The Global Coal Market

Coal has been the world's fastest-growing energy source in absolute terms for over a decade. Coal also emits more CO_2 than any other fossil fuel and contributes to serious air pollution problems in many regions of the world. If we hope to satisfy the demand for affordable energy in emerging economies while protecting the environment, we need to develop a keen understanding of the market that supplies coal. This book offers an in-depth analysis of the key producers and consumers that will most influence coal production, transport, and use in the future. By exploring how countries such as China, India, Indonesia, Australia, and South Africa have developed their respective coal industries – and how these industries link together through the international coal trade – experts shed light on how the global coal market may evolve, and the economic and environmental implications. This book is the most comprehensive treatment of these topics to date and will appeal to a wide readership, including scholars and practitioners working on energy economics and policy.

Mark C. Thurber is Associate Director at the Program on Energy and Sustainable Development (PESD) at Stanford University, where he studies coal, natural gas, oil, electricity, and carbon markets, as well as how to deliver energy to the poorest populations around the world. He teaches on energy markets and policy at Stanford University's Graduate School of Business.

Richard K. Morse is an affiliated researcher at the Program on Energy and Sustainable Development (PESD) at Stanford University. He has published multiple articles and presented at major coal industry conferences on the operation of the global coal market, with a particular focus on China.

The Global Coal Market

Supplying the Major Fuel for Emerging Economies

Edited by

Mark C. Thurber and Richard K. Morse

Stanford University

CAMBRIDGE
UNIVERSITY PRESS

CAMBRIDGE
UNIVERSITY PRESS

University Printing House, Cambridge CB2 8BS, United Kingdom

Cambridge University Press is part of the University of Cambridge.

It furthers the University's mission by disseminating knowledge in the pursuit of education, learning and research at the highest international levels of excellence.

www.cambridge.org
Information on this title: www.cambridge.org/9781107092426

First published 2015

A catalogue record for this publication is available from the British Library.

Library of Congress Cataloguing in Publication data
The global coal market : supplying the major fuel for emerging economies / edited by Mark C. Thurber, Richard K. Morse.
pages cm
ISBN 978-1-107-09242-6 (hardback)
1. Coal trade – Developing countries. 2. Energy consumption – Developing countries. I. Thurber, Mark C. II. Morse, Richard K.
HD9554.D442G66 2015
381'.4224–dc23

2015001500

ISBN 978-1-107-09242-6 Hardback

Contents

Figures

Tables

Boxes

Contributors

JEREMY CARL, Hoover Institution at Stanford University

ANTON EBERHARD, University of Cape Town

CLEMENS HAFTENDORN, DIW Berlin (former)

GANG HE, University of California, Berkeley

FRANZISKA HOLZ, DIW Berlin

BART LUCARELLI, Roleva Energy Limited

ROMAN MENDELEVITCH, DIW Berlin

RICHARD K. MORSE, Stanford University

WUYUAN PENG, China University of Geosciences, Wuhan Campus

VARUN RAI, The University of Texas at Austin

HUAICHUAN RUI, Royal Holloway, University of London

MARK C. THURBER, Stanford University

KEVIN JIANJUN TU, Canadian Industrial Energy End-Use Data and Analysis Centre (CIEEDAC)

CHRISTIAN VON HIRSCHHAUSEN, TU Berlin and DIW Berlin

Acknowledgements

This book grew out of the observation that academic attention to coal markets in recent years has not been even remotely commensurate with the significance of the fuel in meeting energy needs globally. Whether one believes that coal is a significant positive for the world, a significant negative, or a mixed bag, there is no escaping the fact that dealing with reality means accepting coal's current importance and trying to understand where we go from here. We are delighted that we were able to assemble a group of experts with such deep country, modeling, and technology expertise to explore what we can learn from the development thus far of coal industries and coal markets around the world.

This project began at the Program on Energy and Sustainable Development (PESD) here at Stanford University when David Victor was director of the program. We are grateful for his vision in recognizing the importance of the global coal market and the relative dearth of attention that has been paid to it in academic circles. The current director of PESD, Frank Wolak, picked up where David left off in fully supporting this work and helping to bring it to fruition. Kathy Lung, PESD's operations director, was, as usual, extremely supportive throughout the process. PESD staffer Brian Kooiman provided helpful editing and formatting assistance, as did Laura Seaman from our parent institute at Stanford, the Freeman Spogli Institute for International Studies (FSI).

We appreciate our ongoing relationship with Cambridge University Press and the highly capable work of many staff there in turning this manuscript into a book. Chris Harrison was an advocate for the project from the very beginning, and Philip Good was an excellent steward as well. Claire Wood was a pleasure to work with on all aspects of the editing and production process. David Morris did great work on the cover, and Sri Hari Kumar at Integra was responsive and helpful.

We would like to specially thank contributor Bart Lucarelli for valuable comments and insights that helped improve a number of chapters in the book. Mark Hutchinson of IHS CERA also provided helpful support in the form of published price data for Indonesian steam coal.

Staff at the National Bureau of Asian Research, and especially Meredith Miller and Clare Richardson-Barlow, have been supportive of this project for several years, and they helped provide forums for discussing the themes of this book.

Many, many other people contributed significantly to the individual chapters herein. They are thanked at the beginning of each chapter.

We hope you will find the material in this volume to be useful.

Mark C. Thurber
Richard K. Morse

Part I

Introduction

1 The Asia-centric coal era

Mark C. Thurber and Richard K. Morse

1.1 Introduction: coal's dominant role in the world energy mix

The world has witnessed tremendous upheaval in global energy markets in the last decade. Oil markets were roiled by war and uprisings in the Middle East, with economic expansion helping to push real oil prices to near-record high levels. Natural gas production in the United States was revolutionized with the emergence of new techniques for extracting gas from shale deposits. The extension of these techniques to oil created a surge in North American oil production that, coupled with a global economic slowdown, helped bring oil prices far down from their peak. Nuclear power appeared on the verge of a dramatic expansion before its rise was cast into doubt by the tragic disaster at Fukushima in March 2011. And renewable energy began to grow from a small contributor to world energy supplies into one of the largest sources of new electricity generation in many markets. Those narratives, for the most part, are well known. But one of the most important stories in modern energy markets remained largely untold and unappreciated, even as it reshaped the map of global energy use and the pace of climate change. That story is the story of coal's stunning growth.[1]

For more than a century, coal has been a cornerstone of the global energy supply due to its wide availability and low cost of extraction. In recent years, far from seeing its place overtaken by competing energy sources, this catalyst of the Industrial Revolution[2] has grown to have an even more pivotal role in the world's energy mix (see Figure 1.1). Over the decade to 2012, energy generation from coal grew far faster in absolute terms than any other energy source. In fact, coal served nearly as much incremental demand for primary energy as *all other forms of energy combined* (IEA 2013). As of 2011, 29% of total world energy was supplied by coal, second only to oil at 31% (IEA 2013). Most of this coal is used in the power sector, where it accounts for more than 41% of global electricity production (IEA 2013) – a larger share than any other form of energy. If

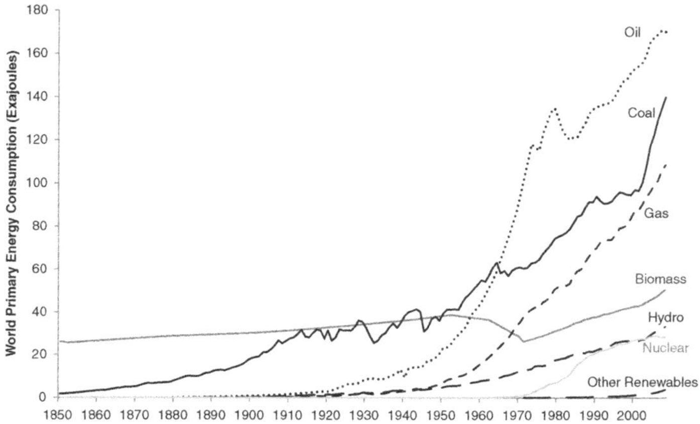

Figure 1.1 Historical trends in world primary energy consumption by energy source. Data sources: IIASA (2012); Grubler (2012).

there is no new policy action on climate change, coal is expected to be the single largest source of energy globally by the early 2020s, overtaking oil (IEA 2013, p. 58).[3]

Coal use in Europe and the United States powered the first global industrial revolution; the second coal era is centered in Asia and is intimately connected to the rise of developing economies. In its "New Policies Scenario,"[4] the International Energy Agency (IEA) projects a reduction in coal demand from developed countries of 362 million tonnes (Mt) per year by 2035, but an increase from non-OECD (Organization for Economic Cooperation and Development) countries of 1,298 Mt/year, of which demand from China, India, and ASEAN nations accounts for 90% (IEA 2013, p. 145). Coal-fired power has helped drive robust economic growth in these and other Asian countries, which in many cases see few affordable alternatives to coal.

As coal is transforming economies, it is also transforming the climate through its CO_2 emissions, and in many cases leaving local pollution problems such as acid rain and smog in its wake. Coal combustion is now responsible for more CO_2 equivalent emissions than any other fossil fuel (Figure 1.2). And while the SO_2, NO_x, particulate, and toxics emissions that degrade local air quality can be mitigated to a substantial degree through well-understood and affordable technological fixes, curtailing CO_2 emissions from coal is a much more difficult task (see Section 1.4.2 of this chapter). The IEA predicts that about half the total increase in global CO_2 emissions from fossil fuel use by 2035 will come from coal use in

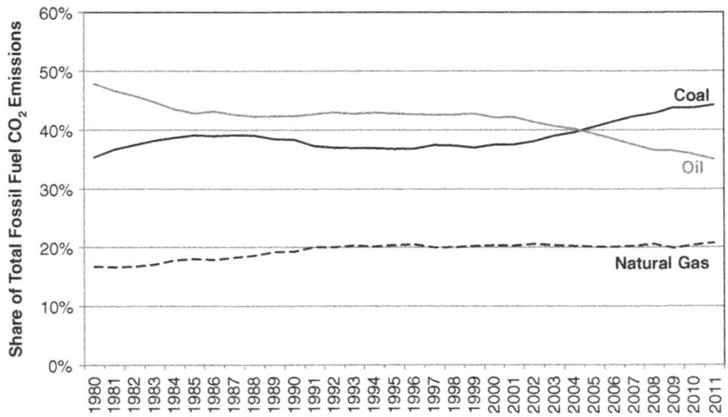

Figure 1.2 Share of total global CO_2 emissions from fossil fuels for coal, oil, and natural gas. Data source: EIA (2014).

non-OECD Asia (IEA 2013).[5] Coal appears likely to persist as the single largest cause of global warming in the coming decades.

But seeing coal solely as "climate enemy number one" misses an important point. Just as coal is the leading *cause* of climate change, so it represents a leading *opportunity* for carbon mitigation. As an illustration of this, we estimate that upgrading the average efficiency of China's fleet of coal power plants by 4%, which is ambitious but not infeasible, would contribute more to CO_2 mitigation on an annual basis than all of the world's wind, solar, tidal, and geothermal energy generation.[6] (Of course, such a reduction in emissions intensity would not be able to counter the increase in emissions if China and other countries continue to build *new* coal power plants at a rapid pace.) This suggests that emerging technologies that might reduce the environmental impact of coal are essential to the fight against climate change. There is another important reason why those who are serious about combating climate change need to understand the world's coal markets – namely, those markets play a key role in determining the competitiveness and viability of alternative energy sources. Coal-fired power represents the cheapest source of electricity in many economies.[7] Thus coal prices help determine the levelized cost of electricity against which "cleaner" generation sources have to compete in the market. Factors that cause coal prices around the world to increase or decrease will have a direct effect on the size of this cost gap, the competitiveness of alternatives to coal, and consequently global carbon emissions.

In short, no climate policy is likely to succeed without a strategy for cleaning up or displacing coal. Efforts by environmentalists to block new coal plants in the US and Europe are an implicit recognition of this fact. However, such efforts risk irrelevancy if they fail to recognize the interconnected nature of the global coal market and the central role of developing economies in determining the future of coal. Moreover, many of these emerging economies feature heavy government control and regulation of markets, necessitating a different understanding of the possible drivers of change.

One of the key objectives of this book is to understand how domestic coal markets in large developing countries will interact with and shape global coal markets. We will consider which domestic factors in key coal-producing and consuming countries have the most impact on the balance of global coal supply and demand, and thus on world prices. As one example, understanding that China, now the world's largest importer of steam coal, imports coal as a form of price arbitrage rather than because of a domestic shortage (Chapter 9) may lead to a surprising conclusion – namely, that the single largest source of incremental demand in world coal markets over the last five years is highly dependent on the particulars of domestic Chinese energy policy. As another example, whether India's domestic coal monopoly, Coal India Limited (CIL), can increase production to meet domestic demand will have major impacts on how much coal India seeks on world markets as well as on the country's CO_2 emissions. Models of the global coal trade that capture these important feedback effects between domestic and global markets are needed to better understand such effects (Chapters 9 and 10).

The China case highlights another key theme of our analysis: that heavy market intervention by governments in key coal countries means that the leverage points over coal production and use are different than in more liberalized economies.[8] For example, it is impossible to understand China's coal industry without understanding its largest, highly regulated customer, the power industry, which is majority government-owned. China's government is very concerned with maintaining low prices for electricity – both to hold down inflation and to fuel industrial development and economic growth. But as coal markets and prices have been partially liberalized over several decades to incentivize production, the power sector has not kept pace and remains dominated by state-owned firms providing electricity at fixed prices (see Chapter 2). The result today is what is known as the "coal-power conflict," in which the mismatch in industrial organization between the coal and power sectors, and the constraint of the monopoly rail network that transports coal, creates significant distortions and bottlenecks in the coal-to-power value chain – especially when coal prices go up. To illustrate how dramatic these distortions have become, for many periods between 2009 and 2012, raw coal was worth more on an energy basis than the final product, electricity, it was used to produce (see Chapter 12). This resulted in supply shortages, the imposition of price controls, and an urgent need for imports. China is looking to vertical integration as one possible solution to these problems, which have resulted from its hybrid of liberalized market and state control (Chapter 3).

We also examine the even more daunting challenges India's government faces around the performance of its coal-producing and consuming industries, which are the result in part of the ways India has involved itself in coal, rail, and power markets (Chapter 4). Due to a low-performing state coal company and rail monopoly, an unsettled legal regime around land rights, and conflict in coal-producing regions, the country's domestic production is falling far short of demand. The government has projected that India's "coal deficit" could reach around 185 Mt by 2017 and over 270 Mt by 2021–2022 (*Economic Times* 2012; Thapar 2015). However, given government control over power prices and numerous other dysfunctions of the Indian power sector, including line loss and theft, there are serious questions about whether India can generate the revenue from electricity sales that would enable it to pay for sufficient coal from the international market. How the interaction between India's power sector and its monopoly coal producer evolves may have a crucial influence on whether massive import demand from the country materializes.

We devote significant attention to major coal-exporting countries, which to varying degrees are facing supply expansion problems of their own. Australia (Chapter 6), Indonesia (Chapter 7), and South Africa (Chapter 5) have contended with challenges of different sorts in the development of new mines and/or transportation infrastructure. At times, competing priorities seem to be making it more difficult than in the past for these countries' governments to maintain a supportive policy environment for the coal export industry. Constrained supply from these key exporters would reverberate throughout the global coal market.

To gain insight into these kinds of issues – how the global coal market works and the economic and environmental implications – this book poses three fundamental questions. First, which kinds of government intervention – such as economic and environmental regulation at central and local levels of government, or direct involvement in coal-producing or consuming enterprises through state ownership – have tended to have the most significant impact on the evolution of coal production and consumption in a country? Second, how do domestic policies and stakeholder processes in key producing and consuming countries collectively impact the global coal trade and shape economic and environmental outcomes? Third, can policy or technological measures reconcile increased coal use with the imperative to reduce greenhouse gas emissions, and, if so, what will be the resulting impacts on global coal markets?

In Sections 1.2 through 1.4 of this introductory chapter, we elaborate on each of these three questions, provide relevant background, and discuss how our country case selection provides insights into these questions. Section 1.2 describes how the nature of government regulation strongly affects the character of the coal value chain and, as a result, the functioning of the coal market itself. Section 1.3 charts the historical evolution of the international coal trade and sets the stage for subsequent discussion of how this trade functions at present and will evolve going forward. Section 1.4 considers how the different instruments of international and domestic climate policy as well as emerging "clean coal" technologies might affect the coal industries of specific producing countries and the global coal market. Finally, Section 1.5 concludes with a description of how this book is structured in order to address the central questions we pose.

1.2 The coal value chain

1.2.1 A brief introduction to the coal value chain

Coal was a key driver of the Industrial Revolution that began in Great Britain in the mid-1700s. It substituted for increasingly scarce biomass in firing blast furnaces for the production of iron, was tapped as a fuel for commercial and residential heating, and, with the refinement of steam engine technology, found growing use in all applications requiring mechanical power, prominently including manufacturing, mining, and transportation (Landes 1969; Gordon 1987). By the 1920s coal had surpassed biomass as the preeminent fuel in the global energy system (Wilson and Grubler 2011).

With the sharp rise after World War II in the availability of petroleum-based fuels and electricity, which have higher energy densities and are easier to transport,[9] coal use became increasingly concentrated in power generation and iron making. In the latter application, coal has unique technological advantages. Specific kinds of low-ash, low-sulfur coals ("coking coals") are used to produce "coke," a solid fuel that meets the special chemical and physical requirements of the iron-making process (Landes 1969; Gordon 1987). Electricity generation or industrial uses do not demand coals with the exacting specifications of coking coals, explaining the characteristic differentiation of coal into "steam" and "coking" types. In this volume we focus on steam coal markets, which now represent the majority of internationally traded coal and are most consequential for today's global energy system.

Before it can be used to efficiently generate electricity, coal requires additional processing. It is typically pulverized prior to burning. It may also be screened and washed to reduce impurities. Less commonly, it is converted into a gaseous or liquid fuel by means of thermo-chemical conversion processes such as gasification and pyrolysis. Because of the required pre-combustion processing and handling, and sometimes the environmental controls that are mandated, coal power plants are significantly more capital-intensive than oil- or gas-fired alternatives. Large power plants provide economies of scale in coal processing and environmental control, making electricity generation in such facilities a preferred application for coal. But coal's only competitive advantages as a fuel for power generation relative to oil and natural gas are low cost and wide availability. In growing economies such as those of China and India, where coal is one of the relatively few domestic energy resources found in abundance, these advantages are of overriding importance.

Transportation of coal from mines to end use sites has always been a crucial link in the coal value chain. Domestic waterways played a central role in transport in early coal markets in Great Britain and the United States (Finch 1973; Miller and Sharpless 1985). The advent of steam-powered locomotives in the 1800s led railroads to assume the central role in coal transport that they maintain to this day.[10] The need to move coal was in fact an important spur to the development of both canal systems and, later, rail networks (Miller and Sharpless 1985; Garnett 2005). In theory, an alternative to barges and trains is to mix coal with water and transport the resulting slurry by pipeline, but disadvantages of this method include its water intensity and the need to dry the coal before burning. The only plant fueled with coal transported by this method, the Mohave Generation Station in Laughlin, Nevada, was shut down in 2005 (SCE 2012).

An alternative to bulk transport of coal itself is "coal by wire," in which electricity from coal-fired power plants near coal mines ("mine mouth" plants) is carried to the location where energy is consumed by high-voltage transmission lines (see, e.g., Gordon 1987; EIA 1998). The relative attractiveness of transporting coal versus electricity depends on a number of factors including the distance between coal mine and end use as well as the respective regulatory environments for coal transportation, power generation, and electricity transmission (including the relative ease of siting new rail lines as compared with new transmission lines). In the Soviet era, for example, the characteristics of Siberian lignite and their distance from population centers might have argued for electricity transmission from mine mouth plants, but underdevelopment in the region made this approach infeasible, whereas transport by rail was encouraged by low tariffs under central planning (Gordon 1987; Victor and Morse 2009).

1.2.2 Government involvement in the coal value chain

A key tenet of our approach throughout this volume is that the coal industry cannot be understood in isolation from the applications that it serves – particularly power generation – and the means by which it is transported.[11] The organization of these three linked industrial segments – coal production, coal transportation, and the electric power sector – is a function both of "natural" factors, such as geology and geography, and of how governments choose to control each respective segment. Regulation of one part of the coal value chain often has powerful effects on another part. Through comparative analysis of the different forms of government involvement along the coal-to-power value chain in these five key countries, we seek insight into which kinds of government policy and intervention have had the most significant impact on coal production and use over the past several decades.

There is significant literature addressing the interactions between coal, rail, and power sector regulation in the United States and Europe, but far less for the world's fastest growing coal markets – a gap that we seek to address in this book. The effect of market restructuring and price liberalization in the power, and sometimes rail, markets has been an important theme in the US and Europe. Effective deregulation of rail transport rates in the US by the Staggers Act of 1980 combined with barriers to entry have allowed railroads to exercise significant market power and extract a large portion of the rents in the coal value chain (Wolak and Kolstad 1988; EIA 1998; Busse and Keohane 2007). In fact, Gerking and Hamilton (2010) argue that the market power of US railroads has blunted the impact of SO_2 regulation on the US coal market, illustrating the potential policy implications of how the coal value chain is structured.

Restructuring of US electricity markets has required generators to compete and in turn to seek lower-cost supplies of coal (although in the last several years low prices for natural gas have exacted an even heavier toll on the coal industry by giving gas an economic advantage over coal in many cases). Generators also face more uncertainty about electricity prices and quantities and thus may prefer to procure coal via shorter, more flexible contracts. The uncertainties flowing from a competitive electricity market translate into risk that needs to be hedged by both coal and power producers, encouraging the creation of derivative financial contracts (EIA 1998).

Table 1.1 *Role of SOEs in the coal value chain*

	India	China*	South Africa	Indonesia	Australia
Dominant coal SOE(s)	✓	*			
SOE transport monopoly	✓	✓	✓		
Dominant power SOE(s)	✓	✓	✓	✓	

* China has one very large and important SOE in coal – Shenhua (see Chapter 3) – but also many other coal producers of various sizes and ownership.
Note: Rail is the primary domestic transportation mode except in the case of Indonesia, which relies heavily on inland waterways.

Restructuring of power markets in Europe has had similar effects on coal markets. By the 1990s, the push for wholesale competition in many of the EU's power markets forced generators to compete on price. Pressure was passed on to coal suppliers to deliver competitively priced fuel inputs, which was not possible under the "cost plus" contract structures that had historically been used in many coal markets. As in the US, spot markets arose along with demand for financial instruments that would help manage fuel procurement risk (Ekawan and Duchêne 2006; Ritschel and Schiffer 2007).[12] Because the EU's coal supply has increasingly come from international markets, the demand for derivatives instruments by European coal buyers played an important part in driving the "financialization" of global coal markets, as we will discuss in the next section.

But studies of the US and European experiences provide only limited guidance for understanding the fast-growing coal markets of Asia, which we focus on in this book. Many governments in our sample involve themselves much more directly in the coal value chain by explicitly setting prices for coal, transportation, and/or electricity, or through state ownership of coal, transportation, and/or power companies. Table 1.1 shows the variation in how and where the governments in our sample use state-owned enterprises (SOEs) as their agents in the coal value chain.[13] We will observe and compare how these different choices affect the functioning of the coal sector itself as well as the downstream sectors (especially power generation) that use coal. We will also devote significant attention to how government control over power prices influences coal market structure and operation. In some of the countries in our sample, electricity prices are not set by bids and offers but rather are held at artificially low levels to fuel industrial growth and keep inflation in check.[14] As described in Chapters 2 and 3, China's tight caps on power prices distort incentives and resource allocation along the coal value chain, a problem the government is hoping to finesse through vertical integration within state-owned firms.[15] Chapter 7 discusses how, in the past, low power prices in Indonesia prevented national power company PLN and independent power producers (IPPs) from making bids for coal supply that were competitive with coal prices in export markets. Today the government of Indonesia allows PLN and domestic IPPs to pay prices that are at or above the export prices for Indonesian coals of similar quality.[16]

Because capital-intensive infrastructure is required at every step of the coal value chain – mining, transport, and electricity generation – governments can also play an important positive role by either making these investments themselves or creating a regulatory and investment climate that encourages the needed private investment. Across our sample of countries, we observe that a common challenge is ensuring adequacy of rail and port infrastructure. In many cases there are shortfalls in coal supply or exports due to inadequacy of infrastructure even when ample reserves exist. In the individual country studies and the conclusion of this book (Chapter 13), we discuss the factors that make some governments more successful than others at attracting the necessary investment and achieving a coordinated build-out of the coal value chain.

1.3 The international trade in coal

1.3.1 The development of the global coal market

Before 1960, traded coal markets rarely extended beyond national borders.[17] The international trade that did exist was mostly within Europe, with little seaborne trade apart from US shipments of coking coal to Japan. In the 1960s, that began to change as Japan sought additional suppliers of coking coal to feed its growing steel-making needs, while other OECD countries – and, starting in the 1970s, newly industrializing nations such as South Korea, Taiwan, and Brazil – started to import significant quantities of coking coal as well (Ellerman 1995; IEA 1997).[18]

But the more important shift for the purposes of this book began in the 1970s. The oil crises of that decade created strong incentives for oil-consuming nations to replace oil use in power generation with coal. In the face of stark demonstrations of OPEC's leverage, governments and firms saw a clear price and security advantage from swapping cheap, abundant coal for expensive, cartel-controlled oil. Starting in 1980, members of the Organization for Economic Cooperation and Development began a concerted effort to do just that (IEA 1997).[19] Between 1980 and 2000, OECD countries collectively increased the use of coal in electricity

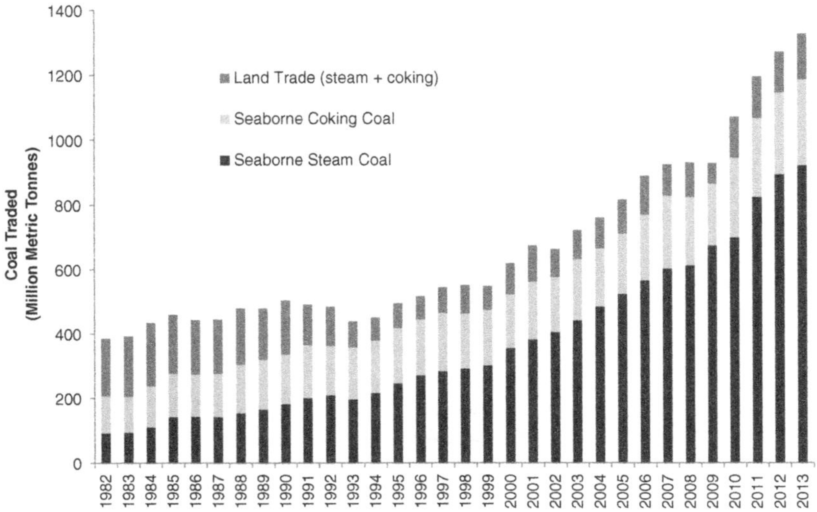

Figure 1.3 Historical evolution of the world coal trade. Data source: IEA's *Coal Information*, various years: 1982–1985 from 2004 edition, 1986–1989 from 2011 edition, 1990–2013 from 2014 edition (IEA 2014a).

generation by 61% (and reduced the use of oil in that sector by 41%). Thus the emergence of the international coal trade was in many ways a response to the evolution of modern oil markets.[20]

The push to establish coal as a globally available oil substitute was effective, and the seaborne trade in steam coal took off, with imports by Japan playing the key role in initiating the steam coal trade in Asia.[21] Between 1982 and 2013, the size of the maritime trade in steam coal grew by roughly a factor of ten, from just over 90 Mt/year to more than 900 Mt/year (see Figure 1.3). While internationally traded steam coal still only constitutes about 17% of global production (IEA 2014a), it has played a pivotal role by linking prices between previously regional markets around the world.

Historical distinctions between the trade in the "Atlantic Basin" and "Pacific Basin," once treated as distinct markets, have for the most part eroded.[22] Since the 1990s, prices have largely moved in tandem across major international hubs for globally traded coal (see the NW Europe, Richards Bay, and Newcastle price indices in Figure 1.4).[23]

The next wave of price integration, which is increasingly in evidence though not yet complete, is the linking of prices in previously isolated domestic markets to prices in international markets (note in Figure 1.4,

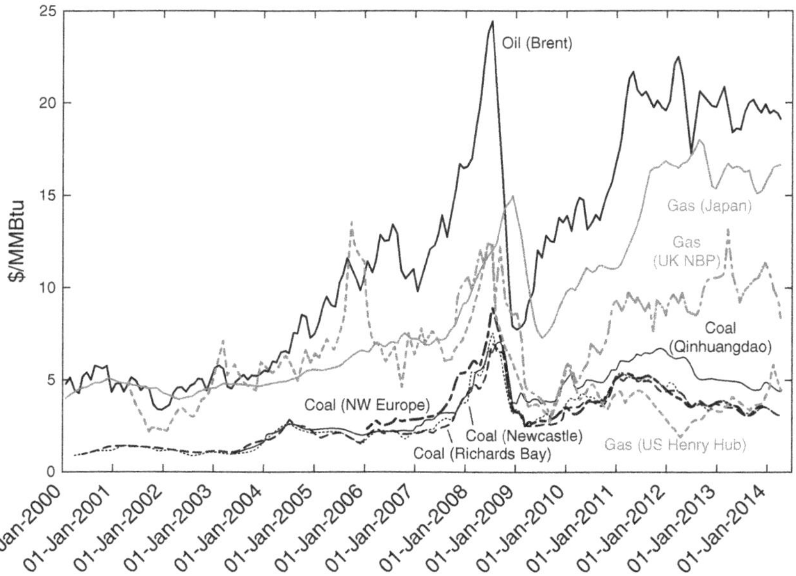

Figure 1.4 Price of international energy commodities, energy adjusted for comparison on a $/MMBtu basis. All prices are spot except where otherwise noted. *Coal*: Qinhuangdao (FOB, net as received, from IHS McCloskey); Richards Bay (FOB, net as received, from IHS McCloskey); Northwest Europe (delivery into ARA hub, net as received, from IHS McCloskey). *Natural gas*: Japan (average imports including freight); UK (National Balancing Point); US Henry Hub. *Crude Oil*: Brent. Data source: Bloomberg.

for example, how China's Qinhuangdao price broadly follows but does not exactly lie on top of prices in major international coal hubs). Geographic coal price differentials which still exist represent an opportunity that coal consumers can attempt to arbitrage, although they may require additional transportation infrastructure and supportive government policy in order to do so (see Chapters 8 and 9).

Growth in traded coal volumes has both spurred and in turn been facilitated by the development of relatively liquid, transparent price indices. In 1980, if you wanted to obtain the price of globally traded coal, it wouldn't have been possible; long-term bilateral contracts constituted the entirety of the market and created informational asymmetries that made price arbitrage either prohibitively difficult or outright impossible. But starting in the 1990s, electricity market liberalization in Europe encouraged utilities to seek shorter-term contracts and spot procurement. Spot market prices gradually began to supplant prices based on long-term, bilateral contracts.[24] The shift toward spot market prices created price risk that utilities had to manage. One way to do this was with derivative financial instruments, but derivatives in turn required information about the price of the underlying asset, which could be provided by price indices. Energy broker TFS launched the first coal price benchmark, API #1, in 1998 and executed the first derivatives contract – an over-the-counter (OTC) coal swap – later in the same year. Follow-on efforts by various players to "commodify" coal led by the 2000s to spot and derivatives markets that were (relatively) liquid, transparent, and standardized.[25] The next major phase in the integration of coal markets was the emergence of coal futures. In 2006, trading platforms globalCOAL and ICE launched futures contracts for delivery at Rotterdam, and by 2009 contracts for Richards Bay and Newcastle followed. (An attempt by the Australian Securities Exchange [ASX] to launch physically settled coal futures for Newcastle failed.) Asia has lagged significantly in the process of "financializing" the coal markets, but 2011 witnessed the emergence of various physical benchmarks and early swaps markets in Asia.[26] In moving from markets based entirely on long-term contracts to those where spot contracts become important and start to set prices, and where derivatives begin to develop, coal seems to be following the trajectory of oil before it, although it remains at a much earlier stage in this evolution. The development of this more sophisticated and connected market for coal matters for several reasons.

First, it yields price and other linkages between previously separate energy markets. For example, Europe, which imports the majority of the coal it uses to generate electricity, now finds itself in competition with India for its traditional coal supplies. Since the domestic price of coal in China helps determine that country's appetite for imports (see Chapter 9), which in turn ripples back to affect prices in India and elsewhere, this means that the price of coal in China now affects the price of power in Berlin (and the demand for coal substitutes such as natural gas). Power markets in other major importing countries such as Japan, Korea, and Taiwan are similarly susceptible to what happens in China's and India's domestic markets, since those developments are transmitted to global prices through spot and derivatives markets. Conversely, the easy availability of coal sourced abroad can impact domestic markets. The ability of China's power companies to buy international coal provides them with enhanced leverage over domestic coal producers. These new linkages through the international coal trade mean that policies and regulations affecting coal production and consumption in one country can end up having repercussions across the world. For example, if India's unsettled legal framework around land rights (see Chapter 4) discourages domestic coal production and forces greater dependence on imports, this will have an important effect on every country that consumes significant amounts of coal.

Second, an integrated global coal market can affect the calculus for environmental protection efforts. All else equal, policies that increase the demand for or restrict the supply of coal available on the international market will push up the global price and thereby reduce CO_2 emissions by encouraging switching to lower-carbon fuels (Wolak and Morse 2010). However, attempts to curb CO_2 emissions by limiting coal combustion in particular countries or regions could be less effective than anticipated if the coal that would have been burned is instead traded and results in incremental coal combustion in another country (Light et al. 1999). The effects of trade on local and regional pollution need to be considered as well. For example, investments that make low-sulfur coal (as from the US Powder River Basin) more available on the international market have the potential to reduce SO_2 emissions in countries with weak environmental regulation by displacing higher-sulfur alternatives.[27]

The fact that today's global coal market functions more like the modern oil market than the coal market of the 1970s will increasingly force policy and investment decisions in the coal space to consider global factors and implications. Traditional notions of coal as the strongest pillar of energy security owing to its previously assumed independence from global developments will need to be reevaluated. What once would have been safely classified as domestic environmental regulation of coal is now likely to have complex international effects that must be considered. In short, energy investments or government policies around coal or natural gas in one country can no longer ignore coal mines in the deserts of China or rainforests of Indonesia.

1.3.2 The major Pacific Basin coal trade

Through the country cases (Chapters 2–7) and trade-related studies (Chapters 8–10) in this book we consider the most important ways in which global coal market linkages may cause policy and market design decisions in a given country to spill over and affect other countries. We also consider how the existence of an integrated global market for coal may affect the outcomes of such policies. To provide context for this investigation, the remainder of this section describes some of the important relationships between key countries in today's coal trade.

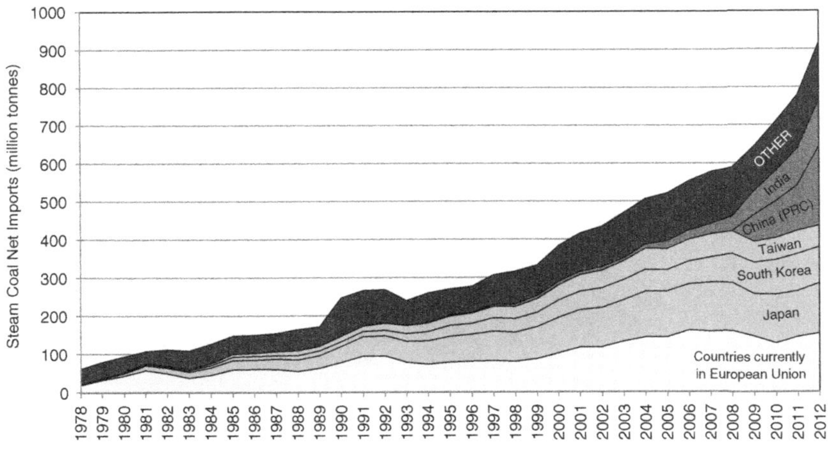

Figure 1.5 Net imports of steam coal for the most significant importing countries. Data source: IEA (2014c).

The countries that imported the most steam coal in 2012 in Asia were China, Japan, India, South Korea, and Taiwan, in that order (see Figure 1.5). We can broadly separate the major importing countries into two groups. The first group consists of developed countries that import coal because they are "energy-short" at home. This group includes Japan, South Korea, and Taiwan. (Western European nations could be considered to be in this category as well; the EU as a block imported over 150 Mt of steam coal in 2012.) The second group consists of major coal importers, namely China and India, which have large domestic coal markets but also import on the margins of their domestic markets.

We focus our analysis of demand in global markets on the developing economies of China and India for three reasons. First, they are the largest and most important sources of future demand growth. China and India are projected to account for 69% of all coal demand growth to 2035 (IEA 2013).[28] Second, they are poised to exert an outsized influence on the global market in the coming decades. In the case of China, this is because imports, which comprise a major share of the global trade, are largely a function of what happens in China's much larger domestic market. As shown in Figure 1.6, China's steam coal production (and consumption) dwarfs that of all other countries. As a result, even very small percentage shifts in domestic supply or demand are highly consequential for the international trade. China is pursuing a restructuring of its coal sector, which, if widely successful, could remake its domestic market (see Chapter 3) – and

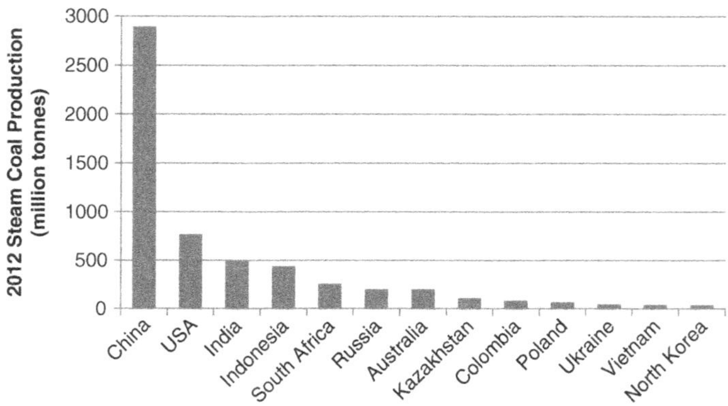

Figure 1.6 Steam coal production by country in 2012. Data source: IEA (2014a).

by extension the international trade as well. India is also particularly pivotal because, despite ample reserves, a major gap appears to be growing between domestic coal production and domestic coal demand, driven by an aggressive (though often underperforming) electrification program that calls for substantial expansion of coal-fired generation capacity. This anticipated gap between Indian supply and demand may cause India to supplant China as the world's largest steam coal importer before 2020.

Third, as discussed in Section 1.2, China and India are especially worthy of further study because of how the balance of production and consumption in both countries depends heavily on the political economy of a complex and mostly nonliberalized coal-to-power value chain.

As of 2012, the world's largest exporters of steam coal were, in order, Indonesia, Australia, Russia, Colombia, and South Africa (see Figure 1.7). We focus our analysis on the two largest exporters, Indonesia and Australia, as well as South Africa, which is the key "swing supplier" between the Atlantic and Pacific Basins. Several main factors that we will explore further in this book have historically had an important effect on how these countries supply the global market.

First, the nature of the resources and the technology that is brought to bear for extraction, processing, and transport has unsurprisingly shaped cost and thus market competitiveness of a given country's coal. For example, Australia has had the benefit of massive reserves but has also devoted significant resources and attention to technological innovation in mining. Additionally, its decision to electrify rail transportation has reduced exporters' exposure to fluctuations in oil and diesel markets.

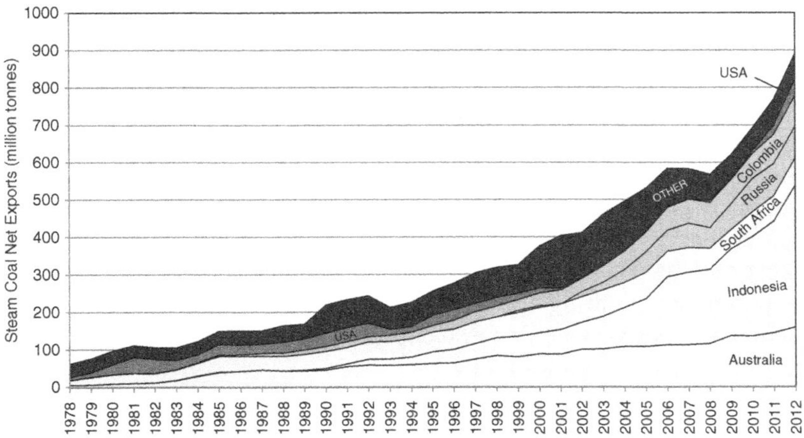

Figure 1.7 Net exports of steam coal for the most important exporting countries. Data source: IEA (2014c).

Second, the ability of a country's government to either directly mobilize public investment or facilitate private investment in both ports and domestic transportation infrastructure plays a key role in determining export capacity.[29] The governments of the Australian states of Queensland and New South Wales, in cooperation with the federal authorities, are generally considered to have done an effective job at building out an efficient transportation network, although port limitations continue to exist (see Chapter 6; Rademacher 2008; Bayer et al. 2009). South Africa has built the world's largest coal export terminal – Richards Bay, with a current annual nameplate capacity of 92 Mt – but capacity limitations of the state railroad Transnet were capping exports around 68 Mt as of 2012 (see Chapter 5).[30] Indonesia, on the other hand, has leveraged its plentiful domestic waterways for low-cost and flexible coal transportation, sidestepping some of the constraints in capital-intensive, land-based infrastructure faced by other exporters (Chapter 7). The development of infrastructure often shapes export capacity and cost structure as much as coal reserves or mining technology.

Third, the tension between exports and domestic use of coal in developing economies can alter the role a country plays in the global trade. South Africa has been a key coal supplier to Europe since the 1960s, but it is now facing the challenge of how to reconcile a vibrant export trade with the demand of state-owned power company Eskom for coal to keep the lights on and spur development. The issue is coming to a head with the expiration of favorable legacy contracts for coal held by Eskom

(Chapter 5). As these contracts expire, Eskom may increasingly have to compete with export markets and the higher prices they offer in order to secure coal for domestic use. As a result, significant discussion has arisen in South Africa around whether to declare coal a "strategic resource" or take other actions (such as nationalization of some coal assets) that would withhold a portion of coal production for use by Eskom and reduce supplies to global markets. Indonesia is building significant coal-fired power to fuel its own growth and has put in place a Domestic Market Obligation (DMO) intended to reserve a share of its coal for domestic use, potentially limiting the growth of exports if implemented in an inflexible way (as we examine in Chapter 7).

1.3.3 Other countries and issues of possible importance for the world coal trade

In this book, we deliberately chose to focus on the countries most engaged in the Pacific Basin coal trade due to their pivotal role in shaping the global market. However, there are of course countries outside of our sample that will also be important to the future of coal. Russia and Colombia are two particularly noteworthy exporters that were not included among our case studies. Among importers, we have not included case studies of Europe (where demand growth is mostly stagnant but nuclear retirements have given a boost to coal), Taiwan, Korea, or Japan specifically, but we have addressed their roles to some extent in chapters on key exporters (Chapters 5–7) and on the international coal trade (Chapters 8 and 10). In addition, our country rather than firm-level orientation means that we have only limited leverage to study the particular role of coal companies, some of which control large volumes of coal and thus may be able to exert significant influence on global markets (Haftendorn 2012).

One current development, which we treat in Chapter 8, is the possible reemergence of the United States as a substantial exporter of steam coal. When this study began, this notion was far from anyone's mind. The US was the largest exporter of seaborne coal up until the early 1980s, when it lost that position to Australia. The US played a role after that as a residual supplier in the world trade due to the size of its domestic market and consequent ability to serve excess global demand when global prices became sufficiently high to make international shipment of the high-cost production in the eastern US economically attractive (Ellerman 1995).[31] Since the financial crisis of 2008, however, changes in US power, coal, and gas markets combined with strong demand in Asia are generating serious investment interest in expanding export capacity for large amounts of US coal that have never before reached international markets in significant volumes. A key factor is that increased US natural gas production from unconventional sources, which drove down US prices to an average of around $3.80 per MMBtu from 2009 to 2011,[32] have combined with increased regulation of pollution from coal combustion[33] to cause precipitous declines in US coal demand. Coal's share of monthly US generation totals has dropped to levels not seen in more than thirty years.

Figure 8.1 in Chapter 8 demonstrates the stark disconnect between steam coal prices in the US Powder River Basin (PRB) and major Asian markets. As can be seen, PRB coal trades at a significant discount to other major coal indices in Asia, indicating that a significant arbitrage opportunity exists if new port infrastructure could be put in place on the West Coast to enable PRB coal to reach Asian markets.[34] New Pacific Northwest ports are being considered that might send over 100 Mt of coal into Asian markets, potentially catalyzing a rebirth of the US coal export industry as a dominant player in global markets. The entrance of the US as a high-volume supplier could have disruptive effects on the Asian trade.

A related issue that we flag here but do not treat in depth is the important interaction between natural gas and coal markets. The very phenomenon that makes US producers so eager to get coal offshore, plentiful shale gas, creates long-term uncertainty for the global coal market as a whole. Coal's primacy in China and India, where in 2012 it fueled 75% and 71% of electricity generation, respectively (IEA 2014b), is in part tied to the constrained supply of natural gas in these countries. And imported LNG was quite expensive in Asia up until late 2014 (see the gas price for Japan in Figure 1.4), which helps explain why natural gas represented only around 2% of China's generation and 8% of India's as of 2012 (IEA 2014b).

In 2011 and 2012, Henry Hub gas in the United States was actually cheaper than coal in Asia on an energy adjusted basis (see Figure 1.4).[35] If the gap between natural gas prices in the US and Asia were to narrow significantly – through LNG exports from the US (which still could not eliminate the price differential due to LNG transport costs),[36] redirection of Middle East LNG cargoes once intended for Europe, Australian LNG, or direct application in Asia of shale gas extraction technologies developed in the US – it could be a game changer for the Asian coal market over the long term. The US Energy Information Administration estimates that China has 50% more unconventional gas under the ground than the US, and that India and Europe also have significant potential for new shale gas development (EIA 2011). It remains to be seen how quickly and to what degree the rest of the world will develop its unconventional gas, but a significant shift in capital investment toward gas-fired generation is entirely possible if sufficient gas supplies can be developed in Asia's domestic markets at low cost.[37] Gas, in many ways, is the biggest wildcard for the future of the global coal market (and for the future of the climate, as we will discuss in the next section).

1.4 Climate change and coal markets

Because coal is the world's leading source of CO_2 emissions, it is impossible to separate the trajectories of coal markets, coal technologies, and the climate. The evolution of coal markets and coal technologies will be one of the most significant factors affecting the pace and magnitude of climate change. Conversely, efforts to address climate change are likely to affect coal markets more than other energy markets. Our selection of China, India, Australia, Indonesia, and South Africa provides excellent leverage to study how countries with different political systems and levels of development are trying to balance economic and environmental concerns in their policy making. By considering the range of climate policy instruments employed – from a (short-lived) carbon tax in Australia to a voluntary carbon intensity reduction directive in China – we will be able to glean broader insights about which kinds of policies are most likely to materially affect coal markets. By studying the path-breaking "clean coal" efforts of China and Australia, we will also be able to make an initial assessment of the potential for technology to reconcile coal use with carbon mitigation.

1.4.1 The impact of climate policy on coal markets

The vision of an all-encompassing climate regime to cap global greenhouse gas emissions has stalled with the wind-down of the Kyoto Protocol in 2012 without a successor agreement to bind the world's top emitters. However, there remain numerous other types of national, regional, and local climate policies that could have an impact on energy markets in general and coal specifically (Keohane and Victor 2010; Victor 2011). We will now describe some of the most important of these, and how they might intersect with coal markets.

The largest, most established carbon market in the world is the EU's Emissions Trading Scheme (ETS). The effects of this market on coal demand are straightforward. Coal combustion typically creates 0.8–0.9 Mt of CO_2 equivalent per MWh of energy generation, and plants must submit "allowances" (EU allowances, or EUAs) for each tonne of

CO_2 they produce. These allowances are obtained either through free allocation by the government or on the traded market, where prices are set based on supply and demand. In 2008, the EUA price ranged between €20 and €30. It dropped to around €15 for much of the period from 2009 to 2011 and then fell below €10, where it remained at the time of writing. The causes of the EUA price decline included reduced demand due to economic weakness, plentiful supply of international offsets, and complementary EU environmental policies. Because of a coal plant's high carbon intensity, it requires approximately twice as many allowances per MWh as an alternative facility fueled by natural gas. Whereas coal-gas fuel price differences have usually remained more salient than EUA prices in dispatch decisions, at the margin Europe's carbon pricing regime has shifted some electricity generation from coal to gas plants in power markets in Europe that dispatch on a least-cost basis.[38]

In addition to their direct impact on the dispatch of coal power plants, policies that create the expectation of a future (and possibly higher) carbon price may indirectly influence coal markets by chilling investment in new coal-fired capacity. In testimony before Great Britain's House of Commons, carbon market analyst Trevor Sikorski argued that the EU ETS has helped decarbonize the EU power sector in part by curtailing investment in new coal plants (Great Britain Parliament 2012). In particular, significant uncertainty as to the nature of carbon price exposure beyond 2020 – for which the rules for the EU carbon market are not finalized – creates an unquantifiable risk that could kill a $2 billion investment in new coal capacity whose payback period may be decades. On the other hand, Germany's recent investments in new coal-fired power (in part to backfill retiring nuclear capacity) suggest that expectations for future carbon prices are not yet sufficiently high to shut off all investment in coal plants (*Spiegel* 2014).

By showing that carbon pricing is feasible, schemes such as the EU ETS could encourage other jurisdictions to follow suit. Right now, carbon pricing is the exception rather than the rule, but if a large number of regions, countries, and subnational jurisdictions start pricing carbon, the impact on coal markets could be significant. In 2012, Australia appeared to be following the lead of the EU when it established its carbon pricing mechanism (CPM) with the passage of the Clean Energy Act. However, it reversed course in 2014 by repealing the CPM after a change in government from Labor to the National-Liberal Coalition (see Chapter 6). (In line with the arguments in the previous paragraph, the repeal of the CPM does not remove the risk that carbon pricing in Australia will reemerge.) Given Australia's status as a major coal producer, user, and exporter, its politically contentious efforts to balance economic and climate change considerations are an especially instructive case study of the policy (and political) tradeoffs around coal. The US government recently launched a plan that would involve state-level regulation of CO_2 emissions from existing power plants by the Environmental Protection Agency. As described in Chapter 8, environmental impact approval processes in the US – notably around proposed new coal terminals – have also become proxy fights about climate change policy. Other countries are certainly watching how climate policy evolves in countries such as Australia and the United States.

The developing countries in our sample – China, India, Indonesia, and South Africa – have declined to sign on to an international agreement with binding emissions caps, wanting to preserve their freedom to grow and develop. However, some of these countries have floated voluntary plans to limit their own greenhouse gas emissions. As will be discussed in Chapter 5, South Africa volunteered at the Copenhagen meeting of the United Nations Framework Convention on Climate Change (UNFCCC) in 2009 to constrain its greenhouse gas emissions below a "business-as-usual" trajectory. China's officials have supported renewable energy, declared their intent to reduce the carbon intensity (carbon emissions per GDP) of the economy by 16%–17% from 2010 to 2015, and expressed a desire to hold coal output below certain levels (see the Appendix). We will consider the degree to which voluntary efforts like these will substantively affect coal markets.

We also consider whether multilateral development institutions such as the World Bank may have leverage over the carbon intensity of energy systems through their role in mobilizing investment in power plants and other major energy projects in developing countries. Indonesia, India, and South Africa all have large (and desperately needed) coal power projects in the works, with substantial involvement by multilateral development institutions.

Taken collectively, the five country studies in our book afford significant insight into the conditions under which climate policy genuinely constrains the growth of coal in the world's key coal producing countries, and when the economic drivers animating the coal market in Asia are simply too compelling to be curtailed by attempts to regulate greenhouse gases.

1.4.2 The potential for technology to reconcile coal use and carbon mitigation

Part IV of the book examines how technological advancement might address the climate challenge by decoupling CO_2 emissions growth from coal demand growth. China and Australia are two bellwether countries for this effort. Both countries are pursuing programs of investments in key technologies that might break the tight correlation between coal combustion and CO_2 emissions – including highly efficient coal combustion, carbon capture and storage, coal bed methane, and underground coal gasification.

Carbon capture and storage (CCS) is a process that could deal with emissions from coal by capturing CO_2 from coal-fired power plants, condensing it into a liquid, and then injecting it into geological formations underground where it would be stored in perpetuity and not emitted into the atmosphere.[39] CCS in theory could eliminate nearly all of the CO_2 emissions from coal combustion. However, the road ahead for the technology is extremely challenging, in large part due to its high cost. In Chapter 12, we consider the prospects for CCS to be a major solution for the single most important target market, the Chinese power sector. We examine the factors that are crucial to determining whether CCS in China can in fact play a major role in stopping climate change, including (1) the true costs of CCS across the entire energy value chain; (2) how CCS might be integrated into the industrial organization of the Chinese coal, power, and transport sectors; and, most importantly, (3) the motivations of China's government to develop CCS. Chapter 12 also considers the clean coal technologies that have so far played the most significant role in decreasing emissions from coal in China: higher efficiency thermodynamic cycles such as ultrasupercritical combustion and integrated coal gasification combined cycle (IGCC).

Other technologies are also emerging that might help achieve significant CO_2 reductions relative to traditional coal combustion. These technologies include underground coal gasification (UCG), in which coal is gasified while still in the coal seam to yield a more concentrated CO_2 stream amenable to lower-cost CCS, and extraction of coal bed methane (CBM) from wells drilled deep into coal beds. Australia has been a leader in CBM and UCG development and also took an early interest in CCS. In Chapter 11 we consider what the progress of CBM, UCG, and CCS in Australia might say about the broader prospects for technology to allow coal resources to be utilized with fewer CO_2 emissions.

1.5 The organization of this book

As previously described, we commissioned a series of studies that would allow us to explore the three overarching questions posed in Section 1.1: (1) Which kinds of government intervention in domestic coal production and consumption are most important? (2) How will the international coal trade be affected by domestic markets going forward? and (3) How will climate policy affect the coal market (and vice versa), and to what degree can technology reconcile coal use with climate mitigation? We focused on the main suppliers and consumers in the Asian coal market, given that consumption and production from this region will largely shape the future of coal.

Following this introduction (Part I), we organize the book in several major parts. Part II consists of country case studies of the two largest producers and consumers of coal in Asia – China (Chapters 2 and 3) and India (Chapter 4) – as well as three critical suppliers to the region, Indonesia (Chapter 7), Australia (Chapter 6), and South Africa (Chapter 5). Because of China's particular importance, the historical evolution and current status of the coal value chain in that country are treated in three separate parts: Chapters 2 and 3 plus an Appendix at the end of the book which richly describes the production, transportation, and use of coal in China.

Part III explores the critical interconnections between individual country markets and the global trade. We start by considering how transportation constraints in coal and other fossil fuel markets shape the global energy trade, focusing in particular on the possible implications of increased US exports of steam coal from the Powder River Basin to Asia (Chapter 8). We follow with a simple conceptual model of why China imports coal (Chapter 9), which is crucially important to understand given how much China's trading patterns influence the global market. We conclude Part III by introducing a model of the entire global coal trade that can be used to investigate how dynamics in individual countries affect the world market (Chapter 10).

In Part IV, we assess the prospects that technology can resolve the coal-climate dilemma, examining technological efforts in Australia (Chapter 11) and China (Chapter 12). Finally, in Part V, we synthesize the collected research to produce new insights into the three questions noted earlier.

This book offers a close look at how coal markets really work, with a focus on the Asian producers and consumers that are most critically shaping their evolution. We hope that it may be of use to government policy makers who want to learn from the experience of other countries and jurisdictions that have attempted to balance the economic, environmental, and political questions around coal; to industry players who seek to learn from patterns in how the climate for coal investments may shift over time, both in specific jurisdictions and globally; to environmentalists who want to better appreciate the levers for environmental protection through an understanding of the interconnections within the global coal market and between coal and other energy; and, given the relative dearth of previous work in this area, to academics who seek a baseline assessment of coal market functioning off of which to build as coal's resurgence transforms global economies and the climate.

Notes

1. Coal can be separated into two main categories: steam coal, which includes thermal and lignitic coals, and coking coal, which includes all forms of metallurgical coals such as hard coking coal, semi-soft coking coal, and PCI. In this chapter, when we use the term "coal," we are referring primarily to steam coal.
2. The Industrial Revolution was supported by both steam coal, which provided energy for steam-driven pumps and locomotives, and coking coal, which provided a reductant for making steel.
3. There are, of course, many factors that could alter this outlook, notably including greater availability of natural gas if the successful development of shale gas in the US is replicated elsewhere. An emerging question is to what extent the shale gas development in the US, which has contributed to lower natural gas prices and consequently an erosion of coal demand in the power sector, can be replicated internationally. If it can, downward pressure on natural gas prices in key domestic markets as well as in internationally traded LNG markets might challenge coal's position in energy markets. The degree to which gas can compete with coal in the coming decades is a critical question for global energy supplies and global CO_2 emissions.
4. The New Policies Scenario was the central scenario used by the IEA in its analysis (IEA 2013). This scenario assumes that governments will follow through on climate policy commitments that they have announced but have not yet implemented.
5. This result holds in both the New Policies and Current Policies scenarios.
6. To make this estimate, we assumed that these renewable energy sources replaced an equivalent quantity of energy generation, with carbon intensity equal to the 2009 global average for electricity generation, from IEA figures. Clearly, long-term solutions to climate change will require power generation technologies with CO_2 intensities far lower than from even the most efficient coal plants currently available. Thus, strategies to achieve mitigation through enhanced coal combustion efficiency need to be optimized in the broader context of a "full portfolio" approach, which also includes alternative technologies and fuels.
7. We do not attempt to represent unpriced externalities from coal combustion here.
8. Of course, even broadly "liberalized" economies typically regulate their railway, power, and coal sectors. However, countries such as China and India, which have a strong heritage of central planning, are distinguished by the degree of government intervention – for example, through strict price controls and/or state ownership of key industrial players.

9. Note that pipelines (especially for natural gas) and electricity transmission lines do have high upfront costs and can be difficult to site in many jurisdictions.
10. Inland waterways nonetheless maintain a key role for coal transport in a number of countries. Indonesia is probably the foremost example of a country whose geography heavily favors transport of coal by water (see Chapter 7).
11. Significant differences within the segments of the value chain are common. For instance, the physical character and cost structure of underground and open-pit mining can vary significantly. In the transport sector, the economics of overland transport via rail networks will differ significantly from transport using navigable waterways. It is also important to note that international shipping of coal in seaborne markets utilizes dry bulk ships, for which there exists an independent commodity market where prices are governed by numerous factors exogenous to the coal market. Finally, as we will explore in our cases, power market structure and regulation varies greatly across countries.
12. In addition to this effect on pricing and contract structure, Parker suggests that, for the UK case, privatization of the electricity sector at least indirectly contributed to privatization of the coal industry (1996, p. 122).
13. It should be noted that this table focuses on SOE participation in the value chain and does not indicate where governments may take a strong role in controlling prices or other market outcomes without directly taking ownership of key assets.
14. By "artificially low" we mean prices that are below those which a competitive market would deliver. In some cases, electricity prices may not even allow generators to recoup the cost of coal (see Chapters 2 and 3).
15. This case illustrates how governments sometimes seek state ownership as an attempt to mitigate the negative impacts of market distortions without changing the underlying price and market frameworks that caused them.
16. Specifically, PLN and domestic IPPs now pay the government-mandated coal reference prices used to determine royalty payments to Indonesian coal suppliers.
17. Among other factors, subsidies for domestic coal producers, such as those in Europe, helped ensure that regional markets remained segmented and that countries could continue to develop domestic supplies even where normal market conditions would have given rise to international trade.
18. Sometimes the trade in coking coal facilitated the growth in the trade of steam coal by catalyzing the development of infrastructure that could serve both markets (see Chapter 6).
19. In 1978, the Carter administration passed the Power Plant and Industrial Fuel Use Act, which restricted construction of new oil and natural gas power plants and favored nuclear, coal, and alternatives. The prohibition on building natural gas plants was later repealed in 1987, leading to a dramatic increase in the use of natural gas in the power sector.

    Interestingly, writing in 1979, IEA executive director Ulf Lantzke acknowledged that CO_2 emissions would be a challenge in the future for expanding coal use due to their contribution to the "greenhouse effect" (Lantzke 1979). This is evidence that policy makers and IEA governments were broadly aware of coal's climate risks (although climate science was then at a far less advanced stage) even as they sought to triple the size of the coal trade. Despite this awareness dating back more than thirty years, few serious efforts to constrain coal's use on climate change grounds have been put forth to date.

20. The IEA's Ulf Lantzke put forward a vision in which coal would not only replace much oil use in power generation, but coal markets would also grow to function much as oil markets did: "The concept of an international free-trading market in coal analogous to other major commodity markets must be present throughout the planning process. Such a market could ultimately develop in a fashion similar to today's international oil market" (Lantzke 1979, p. 365).
21. Japan led the way through a targeted expansion of coal consumption in reaction to the oil crisis. Japanese policy explicitly targeted expanding coal use through policies aimed at increasing imports including loans to coal producers in return for off-take agreements, favorable loans for the construction of new coal-fired plants, and the promotion of the construction of coal import infrastructure. See Eguchi (1980) and Chapters 6 and 7.
22. We are mainly referring to price integration here. Physical flows of coal between the Atlantic and Pacific basins are also more significant now than they were in the 1990s. This "physical integration" is a function to a large degree of low charter rates for dry bulk vessels in recent years, and the extent of physical integration could change, for example, if these freight rates go up.
23. Ellerman argued in 1995 that a unified global market had developed for coal (Ellerman 1995). Several more recent studies using price co-integration analysis have concluded that the steam coal market is mostly, though not completely, integrated on price (Li et al. 2010; Zaklan et al. 2012). One study concluded that it was not possible to confirm co-integration in the 1990s for steam coal (Wårell et al. 2006). Considering the balance of the evidence, it is reasonable for practical purposes to assume that global coal prices in today's market are linked to a significant degree.
24. In 1997, the IEA described an emerging spot market that had begun to influence a market still defined by illiquid, bilateral pricing: "There is no single posted price. Annual negotiations for contract supplies are important, and partly fulfill this function, but also depend on continue market activity for spot supplies. Spot markets are limited in their operation because the variable part of demand is limited but do influence contract prices in the Europe-Atlantic market" (IEA 1997, p. 42).
25. Enron, GlobalCoal, and TFS were some of the early key players in the coal derivatives space. Firms such as Argus, Platts, and McCloskey, many of which have long offered price indices for physical oil markets, also became providers of the major coal price indices (though TFS first created some of the important indices).
26. A key reason for the slower adoption of index-linked contracts and coal derivatives in Asia is that utility markets in Asia have lagged in terms of

liberalization and introducing competition. Additionally, business culture in Asia has traditionally been slower to adopt derivatives as risk management tools.

27. Conversely, high-sulfur exports from US regions such as the Illinois Basin might increase SO_2 emissions in countries with weak environmental controls.
28. Virtually all of the growth in demand for coal will come from non-OECD countries, and the projected demand increase from China and India accounts for 69% of the incremental non-OECD demand. ASEAN countries account for another 21% of this projected non-OECD demand increase (IEA 2013).
29. As pointed out by Kolstad and Abbey (1983), other researchers have used transport infrastructure limitations – caused presumably by the inability of government institutions to create an attractive climate for such investments – to reconcile differences between actual trade patterns and those that would be expected based only on the supply and demand in different countries. Kolstad and Abbey show that, for the market of 1980, trade patterns could be explained equally well through models in which major participants exercised market power – especially South Africa and Australia as a supply duopoly and Japan as a monopsony buyer. In most of this book, we focus on the possible institutional drivers of trade patterns, but Chapter 10 also discusses market conduct explanations.
30. Though as of 2012, new agreements between Transnet and coal producers seemed to be paving the way for future export capacity expansion to 98 Mt by 2019.
31. An intriguing question is whether China could evolve to play a similar role on the world market – see Chapter 9 for relevant discussion.
32. Computed as the average of daily Henry Hub spot prices, traded on NYMEX and as reported by EIA.
33. The Cross-State Air Pollution Rule (CSAPR) and Mercury and Air Toxics Standards are of particular significance for existing coal assets.
34. Of course a more complete calculation of this arbitrage opportunity would have to include transportation costs and assessment of infrastructure costs.
35. While this price differential does not capture the very substantial infrastructure and transportation costs that would be incurred to arbitrage that differential, it may portend a looming challenge to coal's dominance as the cheapest source of energy traded in international markets.
36. Even in the absence of direct LNG exports to Asia, US shale gas could start to transform global gas markets in a way that makes gas more competitive in Asia. The emergence of US shale supplies has depressed US gas prices, causing LNG cargoes once intended for the US to be diverted to other markets. This offers a new source of cheaper supply (often through spot LNG cargoes) that is beginning to chip away at the oil-indexation regime in places such as Europe. As gas supplies become more abundant and spot markets proliferate, pressure on the oil-indexation regime could change energy economics in Asia, which also still relies on the oil-indexation regime. This would be likely to make gas cheaper in countries such as Japan, Korea, Taiwan, China, and India that now depend heavily on coal.

37. The UK's "dash to gas" in the 1990s as North Sea production ramped up is a possible analog.
38. Insight into how carbon prices impact dispatch economics in the EU is provided by the "clean dark spread" and the "clean spark spread."
39. Carbon capture and utilization is a related process in which CO_2 would be used as an input to the manufacture of useful products, such as cement. In theory, it would avoid the need for geological storage and derive value from the CO_2, but it has not yet been demonstrated to be economically viable.

References

Armstrong, Mark, Simon Cowan, and John Vickers (1994). *Regulatory Reform: Economic Analysis of the British Experience*. Cambridge, MA: MIT Press.

Bayer, Arne K., Maggi Rademacher, and Andrew Rutherford (2009). "Development and Perspectives of the Australian Coal Supply Chain and Implications for the Export Market." *ZfE Zeitschrift für Energiewirtschaft* 03/2009, 255–267.

Busse, Meghan R. and Nathaniel O. Keohane (2007). "Market Effects of Environmental Regulation: Coal, Railroads, and the 1990 Clean Air Act." *RAND Journal of Economics* 38(4), 1159–1179.

Eguchi, Yujiro (1980). "Japanese Energy Policy." *International Affairs* 56(2), 263–279.

Economic Times (2012). "India's Coal Imports Projected at 185 Mt by 2017." *Economic Times*, July 16, 2012. Available at http://articles.economictimes.indiatimes.com/2012-07-16/news/32698225_1_production-target-coal-imports-coal-india (accessed March 13, 2015).

EIA (1998). "Challenges of Electric Power Industry Restructuring for Fuel Suppliers." U.S. Energy Information Administration website. Available at http://www.eia.gov/pub/electricity/chg_str_fuel.pdf (accessed July 17, 2012).

EIA (2011). *World Shale Gas Resources: An Initial Assessment of 14 Regions Outside the United States*. Washington, DC: U.S. Energy Information Administration.

EIA (2014). "International Energy Statistics," U.S. Energy Information Administration website. Available at http://www.eia.gov/cfapps/ipdbproject/IEDIndex3.cfm (accessed August 20, 2014).

Ekawan, Rudianto and Michel Duchêne (2006). "The Evolution of Hard Coal Trade in the Atlantic Market." *Energy Policy* 34(13), 1487–1498. ISSN 0301-4215, 10.1016/j.enpol.2004.11.008.

Ellerman, A. Denny (1995). "The World Price of Coal." *Energy Policy* 23(6), 499–506.

Ellerman, A. Denny and Juan-Pablo Montero (1998). "The Declining Trend in Sulfur Dioxide Emissions: Implications for Allowance Prices." *Journal of Environmental Economics and Management* 36, 26–45.

EPA (2012). "Standards of Performance for Greenhouse Gas Emissions for New Stationary Sources: Electric Utility Generating Units." *Proposed Rule, Docket Number EPA-HQ-OAR-2011-0660*. U.S. Environmental Protection Agency, March 27, 2012. Available at http://epa.gov/carbonpollutionstandard/pdfs/20120327proposal.pdf (accessed August 1, 2012).

Finch, Roger (1973). *Coals from Newcastle: The Story of the North East Coal Trade in the Days of Sail*. Lavenham: T. Dalton.

Garnett, A. F. (2005). *Steel Wheels: The Evolution of the Railways and How They Stimulated and Excited Engineers, Architects, Artists, Writers, Musicians, and Travellers*. Waldenbury: Cannwood.

Gerking, Shelby and Stephen F. Hamilton (2008). "What Explains the Increased Utilization of Powder River Basin Coal in Electric Power Generation?" *American Journal of Agricultural Economics* 90(4), 933–950.

Gerking, Shelby and Stephen F. Hamilton (2010). "SO_2 Policy and Input Substitution under Spatial Monopoly." *Resource and Energy Economics* 32, 327–340.

Gordon, Richard L. (1987). *World Coal*. Cambridge: Cambridge University Press.

Great Britain Parliament House of Commons Energy and Climate Change Committee (2012). *The EU Emissions Trading Scheme: Tenth Report of Session 2010–12*. London: The Stationery Office.

Grubler, Arnulf (2012). Personal communication.

Haftendorn, Clemens (2012). *Evidence of Market Power in the Atlantic Steam Coal Market Using Oligopoly Models with a Competitive Fringe* (January 26). DIW Berlin Discussion Paper No. 1185.

IEA (1997). *International Coal Trade: The Evolution of a Global Market*. Paris: International Energy Agency.

IEA (2010). *CO_2 Emissions from Fuel Combustion, 2010 Edition*. Paris: International Energy Agency.

IEA (2013). *World Energy Outlook 2013*. Paris: International Energy Agency.

IEA (2014a). *Coal Information 2014 (with 2013 Data)*. Paris: International Energy Agency.

IEA (2014b). *Electricity Information 2014 (with 2013 Data)*. Paris: International Energy Agency.

IEA (2014c). "IEA Coal Information Statistics." International Energy Agency. Available at http://www.oecd-ilibrary.org/energy/data/iea-coal-information-statistics_coal-data-en (accessed April 2, 2014).

IIASA (2012). *Global Energy Assessment: Toward a Sustainable Future*. Cambridge: International Institute for Applied Systems Analysis/Cambridge University Press.

Keohane, Robert O. and David G. Victor (2010). "The Regime Complex for Climate Change," Discussion Paper 2010–33, January 2010. Cambridge, MA: The Harvard Project on International Climate Agreements.

Kolstad, Charles D. and David S. Abbey (1984). "The Effect of Market Conduct on International Steam Coal Trade." *European Economic Review* 24, 39–59.

Kolstad, Charles D., David S. Abbey, and Robert L. Bivins (1983). "Modeling International Steam Coal Trade." Report from Los Alamos National Laboratory LA-9661-MS/UC-98f, January.

Landes, David S. (1969). *The Unbound Prometheus: Technological Change and Industrial Development from 1750 to the Present*. Cambridge: Cambridge University Press.

Lantzke, Ulf (1979). "Expanding World Use of Coal." *Foreign Affairs* 58(2), 351–373.

Li, Rayond, Roselyne Joyeux, and Ronald D. Ripple (2010). "International Steam Coal Market Integration." *The Energy Journal* 31(3), 181–202.
Light, Miles K., Charles D. Kolstad, and Thomas F. Rutherford (1999). "Carbon Markets, Carbon Leakage and the Kyoto Protocol." Working Paper No. 99–23, October. University of Colorado at Boulder, Center for Economic Analysis.
McCarthy, James E. and Claudia Copeland (2011). "EPA's Regulation of Coal-Fired Power: Is a 'Train Wreck' Coming?" *CRS Report for Congress R41914*, August 8, 2011. Washington, DC: Congressional Research Service.
Miller, Donald L. and Richard E. Sharpless (1985). *The Kingdom of Coal: Work, Enterprise, and Ethnic Communities in the Mine Fields*. Philadelphia: University of Pennsylvania Press.
Parker, Mike. (1996). "Effects on Demands for Fossil Fuels," in John Surrey, ed., *The British Electricity Experiment. Privatization: The Record, the Issues, the Lessons*. London: Earthscan, 120–137.
Rademacher, Maggi (2008). "Development and Perspectives on Supply and Demand in the Global Hard Coal Market." *ZfE Zeitschrift für Energiewirtschaft* 02/2008, 67–87.
Ritschel, Wolfgang and Hans-Wilhelm Schiffer (2007). *World Market for Hard Coal*. Available at http://www.rwe.com/web/cms/mediablob/en/247566/data/235584/1/rwe-power-ag/hard-coal/blob.pdf (accessed March 13, 2015).
SCE (2012). "Mohave Generation Station," Southern California Edison website. Available at http://www.sce.com/PowerandEnvironment/PowerGeneration/MohaveGenerationStation/default.htm (accessed July 21, 2012).
Schernikau, Lars (2010). *Economics of the International Coal Trade: The Renaissance of Steam Coal*. Dordrecht: Springer.
Spiegel (2014). "Green Revolution? German Brown Coal Power Output Hits New High." *Spiegel Online*, January 7, 2014. Available at http://www.spiegel.de/international/germany/researchers-alarmed-at-rise-in-german-brown-coal-power-output-a-942216-druck.html (accessed August 25, 2014).
Thapar, Kapil (2015). "India: Expectations and Projections for Coal Demand." Presentation at 13th Annual Coal Markets conference, Singapore, February 3–6, 2015.
Victor, David G. (2011). *Global Warming Gridlock*. Cambridge: Cambridge University Press.
Victor, David G. and Richard K. Morse (2009). "Living with Coal: Climate Policy's Most Inconvenient Truth." *Boston Review*, September/October, 7–14.
Wårell, Linda (2006). "Market Integration in the International Coal Industry: A Cointegration Approach." *The Energy Journal* 27(1), 99–118.
Wilson, Charlie and Arnulf Grubler (2011). "Lessons from the History of Technological Change for Clean Energy Scenarios and Policies." *Natural Resources Forum* 35, 165–184.
Wolak, Frank A. and Charles D. Kolstad (1988). "Measuring Relative Market Power in the Western U.S. Coal Market Using Shapley Values." *Resources and Energy* 10, 293–314.
Wolak, Frank A. and Richard K. Morse (2010). "China's Green Gift to the World." *The Guardian*, December 30. Available at http://www.theguardian.

com/commentisfree/cifamerica/2010/dec/30/coal-energy-industry (accessed March 13, 2015).

Xinhua (2011). "China to Cap Energy Use at 4 Bln Tonnes of Coal Equivalent by 2015: Former Energy Head." Xinhua News Agency, March 4, 2011. Available at http://news.xinhuanet.com/english2010/china/2011–03/04/c_13761106.htm (accessed August 1, 2012).

Zaklan, Aleksandar, Astrid Cullmann, Anne Neumann, and Christian von Hirschhausen (2012). "The Globalization of Steam Coal Markets and the Role of Logistics: An Empirical Analysis." *Energy Economics* 34, 105–116.

Part II

Case studies of key coal countries

2 The evolution of China's coal institutions

Wuyuan Peng

2.1 Introduction

Coal is the major primary energy source that fuels economic growth in China. China has abundant coal but far less oil and natural gas. Alternatives, such as hydropower and nuclear power, cannot keep pace with the speed of economic growth due to their high capital cost and the controversy regarding issues such as relocation of populations, ecological degradation, and safety concerns. Therefore, almost 70% of primary energy in China comes from coal.

When the People's Republic of China (PRC) was founded in 1949, its annual coal production was only 32.4 million metric tons (Mt). China's coal production has since grown to be the largest in the world. In 2012, coal production was approximately 3.65 billion metric tons (Bt), accounting for about 47% of the world's production of coal.

Soviet-style central planning of the coal sector was adopted after the PRC was founded. But since the end of the 1970s, there have been major changes to the institutions in the coal industry: market mechanisms were introduced to the coal sector, and the major state coal mines were transferred from central to local governments (Table 2.1). In 1985, the central government initiated a general contract system to regulate the input and output of unified distribution coal mines, which are under central planning and have access to railway transportation.[1] Construction of key mines belonging to the Shenhua Group,[2] the largest coal project in China, also started in 1985. During the same period, the permitting procedure for small-sized coal mines was transferred to local governments and made much simpler than before. In 1993, except for power

The author would like to thank David Victor, Mark Thurber, Varun Rai, Gang He, and Richard Morse for their contributions to this chapter. Kathy Lung, Alexander Slaski, and others also contributed significantly. Catherine Vogel spent a lot of time editing this chapter. The author would also like to thank Jiahua Pan who introduced him to PESD in 2004. Most importantly the author is grateful to the China University of Geosciences (Wuhan Campus) for granting a one year sabbatical to complete this project.

Table 2.1 *Trend of coal institutions' evolution in China*

	Applying central planning (1949–1978)	Introducing a market system (1979–1992)	Establishing a market system (1993–2002)	Perfecting a market system (2003–present)
Administration	Hierarchical control by central agency	Two-level system, by central and local governments	Local governments after 1998 except Shenhua, China Coal, and Yimin*	Regulation by central government
Finance**	State budget	State budget switching to state loan	State creditor switching to shareholder	Self-raised
Pricing**	Regulated and underpriced	Two-track system	Relaxed except for power generation and agricultural use	Relaxed

* The total raw coal production share of Shenhua, China Coal, and Yimin is between 10% and 15% nationwide.

** This refers to the Shenhua Group and unified distribution coal mines, which are the mainstays of the coal sector.

generation and agricultural use, the price of coal was liberalized, and the unified distribution coal mines were pushed to market competition. In 1995, the Shenhua Group was founded as a central firm under the direct administration of the State Council. In 1998, the administration of the unified distribution coal mines was transferred to local governments, further decentralizing control of the coal sector. By 2002, all coal pricing, even for power generation, had been relaxed. Over the course of about twenty years, the coal sector had shifted from central planning to a market-based and decentralized system.

This chapter explains these remarkable market-oriented and decentralizing trends and explores their implications for the power sector, now the largest single consumer of coal. There is a lot of existing research about the coal industry in China but little on the institutional evolution of the coal sector. The existing research on coal in China can be grouped into the following main categories.

First, the issue of sustainable development of the coal industry has received attention from many institutes and researchers. Pan et al. (2002) compared the coal market concentration between China and the United States and argued that the market concentration in China was too low. The World Bank Energy Sector Management Assistance Program

(ESMAP 2004, 2008) identified many of the problems that China's coal mining industry needs to address on the path toward sustainability, including safety, health, and environmental problems. The China Energy Group at Lawrence Berkeley National Laboratory has researched energy efficiency and strategy (Sinton, et al. 2005; Zhou et al. 2007). The Industrial Performance Center at MIT published several working papers related to China's coal sector (Lester and Steinfeld 2006, 2007), with an emphasis on patterns of coal use. The International Energy Agency (IEA 2007, 2009) has addressed coal supply and clean coal technology in China. Clean coal technologies are needed to utilize coal in an environmentally acceptable way and to improve coal utilization efficiency (Chen and Xu, 2010). Xiao et al. (2008) used a vector auto regression (VAR) model to analyze the effect of coal mine safety regulation, finding the long-term effect to be significant. Wang (2010) discussed the challenges and opportunities faced by China's coal industry, including production capacity, safety, technology options, and pollution control.

Second, a number of studies have examined China's role as an important player in the global coal market. The Australian Bureau of Agricultural and Resource Economics and the Japanese Institute of Energy Economics have conducted a series of research studies focusing on coal import and export projections for China (Ball et al. 2003; Schneider 2004; Sagawa and Koizumi 2007, 2008). Lin and Liu (2010) estimated China's coal production peak and analyzed trends in coal imports.

Third, many researchers analyzed the coal sector from an enterprise level or industrial perspective. Thomson (2003) treated the coal industry in China as a case study with lessons for the country's economic development as a whole. Rui (2005) analyzed different types of Chinese coal mining enterprises and their implications for China in the context of global changes in the coal industry. Rosen and Houser (2007) analyzed China's energy demand and supply system and its global impacts. Shi (2008) evaluated the opening up and reform program of the energy industry in China, but did not explore the driving forces and implications of the reforms.

Fourth, researchers have explored the relationship between the coal and power sectors in China. Yu and Yu (2006) proposed the vertical integration model between coal and power to solve the conflict between the two sectors (see Chapter 3, this volume). Wang (2007) explored the unbalanced development of the coal and power sectors, suggesting that the imbalance was caused by government intervention. Zhang and Heller (2007) explained why China's power sector reform stalled. Zhu (2009) identified coal as the pillar of electric power security in China.

Most of the research has put an emphasis on coal mine safety, clean coal technology, coal demand projections, international trade, and vertical integration, but very little has considered the cause and effect of the evolution of coal institutions in China.

The hypothesis of this chapter is that the market-oriented and decentralizing reforms in the coal sector were influenced by the changes of state energy development priority as well as the relationship between the central and local governments in the context of broader reforms within China's economy. However, those reforms have not influenced all related sectors equally. Although the central government has allowed substantial market-oriented reforms in coal, those reforms have not easily extended to the power sector. The fragmented character of these reforms has had significant implications. Since power sector reform lags coal sector reform, coal prices have become market-oriented while power prices remain tightly regulated. Because coal is the primary input into Chinese power generation, the tension between the power and coal sectors is unavoidable and has raised concerns about electricity shortages.

The rest of the chapter is organized as follows: Section 2.2 illustrates the initial conditions of coal institutions' evolution in terms of administration, finance, and pricing in China; Sections 2.3, 2.4, and 2.5 discuss the coal institutional evolution in the subsequent stages of reform; and Section 2.6 concludes the study.

2.2 Applying central planning to the coal industry (1949–1978)

After the People's Republic of China was founded in 1949, Soviet-style central economic planning was adopted. From the beginning of socialism in China, it had been regarded as an unalterable principle to abolish the market system and establish a planned economy characterized by highly centralized administrative coordination, following the example of the Soviet Union. After the outbreak of the Korean War in 1950, China had to give top priority to national defense-related industries. Chinese leaders chose an institutional arrangement that would mobilize and allocate resources through central planning so that limited resources could be used to build up heavy industries, especially the country's core military industry.

With the goal of heavy industry development, all of the key energy sectors – coal, electric power, and petroleum – were put under central planning and given priority to develop.[3]

The local governments under the central planning system had little authority and instead had obligations to implement the planned tasks. In the Soviet economic planning system, production was organized hierarchically, with targets set by the central authorities. In the final quarter of each year, the State Planning Commission (SPC)[4] prepared the following year's preliminary balances for all commodities and capital goods under state control. These goods were referred to as being "under unified distribution" or simply "under plan."

2.2.1 *Administration*

In 1949, the new government nationalized most coal mines, and from 1953 the first five-year plan was implemented. The foreign coal mines and nationalist (or Kuomintung) government coal mines were taken over by the central government, and the remaining coal mines were transferred to local state coal mines and township and village coal mines,[5] with compensation according to the size of coal mines. In terms of coal production, the shares of state-owned companies, public/private joint ventures, and private coal mines were 68.2%, 3.5%, and 28.3% respectively in 1949. Five years later, the share of private coal mines had decreased to 6.1% (Zhang 1989).

As coal was given one of the top priorities to develop, it was necessary to set up an independent ministry in charge of the coal sector. The Ministry of Coal Industry was separated from the Ministry of Fuel Industry as an independent body in 1955, with greater autonomy and a more coherent planning structure.[6]

Each ministry under the planned economy was given proposed output targets that were communicated to all the various sub-units. Additionally, balance tables were drawn up to schedule the mode and route of transportation for every commodity under plan.

Monopoly administration of the coal sector under central planning was used to boost downstream heavy industries. The state tried to tailor both forward and backward economic linkages through the coal industry. The national agencies (State Bureau of Material, Ministry of Railways, and others) arranged the supply of input materials, such as timber, explosives, and electric power, and also directly coordinated the sales, allocation, and transport of all coal under plan through the annual National Coal Ordering Conference[7] in order to match the coal supply to the demand from key sectors (see Chapter 3, this volume, for more discussion of this conference). The state also tried to balance the total amount of coal produced and consumed by each region.

Unified distribution coal mines[8] dominated the coal sector during the central planning period. China's coal mines were divided into three main categories according to the type of ownership: the major state coal mines, the local state coal mines, and township and village coal mines.[9] The former two were state-owned enterprises. The major state coal mines were controlled by the central government through the administration of national agencies, with most production being allocated under central plan. The local state coal mines were operated either by provincial, prefectural, or county governments, while township and village coal mines were operated by townships, communes, or collectives. Both the local state coal mines and township and village coal mines were administrated by a Provincial Coal Industry Bureau. The output from the local state coal mines was also generally under unified distribution, whereas the output from the township and village coal mines was not. Because the production from township and village coal mines was not put under central planning, it could not be sold freely, even to the neighboring communities.

2.2.2 Finance

The leading characteristic of the financial system under the planned economy was the integration of public finance and business finance. Important economic decisions of firms were never based on financial considerations of the firm. Rather, decisions about which investments to undertake were made by government planners and financed by the government budget. Firms that possessed a monopoly position forwarded their cash surplus to the government, which made up the main source of government revenues.

The state bank was considered the cashier for the state fiscal system under central planning, and a single institution played the roles of both central bank and commercial bank. From 1949 to 1978, the People's Bank of China (PBC) was the only state bank in China, integrating the functions of the central bank, such as financial supervision, with those of policy banks and commercial banks, such as savings and loans.

The state coal mines were financed via the state budget allocation. During the central planning period, the government provided over 90% of the funding in the coal sector. State-owned enterprises did not function as independent firms with management responsibility, the ability to allocate assets, or control over how to spend their earnings. A year-end plan target for total capacity specified where and how their goals should be achieved, whether by new mine construction or by renovation and expansion of existing mines. Prospecting, surveying, and designing teams were assigned to each project. The amount of new capacity created each year was directly related to the investment funds allocated to the Ministry of Coal Industry from the state budget allocation. The Ministry of Coal

Industry also oversaw investment in coal washing plants, machinery plants, and railway branch lines. Between 1953 and 1978, China's coal industry received a total of RMB 349.5 billion[10] in state investment compared to RMB 457.7 billion for the electricity industry (National Statistical Bureau, various years).

2.2.3 Pricing

In the Chinese centrally planned economy, prices had several main functions. First, price was used as an accounting device in the tabulation of the annual and five-year plans. The inputs and outputs of each enterprise, as well as targets for sales, profits, taxes, etc., were calculated by the central planners and expressed in value terms; enterprise managers were evaluated on the basis of their fulfillment of these targets. Prices were based on average production costs and may have included a turnover tax or subsidy and a wholesale and/or retail profit margin. There was no consistent relationship between prices and demand. Second, price was used for mobilizing resources, with the prices of agricultural and upstream products kept artificially low and the prices of final industrial and downstream products kept high. The distorted price system meant that state-owned downstream industrial enterprises were extremely profitable, thus giving the government the fiscal capacity to mobilize economic resources in pursuit of its goals.

The price of coals from the unified distribution mines were tightly regulated and set very low by the central government in order to boost downstream activities, such as electricity production, as well as to prevent inflation. The "producer" or "supply" price of coal in each of the provinces was strictly a function of production costs plus local transportation costs. The production costs were calculated as the average of annual operating costs per tonne incurred by the state underground mines operating at peak levels, and excluded any investment costs sustained in the preliminary stages of mine development. From 1949 to 1985, there were only four nationwide increases in average state mine producer prices (Table 2.2).

The underpricing of coal resulted in a severe shortage of coal supply. Thus, from roughly 1972 onward, the non-state mines were allowed to sell their coal at prices negotiated in the market in order to incentivize output and mitigate the coal supply shortage.

2.2.4 Performance

Productivity improvement was slow even though in 1978 China's coal production had reached nearly 618 Mt, placing it third in the world after

Table 2.2 *Coal price adjustments nationwide (RMB/tonne)*

	1958		1965		1979		1985	
Raw coal	Before adjustment	After adjustment	Before adjustment	After adjustment	Before adjustment	After adjustment	Before adjustment	After adjustment
Selling price	10.99	13.44	15.33	17.34	15.91	20.98	23.81	26.86
Cost	9.28	10.20	14.34	15.78	13.54	17.80	17.80	23.37

Note: Some local price adjustments in interim periods are not shown here.
Source: Zhang (1989).

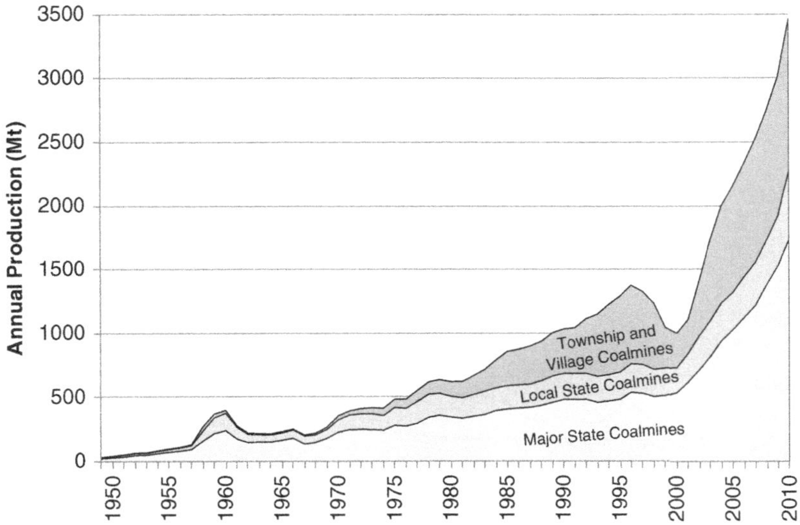

Figure 2.1 Raw coal production in China by ownership. Data source: China Coal Industry Association (2011).

the US and the USSR.[11] Output per man shift rose from an average of 0.36 tonnes in 1949 to 0.93 in 1978, still a fraction of what it was in the other large coal-producing countries. For example, in 1974, output per man shift was 2.44 in the USSR, 3.41 in the UK, 4.06 in West Germany, 4.15 in Austria, and 9.60 in the United States. Output per man shift in China was closer to India's 1970 output of 0.73. The growth of coal production was mainly attributable to higher labor input rather than capital investment. From 1957 to 1978, total employment at the state coal mines increased from 0.3 to 4.1 million while the mechanization level (extraction and tunneling) grew only from 4.1% to 32.8%.

Although production was dominated by the central coal mines, local coal mines also developed steadily in number and capacity. In 1952, 1.4% of the national output came from the township and village coal mines; by 1978, this total was 15.4% (Figure 2.1). This was partially due to price controls on non-state mines being lifted in 1972, as mentioned earlier.

The Ministry of Coal Industry was charged with the task of balancing the regional production and consumption of coal. China's highest quality coal – high in calorific value and low in sulfur and ash level – is found in western China: Shanxi, Shaanxi, and Inner Mongolia. However, consumption was (and still is) much higher in eastern China.[12] In 1952, eastern China's production share of coal nationwide was 13.0% while its

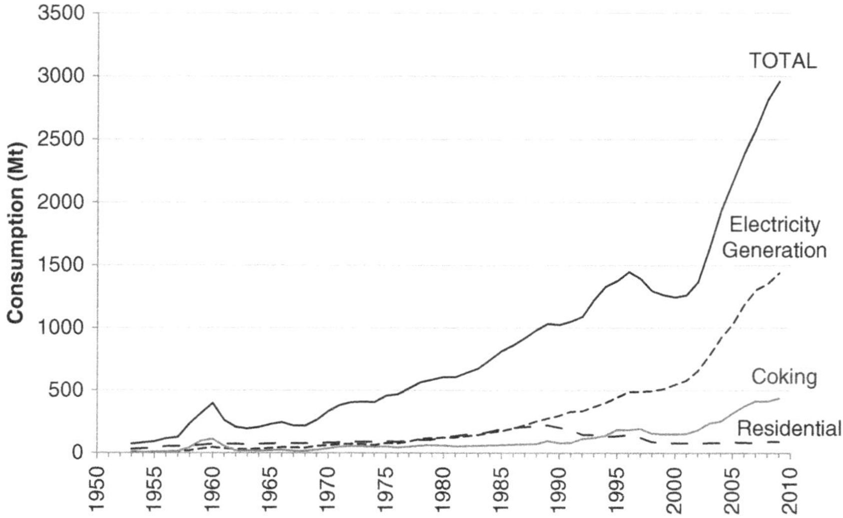

Figure 2.2 Coal consumption in China by end use. Data source: China Coal Industry Association (2011).

consumption share was 36.3%, leaving a deficit of 23.3%. In 1978, eastern China's production share increased to 17.2% while its consumption share decreased to 34.9%, for a reduced but still significant deficit of 17.7%. This regional imbalance required the central government to carefully coordinate the production, transportation, and consumption of coal.

Under central planning, it was a priority to allocate as many resources as possible to develop the heavy industries. Steel production was initially prioritized, but afterward the development of power generation was deemed more important (Figure 2.2).

Despite its growth, the coal industry was afflicted by the unavoidable flaws of central planning, such as economic inefficiency and chronic shortage. These inefficiencies existed because the coal sector faced soft budget constraints under the planned system. The state coal mines got their investment capital funding from the state budget allocation without the need to repay it, and the allocation for miners' salaries was based on the number of employees. The chronic shortage in coal production was due to underpricing of coal. In 1957, 50% of China's major mines had financial deficits, and as of 1977, the proportion of major mines nationwide with deficits had reached 73%. Instead of a steady increase in profit per ton, there were large fluctuations in output and profits. The volatility in profits greatly affected production in that there was never enough capital available on a sustainable basis to reinvest and to increase the levels of mechanization, processing, and storage capacity. In turn, the chronic shortage in coal production led to a power supply shortage.

2.3 Introducing a market system to the coal industry (1979–1992)

Even though the central planning system and heavy-industry priority strategy created an important heavy industrial base, efficiency was very low. The industrial growth rate was high in the 1950s but slowed during the 1970s. The combination of low industrial growth and high unemployment generated profound dissatisfaction with the Soviet central planning system. In response, markets were accorded a greater role in the planning process after the "reform and opening up program" was initiated at the Third Plenary Session of the 11th Central Committee on December 18, 1978. The assumption was that system transformation from central planning to a market economy would have to take place concurrently with economic development. This reflected China's approach to economic transition; market incentives were introduced and combined organically with central planning. The overarching vision was that the process of economic development would drive market transition forward and guarantee its eventual success.

After the initiation of market-oriented reforms, the priority of state energy development was changed. The focus of the Chinese government shifted away from national defense toward overall economic development. The program of prioritizing heavy industry development was phased out; in turn the state's direct investment in the coal sector was reduced. Electric power became the center of state energy development. Hydropower was given priority to develop because hydropower resources are abundant in China (Zhou and Wang 2001).[13] By the initiation of the reform and opening up program in 1978, electricity shortages had become the main constraint on economic growth; 20% to 30% of industrial production capacity could not operate normally due to the shortage of electricity. The consequence of the adjustment of state energy development priority was that the strategic position of coal in the central planning system was diminished. At the same time, coal production had to be increased very rapidly to meet the new demands of economic development. In 1982, the Twelfth Plenary Meeting of the Chinese Communist Party decided that GDP should be tripled by 2000 in light of Deng Xiaoping's slogan "development is the paramount force" (Wu 2005). In order to fuel economic growth, coal production was to be doubled by 2000, from 600 Mt to 1.2 Bt. This brought new challenges to the coal sector.

Local governments became important players after the initiation of the reform and opening-up program. In addition to delegating power to and sharing profit with state-owned enterprises, the central government delegated power to and shared profit with local governments in order to give local governments more incentive to develop the local economy. As a consequence, the central government's monopoly over the industry was reduced. In the mid-1980s, the guiding principle of "keeping public ownership as the mainstay of the economy and allowing diverse forms of ownership to develop side by side" was put forward (Wu 2005). This provided opportunities for non-state enterprises, especially rural industries, to take advantage of large potential profits in the industrial sector. Their activities were usually unregulated and lightly taxed.[14]

A market system was introduced to the coal sector and regarded as the necessary way to increase coal output while mitigating the state fiscal burden. The "two-leg walking" strategy, which encouraged growth of both the central mines under central planning and local mines outside central planning (in particular township and village coal mines) to develop, was adopted even though the central coal mines remained much more important than local coal mines in terms of scale of resources and application of technology.[15] These reforms caused a series of changes in the areas of administration, finance, and pricing in the coal sector.

2.3.1 Administration

A two-level administration system, central and local, was adopted in accordance with the two-leg walking strategy (Figure 2.3). At the central

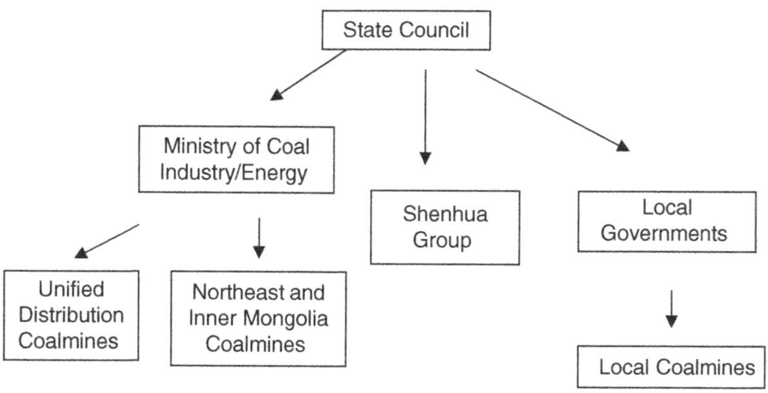

Figure 2.3 Administration of the coal sector in China.

level, the Northeast and Inner Mongolia coal mines were separated from the unified distribution coal mines because in the Northeast region the coal resource was depleted, whereas in Inner Mongolia the coal resource was rich but the public infrastructure (including transportation and power) was poor. The Ministry of Coal Industry, which was originally responsible for the unified distribution coal mines, lasted until 1988 and then merged into the Ministry of Energy. In 1982, the central government started the construction of the Shanxi Coal Base[16] and created the China Coal Import and Export Corporation (China Coal) to absorb foreign direct investment.[17] Even more important, it initiated the general contract system for the input and output of unified distribution coal mines in 1985. The general contract system in the coal industry is a policy of delegating powers to and sharing profit with lower levels; the Ministry of Coal Industry entered into contracts with the state first, then subcontracted with the unified distribution coal mines. The contracts covered both production output and planned loss. The general contract approach abandoned the planned system of coal distribution, enhanced the enterprises' power of self-determination, and increased their incentive to produce.[18]

The Shenhua Group (discussed in more detail in Chapter 3 of this volume) remained outside of the Ministry of Coal Industry and was directly administered by the State Council. The central government initiated production at the Shenhua Group in 1985. Shenhua's primary resource was the Shenfu Dongshen Coal Fields located on the border of Inner Mongolia and Shannxi, which had a recoverable reserve of 22.3 Bt, out of 110 Bt of reserves nationwide. Because of these incredible resources, Shenfu Dongshen was planned as China's most important future coal base, with the Shenhua Group overseeing this project.

At the local level, the two-leg walking strategy transferred the administration of small-sized coal mines from the central to local governments in order to increase overall output. Local governments or coal departments could approve mining permits to small mines. In April 1983, the State Council approved the "report on eight measures for accelerating the development of small scale coalmines" and ordered all jurisdictions to implement it (Zhou and Wang 2001). Communes, brigades, and all trades and industries were encouraged to operate mines, and citizens were encouraged to pool finances to support the mines. Areas which were short of coal could operate mines in partnership with other areas that had abundant supplies. Small mines could extract resources belonging to the state mines that could not be mined by the large mines. The operators of locally produced coal could haul and sell their coal anywhere at a negotiated price. However, if rail transport was required, the coal would be put under unified distribution.

2.3.2 Finance

In order to improve the efficiency of capital usage, the fiscal budgetary system of China was switched in 1980 from the "unitary system," under which state-owned enterprises were funded out of the government budget and forwarded all of their profits to the government, to the "responsibility system," under which SOEs had to repay the loans that funded investments but also had more autonomy to choose their own projects. The source of investment in fixed assets of state-owned enterprises shifted from the state budget allocation to bank loans. A revenue-sharing system between central and local governments was introduced in order to generate enthusiasm on the part of local governments.[19] The local governments were given incentives to boost local firms, including local coal mines. These various reforms altered the relationships between the central and local governments.

After the initiation of market-oriented reform, a series of financial reforms were set in motion. From 1979 to 1985, specialized banks[20] were created, and the People's Construction Bank of China became the bank specializing in financing investment in fixed assets. In 1983, the State Council decided that the People's Bank of China (PBC) should focus on its role as the central bank.

The fiscal and financial system reforms extended to the unified distribution coal mines. Under the "general contract of input and output," the financing of basic construction investment in unified distribution coal mines shifted from the state budget allocation to state loans and was operated by the Ministry of Coal Industry and the Energy Investment Corporation, which was established by the central government to deal with state loans for energy projects (Figure 2.4). At the same time, the profit-forwarding system shifted to a tax-collecting system, and the state coal mines were given the authority to retain a share of profit for reinvestment. From 1985 to 1993, the amount of state loans awarded to mines totaled RMB 35.2 billion. Switching of financing for basic infrastructure construction from the state budget to state loans caused a heavy debt burden on major state coal mines, mainly because coal prices were not liberalized. During this period, multiple sources for financing outside of the state budget plan were created, for example, "coal replacing oil" loans (with an interest rate of 2.4%), loans from the People's Construction Bank of China (interest rates between 6.48% and 9.9%), construction bonds (interest rates from 6% to 11%), foreign

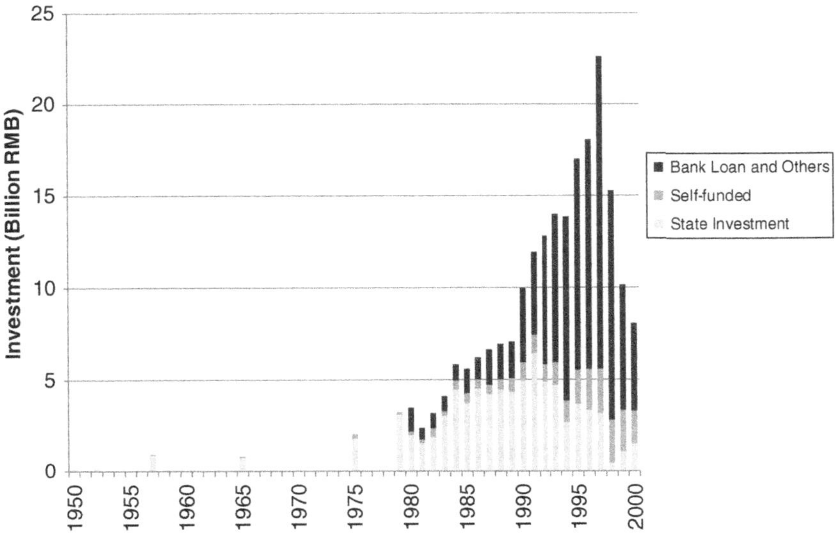

Figure 2.4 Annual investments at major state coal mines. Data source: China Coal Information Institute, various years.
Notes: (1) State investment includes state loan under budget plan. (2) Data after 2000 is not available due to the change of statistical method.

government loans, and World Bank loans (interest rate of 8%). From 1985 to 1993, the total amount of bank loans exceeded RMB 31.4 billion.

Only Shenhua Group[21] was the beneficiary of continued state budget allocations (Figure 2.4). The total investment in Shenhua was targeted for RMB 89 billion from 1985 to 1995. By this time, the group was made up of the Shenfu Dongshen mines, pit mouth power plants, the Shenhuang double tracking electrified railway, and the Huanghua coal transport sea port.

After market access was made easier under the two-leg walking strategy, millions of local coal mines entered the market, receiving funding mainly from local communities.

2.3.3 *Pricing*

The unified distribution coal mines under the general contract adopted a dual-track pricing system: planned price and "floating price." The non-unified distribution coal mines sold all their coal at a market price (negotiated or spot).

The dual-track pricing system was applied as an incentive to increase production. Above-quota output that the state mines produced could be sold to the state at 50% (later changed to 70%) above the plan price, or on the open market at a floating price. However, during the years the general contract was in effect, the average selling price of raw coal was lower than the production cost because the plan price was fixed while production costs increased rapidly due to severe inflation,[22] causing most unified distribution coal mines to lose money. The general contract included clauses on losses. In order to make up for the difference between production costs and selling prices, the Ministry of Finance subsidized the industry by RMB 300 million per year between 1985 and 1990, with the provision later extended to 1992. If the enterprises incurred losses beyond those expected, the subsidies would cover these losses. If, on the other hand, the losses were less than anticipated, the subsidy was to be shared, with 60% going to the enterprises and 40% to the state. However, the subsidies could not compensate for the losses, and the mines did not have the capacity to invest and develop on their own. By 1988, 80 out of 86 unified distribution coal mine bureaus were operating at a loss.

Since 1972, non-unified distribution coal mines had been permitted to sell their coal at market prices, and as a result their coal selling prices were much higher than the planned price in the main coal consumption markets. For example, the market prices of coal were usually 50% to 100% higher than planned prices in Shanghai, Nanjing, and Suzhou during this period (Tian and Qiao 1991).

Already, the central government's reliance on limited market pricing to increase coal output was shifting production incentives away from the central government–owned unified distribution mines and toward locally controlled mines. The consequence was that a larger share of production was controlled at the local level, and the local governments and miners became important players in subsequent reforms of the coal sector.

2.3.4 Performance

Coal production in China witnessed a great expansion in response to the new incentives (the combination of output-based incentives and market signals) at both unified distribution mines and local mines. The two-leg walking program helped increase coal production from 635.5 Mt in 1979 to 1.1 Bt in 1992, close to the objective of 1.2 Bt by 2000. Production at major state coal mines increased from 357.8 Mt in 1979 to 482.5 Mt in 1992; production at local coal mines (including local state coal mines as well as township and village coal mines) increased from 277.8 Mt in 1979 to 632.0 Mt in 1992, surpassing the major state coal mines[23] (Figure 2.1).

Local governments had many incentives to develop local mines: developing the local economy, creating jobs, and improving living standards of local residents.

Productivity improvement was faster than before. Under the general contract system, the gross revenue of unified distribution coal mines was linked to the coal output, so the operators of the unified distribution coal mines had the incentive to increase the use of machinery and equipment and decrease the use of laborers. In terms of extraction, tunneling, and loading, the mechanization level of major state coal mines increased from about 30% in 1979 to more than 70% in 1992. However, the size of the work force decreased; in 1985, the number of employees decreased by 110,000 while production increased by 11.6 Mt. Productivity increased because each unit increase in output required less labor input. Output per man shift at the major state coal mines improved from 0.912 tonnes in 1980 to 1.33 tonnes in 1992. Output per man shift at local state coal mines also increased from 0.583 tonnes in 1985 to 0.795 tonnes in 1994.[24]

The favorable policies for small coal mines and small thermal power facilities increased coal consumption for power generation, particularly in rural areas.[25] The share of coal for power generation increased from 20.3% in 1980 to 28.9% in 1990. By the mid-1990s, local coal-fired power generation was one of the main sources of rural electrification (Peng and Pan 2006).

Regional coal imbalances widened as eastern China became the manufacturing base and coal consumption center while northern China grew into the coal production base. In 1979, the central government began wider market-oriented reforms that spurred economic growth, in particular along the east coast. In 1985, eastern China imported 63.8 Mt of coal from other regions; by 1990, 109.5 Mt of coal was imported. The locations of coal production and consumption were now driven by market forces instead of central planning. Coal resources were much richer and mining costs were much lower in Shanxi, Shaanxi, and Inner Mongolia. (Shanxi and Inner Mongolia are in the north; Shaanxi is in the northwest.) For example, in 1985, the production cost of local coal mines was RMB 16.8 per tonne in Shanxi, RMB 24.7 in Hebei, RMB 25.5 in Henan, and RMB 32.6 in Shandong. There was a rush of mine development in Shanxi, Shaanxi, and Inner Mongolia once the restrictions on local mines' development were lifted in 1983 (Figure 2.5). This regional imbalance affected internal coal flows and rail networks. In 1986, the coal transported by railway was 531.1 Mt; in 1992, this increased to 641.1 Mt. This increased the importance of transportation as well as the exposure to transportation constraints.

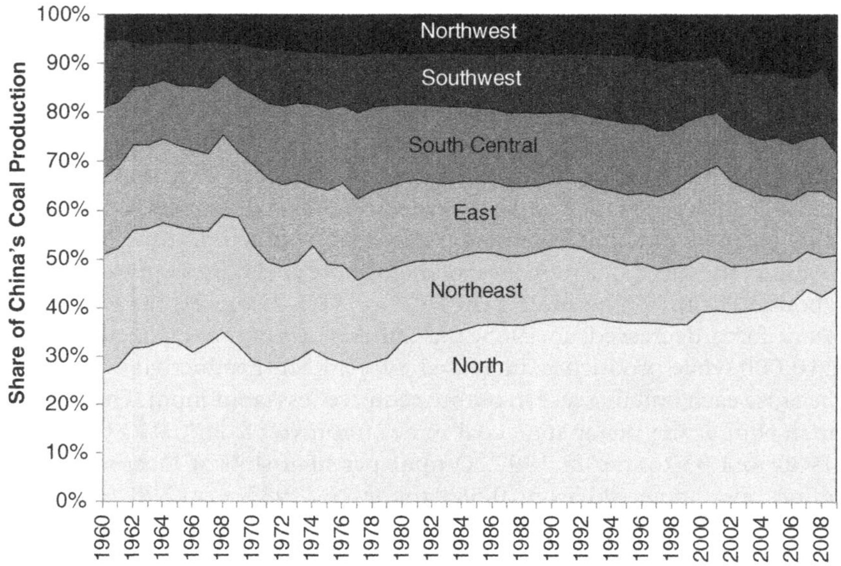

Figure 2.5 Share of raw coal production in China by region. Data sources: China Coal Industry Association (2011); regional shares calculated by author.

The market reforms did not influence all sectors equally. By the late 1980s, the permitting procedure for small-sized coal mines belonged to the local administration and was a much simpler process. But railway access was still limited to unified distribution coal mines, which constrained the ability of local mines to bring coal to market.

Since the unified distribution coal mines obtained access to rail transportation while local coal mines did not, China's coal market divided into two distinct markets, national and local, operating in fundamentally different ways. The national market is connected by railway transportation and populated by larger producers supplying higher quality coal. Often this coal is intended for consumption by the big end-users such as key national electric power generators. Price in the national market is still strongly influenced by the central government. The local market is usually populated by smaller producers, with transportation generally by truck and limited to shorter distances. The average delivery distance is about 80 km by road (compared to 550–570 km by rail in the national market). Cost structures and pricing mechanisms for these local coal producers are sensitive to local political and economic factors, with market oversight occurring at the provincial or local level.

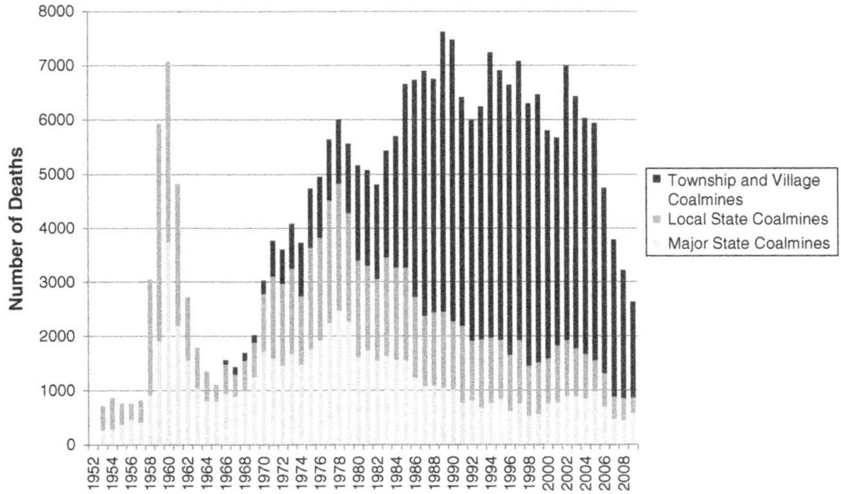

Figure 2.6 Number of deaths from coal mine accidents in China. Data source: China Coal Industry Association (2011).

Even though the development of the township and village coal mine operations spurred production, these mines had a harmful impact on other aspects of the Chinese coal industry. The township and village coal mines extracted coal resources less efficiently and caused frequent accidents. The average production capacity of a township and village coal mine was just 1% of a typical major state coal mine, and its recovery rate was 10% to 15%, compared to 70% to 80% for major state coal mines. Furthermore, the safety record of the township and village coal mine was poor (Figure 2.6).

2.4 Establishing a market system in the coal industry (1993–2002)

By the 1990s, broader reforms in the Chinese economy were continuing to replace central planning with market mechanisms. Even though the combination of market and central planning increased output, it also caused market distortions. Each commodity had two different prices: the planned price and the market price. Furthermore, the state-owned enterprises (SOEs) had a much lower selling price and a much heavier tax burden compared to the township and village enterprises (TVEs). In order to reduce some of these obvious distortions in the economy, the Fourteenth Central Committee of the Chinese Communist Party in 1992

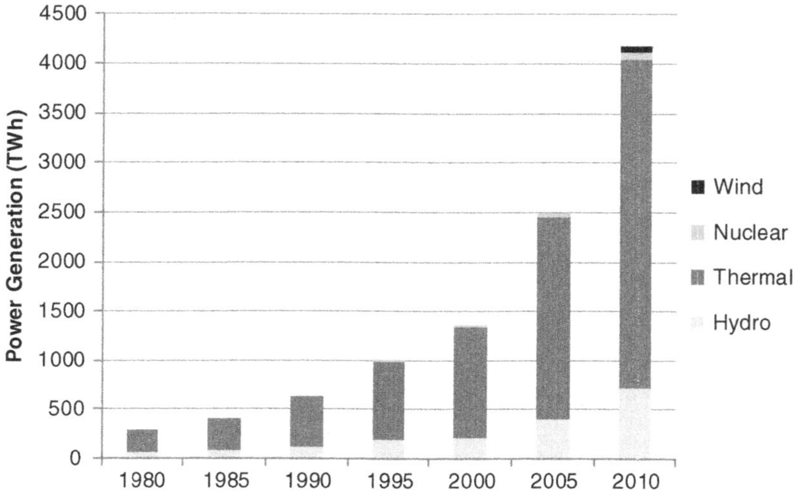

Figure 2.7 Power generation by source in China. Data source: National Statistical Bureau, various years.
Note: Thermal power in China is mostly coal-based.

decided that China should work toward becoming a "socialist market economy" (Wu 2005). The objective of these reforms was to separate government administration from business operations, create enterprises that were truly market players, and unify the market so that different enterprises could compete fairly in the market.[26]

Since the initiation of reform, energy production has developed rapidly to fuel economic growth, and as a consequence has caused serious environmental pollution. China's sulfur dioxide emissions in 1995 amounted to 23.7 Mt, more than any other country. Hydropower could not keep up with the pace of economic development, and China relied more on coal than before, even for power generation (Figure 2.7). Large-scale exploitation and utilization of coal is one of the major causes of ecological destruction and environmental pollution. Environmental protection was increasingly given priority in state energy development, and a series of measures were adopted to counter the environmental damage. First, clean coal technologies were encouraged to improve environmental quality. Second, the development of cleaner forms of energy such as natural gas and renewables was encouraged. Third, efforts were made to consolidate small coal mines because they tended to produce poor quality coal (and lacked the coal-washing equipment that could help improve coal quality) and in general had little capacity to mitigate ecological damage.

Management responsibilities between central and local governments were divided more clearly after the initiation of a program of separating government administration from business operations in 1993. The central government moved toward regulation and macroeconomic management, and the local governments played a pivotal role in implementing the central government's policies, with greater authority within their jurisdiction.

According to the objective of establishing a "socialist market economy" and the requirement under the new state energy development priorities to protect the environment, the market was to be unified and the two-leg walking strategy was to be phased out in the coal sector; the unified distribution coal mines were to be pushed toward market competition; and the local coal mines were to be consolidated. Market reforms now extended to the entire coal sector. All of these reforms caused significant changes in administration, finance, and pricing in the coal sector.

2.4.1 Administration

In 1993, the Ministry of Coal Industry was reconstituted when the Ministry of Energy was phased out. The two-level (national and local) administrative system continued until the first half of 1998 and was then dissolved. Later in 1998, the Ministry of Coal Industry was phased out and the relevant macroeconomic management functions were transferred to the State Economic Trade Commission and later to the National Development and Reform Commission (NDRC) because it was not considered necessary to have an independent coal ministry after the coal sector had been pushed to market competition. Following this, the administration of the ninety-four unified distribution coal mine bureaus was decentralized and transferred to the local authorities, and the administration of the Northeast and Inner Mongolia coal mines was also transferred to local governments. With the exception of Shenhua Group, the administration of the coal sector had been decentralized, and the central government focused more on administration of environmental protection and safety.

The key factor that led to the transfer of major state coal mines from central to local governments was the change in the relationship between central and local governments. Even though the central government wanted to reduce the output of local coal mines and give more market space to the major state coal mines, the local governments had every incentive to support the local coal mines, which competed for resources and market share with major state coal mines. During the Asian financial crisis of 1997, excess coal supply capacity started to appear. The brunt of the decline in demand was shouldered by the central mines because the local mines refused to roll back production. Sixty percent of major state coal mines stopped production partly or completely. By the beginning of 1998, the unused production capacity had reached 90 Mt. Major state coal mines lost RMB 3.75 billion during the first three quarters of 1998. Because the township and village coal mines competed for coal resources and market share with the major state coal mines, local governments had little incentive to discourage their development. Out of a total of 61,000 small coal mines in 1998, 51,200 were illegal mines, producing 430 Mt of coal in total. Of these illegal coal mines, 13,000 coal mines producing 114 Mt were located in the major state coal mines' fields. From 1993 to 1997, 639 floods and 57 gas explosions caused by local coal mines occurred in the fields of the major state coal mines and led to direct economic losses reaching RMB 2.5 billion. These circumstances drove the State Council's policy of transferring major state coal mines to local governments and closing small pits. In 1998, the State Council proposed to close illegal and "wrongly-located" coal mines (document no. 43). Pursuant to the policy, from the end of 1998 to the end of 1999, 25.8 thousand illegal and wrongly located small-sized coal mines were to be closed, and coal production capacity was to be decreased by 250 Mt.

In 1999, an experimental bankruptcy of major state coal mines was initiated. Of the 600 major state coal mines in operation at that time, more than 120 mines were running out of resources and facing large financial deficits while producing high sulfur and high ash coal. The annual production capacity of these 120 mines was 90 Mt, but in 1998, the actual production was only 50 Mt. Their deficit was RMB 3.5 billion, which accounted for 88% of the total deficit of major state coal mines. When the central government wanted to transfer these coal mines to local governments, the local governments refused, thus forcing the mines into bankruptcy. Pilot projects including the Benxi, Longfeng, and Jixi coal mines went bankrupt, loans were canceled, and employees and retired workers were resettled.

Pushing the major state coal mines into the hands of the local governments did not mean that the mines were now independent. The local governments exerted strong influence over the behavior of the mines in the market, and the major state coal mines were now used by the local governments to boost the local economy and maximize local tax revenues. The consequence of transferring major state coal mines was that these mines were now treated more favorably by the local governments.

The Shenhua Group started coal production in 1995 and has stayed as a central firm up to now. The first reason that Shenhua Group was not decentralized is that Shenhua Group is a vertically integrated firm requiring approval and coordination from more than four ministries and six provincial governments, each responsible for either the coalfield or local land or infrastructure. It would be very difficult for a local firm to coordinate these complex interests. The second reason is that Shenhua Group faces little competition from local coal mines. There are far fewer local coal mines in the Shenfu Dongshen coalfields because the public infrastructure is poor.

2.4.2 Finance

New tax rules that applied equally to all economic participants were introduced in order to establish the market. In 1994, tax systems were unified, the revenue-sharing system was phased out, and the tax-sharing system between central and local governments was implemented nationwide.[27] These reforms changed the coal tax from a product tax with a tax ratio of 3% into a value-added tax with a tax ratio of 13%. They had a major impact on the coal mines,[28] with RMB 337 billion of the major state coal mines' former income becoming value-added tax remittance.

New financial rules were introduced in order to increase financial prudence. During the second half of the 1990s, the banking system underwent fundamental restructuring with the objective of separating government policy involvement from banking operations.[29] In 1998, the PBC was finally given a new organizational structure with a renewed mandate to conduct monetary policy. State-run commercial banks soon found themselves facing tougher budget constraints, which created a ripple effect among clients in SOEs. In 1993, the National Development Bank was established and started offering policy loans to spur infrastructure investment.[30]

Until the major state coal mines were transferred to local governments, the central government continued the pattern of investing in major state coal mines. From 1993 onward, the National Development Bank started offering the major state coal mines policy loans for infrastructure investment since most major state coal mines operated at a loss, mainly due to the tax policy change and fierce competition from the local coal mines. When the ninety-four major state coal mine bureaus were transferred to local governments in 1998, the central government shifted from the role of creditor to that of stockholder because the coal mines did not have the capability to repay state loans. After this transfer, the central government only invested in Shenhua Group and provided some loss subsidy to the major state coal mines.

2.4.3 Pricing

In order to push major state coal mines into market competition and phase out the subsidy, the coal price exception for power generation (power coal) and agricultural uses was liberalized in 1993. After the liberalization, coal price rose sharply until the outbreak of the Asian financial crisis in 1997. For example, the average selling price of raw coal was RMB 76.95 per tonne in 1993 compared to RMB 46.15 in 1992 (Zhou and Wang 2001). But most benefits from liberalization went to the local coal mines. The output of township and village coal mines increased by 14% in 1993 while major state coal mines' production fell by about 5%; the average selling price of coal for major state coal mines increased 16.3% in 1993, much less than the national average selling price increase of 66.7%.

The central government initiated several programs to slow local coal mines. In March 1994, the State Council proposed a program of consolidation, renovation, adjustment, and upgrading for township and village coal mines. In December 1994, the State Council issued a management statute intended to slow the production of local coal mines. In 1996, the "coal law" was put into effect to control the mining activities of township and village coal mines. However, these measures were not successful in slowing the boom in local mines, and the township and village coal mines' production of raw coal kept increasing (Figure 2.1).

2.4.4 Performance

The reforms to decentralize control and establish a vibrant market for coal improved productivity while also apparently enhancing safety. Under the pressure of market competition, the major state coal mines managed to solve the overstaffing situation by relocating employees. From 1993 to 2001, the number of employees in major state coal mines decreased from 3.5 million to 2.8 million. The mechanization level of major state coal mines increased from 72.3% in 1992 to 77.8% in 2002, and the total number of deaths by accident in the coal sector decreased from 7,239 in 1994 to 5,670 in 2001 (Figure 2.6).

In 2000, the central government announced that the policy to close small coal mines had been successful. In 1998 and 1999, approximately 31,000 small mines with about 19% of the annual production capacity of China's coal market were closed. But throughout this period, some closed coal mines reopened surreptitiously, and many analysts argue a certain amount of coal output went unreported (Appendix, this volume; IEA, 2009).

Although the central government has allowed substantial market reforms, these reforms have not easily extended to the power sector – the largest single consumer of coal to this day. In 1998, the Ministry of Electric Power Industry was eliminated and the State Power Company was created, with the goal of further separating government administration from business operations. In December 2002, the central government announced its intent to create more competitive power markets. The State Power Company was dissolved and five big power companies, two grid companies, and the State Electric Power Regulatory Commission were established. This unbundling and attempt to invoke the competitive market model fit with the country's general effort to replace central planning with market mechanisms. However, a "true" power market has not been established because the power price is still tightly regulated by the central government.

2.5 Perfecting a market system in the coal industry (2003–present)

The gradual nature of China's economic transition has implied uneven progress in many areas of reform. Reforms implemented early, including incentive schemes, enterprise autonomy, liberalization of output prices, and opening to foreign trade and investment, have accomplished many of their intended objectives. But in other areas, including the development of factor markets, protection of the environment, and provision of social security, reforms are far from complete. At the Third Plenary Session of the Sixteenth Central Committee in 2003, construction of a "modern market system" was proposed, and the fundamental role of the market in resource allocation was emphasized. The state-owned Asset Management Committee was created to act as the government's investor in SOEs, and improving the corporate governance of SOEs was designated as a priority. Social management and public service were identified as the core function of government.

China entered into the World Trade Organization (WTO) on December 11, 2001, which created a further push for investment liberalization and ownership diversification. After its accession to the WTO, China became the world's manufacturing base, its second-largest foreign direct investment (FDI) recipient, and its third-largest trader. China is the world's largest producer (and consumer) of several major commodities, including iron, steel, cement, and ammonia. Economic growth has been very high and driven mainly by industry, which now accounts for almost 50% of GDP.

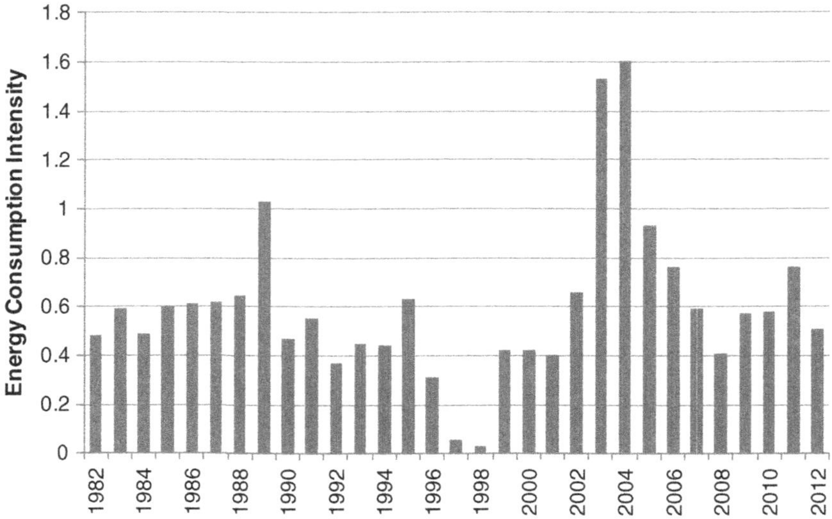

Figure 2.8 Energy consumption intensity in China (defined as the annual growth rate of energy consumption divided by the corresponding growth rate of GDP). Data sources: National Statistical Bureau, various years.

In less than a generation, China has changed from a minor and largely self-sufficient energy consumer into the world's fastest-growing energy consumer and a major player on the global energy market. Demand for all forms of energy has risen very fast in the last decade. After falling during most of the 1980s and 1990s, energy intensity rose between 2002 and 2004 (Figure 2.8), threatening the sustainability of economic expansion. Despite some successes in countering environmental damage from energy production and use, emissions of air pollutants in China remain very high, and energy-related greenhouse-gas emissions are rising very rapidly. Strong policy action is needed to address these issues, in particular by encouraging more efficient energy use throughout the energy system.

Energy conservation became a top goal of policy makers, and a series of priorities were adopted to improve energy efficiency. First, energy production and consumption were to be restructured so as to increase the size of enterprises, with the goal of achieving economies of scale. Second, advanced technologies were to be disseminated in an effort to increase the efficiency of the energy system. Third, new government entities were created to reinforce the design and implementation of energy policy. For example, the National Energy Leading Group of the State Council, supported by the Office of the National Energy Leading Group (ONELG), was established in 2005 and headed by the premier in order to coordinate among ministries to achieve energy policy goals. The Energy Bureau within the NDRC was created in 2008 and takes the lead in formulating energy supply policy.

2.5.1 *Administration*

The administration of the coal sector is shared among a number of different bodies. The State Council is the country's highest governing body. Energy policy-making responsibilities are held by the NDRC. The State Administration of Coalmine Safety governs coal mine safety, and the Ministry of Land and Resources oversees the exploitation of fossil fuel resources and reserves.

The adjustment of state energy development priorities was favorable to the major state coal mines but not favorable to the small coal mines, since small coal mines used more primitive technologies.[31] The State Council argued in 2005 for promoting the healthy development of the coal industry by shutting down the small-sized coal mines (document no. 18).[32] Following this, the Ministry of Land and Resources and the NDRC proposed the integration of mineral resources development, usually with small-sized coal mines being acquired by large ones. In 2007, the NDRC issued the eleventh five-year plan, which stipulated that the number of small-sized coal mines should be no more than 10,000 and their combined output should not exceed 700 Mt in 2010.

2.5.2 *Finance*

Restructuring and ownership diversification of the coal sector was initiated, especially after China's entrance into the WTO. The major state coal mines were transformed into limited liability companies, some of which had diversified ownership on domestic and foreign stock exchanges, although the state retained a controlling interest (Table 2.3). By the end of 2011, there were more than thirty listed coal corporations. Local mines were absorbed into state-owned and/or private company structures.

2.5.3 *Pricing*

By 2002, all coal pricing, even for power generation, had been relaxed. After China's economy recovered from the Asian financial crisis in

Table 2.3 *Coal mining enterprises by ownership in 2005*

Ownership	Number of enterprises	Share of national total, %	Design capacity in 2005, Mt/year	Share of national capacity, %
State-owned or state-held shares	4,185	16.9	1,340.5	59.2
Collective and co-operative	11,695	47.1	642.9	28.4
Private	8,919	35.9	278.8	12.3
Foreign joint ventures	14	0.1	2.1	0.1
TOTAL	24,813	100	2,264.3	100

Source: IEA (2009).

2001, power coal demand increased rapidly. But the local governments and power coal producers had no incentive to increase the power coal supply because the price of power coal was held at artificially low levels. In order to increase the production and delivery of power coal, the central government further relaxed price controls over power coal and instead set a guidance price for term contracts between unified distribution coal mines and key national power generators[33] at the annual Coal Ordering Conference. In 2002, about half of China's power coal was sold by term contract through the annual National Coal Ordering Conference and half was sold in spot markets with the Qinhuangdao port increasingly acting as the leading benchmark. Since 2002, the central government has no longer directly controlled any coal price (Figure 2.9). Because the central government no longer controls the power coal price but still controls the power price, this creates a conflict between the coal and power sectors.

2.5.4 *Performance*

Because coal is the primary input into Chinese power generation and coal prices have become market oriented while power prices are tightly regulated, the tension between the power and coal sectors is unavoidable and has raised concerns about electricity shortages. Since 2002, the government has been dealing with the consequences of reforms not being extended to all sectors. As the gap between term contract power coal prices and spot market prices has widened (Table 2.4), the tension

Table 2.4 *Average production cost, term contract price, and spot price of power coal for major state coal mines (RMB/tonne)*

	Average production cost	Term contract price (FOB)*	Spot price in Qinghuangdao
2002	121.4	132.0	270.0
2003	123.9	138.6	280.0
2004	169.0	162.5	370.0
2005	208.0	212.8	440.0
2006	214.2	227.8	470.0
2007	215.4	233.6	500.0
2008	339.9	320.0	760.0

* (1) Railway transportation cost is not included in the term contract price; the average rail transportation cost is about RMB 100/tonne nationwide. (2) The term contract price has been unavailable since 2009 because the central government no longer regulates it.

Source: Term price is from China Coal Information Institute, various years; Qinghuangdao spot price and railway transportation cost come from the China Coal Market network (CCTD).

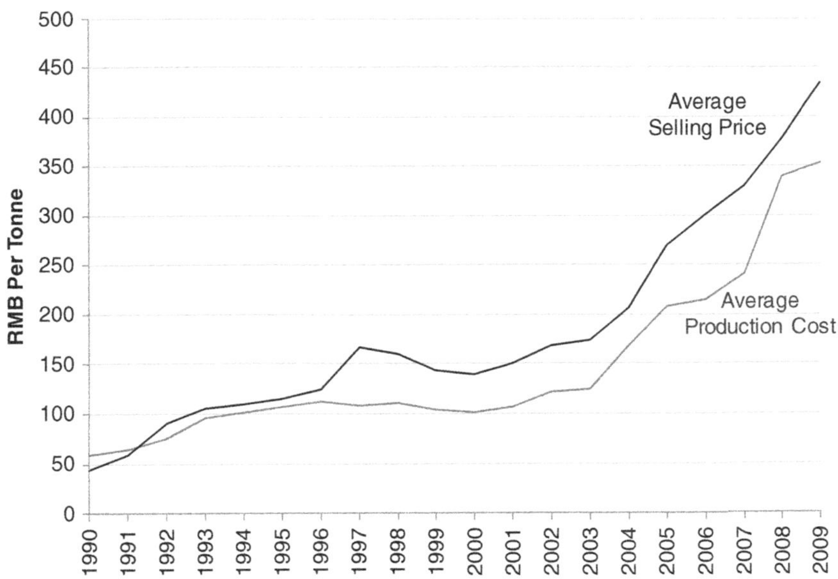

Figure 2.9 Average production cost and selling price for major state coal mines in China. Data source: China Coal Information Institute, various years.

between coal producers and power generators has deepened. Coal producers become increasingly reluctant to enter into contracts, and key power producers have been pinched and are reluctant to pay the market price. In the National Coal Ordering Conference for 2003, coal sellers wanted to use the Qinhuangdao spot price as the benchmark for term contracts, but coal buyers refused to accept it; eventually, the central government set a 5% price increase. Despite the price increase, 50% of coal term contracts were not signed, which resulted in electricity shortages in most provinces. In the following years, the government set a "price linkage mechanism" that allowed power plants to apply for higher prices based on coal price increases. Because the price linkage mechanism encouraged coal suppliers to boost coal prices and thereby contribute to inflation, it was later canceled. Power prices are still strictly controlled by the government.

The tension between the coal and power sectors is in reality between the central and local governments. Coal mines with about 85% of China's coal production are controlled by local governments. For example, Shanxi, Shannxi, and Inner Mongonia together have more than 50% of term contracts for power coal. Shanxi province even established a coal distribution group to manage the coal selling price. The key national power companies consume about 50% of power coal and are owned by the central government. The central and local governments need to be carefully coordinated so as to avoid disruptions to coal supply.

The tension between the coal and power sectors has had repercussions in the international coal market. Under the pressure of price increases from domestic coal suppliers, the power sector responded by going to overseas markets to purchase coal. The option to acquire coal in the international coal market is an important source of leverage for the power companies because it reinforces the negotiating ability of power companies in the domestic coal market. In 2009, the amounts of imported coal increased rapidly, surpassing the amount of exports by more than 100 Mt in 2009 (Figure 2.10) with substantial effects on the international market. Chinese power firms will continue to have this incentive to enter the international market for coal in response to impasses in domestic negotiations. China's reform imbalance, which can cause Chinese power firms to enter and exit the global market in an erratic way, therefore, can translate into volatility in international markets for coal, especially given that even small residuals of the Chinese market can be large relative to the international coal trade. (See Chapter 9 for further discussion of China's coal trading behavior and its implications for the international market.)

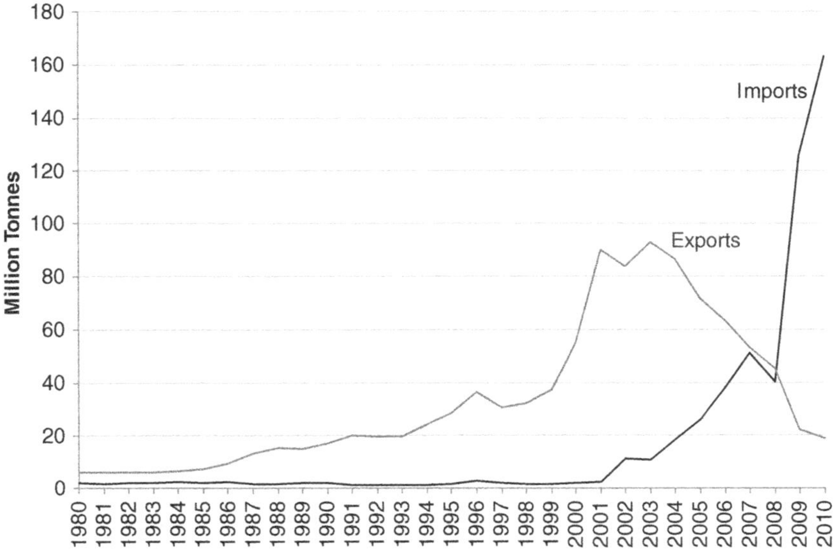

Figure 2.10 China's international coal trade. Data source: China Coal Information Institute, various years.

2.6 Conclusion

The coal sector is integral to the Chinese economy. Coal provides almost 70% of the nation's primary energy, generates about 75% of the electricity, and supports millions of employees.

After the PRC was founded in 1949, the coal sector was put under central planning and given priority to develop in order to boost downstream heavy industries. The market-oriented reforms initiated by China's central government influenced the coal sector directly, as did changes in the state's energy development priority and the relationship between central and local governments. Under the influence of the broader reforms in the Chinese economy, the focus of state energy development changed from being oriented to the entire energy sector to being directed in particular at electric power, and then later to environmental protection and energy conservation. Local governments were given more and more independent responsibility from the central government to boost the local economy. All of these objectives translated into market-oriented and decentralizing changes in the areas of administration, finance, and pricing in the coal sector.

However, these market-oriented and decentralizing reforms have not extended equally to the power sector, now the largest single consumer of coal, and to the railway sector, which connects the coal and power sectors. The fragmented nature of these reforms has had significant impacts. Because coal is the primary input into Chinese power generation, the tension between the power and coal sectors is unavoidable and has raised concerns about electricity shortages. Since 2002, the government has been dealing with the consequences of these differential reforms.

Even though the central government now encourages the vertical integration of coal and power to mitigate the conflict, it is important that the vertical integration process be very deliberate. The ownership of key national power companies is with the central government, which has control over electricity prices and railway transportation, while major state coal mines – with the exception of Shenhua Group and China Coal – belong to the local governments, which have control over coal prices. This creates a new tension between the central and local governments, which needs to be carefully managed. The reform of electric power, which is arguably the most lasting and effective solution to the tension between the coal and power sectors, is therefore critical.

Notes

1. The central government delegated power to and shared profit with state-owned enterprises through the general contract system.
2. As one of the major state coal mines, Shenhua Group does not belong to the unified distribution coal mines because it has captive rail lines.
3. There were 24 coal projects within the 156 key projects designed with aid from the Soviet Union during the 1950s.
4. The SPC was later reconstituted as the National Development and Reform Commission (NDRC).
5. Township and village coal mines are also named small coal mines since they are typically much smaller.
6. The ministry responsible for the coal industry during this period was variously designated as the Ministry of Fuel Industry (1949–55), Ministry of Coal Industry (1955–70), Ministry of Fuel and Chemical Industry (1970–75), and Ministry of Coal Industry (1975 onward). For consistency, the Ministry of Coal Industry is used throughout this chapter.
7. The National Coal Ordering Conference was hosted by the central government to arrange the following year's coal production, consumption, and transportation under unified distribution. It was renamed the National Coal Negotiation Conference in 2002. For consistency, National Coal Ordering Conference will be used throughout this chapter.
8. The unified distribution coal was usually transported by railway.
9. These three categories were formed under central planning. For consistency, their original titles will be used throughout this chapter even though they were renamed later.

10. RMB is the currency unit in China. The exchange rate of RMB was overvalued by the Chinese government during the central planning period. In 1993, a market-based exchange rate of RMB was initiated. In 2008, USD 1 was equal to about RMB 6.8.
11. In 1949, China's coal production was ninth in the world.
12. There are six administrative regions in China. East China includes Shanghai, Jiangsu, Zhejiang, Anhui, Fujian, Jiangxi, and Shandong; North China includes Beijing, Tianjin, Hebei, Shanxi, and Inner Mongolia; Northeast China includes Liaoning, Jilin, and Heilongjiang; South Central China includes Henan, Hubei, Hunan, Guangdong, Guangxi, and Hainan; Southwest China includes Sichuan, Chongqing, Guizhou, Yunnan, and Xizang; Northwest China includes Shaanxi, Gansu, Qinghai, Ningxia, and Xinjiang.
13. The Ministry of Electric Power and the Ministry of Water Resources were merged in 1982 to facilitate the development of hydropower.
14. During the course of initial reform, China's fiscal position eroded significantly, dropping from 33.8% of GDP in 1978 to only 10.8% at the low point in 1995. The decline of budgetary revenues was driven primarily by the inexorable erosion of the old system in which state enterprises raised revenues from their monopoly position. More generally, fiscal decline was also the logical result of a transition strategy that stressed decentralization of authority and benefits, along with releasing resources from government control to the marketplace. From 1995 onward, a modern tax system was gradually built up and led to a steady revival of government budgetary collections.
15. Another way to spur coal production was to increase the coal price, but this was not feasible because inflation was severe in the 1980s (the growth rate of the retail price index was 8.8% in 1985 and 18.5% in 1988).
16. The Shanxi Coal Base includes Shanxi, the northern part of Shaanxi, the western part of Inner Mongolia, Ningxia, and the western part of Henan, which are all neighbors.
17. The coal resource for opencast mines is mainly in Inner Mongolia. The use of foreign technology was stressed because it would be more expeditious to use the latest technology available elsewhere rather than develop China's own. The opencast mines were selected because they were suitable for large-scale mechanization technology. The production per year of opencast was generally twice as much as from underground mines, construction time was at least 25% less, construction and operation costs were lower, and working conditions were safer.
18. The Ministry of Coal Industry took marketing authority back from the National Material Bureau. Unified distribution coal mines could sell coal directly to end users at the regulated prices.
19. Under this new fiscal administration system, revenue usually was shared between the central and local governments according to predefined formulae.
20. There were four specialized banks: the Industrial and Commercial Bank of China, the Agricultural Bank of China, the People's Bank of China, and the People's Construction Bank of China. These banks later were converted to commercial banks.

21. Another coal project, Yimin, was set up by the Huaneng Group, which is the biggest electric company owned by the central government, and was planned to be put into operation in 1993. The planned capacity of coal production was 5 Mt per year, to be used to generate electric power.
22. The annual growth rate of the retail price index was 8.8% in 1985 and 18.5% in 1988, respectively.
23. It can be thought of as "growth out of plan."
24. There is no nationwide data about the productivity of township and village enterprises, but it is likely to be very low due to the use of primitive technology.
25. Because the central government was unable to supply the financial resources required to satisfy the surging power demand, it issued "provisional regulations on building power plants by fundraising and practicing multiple electricity prices" and "regulations on the electricity price of small thermal power plants." The central government also transferred the administration of small thermal power to local governments. However, in 1999 the State Council put forward a "notice of issues concerning the closing down of small thermal power units" in order to improve the efficiency of coal-fired power generation (Zhou and Wang 2001).
26. The Company Law was put into effect at the beginning of 1994 and guaranteed the separation of the government administration function from business operations.
27. This fiscal reform was designed to broaden the tax base by implementing the value-added tax and other business taxes and also to stabilize the tax revenue relationship between central and local governments.
28. From this period, "coal mine" is also named "coal enterprise."
29. The PBC had been nominally established as a central bank in 1983, but at that time it remained beholden to government officials at both central and provincial levels.
30. A policy loan is usually used for public goods production or loss subsidy due to policy changes. The interest rate is very low.
31. Small-sized coal mines usually have little capacity to make use of coal-bed methane.
32. The "township and village enterprise" was renamed "small-sized enterprise" in the second half of the 1990s.
33. There are five key national power generators in China: Huaneng, Huadian, Datang, Guodian, and Zhongdiantou.

References

Ball, A., A. Hanstead, R. Curtotti and K. Schneider (2003). *China's Changing Coal Industry: Implications and Outlook*, eReport 03.3. Canberra: Australian Bureau of Agricultural and Resource Economics.

Chen, Wenying and Xu Ruina (2010). "Clean Coal Technology Development in China." *Energy Policy* 38, 2123–2130.

China Coal Industry Association (2011). *Compilation of China Coal Industry Statistical Data 1949–2009*. Beijing: China Coal Industry Publishing House.

China Coal Information Institute (various years). *China Coal Industry Yearbook*. Beijing: China Coal Industry Publishing House.

ESMAP (2004). *Toward a Sustainable Coal Sector in China*, June. Available at https://www.esmap.org/node/928 (accessed May 5, 2015).

ESMAP (2008). *Economically, Socially and Environmentally Sustainable Coal Mining Sector in China*, December. Available at http://documents.worldbank.org/curated/en/2008/12/10177734/economically-socially-environmentally-sustainable-coal-mining-sector-china (accessed May 5, 2015).

IEA (2007). *World Energy Outlook 2007: China and India Insights*. Paris: International Energy Agency.

IEA (2009). *Cleaner Coal in China*. Paris: International Energy Agency. Available at https://www.iea.org/publications/freepublications/publication/coal_china2009.pdf (accessed May 5, 2015).

Lester, Richard and Edward Steinfeld (2006). *China's Energy Policy: Is Anybody Really Calling the Shots?* Working Paper, MIT-ICP-06–002.

Lester, Richard and Edward Steinfeld, (2007). *The Coal Industry in China and Secondly India*. Working Paper, MIT-ICP-07–001.

Lin, Boqiang and Liu Jianghua (2010). "Estimating Coal Production Peak and Trends of Coal Import in China." *Energy Policy* 38, 512–519.

National Statistical Bureau (various years). *China Yearbook of Statistics*. Beijing: China's Statistical Press (in Chinese).

Pan, Kexi, Pu Jin and Xiang Tao (2002). "A Study on Market Concentration in China's Coal Sector," *Management World* 12, 77–88 (in Chinese).

Peng, Wuyuan and Jiahua Pan (2006). "Rural Electrification in China: History and Institution." *China and World Economy*, 14(1), 71–84.

Rosen, Daniel and Trevor Houser (2007). *China Energy: A Guide for the Perplexed*. Washington, DC: Peterson Institute.

Rui, Huaichuan (2005). *Globalization, Transition and Development in China: The Case of the Coal Industry*. London: RoutledgeCurzon.

Sagawa, Atsuo and Koichi Koizumi (2007). *Present State and Outlook of China's Coal Industry*. Tokyo: Institute of Energy Economics, Japan.

Sagawa, Atsuo and Koichi Koizumi (2008). *Trends of Import and Export of Coal by China and Its influence on Asian Market*. Tokyo: Institute of Energy Economics, Japan.

Schneider, Karen (2004). *China's Energy Policy: Recent Developments and Outlook*. Canberra: Australian Bureau of Agricultural and Resource Economics.

Shi, Dan (2008). "Review and Survey of 30 Years Reform and Opening Up in Energy Industry," *China Energy*, 30(6), 5–12.

Sinton, Jonathan, Rachel Stern, Nathaniel Aden and Mark Levine (2005). *Evaluation of China's Energy Strategy Options*. China Energy Group in Lawrence Berkeley National Laboratory. Available at https://eaei.lbl.gov/sites/all/files/LBL_56609_China_Energy-Strategies.May_.2005_0.pdf (accessed May 5, 2015).

Thomson, Elspeth (2003). *The Chinese Coal Industry: An Economic History*. London: RoutledgeCurzon.

Tian Yuan and Qiao Gong, eds. (1991). *Research on Price Reform in China, 1984–1990*. Beijing: Electronic Industry press, 390 (in Chinese).

Wang, Bing (2007). "An Imbalanced Development of Coal and Electricity Industries in China." *Energy Policy* 35, 4959–4968.

Wang, Xianzheng (2010). "Challenges and Opportunities China Coal Industry Encounters." *Shanxi Energy and Conservation* 5, 4–6 (in Chinese).

Wu, Jinglian (2005). *Understanding and Interpreting Chinese Economic Reform.* Singapore: Thomson.

Xiao, Xinzhi, Qi Yingfei, and Li Hongjuan (2008). "An Empirical Research on Effect of Coalmine Safety Regulation in China." *China Industrial Economy* 5, 67–76 (in Chinese).

Yu, Lihong and Yu Yihong (2006). "A Study on Vertical Regulation Model between Coal and Power Based on Value Chain Efficiency." *China Industrial Economy* 6, 5–13 (in Chinese).

Zhang, Chi and Thomas Heller (2007). "Reform of the Chinese Electric Power Market: Economics and Institutions," in David Victor and Thomas Heller, eds., *The Political Economy of Power Sector Reform: The Experiences of Five Major Developing Countries.* Cambridge: Cambridge University Press.

Zhang, Mingli ed. (1989). *Coal Industry in Contemporary China.* Beijing: Social Science Press in China (in Chinese).

Zhou, Fengqi and Wang Qingyi (2001). *50 Years of Coal Industry in China.* Beijing: China Power Press (in Chinese).

Zhou Nan, M. A. McNeil, D. Fridley, Jiang Lin, L. Prices, S. de la Rue du Can, J. Sathaye and M. Levine (2007). *Energy Use in China: Sector Trends and Future Outlook,* Paper No. LBNL-61904, Lawrence Berkeley National Laboratory.

Zhu, Chengzhang (2009). "Coal Is the Vitals of Electric Power Security in China." *Sino-global Energy* 14(7), 15–18 (in Chinese).

3 Developing large coal-power bases in China

Huaichuan Rui, Richard K. Morse, and Gang He

3.1 Introduction

China's coal market – already the largest in the world – is increasingly a driving force in the global trade and pricing of coal. China's imports accounted for more than 15% of all globally traded coal in 2010. The coal market has entered an era in which even slight shifts on margins of China's internal market directly impact coal trade and prices globally (see Chapter 9, this volume).

China is now in the midst of a radical restructuring of its coal and power sectors that has the potential to change how coal is produced, traded and consumed both in China and the rest of the world. That restructuring aims to integrate the coal and power sectors at giant "coal-power bases" that combined would churn out more coal annually than all the coal produced in the United States. This chapter analyzes the conditions in China's coal market that catalyzed these massive reform and integration efforts, their prospects for success, and the implications for both Chinese and global coal markets.

Coal plays a central role in China's economy, and has been a key ingredient of the economic growth that catapulted the country toward its present position as the world's second largest economic power. To this day, few countries rely on coal to the extent that China does. Despite a dramatic increase in oil and gas production and an increasing share of non-fossil energy sources, including hydropower and nuclear power, coal as of 2011 still provided about 68% of China's primary energy supply (IEA 2013b) and generated 78% of China's electricity (IEA 2013a). Moreover, despite large-scale industrial restructuring and government attempts since the 1990s to close down small and inefficient coal mines, coal still provides no less than 4 million jobs directly in the coal sector and millions more in coal-related industries.

Since 1978, China's coal industry has experienced the fastest pace of growth in the history of the world's coal industries, with production multiplying nearly five times from 620 million metric tons (Mt) in 1978 to 2.79 billion metric tons (Bt) in 2008 (see Figure 2.1 in Chapter 2). China's coal output accounted for 27% of the world total in 1998, increasing to 39% in 2007. Among the industry's most significant achievements is that it has ended China's decades long supply shortage of coal and is able to meet the demand of the rapidly developing national economy. Several major developments enabled this accomplishment. Regulation and administration of the sector has shifted over several decades from a command-economy model to a market-oriented model (see Chapter 2, this volume). Competition has been introduced to the industry. Production methods have been improved considerably, with adoption of modern technology and efficient equipment in many large coal mines. The integration of Chinese coal into the globally traded coal market has accelerated rapidly (see Chapter 9, this volume). Chinese import and export patterns now have significant impacts on international coal prices.

But despite coal's paramount importance to the economy and impressive achievements during the reform era, the coal industry remains among the most troubled industries in China. It suffers from declining government investment in the sector, a poor safety record, environmental pollution problems, bottlenecks in coal transportation (rail and port), and wide persistence of outdated technology. The policy environment has been unstable, with the government pushing aggressively to close small mines when coal supplies are adequate and condoning or encouraging them when production is lagging demand. Above all, the sector's development is significantly constrained by the regulatory structure of the related railway and power sectors. Electricity prices for the power sector are not liberalized and the railway networks are still under monopoly control. This situation creates conflicts between the goals of the coal, power, and railway industries, and impedes the healthy development of the coal industry. While state-owned coal companies have for more than a decade been expected to be responsible for their profits or losses, they are not allowed to sell coal to power-generating companies at market prices. These power companies, which represent the largest source of coal demand, are in turn not allowed to pass their costs through to electricity consumers, with the result that revenues from electricity production at times do not cover coal input costs, forcing either coal or power companies to accept losses. The struggle over who will accept these losses is commonly called the "coal-power conflict" and plays out in negotiations over the coal price. Meanwhile, the monopolized rail network is able to take advantage of its dominant position to extract rents from the coal producers, which rely on the rail networks to transport and sell their product. These various conflicts have resulted in numerous problems for China's coal sector that will be explored in this chapter.

Various reforms have been implemented or attempted by the Chinese central government in an effort to resolve the conflicts between coal, power, and rail as previously described. One of the most significant reform efforts now under way is the radical restructuring of these sectors through the integration of coal production and power generation businesses at key sites designated as "coal-power bases." Per the 11th Five Year Plan of China's Coal Industry issued by the National Development Reform Commission (NDRC 2007), thirteen large coal-power bases were planned (and a fourteenth, Xinjiang, was added more recently). The policy envisioned that each of the mega-bases would produce more than 100 Mt of coal and generate massive amounts of power. These bases involve fourteen provinces and prefectures, covering a total area of 287,000 sq. km and comprising more than ninety-eight coalfields. The available coal resources of the bases are to be 690.8 Bt, representing 70% of the total national coal resources. These bases can be developed utilizing three primary "modes" of integration across the existing coal and power value chain: coal firms extending into the power sector, power firms extending to the coal sector, or coal and power (or other firms) acquiring shares of each other.

This report provides a detailed account of the coal-power base policy as a key strategy of the Chinese government and assesses its potential to dramatically alter how coal is produced, traded, and priced in the Chinese market. We analyze both broad-based drivers of the development of large coal-power bases in China as well as the specific factors that may influence particular cases. We consider whether the potential improvements in coal and power sector functioning envisioned by central planners can be realized, and if so, in what form and at what pace. We make the following major arguments in this chapter:

First, despite the coal industry's paramount importance to the economy and its impressive production growth and restructuring in response to previous reforms, it nevertheless faces considerable challenges. One of the most significant is the conflict between the coal, power, and railway sectors, which are all essential to expanding Chinese energy supply, but whose regulatory and incentive structures are often in direct opposition to each other. Each sector evolved as it did in response to specific policy drivers (see Chapter 2, this volume), but the resulting "coal-power conflict" has created major problems for China and is the underlying driver of current coal reform efforts.

Second, the attempt to radically restructure the sector by integrating coal and power businesses at key coal-power bases is one of the most important reform efforts currently under way. Grounded in the desire to redistribute risk and profits across the coal and power sectors, remove current bottlenecks, and enhance central government control of China's energy system, this reform has the potential to fundamentally alter the structure of the Chinese coal and power markets.

Third, the coal-power base strategy is typical of China's policy making in its emphasis on consolidation of industrial activities within a limited number of state-owned enterprises (SOEs), which the government believes can be more easily controlled. Rather than liberalizing power prices and opening the rail sector to competition, it attempts instead to resolve conflicts by integrating small and private coal firms into large state-owned coal corporations and bringing all steps of a given coal-to-power value chain within limited numbers of large SOEs. The government hopes thereby to remedy problems in the market while still achieving policy objectives such as a low power price that the market might not otherwise guarantee.

Fourth, implementation of the coal-power bases has proceeded most rapidly where it creates new, alternative coal-to-power value chains rather than dismantling or reshaping existing ones. In such cases, the new coal-power bases have not directly challenged powerful incumbents such as the Ministry of Railways. The "Shenhua model," which we will discuss in detail, was so effective because it involved the creation of an entirely new company that was granted vast untapped resources. Other successful cases were not as dramatic but have also involved finding specific convergences of interests that are minimally threatening to incumbents.

Fifth, this case-specific aspect of developing a new coal-power base suggests that the status quo in China's coal market will not be drastically changed in the near future (the next three to five years) by coal-power bases but that their effects will be more important over the long term. As this chapter will discuss, the characteristics and involvement of various actors – including the central government, local governments, the financial sector, and the coal, rail, power, and chemicals industries – as well as the nature of local coal resources and energy needs all influence how quickly bases can be constructed and the sector can be restructured.

The remainder of this chapter is organized as follows. Section 3.2 analyzes the major rationales for building up coal-power bases and explores the underlying drivers for the current reform policy. Section 3.3 describes China's policy for creating coal-power bases and the government's key objectives in doing so. Section 3.4 discusses the details of the Shenhua model, which is the template that many other Chinese firms hope to emulate in creating coal-power bases. Section 3.5 analyzes the various "modes" (models) of building up coal-power bases. Section 3.6 examines the feasibility and difficulties of "cloning" the Shenhua model. Finally,

Section 3.7 concludes with a broad look at the issues that will shape the success of coal-base policy going forward.

3.2 Underlying drivers for building large coal-power bases in China

This section presents a brief history of the reform efforts in the coal industry that have led to the current industry structure. Reform efforts since 1978 have at different points emphasized the transition to a competitive coal market and the consolidation of the coal industry, which are sometimes in tension. Critically, reforms in the coal industry have not moved in tandem with reforms in the power and rail sectors. The resulting problems across the coal-to-power value chain have created the underlying rationale for the establishment of coal-power bases.

3.2.1 Reform in the coal sector

Since the establishment of the People's Republic of China (PRC) in 1949, the government has considered the coal industry to be one of the most strategically important industries. The government maintained a tight grip on the industry under the command economy between 1949 and 1978 by controlling the investment, output, price, transportation, and wage structure of state-owned coal mines (Thomson 2003; Nolan 2001). By the early 1980s, many industries in China had undergone market-oriented reforms, but the coal industry was still under tight government control because of its strategic importance as a primary input to the economy and a fear that reform could cause unpredictable outcomes that would harm other industries. For example, given coal's position as the major primary energy source in China, any change in coal prices would lead to corresponding impacts on prices of other industrial products that use coal as a raw material or input fuel. Instead of following the reform trend taking place in other industries, the coal industry maintained the old command structure, which was dominated by state-owned coal companies. In 1978, there were 104 key SOE coal bureaus ("major state coal mines" in Chapter 2, Figure 2.1), which were large and had connections to the central and provincial governments, and roughly 2,000 local SOE coal producers ("local state coal mines" in Chapter 2, Figure 2.1). Together the key SOE and local SOE mines produced more than 85% of the national output. The remaining 15% was produced by commune and brigade (CBE) coal mines, which were officially renamed township- and village-owned coal mines (or TVEs) in 1984 (Rui 2005). However, as in most former centrally planned and SOE-dominated

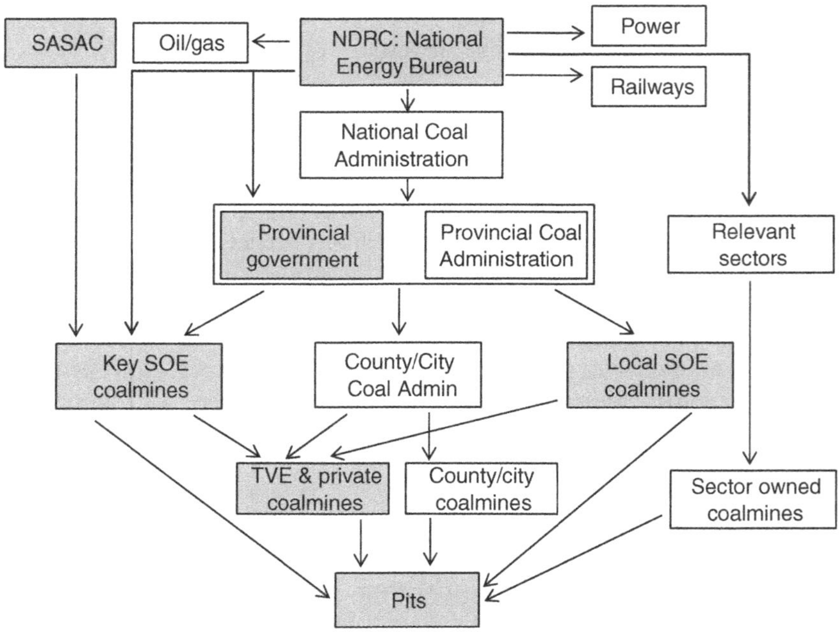

Figure 3.1 Illustration of the administrative structure of China's coal industry. The boxes highlighted in gray refer to levels of administration that are still functional. Those not highlighted refer to levels of administration that used to play an important role but are no longer functional. "SASAC" stands for the State Asset Supervision and Administration Commission. Source: Authors' analysis.

industries, shortages of the most wanted commodities were a common phenomenon. To meet higher and higher demand for coal that arose from the nation's economic boom, the government had no option but to reform the industry starting in the mid-1980s. Since the Ministry of Coal Industry was abolished in 1998, reforms have mainly been designed and implemented by the National Development and Reform Commission (NDRC) and facilitated by provincial governments and the State Coal Administration at its central, provincial, and county levels. Figure 3.1 is a simplified illustration of both the evolution and the current status of coal sector administration in China.

Reforms in China's coal industry since the 1980s have involved both ownership privatization and price liberalization. In 1978, key and local SOEs produced 85% of all coal whereas the share of production by TVEs (non-SOE mines) was only 15%. This industrial structure began to shift in the 1980s. TVEs and private coal mines were gradually allowed to develop by central planners starting in 1983 because they helped alleviate coal shortages, contributed to the eradication of rural poverty through job creation, and improved tax revenue. This created a situation in which "the state, collectives, and individuals all work together to develop large, medium, and small mines" (Ye and Zhang 1998, p. 2). The number of TVE mines jumped from 16,000 in 1982 to 63,000 in 1985 and then peaked at 72,000 in 1997, when TVE mines produced more than half of the total national output – up from 20% in 1981 – finally ending the country's decades-long history of coal shortages (Rui 2005).

By 1997, the Chinese government was determined to restructure the coal industry, for several reasons. First, state-owned coal mines suffered large losses from competition with TVE mines – losses that were ultimately shouldered by the government. Having been near-monopoly coal producers under the planned economy for decades, SOE coal mines were unfamiliar with market mechanisms and unprepared to deal with competition from the TVE mines. Crucially, their higher production costs and inefficient management meant that SOE mines were also unable to compete with TVE mines on price. The resulting financial losses of SOE coal mines peaked in 1998, when the state-owned coal sector was the second largest loss-maker among China's industrial segments (Rui 2005). Second, the increase in TVE mines' share of production led to other problems, including poor safety performance and inefficient use of coal reserves. Both the poor safety record of the industry and the financial losses of SOE mines pushed the government to reform the industry.

Restructuring efforts focused on three major categories of coal producers: tens of thousands of TVE or private mines were closed down; dozens of state-owned loss-making coal companies were declared bankrupt; and large, modern coal corporations such as Shenhua were founded in order to consolidate the industry. The restructuring effort also abolished the Ministry of Coal Industry in 1998 and passed its administrative power over key SOE coal mines (excepting a few large coal companies including Shenhua and China Coal) to local provincial governments. The elimination of the Ministry of Coal Industry and the encouragement of the development of non-SOE mines led to the existence of tens of thousands of coal companies, an industrial structure that was far more decentralized than it had been under the planned economy.

Price reforms were required alongside ownership changes in order to incentivize the coal production that China needed. Moving from a fully planned coal price to a market-based coal price was one of the most difficult reforms in China's coal industry. Because of the extreme importance of coal to the national economy, the government for decades had controlled coal prices, maintaining them at low levels (and predictably causing shortages). For most of the period of 1949–1993, the cost of coal production was actually higher than the sales price fixed by the government. To mitigate shortages and encourage competition while minimizing the possible negative impacts of full coal price liberalization, a dual-track price system was adopted in the coal sector starting in 1985 (Rui 2005). The new system took the planned output of 1984 as a baseline quota. Coal that was produced up to this quota would be sold at the planned price, but production volumes above the 1984 quota could be sold at the market price. By 1992, most of the commodities and products in China could be traded at market prices, but coal prices were still controlled by the government and did not reflect the real cost of producing coal. Coal from state-owned companies was eventually allowed in July of 1992 to fetch market prices but, crucially, electricity coal (coal sold to power companies to generate electricity) was still subject to the price set (or, after 1996, "guided") by the government.

3.2.2 Power sector industrial structure and pricing

Before 2002, China's power generation sector was dominated by the State Power Corporation (SPC), which owned 46% of the country's generation assets and 90% of the electrical transmission and distribution assets. In December 2002, China's State Council dismantled the corporation to introduce competition into the power industry, separating power grid and generation functions and allowing for increased private investment in generation to speed up the development of the industry. After this restructuring China's power industry comprised eleven state-owned companies, including two electric power grid operators, five electric power generation companies, and four companies in other power-related businesses.[1] The five power-generating firms – known as the "Big Five" (*wu da fa dian*) – are Huaneng Group, China Power Investment, Datang Group, Guodian Group, and Huadian Group. Each of the Big Five owned less than 20% (equivalent to 32 GW of electricity generation capacity in 2002) of China's market share for electric power generation. These reforms were intended to push the power industry toward greater market orientation and competition. But unlike coal sector reforms that allowed coal companies to independently determine some prices, the Big Five were not allowed to determine prices for electricity sold to the grid or to final customers. The NDRC's "provisional regulations on managing the selling price of electricity" (*xiaoshou dianjia guanli zanxing banfa*) in April 2005 clearly stated that the "selling price of electricity is set by the government." The government keeps such tight control on power prices mainly because changes in power prices have direct impacts on industrial and residential consumers as well as on the large number of foreign and domestic manufacturing firms in China that prize cheap electricity as a competitive advantage.

3.2.3 The coal-power conflict

Coal industry restructuring and price reform put an end to the persistent coal supply shortages in China and drove SOE coal companies to improve their management, efficiency, and competitiveness, which benefited the entire economy. However, the differences in the industrial structures and pricing regimes that had developed in the coal and power sectors created a situation of "market coal versus planned electricity" (*shichang mei miandui jihua dian*) – the coal-power conflict.[2] While power plants were able to purchase coal in large quantities from non-SOE coal companies, the newly fragmented coal sector was forced to trade with the highly consolidated power sector, which was made up of just five major power-generating companies and negotiated with coal suppliers as an integrated body through China Fuel Ltd. This mismatch of negotiating power led to serious difficulties for coal companies in negotiating and determining coal prices, which in turn caused endless disputes between coal and power firms. (As we shall see, the idea of cloning the Shenhua model in coal-power bases is in many ways about re-consolidating those coal producers and establishing long-term relationships with power companies that resolve this conflict – for example, through acquisitions that put coal and power under the same ownership.)

Against a backdrop of rising domestic and international coal prices after 2003, coal supply firms naturally sought to raise the price for electricity coal. However, this would imply rising costs of coal procurement and thus electricity generation for power companies. For the reasons discussed previously, the government did not allow a significant increase in electricity prices. Power companies were reluctant to accept coal price increases that could not be passed on to their customers.

The coal-power conflict played out at the Annual Conventional Coal Conference (hereafter, "Coal Conference"), which has served as the most important mechanism over the past half century for pricing and allocating electricity coal and transportation contracts. During the conference, coal and power companies would sign supply contracts for the next year based on the price set by the government, and the Ministry of Railways (MOR) would then allocate transportation quotas for the contracted coal. The State Planning Commission (the predecessor of the NDRC) together with the coal, power, and railway authorities supervised the event to ensure that important customers had secured coal supply and transportation.

The task of the Coal Conference became much more complicated starting in 1984, when SOE coal producers were granted the right to sell part of their coal production at market prices. The Coal Conference became even more difficult to manage when the hybrid coal-pricing regime for electricity and non-electricity coal began in 1992. Coal companies and power generators were supposed to form coal-power contracts on both planned and unplanned terms, but fluctuations in coal prices caused recurrent problems. Fewer and fewer contracts were made at the Coal Conference between coal and power producers, and those that were made were more and more difficult to enforce. During both 1998 and 2003, market coal prices were below the electricity coal contract prices, so power companies all favored market coal and did not fulfill their contracts. In 2004 and 2007, the situation was the opposite, but contracts were also not fulfilled as coal firms preferred to sell coal on the free market. By 2008, market coal prices had risen dramatically; electricity coal contracts became useless as the resulting conflict reached its peak in December 2008.

Due to the rising prices after 2003 and the strong desire of the coal producers to increase the price of electricity coal, the NDRC decided not to set ceiling prices for electricity coal in 2005 but then was called back to intervene in Conference negotiations after coal and power firms were unable to reach an agreement a full four months after the Coal Conference. Eventually, the NDRC implemented a new policy called "coal-power linkage" (*meidian liandong*), which allowed electricity prices to adjust to match fluctuations of the coal price. However, this new policy also failed because coal prices rose much faster than policy makers were willing to allow electricity prices to rise. Electricity coal prices rose by 77% from 268 yuan/ton in 2004 to 476 yuan/ton in 2008, but electricity prices were not allowed to increase to a similar extent, as cheap electricity is fundamental to the Chinese government's priority of maintaining rapid economic development, low inflation, and social stability. Under the policy of linked coal and power prices, power-generating companies increased electricity prices only twice, by 0.03 yuan/kWh and 0.015 yuan/kWh, while also obtaining an additional 0.02 yuan/kWh passed back from the grid companies. As a result the electricity price rose by 0.066 yuan/kWh in total, which was able to compensate for about 130 yuan/tonne of increased coal price, far below the actual increase in the electricity coal price, which exceeded 200 yuan/tonne in some cities.

Power-generating firms made a total loss of 70 billion yuan on coal-fired power generation in 2008 because they could not pass on the high price of coal to electricity consumers. Although this loss was also related to the declining demand for electricity after the financial crisis, it mainly resulted from the rising cost of coal. Costs for coal accounted for more than 70% to 80% of the total cost for power generators, compared with about 60% in previous years. Power companies therefore classified this loss as a "policy loss," indicating that it resulted from the central government policy of not allowing an increase in electricity prices. (Even after mid-2009, the Big Five were reluctant to update their financial statements by adding this loss to their 2008 performance.)

At the 2009 Coal Conference in Fuzhou, therefore, power generators wanted a 10% cut from the 2008 contract price. With the market having seen record high coal prices in 2008, however, coal producers aimed to charge power producers 10% more than the contract price of 2008. The difference in the prices demanded by the negotiating parties was as high as 150 yuan/tonne. Coal and electricity firms were unable to reach any agreement on coal prices at the Coal Conference, and no formal contracts were signed. Even the NDRC admitted it was unable to mediate when the expectations from the two sides were so different. There was no government policy that could resolve the dispute and satisfy both coal and power interests. A fundamental imbalance of reform in the two sectors had created a conflict that could not be resolved by the historical means of the conference.

Yet a very interesting phenomenon emerged from the breakdown of coal-power negotiations in 2008: Although no agreement was reached on coal prices at the Coal Conference, power-generating firms did not stop generating power and coal storage levels were at a record high (Xie 2009). This reveals another side of the story. Although it was reasonable for coal producers to ask at the conference for market prices and also understandable that power companies disagreed, both sides were pragmatic enough to make individual deals either directly or through traders, at prices which were typically higher than the controlled price at the Coal Conference but lower than the coal companies' asking price at the conference (Xie 2009). This dynamic indicated that China did not in fact have a fundamental coal supply shortage.

The coal-power conflict also began to influence where coal supply was coming from. It incentivized many thermal power producers to purchase cheaper coal from overseas, especially in the case of coastal customers closer to coal markets in Indonesia, Vietnam, and Australia (Chapter 9, this volume). Import levels of 40 Mt in 2008 and 126 Mt in 2009 were observed (CCIA 2010) – quantities that were small relative to China's total consumption but large relative to the international coal trade.

Ultimately, despite the difficulties arising from the coal-power conflict, China's coal market has found sufficient supply to allow power generators to keep the lights on (Xie 2009).

Yet is not clear whether such a situation of sufficient supply will persist. Overseas resources can only play a supplementary role in supplying Chinese generators because available volumes for import are much less than Chinese demand; import markets are often too volatile for Chinese power generators; and imports can be infeasible or uneconomic for non-coastal power generators. The conflict between coal and power cannot be solved by an increase in imports alone. The domestic coal market, which must satisfy the bulk of Chinese demand, may witness shortages as government policies of shutting down small coal mines become tougher.[3] (Of course, it is debatable whether China's government would continue to press these policies in a true crisis of supply.) Most importantly, with the Coal Conference having lost its effectiveness at facilitating long-term, stable supply contracts between coal and power companies, both large power generators and coal producers are forced to operate in a more risky business environment.

These concerns are important motivations for coal companies, power producers, and the government to push for the integration of coal and power businesses under the government's plan for building large coal-power bases.

3.2.4 The role of transportation in the coal-power conflict

A key question arises from the discussion of the coal-power conflict: Why would coal companies still come to the Coal Conference to negotiate with power companies, even though they could sell their coal to the market at a higher price? The short answer is that they came to obtain railway transport capacity. For half a century the Coal Conference has had two functions: to seal contracts between coal and power companies, and to negotiate and allocate limited rail transportation quotas based on the contracts signed. Coal purchased under these contracts could only be transported via rail after the "special stamp for transportation contract" (later changed to "check stamp for supply of goods") was awarded by the MOR. The state railway (the MOR itself was officially disbanded in 2013) gives priority to this sort of contracted, or "inside plan" (*jihua nei*), coal over coal transacted in the market. It is the prospect of guaranteed transportation that attracts the thousands of coal and power companies that attend the Coal Conference.[4]

Over the last half century, the MOR played a crucial role in China's economy as the body with sole responsibility for investment, technology upgrading, construction, operation, and maintenance for China's

Table 3.1 *China's national freight via railways in 2008*

Indicators	Units	Achieved	Over the previous year (±%)
Volume of freight traffic	Mt	329,035	4.6
National Rail	Mt	273,932	3.7
Non-controlling joint-venture railways	Mt	27,975	8.2
Local railway	Mt	27,128	11.2
Turnover of freight traffic	100 Mt-km	24,817	3.6
National Rail	100 Mt-km	23,360	3.4
Non-controlling joint-venture railways	100 Mt-km	1,306	5.9
Local railway	100 Mt-km	151	13.7

Source: Ministry of Railways (2009).

Table 3.2 *Volume of freight traffic for key goods and commodities in 2008 (10,000 metric tons)*

Category	Achieved	Over the previous year (±)	Over the previous year (±%)
Coal	169,146	9,257	5.8
Smelting materials	72,800	563	0.8
Food	12,462	1,422	12.9
Oil	14,886	528	3.7
Cotton	388	−3	−0.8
Chemical fertilizer and pesticide	8,398	−387	−4.4
Container	6,863	−487	−6.6

Source: Ministry of Railways (2009).

90,000 km of railways (Ministry of Railways 2009). This near monopoly position is clearly shown in Table 3.1 (as "National Rail"). The national rail network is still the main system for transport of China's most important goods and commodities (Table 3.2), even now that highways are highly developed in China. The pace of railway construction since the 1980s has been impressive, and capacity has been considerably improved (Table 3.3), but rail capacity growth still cannot match the pace of economic development and associated growth in demand for transport of energy commodities.

The importance of the state railway is magnified by the tight supply of rail capacity in China, because it has the power to allocate the limited capacity to transport goods and commodities to customers. The operation, reputation, and profits of hundreds of thousands of Chinese

Table 3.3 *The coal transport capacity of China's railways*

	Cargo transportation capacity (10,000 t/y)			Coal transportation capacity (10,000 t/y)			Actual (10,000 t)
	2005	2006	2010	2005	2006	2010	2005
Northern route	**43,900**	**48,900**	**75,500**	**40,100**	**45,100**	**70,000**	**35,315**
Daqin	20,000	25,000	40,000	20,000	25,000	40,000	20,200
Fenshada	6,800	6,800	10,000	4,300	4,300	6,300	3,849
ShenShuo	13,000	13,000	20,000	13,000	13,000	20,000	9,345
Jingyuan	2,200	2,200	3,000	1,600	1,600	2,100	1,445
Jitong	1,900	1,900	2,500	1,200	1,200	1,600	467
Middle route	**10,500**	**10,500**	**18,000**	**6,800**	**6,800**	**11,600**	**7,356**
Shitai	7,500	7,500	10,000	5,000	5,000	6,700	6,817
Hanzhan	3,000	3,000	8,000	1,800	1,800	4,900	539
Southern route	**24,300**	**24,300**	**41,600**	**14,000**	**14,000**	**22,200**	**10,927**
Taijiao	5,000	5,000	9,000	3,900	3,900	7,100	4,005
Houyue	10,000	10,000	10,000	6,500	6,500	6,500	5,212
Longhai	5,000	5,000	10,000	1,900	1,900	3,800	876
Ningxia	2,300	2,300	5,000	900	900	1,900	377
Xikang	2,000	2,000	7,600	800	800	2,900	457
TOTAL	**78,700**	**83,700**	**135,100**	**60,900**	**65,900**	**103,800**	**53,595**

Source: Beijing Xinhua InfoLink Development Co., Ltd. Quoted from Sagawa and Koizumi (2007).

companies depend on the rail transportation quotas allocated to them and on whether the allocation occurs in a timely fashion.

The MOR could authorize "outside plan" (*jihua wai*) transportation, which refers to goods and commodities to be transported via rail without an official stamp and at a lower priority. Fees charged for transporting inside plan goods were fixed by the government and the MOR could not make a profit from them, but fees charged for outside plan goods were decided by the MOR and it could make a profit from them.[5] As rail capacity expanded rapidly in the past few decades, the MOR gained greater ability to allocate goods and commodities to outside plan transportation. This motivated the MOR, mainly through its departments and subsidiary companies, to emphasize outside plan and charge high fees that took advantage of limited inside plan capacity. Numerous coal companies, especially small ones, had to seek outside plan capacity and pay the higher fees.

The state railway plays a particularly crucial role for the coal and power industries. With more than half of China's coal needing to be transported by rail, the coal industry is and always has been the state railway's number one customer. In 2008, 1.7 Bt of coal were transported by rail (Ministry of

Railways 2009). But railway capacity has never been sufficient. Coal transportation has long suffered from what is referred to as the "transportation bottleneck" problem – the fact that coal from production regions (mainly northwest China) could not always get access to rail networks to be transported to consumer regions (mainly southern and eastern China) due to the short supply of rail capacity. Given the improvement in rail capacity over the last three decades, the rail bottleneck now manifests more in the form of high fees for transportation rather than insufficient capacity per se. A very large proportion of delivered coal cost is in fact attributable to intermediate transport links, mainly railway linkages. In 2008, the average distance from coal production to consumption regions was 635 km. The average price of rail transport was 0.08 yuan/ton-km (Pu 2009). As a result, logistical costs associated with coal transportation accounted for 30% to 60% of total costs incurred by power-generating firms (Fan 2009).

Many of the fees and taxes levied by the state railway and local governments are non-transparent. In 2007 inside plan rail transportation capacity for coal should have been 840 Mt, of which 640 Mt should have been electricity coal. However, in fact only 70% to 80% of inside plan coal was transported; with total coal transport of 1.4 Bt in that year (Fan 2009), this means that at least half of the coal was carried outside plan. The well-known "secret" is that customers can prioritize their shipments with the state railway by going outside plan at a much higher price. The income from outside plan coal transportation goes to the treasuries of the various companies run by railway staff. These "intermediate costs" increase the delivered cost considerably. In 2000 the intermediate fees paid by a private coal company for transporting coal from Shanxi to Tianjin were twice as much as the coal's pithead cost (Rui 2005). In 2008 a state-owned company doubled its cost for coal after adding the intermediate cost of transportation to Qinhuangdao port from the Shanxi region where the coal was produced (Fan 2009). This shows the degree to which transportation costs have been a burden for coal companies, especially the small producers that do not own any rail capacity and have to rely completely on the national rail network, thus exposing themselves to whatever charges accrue.[6] SOE coal companies suffer less from transportation costs because most large state-owned coal companies own some rail capacity or are guaranteed transportation quotas, and they can even sell capacity to small mines or buy coal from small mines at a cheaper price in exchange for extra capacity.

The current transportation structure, which dramatically inflates the coal price, is another reason for companies to move toward coal-power bases.

3.3 Coal-power base policy and the motivations behind it

As described previously, there are severe challenges facing China's coal industry: incomplete price reform, fragmented production structure and diseconomies of scale, a poor safety record, mismatched market structures for coal and power, and the market power of the railways. All of these factors make China's coal market inefficient and contribute to higher costs and prices. But above all, the coal, power, and railways conflict stands out as a key challenge.

In response to this dysfunctional situation, Chinese policy makers have proposed creating large coal-power bases that are intended to end the disputes among the coal, power, and railways and optimize the structure of the coal industry. The coal-power base policy – and its origins and objectives – are described in the following sections.

3.3.1 Government policies on building up large coal electricity bases

The government first launched a campaign to vertically integrate coal and power in 2004. On September 6, 2005, the State Council issued "Some Opinions of the State Council on Promoting the Healthy Development of the Coal Industry" (*Xinhua News*, 2005), which formally stated the idea of building up large coal-power bases. This document for the first time announced thirteen bases rather than the previous figure of seven bases and clearly called for coal-power bases rather than coal-only bases, as had been suggested before. More importantly, this document raised the important question of how to promote the coordinated development of the coal industry along with related industries. It stressed that the construction of large-scale coal-producing bases should be linked with integrated development of shipping and water resources, rail and ports, and local economic development. The government particularly emphasized solving the rail bottleneck issue by means of large coal-power bases and coordination among coal and its related industries. Building large-scale thermal power stations near coal mines to relieve the pressure on coal transportation would be one example of such coordination; encouraging large-scale coal enterprises to have joint operations with metallurgical, chemical, construction material, communications, and transport industries would be another.

The State Council's 2005 guidance also emphasized that the state should apply capital construction funds from the central budget or government loans to support the construction of large-scale coal-production bases. Government policy banks, state-owned commercial banks, and joint-stock commercial banks were all asked to actively improve their financial services and to effectively support the development and construction of such bases.

The government did not underestimate the difficulties of building coal-power bases across disparate regions and sectors. It called for all the relevant local governments and industries to follow the guidance of the central government and voluntarily participate in the integration effort. The government underscored the importance of breaking away from regional, industrial, and ownership constraints in order to accelerate the development of large-scale coal enterprises and enterprise groups while also regulating small coal mines.

The key government document that describes the thirteen coal-power bases plan is the 11th Five Year Plan of the Coal Industrial Development issued by the NDRC in January 2007 (henceforth "the Plan"). The thirteen coal production bases were listed as follows: Shendong, Shanbei (North Shaanxi), Huanglong (Huating), Jinbei (Northern Shanxi), Jinzhong (Central Shanxi), Jindong (East Shanxi), Luxi (West Shandong), Lianghuai (Huainan and Huaibei), Jizhong (Central Hebei), Henan, Yungui (Yunnan-Guizhou), Mengdong (East Inner Mongolia and North East of China), and Ningdong (East Ningxia). Figure 3.2 shows the locations of the thirteen original bases, as well as Xinjiang, which now represents the fourteenth coal-power base.

The Plan set forth five principles for building large coal electricity bases. All of these principles show a preference for large companies and central control. The first principle is orderly and centralized development (*jizhong youxu kaifa*). Essentially, one base is developed by one company, which is responsible for making a comprehensive plan and controlling the pace of development. The second principle is innovation. With the characteristic orientation of China's government, this plank identifies large coal and power group corporations as the best types of firms to drive technology investment and advancement. The third principle is development of an optimal coal production structure based on large-scale, modern opencast mines, improved resource recovery rates, and closure of small mines. (This consolidation of coal companies into larger entities is referred to as horizontal integration.) The fourth principle is comprehensive vertical integration of coal and power, coal and chemical, and coal and railway industries. The fifth principle is the development of "the recycling economy," in which the close connection between upstream and downstream industries reduces environmental impacts (NDRC 2007).

The thirteen coal bases will cover ninety-eight coalfields of around 287,000 sq. km, distributed across fourteen provinces. Across these thirteen bases, the government will allow six to eight coal-power corporate

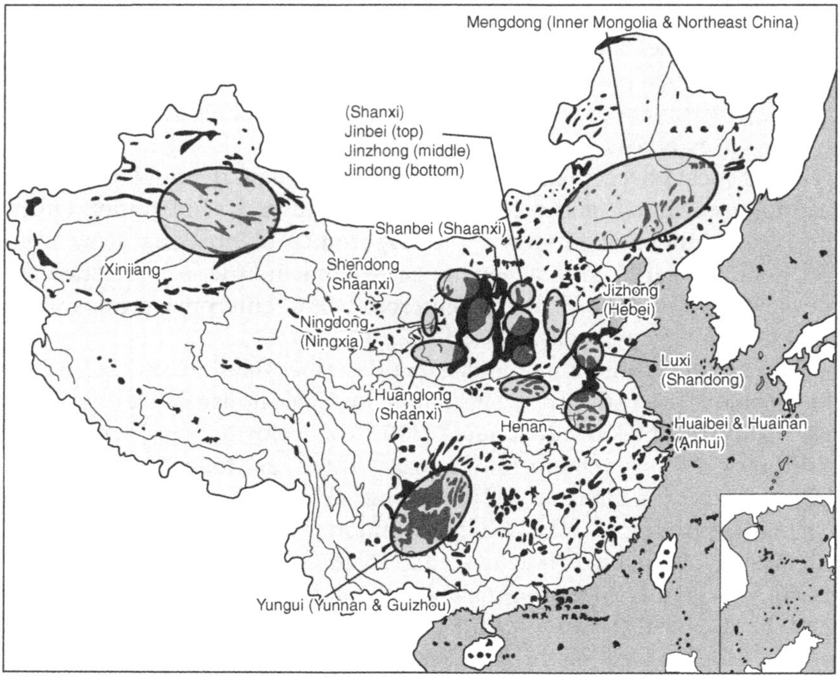

Figure 3.2 Map of coal-power bases in China. Dark areas represent coal resources. Source: Originally from Ministry of Land Resources of China; adapted by Gang He.

groups with 100 Mt annual output and eight to ten companies with 10 Mt annual output. The output of thirteen coal-power bases was hoped to reach 2.24 Bt by 2010 (NDRC 2007), such that output from these thirteen bases would account for about 86% of the national total, assuming 2.6 Bt of national output in 2010 (NDRC 2007).[7] However, actual development of coal-power bases has thus far fallen significantly short of this target.

The contours of the thirteen coal-power bases set out in the 11th Five Year Plan have changed considerably since the initial plan and will be subject to further changes in the future. This is primarily due to the fact that the original coal-power bases did not take all relevant factors into account, or new coal resources were discovered. Fang Junshi, head of coal at the National Energy Administration, explained that many of the planned bases were "not practical and scientific" because these bases were found to lack water resources, a crucial input for extracting coal and washing coal (*Shanghai Stock Exchange Report*, December 5, 2008). As we will discuss later in this chapter, Xinjiang is a massive new coal province not part of the original thirteen bases that could replace some of the bases that proved unsuitable.

3.3.2 Government objectives in building up the coal-power bases

In this section we lay out, and in a few cases critique, the government's logic in promoting the coal-power bases.

3.3.2.1 *Reduce price risk and conflict between coal and power firms*

Price volatility and transportation dependency are the main risks at present for both coal and power firms. As the government sees it, coal-power bases could address both issues, in the following ways. First, creating coal-power bases could reduce transaction costs for coal and power companies when they sell or purchase coal. Second, coal-power bases could help spread market risk across the whole value chain. Coal profits could supplement power losses when the coal price is high, and vice versa. Power company performance would be expected to improve because "if China's utility companies could supply 30% of coal demand with their own source, they would be less vulnerable to coal price fluctuations"[8] (interview with Xie, April 21, 2009). Establishing more coal-power bases would also help reduce risk by providing coal companies with more secured customers and power companies with a long-term secured coal supply.[9] Third, integrated coal-power bases with dedicated transportation would shift rents that are currently accruing to the state railway back to the coal and power industries.

3.3.2.2 *Nurture large, modern coal corporations for better performance*

Chinese energy policy has made building large corporations and enhancing the industrial concentration of the coal sector a top priority. One of the five principles in the 11th Five Year Plan of the Coal Industry is that large corporations will be chosen as the principal enterprises to develop coal-power bases, in order not only to provide a comprehensive plan for the broader industry's development but also to secure the best technology and equipment in developing the bases. The government seeks both horizontal integration of coal production through the consolidation of small mines into larger ones and vertical integration between coal producers and coal consumers.

Coal-power base policy envisions several key gains from optimizing the industrial structure of coal and power through integration. First, integration will ensure that coal resources can be extracted, processed, and transported in a manner that maximizes the value of coal and reduces inefficient resource use through economies of scale. Second, a limited number of large coal corporations should find it easier to make long-term supply agreements or cooperate in exploration and R&D.

Companies that are part of coal-power bases are required to build modern, safe, and high-efficiency coal mines or power plants. This requirement is expected to stimulate the application of advanced technology and high-efficiency equipment and promote R&D. Among other potential benefits, the use of the latest technologies should reduce local pollution and carbon emissions associated with coal extraction and use.

While some of the government's reasoning in favoring large state-owned companies is sound, the history of China's coal sector (not to mention the broader literature on state-owned enterprises) should sound a few cautionary notes for government reformers who expect that consolidation of industrial activities within large SOEs will automatically improve performance. After all, the coal sector was originally liberalized precisely because SOEs could not keep up with production needs (see Section 3.2.1 as well as Chapter 2, this volume). Large SOEs may be more controllable in certain respects (see Section 3.3.2.3), including when it comes to safety and environmental performance, and they offer economies of scale and a valuable platform for the introduction of advanced technology. At the same time, they do not face the same market drivers and competitive threats that can drive cost efficiency and responsiveness in firms without central state backing. Time will tell whether vertically integrated SOEs alone are adequate to keep coal production at necessary levels, or whether China will need to continue to rely heavily on smaller players that are more removed from central government control.

3.3.2.3 Reinforce the central government's control over energy markets

The Chinese government has a strategic vision to use large coal and power bases to reinforce its control over the energy sector. Zhang Guobao, the head of the National Energy Bureau, published an influential article in the *People's Daily* (December 29, 2008) entitled "The current situation: Opportunities among the danger" (*dang qian de xingshi: wei zhong zhi ji*). Zhang clearly stated in this article that "large energy enterprise groups are nurtured with the integrative and comprehensive development of coal, power, railway, port, chemical, and other related industries, so as to toughen the government's influence and control on energy."

The government sees several mechanisms through which the coal-power bases could help enhance its control over the energy sector. First, if the coal-power bases are a dramatic success, large state-controlled firms could be at the reins of up to half or more of national coal output. The government believes it will be able to coordinate energy and industrial policy more easily through its direct control over these firms. Second, in the government's view, fewer large coal producers would imply a more stable price for coal that would pose fewer challenges for government policy. Third, the government would be able to more easily predict how much coal is available for export by keeping tabs on a limited number of large firms and therefore would be more easily able to control export volumes, which are strictly regulated under a quota system. Fourth, given the smaller number of producers, the government would be better able to supervise each corporation's investment in environmental protection, technology and equipment, R&D, and acquisition of new resources both in China and abroad. Adequate supervision is perceived to be difficult if not impossible in a country with tens of thousands of small coal mines protected by local governments aiming to maximize short-term profits.

3.3.2.4 Transform energy transport and the distribution of environmental impacts in China

One major motive for building coal-power bases is to transform the overall way energy is moved in China, from the transportation of coal via rail to the transmission of coal-generated electricity via wire (also known as "coal by wire" or *gai shumei wei shudian*). If the coal-power bases facilitate successful coal by wire projects at scale, they could make a significant contribution toward assuring the energy supply and improving the local environment in population centers far from the coal mines. The following example of energy transport between Shanxi and Shandong provinces illustrates this point.

Shandong's provincial government projected its coal import demand from other provinces to be 190 Mt, 290 Mt, and 350 Mt in 2010, 2015, and 2020, respectively. It also foresaw a shortfall in electricity of between 3000 MWh and 6000 MWh in 2009 and 2010. The provincial government is therefore working hard to satisfy its demand for both coal and power. Meanwhile, neighboring Shanxi province has plenty of coal but needs guaranteed customers as well as roads and ports to transport its coal. After negotiations, the two provinces signed an unprecedented "agreement framework on strategic cooperation in energy and transportation" in early 2009. According to the framework, from 2009 to the end of the 12th Five Year Plan, Shanxi will transmit 10,000 MWh every year to Shandong by building coal-power bases and constructing transmission capacity that enables energy to be exported as electricity via wire rather than coal via rail. As a result, Shandong will obtain electricity from Shanxi at a price 0.10 yuan/kWh lower than the local Shandong grid price. Shanxi will gain access to new markets for its coal after it builds railways from its central and southern regions to Shandong's port of Rizhao. Most experts believe this will be a win-win agreement. Due to the significance of the power grid in this agreement, the State Grid Corporation also joined the partnership. In November 2008, Shandong signed an agreement with the State Grid Corporation in which the latter will build six super high-voltage transmission lines by 2020 in order to transmit electricity from 52,000 MW of dedicated generation capacity to Shandong, a quantity equivalent to 20% of Shandong's total installed capacity. In many settings, there is still a debate over whether it is cheaper to construct new railways to transport coal or transmission lines to transport electricity.[10] However, in the case of Shandong and Shanxi, the cost of transporting electricity will be lower due to the use of existing transmission lines and the network of the State Grid Corporation.

If the goal of "outside electricity entering Shandong" can be realized, there should be significant energy and environmental benefits for Shandong. The scheme will save the province 128 million tonnes per annum (Mtpa) of coal consumption and reduce local SO_2 emissions by 2.06 Mt. (The net environmental effects are less clear, as pollution in Shanxi will presumably rise; however, there is likely to be some net environmental benefit from improving transportation efficiency and in this case from eliminating the emissions associated with rail transport.) The electricity supplied from Shanxi could create much more space for Shandong's industrial development in the future. The framework is regarded as "unprecedented" because it crosses multiple provinces and industries. It will optimize the energy structures of the two provinces and open the previously closed grid system of Shandong, thereby balancing the power supply shortage in Shandong with the surplus in Shanxi (*Xinhua News* 2009).

3.4 The "Shenhua model": its origin and future prospects

The only fully realized demonstration case at large scale that we can use to evaluate the coal-base strategy and its possible impacts is China's first major, vertically integrated coal-power business: Shenhua. Shenhua is the basic model upon which the coal-power base plan is patterned and is therefore a highly instructive example, although as we shall see its experience may not be fully transferable to other coal-power bases.

Shenhua Group Corporation Limited ("Shenhua" hereafter) was born in 1995; today it is China's top coal producer (Table 3.4) and one of the largest coal companies in the world. It is a state-owned enterprise that was founded at the direction of the State Council. Shenhua is a diversified energy company with business segments that include coal production and sales, electricity and thermal generation, coal liquefaction and coal chemicals, and railways and ports. As of 2009, Shenhua owned fifty-four coal mines with a total annual production capacity of 328 Mt, 1,369 kilometers of dedicated railways with a total transportation capacity of 128 Mt/km, and power plants with a total installed capacity of 16,080 MW.

Shenhua established its leading position in China's coal industry by taking advantage of a constellation of uniquely favorable conditions never before enjoyed by a coal producer in China.

First, unlike most large coal companies in China, which had existed for decades, Shenhua was only established in the mid-1990s under a government mandate to shift the locus of China's coal production from the east to the west of the country (*dong mei xi yi zhanlue*). Shenhua was envisioned as China's most important future production base, as northern and central China's coal had already been substantially depleted. Following this plan, Shenhua was granted by far the largest coal reserves that had ever been given by the government to a company. It was granted a mandate to extract coal from the Shenfu Dongshen coalfield, which had total recoverable reserves of 16 Bt, geological reserves of as much as 223 Bt, and possible resources of 1,000 Bt according to Chinese official survey figures in 1998 (Shenhua 1998). This compared with a national total of 115 Bt of recoverable coal reserves.

Second, Shenhua was granted the largest loans ever offered to a company by the government. The central government, through the State Development Bank, extended to Shenhua preferential low interest rate loans of more than $9.2 billion from 1985 to 2005. This made the Shenhua construction project the third largest investment project ever undertaken in China and one of the world's largest overall. As a brand-new entity with access to such vast amounts of capital, Shenhua was able to closely follow the model of the world's leading coal producers and avoid some of the legacy problems that still plagued the industry in China. In a sharp break from the industrial arrangements that caused the coal-power conflict, Shenhua was given a dedicated rail line to transport coal from the Shenfu Dongshen coalfield to its dedicated port facility at Huanghua. It was also assigned dozens of power plants from the State Planning Commission to consume millions of tons of coal from its own field.

Table 3.4 *China's top 10 coal companies in 2006 and 2008 (tonne)*

		Production (2006)			Production (2008)	
Rank	Coal companies	Mt	% of national total	Coal companies	Mt	% of national total
1	Shenhua Group	149.7	6.3	Shenhua Group	281.3	10.1
2	China National Coal	71.9	3.0	China National Group	114.1	4.1
3	Shanxi Coking Coal	60.8	2.6	Shanxi Coking Coal	80.3	2.9
4	Shanxi Datong Coal Mining Group	56.7	2.4	Shanxi Datong Coal Mining Group	68.9	2.5
5	Heilongjiang Longmei Holdings Co., Ltd.	52.7	2.2	Shaanxi Coal and Chemical Industry	60.4	2.2
6	Yan Mining Group	37.0	1.6	Anhui Huainan Mining	56.7	2.0
7	Shanxi Yanquan Coal (Group)	32.5	1.4	Heilongjiang Longmei Holdings Co., Ltd.	55.0	2.0
8	Anhui Huainan Mining	32.4	1.4	Henan Coal and Chemical Industry	44.7	1.6
9	Henan Pingdingshan Coal (Group)	32.1	1.4	Shanxi Lu'an Mining	42.1	1.5
10	Shaanxi Coal Group	30.8	1.3	Pingmei Shenma Energy Chemical Group	41.2	1.5
	National total	**2,373**	**100**	**National total**	**2,793**	**100**

Source: China's Coal Industry Association, various years.

Third, Shenhua utilized advanced exploration technology and equipment from the very beginning (more details in Rui 2005). Much of China's coal industry is relatively backward by worldwide standards, but the government imported more than 98% of Shenhua's equipment (Rui 2005). Shenhua also installed a reliable automated dispatching system for underground mining. Much labor has been saved by Shenhua's high level of mechanization. A comparison with other Chinese SOE coal mines illustrates the resulting productivity improvement. In 1999, the 350 employees in Shenhua's Daliuta Mine produced 8 Mt of coal, while more than 10,000 employees in Jixi's Xiao Henshan Mine, a traditional state-owned coal company, produced only 3.5 Mt of coal.

Fourth, the Shenhua model built in a high degree of horizontal and vertical integration of the coal, power, chemical, and transportation sectors. The company achieved horizontal integration by creating consolidated entities in mines, railways, and ports and vertical integration by linking coal production with power, coal-to-liquids (CTL), and coal-based chemical businesses. Shenhua is unique among Chinese coal companies in the extent of its horizontal and vertical integration, which has provided its most important competitive advantage. To a large extent, building coal-power bases is an attempt by other Chinese firms (and the government) to clone the Shenhua model and achieve this kind of integration in other settings.

3.4.1 Horizontal integration

Horizontal integration is defined here as the consolidation of coal producers. Under the leadership of Ye Qing since 1998, Shenhua has been actively acquiring existing coal mines, expanding its own mines, and building new mines.

First, Shenhua Group successfully expanded the Shendong production base to more than 100 Mt of annual output by adopting advanced technology and equipment and improving productivity.

Second, it restructured inefficient small mines that extracted coal at Shenfu Dongshen coalfield by "renting" them and putting in place improved processes and coordination (more details in Rui 2004). This not only helped achieve economies of scale in Shenhua's coal output but also ensured that modern and high efficiency techniques were utilized in the coalfield.

Third, Shenhua has been acquiring other coal mines, coalfields, or equity stakes in coal assets across all regions in China and even internationally. Shenhua's listed company Shenhua Energy produced 281 Mt and sold 311 Mt of coal in 2008, and it planned to invest $39.5 billion through 2013 to expand production capacity, with the intent of doubling its annual capacity to 500 Mt. The company plans to develop strategic coal reserves of 30 Mt in the eastern, southern, and central parts of China. Outside of China, Shenhua has invested in coal mines in Indonesia and Australia and was one of ten bidders that pursued a stake in Mongolia's Tavan Tolgoi deposit in the Gobi Desert, which with a coal reserve of 6.5 Bt is generally considered the world's largest untapped coking coal deposit (*China Daily* 2009).

3.4.2 Vertical integration

Vertical integration is the consolidation of coal consuming industries with upstream coal providers and coal transport networks. Shenhua's vertical integration is observed across multiple sectors.

First, vertical integration in Shenhua is exemplified by its coal and power business integration. Shenhua "acquired" many power assets in the 1990s, such Guohua Power and others, to vertically integrate into its coal value chain. These power plants were previously owned by the NDRC (then known as the State Planning Commission) but divested under the policy of "separating business from government." Many assumed at the time that taking over these power plants would be a burden for Shenhua, given that the plants were state owned and very inefficient. Ye Qing, however, considered acquiring these power plants to be an opportunity for Shenhua because the coal market was in surplus at that time and the power plants could consume Shenhua's coal while minimizing transaction costs (Ye Qing, interviewed in 2001 and 2003). As indicated at the beginning of this section, Shenhua's power business had expanded to a total installed capacity of 16,080 MW as of 2009.

Second, Shenhua has integrated its coal production and coal liquefaction businesses. Coal liquefaction is a technology through which coal is transformed into a liquid fuel (for more detail see Nolan et al. 2004). In 1993 China became a net oil importer, and the importation of oil has increased year by year since then, leading the Chinese government to focus on reducing its oil import dependency. It therefore strongly supported coal-to-liquids (CTL) R&D that could potentially contribute to security of liquid fuel supply. Shenhua was awarded almost all of the available government funding for "substituting coal for oil" to implement its coal liquefaction program. Ye Qing predicted in 2003 that by 2020 Shenhua would be more of an oil-producing company than a purely coal-producing company (Nolan et al. 2004).

Third, Shenhua has integrated its coal and chemicals projects. For example, Shenhua Ningxia Coal (Group) Industry Corporation is developing a methanol production facility with a capacity of 250,000 tons per year. All of the civil engineering and most of the installation work have been completed, with the 220-ton boiler commissioned in 2009. A number of other vertically integrated projects in the olefins, tar, transport fuel, and chemical sectors are also underway.[11]

3.4.3 The key role of transport networks in enabling vertical integration

Shenhua's successful vertical integration – in particular of coal and power – would not have been possible without its development of an independent, consolidated transport network that allowed it to bypass the MOR's stranglehold on rail capacity. When Ye Qing was appointed president of Shenhua in 1998, the company was concentrating on its familiar activity of mine construction while delaying rail and port construction. Considering that China's coal market was at the time in a state of surplus but rail transportation was still a bottleneck, leading to high transport costs for Shenhua's coal, Ye suspended mine construction and accelerated the company's efforts to build dedicated railways and ports. The effect of such a shift in strategy was very clear: Shenhua's railway and port construction was completed in 2001, two years ahead of schedule, which immediately saved 40 yuan per ton of coal in freight costs (Ye Qing interview, 2001) while significantly reducing the risks of future transportation bottlenecks due to dependency on state rail. Despite already owning 1,369 km of dedicated railways,[12] Shenhua is planning a further huge expansion of its rail network to integrate its mines fully with Huanghua and Qinhuangdao ports. Work on a new railway from Batuta, where Shenhua operates small mines, to Diandaigou in Inner Mongolia began on April 8, 2009 with a 2.7 billion yuan investment and was expected to be completed in 2012. Shenhua also has four additional railway projects that are included in a railway network plan issued by the NDRC in October 2008[13] (China Coal 2009). Although other coal-power bases are allowed some captive rail capacity – for example, Guodian and Huaneng are permitted to build railways to bring coal to power plants or ports – Shenhua is the only company that owns more than 1,000 km of railways as well as a dedicated coal port.

3.4.4 The challenge of replicating Shenhua's integrated model

As previously described, Shenhua is both horizontally and vertically integrated. This integration was enabled by unique historical circumstances.

While coal and power companies in China have stated their ambition to replicate the Shenhua model, they may not be fully aware of the difficulties involved in doing so. (In Section 3.6, we will further consider the obstacles to replicating the Shenhua model as we assess the prospects that coal-power bases will have a significant near-term impact on China's industrial landscape.)

Few, if any, other companies could acquire a coal asset such as the Shenfu Dongshen coalfield, which is blessed with abundant low-cost and high-quality coal. This coal asset is the foundation of Shenhua's competitiveness. Even though other companies may acquire coal assets from various owners, they are unlikely to be able to put together such a large and integrated resource base as Shenhua, which was specially designed from the outset as a new national champion in coal. Strong central government support factored strongly in Shenhua's experience from the beginning. While the government has indicated that other companies involved in coal-power bases may be allowed to build railways and receive loans, the exact preferential treatment they will receive has never been clear.

Other companies may also lack the particular type of leadership Shenhua had during its key period of integration. Management capability and continuity plus a healthy dose of connectedness within the government are crucial to efforts to build large integrative corporations in China. For more than a decade, from 1986 to 1998, Ye Qing was deputy director of the State Planning Commission (the predecessor of the NDRC) in charge of energy and communication. This post was perfect background for managing Shenhua because the structure of the conglomerate demands both the ability and the authority to coordinate at least six sectors (coal, power, railways, port, coal liquefaction, and coal chemical) in at least five provinces (Shaanxi, Inner Mongolia, Shanxi, Hebei, and Beijing). Power-generating companies such as China Datang Corporation have admitted that extending their businesses into coal is a big challenge for them because the leadership must understand not only electricity but also the coal industry, as well as how to coordinate between the two.[14]

In short, even though many firms possess both coal and power assets (or other coal-related businesses), they may struggle to achieve the degree of integration of Shenhua.

3.5 "Cloning" the Shenhua model in China: types of coal-power bases

This section examines the applicability of the Shenhua model of horizontal and vertical integration to the development of other coal-power bases now planned or underway in China. This analysis will help in assessing how quickly and completely the vision of thirteen integrated coal electricity bases can be realized and what impacts they might have on the coal market.

There are three possible modes of setting up a coal-power base.

The first mode is "coal to power": coal companies expanding downstream into the power business. This is the Shenhua approach. Other major coal companies, including China Coal, Yanzhou, Datong, and Lu An, have all started to expand their businesses not only into power but also into coal chemical and other related businesses.

The second mode is "power to coal": power-generating companies expanding their businesses upstream into coal. The Big Five power companies – Datang, Huaneng, Huadian, Guodian, and China Power Investment – are all actively acquiring coalfields and even old coal bureaus[15] such as Hulinhe, Dayan, and Xinzhuang. Many independent power-generating companies such as China Resources Power Holdings have spent large amounts of capital to acquire coal mines near their plants during the last few years.

The third mode is the "equity sharing mode": coal and power firms acquiring equity stakes in each other and/or creating joint ventures with various shareholding agreements. For example, Huainan Mining Group has set up joint-venture power plants with China Power Investment and Zhejiang Power, respectively.

Thus far, the "power to coal" mode has been more commonly attempted than the "coal to power" mode. The major cases of each of these modes of integration and the motivation for the firms involved to pursue these respective modes are explored below in order to draw inferences about the probable development of the coal-power bases and thus the broader Chinese coal sector. We shall see that coal and power companies do have distinct motivations for pursuing particular modes.

3.5.1 Coal to power mode

What follows are brief discussions of the major Chinese coal firms that are pursing vertical integration and coal power bases through the "coal to power mode." Following the example of Shenhua, the investment and growth strategies of China's large coal companies aim at expansion not only into power but also into other value-added businesses related to coal.

China Coal Group, the second largest coal company in China, has achieved significant growth since its major restructuring ten years ago. The company's coal production volume increased by 4.9 times over this period to 125 Mt; total sales revenue increased by 4.7 times; and total asset value increased by 4.8 times. The company set as its development goal to expand not only its coal business but also its activities in the areas of coal to chemical conversion, power generation, and manufacturing of coal mining equipment. It aims to reach 200 Mt of coal production volume by building five major coal production bases[16] in Shanxi, Jiangsu, and Heilongjiang provinces and Inner Mongolia and Xinjiang Autonomous Regions. It then intends to establish integrated coal and power businesses through the development of pit-mouth power plants. It has demonstrated circulating fluidized bed (CFB) combustion technology and built environmentally friendly pit-mouth power plants that consume coal refuse, coal slurry, and middling. The company's installed power generation capacity as of 2009 was 526 MW (Wang 2009).

Datong Coal Group was China's largest coal producer during the era of the planned economy. Despite this notable position, Datong, like most other state-owned coal companies, suffered from financial difficulties during the reform era, given its large social welfare burden and unfamiliarity with market competition. Its situation has improved in recent years owing to China's rapid economic growth and high demand for coal. However, Datong's management is aware that its current prosperity may not last long given the uncertainty of market demand for coal and the possibility that coal prices may drop over time. The company sees its long term success as dependent on "adding a higher value to coal," for example by extending Datong's business from coal to power and chemical operations (Wu 2009). Consistent with this long term strategy, Datong launched Tashan Cyclic Economic Park, the most important expansion project of the company, with a total investment of RMB 17 billion. This project aims for "sustainable" and "comprehensive" use of resources. The company is expanding into power and chemical businesses through cooperation with power, coal to chemical, and other firms with expertise outside of coal (Wu 2009). In fact, the twenty-nine key projects under construction in Datong are all developed via such modes of cooperation, including joint ventures, project contracts, and other approaches.

Lu An Group is ranked eighth among China's coal companies in terms of its coal output and profit. Lu An's Five Year Plan (2008–2012) aimed to build the company into a 100-Mt coal enterprise so that it could develop greater power in the coal market, including enhanced price negotiation leverage with customers. In addition, Lu An has made considerable efforts to build out other coal-related businesses. It is developing coal-power-chemical, coal-coke-chemical, and coal-oil-chemical projects; building four recycling economic parks focused on power-chemical, coke-chemical, coal-power, and coal-oil; and developing broad capabilities in the coal, power, oil, chemical, and silicon sectors.

The company also aims to strengthen international cooperation and to develop overseas coal reserves and an IGCC project (Ren 2009).

Among all of these efforts, building up the company's coal-power base in central Shanxi province is the most important one. As of April 2009, Lu An Group had a total installed power-generating capacity of 540 MW in the region. It sped up the construction of the 2,400 MW Gaohe Power Plant and the 1,200 MW Xiangyuan Power Plant so they could form the foundation of the coal power base. The company has also successfully produced the first barrel of oil from its coal liquefaction program (although the scale of Lu An's program pales besides Shenhua's, as it did not receive the significant government support for CTL that Shenhua did).

3.5.2 Power to coal mode

Power-generating firms can enter the coal business using two primary models. One is a "green field mode" of developing new coalfields and coal mines; the other is to acquire existing, non-state-owned coal mines.

The green field mode has the advantage that the power-generating firm maintains a high degree of control over the coal resources and management of the coal mines. But it has the disadvantage of long waiting periods to obtain the proper permits and rights to start a mine. In contrast, while coal prices remain high, power companies can obtain permissions to acquire existing mines quickly, which makes the acquisition mode comparatively attractive because ownership of coal resources can be achieved much faster. However, the limited availability of coal resources and mining companies for acquisition forces power companies to use whichever option is available.

What follows are brief discussions of the major power generating firms that are pursuing vertical integration by acquiring assets in the coal sector.

Huaneng Group, the oldest power-generating company among the current Big Five, was established in 1985. By June 2009, it had invested in twenty-six provinces and autonomous regions as well as overseas. It has 130 wholly owned or controlled power plants, with 88,967 MW of installed capacity. It aimed to reach 100,000 MW of installed capacity by 2010, with 60 Mt of coal output per year.

Huaneng acquired a 40% share in Huating Coal Group in October 2008 from the Gansu provincial asset administration commission. It acquired another 9% in June 2009, becoming the largest shareholder and gaining control over the company. Huaneng planned to invest as much as 30 billion yuan in Gansu to build up a comprehensive base with coordinated coal, power, and transportation activities. Many experts interviewed argued that this is an ideal model, with power companies providing capital and advanced management while local governments provide access to resources. (We will further discuss the base assumption of China's policy makers that the management quality of power companies is superior to that of coal companies.) This model can also allow power companies to better manage the price risk associated with the highly volatile coal market.

Huadian Group is China's fourth largest power generator. Its Shanghai-listed subsidiary Huadian International Power Company has mainly adopted the acquisition strategy to improve coal supply. Huadian has 69,020 MW of installed capacity, of which 59,880 MW, or 86.8%, is thermal power, and it consumed 128 Mt of coal in 2008. The company aimed to have 100,000 MW of installed power generation capacity by 2013. Huadian announced in July 2009 that it spent 760 million yuan to acquire a 70% stake in both Shanxi Shuozhou Wantongyuan Erpu Coal Mining Company and Shanxi Dongyi Zhonghou Coal Industry Company, through its subsidiary Shanxi Maohua Energy Investment Company. Both coal mines were expected to supply Huadian's power plants in Shandong. In June 2009, Huadian International also announced plans to buy a 70% stake in the Xingbian mine in Heilongjiang province for 479 million yuan, not long after it had acquired Xinwen Mining Group's Changcheng and Fucheng mines in Inner Mongolia.

Huadian International has also adopted other methods to augment its coal resources. In September 2009, it announced its intention to spend 600 million yuan to increase its stake in Yinxin Coal to 45%. According to the agreement, Huadian International has priority rights to acquire 45% of coal output at market price after the coal mine is put in use. Overall, Huadian planned to have 30 Mt of coal output by 2013.

China Datang Corporation (CDT) is now the largest power-generating firm not only in China but also in Asia. It was established in 2002 by combining some of the assets of the State Power Corporation of China with major businesses in northern and northwest China. By 2008, CDT had total assets of 407.6 billion yuan and revenue of 101 billion yuan. The installed capacity of the company reached 90,061 MW by June 2009, up from 23,847 MW in 2002. It generated 352 billion kWh of power in 2008, up from 117 billion kWh in 2002. Datang's share of national power generation increased from 7.1% in 2002 to 10.3% in 2008.

In 2005, CDT set as a strategy the extension of its value chain upstream and identified coal production as a core business. In 2009, after being permitted to expand into coal mining by the State-Owned Assets Supervision and Administration Commission (SASAC), Datang Coal was established as CDT's subsidiary responsible for managing the coal business. By June 2009, CDT owned or managed six coal mines, owned or shared 9.69 Bt of coal reserves, and had 13 Mt of annual coal output.

China Power Investment Corporation (CPI) was also constituted from some of the pieces of the former State Power Corporation of China. As of the end of 2008, its controllable installed capacity was 51,990 MW, equity installed capacity was 40,116 MW, and total assets stood at 284.2 billion yuan. CPI's subsidiaries include six A-share listed companies, notable among them being Shanghai Electric Power Company. (Shanghai Electric Power's coal-power project is discussed further in Section 3.5.3.)

Of CPI's total power capacity in 2008, 41,123 MW, or 79.1%, was from thermal units, which consumed a total of 103 Mt of standard coal. Because of coal's significance to CPI's business and profits, CPI has pursued strategic acquisitions and joint development of coal mines since 2005. The company's strategy is to prioritize electric power as the core business, use coal mine development to support the power business, and pursue further diversification going forward. Following this strategy, CPI acquired Mengdong Group Company, a large coal operation in Inner Mongolia. CPI also owns Huolinhe Coal Power Company, which enjoys good coal production and sales, notably from one of the largest open cast coal mines in Inner Mongolia, and also owns railways and water reservoirs. CPI also cooperates with other companies to operate the Baihuayin Coal Power Company.

Many power-generating companies including CPI have invested in rail to support the "power to coal" mode of coal-power base formation. In March 2008, CPI's Chifeng-Jinzhou Railway Project was ratified by the NDRC. The main line is 282 km long, and the total investment is 7.185 billion yuan, of which 52%, 16%, 16%, and 16% are contributed respectively by Baiyinhua Coal Power Company, Datang International Power Generation Company, China Guodian Corporation, and China Resources Power Holdings Company. Chifeng-Jinzhou Railway will mainly be used to transport coal produced from Baiyinhua and surrounding coal mines in Xilinguole of Inner Mongolia to the West Liaoning area and Jinzhou port. Although relatively modest in scope, this rail project contributes to the slow erosion of state monopoly control over the nation's rail network.

3.5.3 Equity sharing mode: the case of the Huainan and Huaibei coal-power base

Under the "equity sharing mode," coal and power companies exchange shares with each other and/or acquire shares from railway operating authorities.[17] The main advantage of this mode is that two or more entities reap mutual gains from tapping into each other's comparative advantages. Analysis of the necessary conditions for developing such arrangements points to how future coal and power bases might develop under this third mode.

China's first coal-power base, the Huainan and Huaibei coal power base in East Anhui, was initiated under the equity-sharing mode. In 2003, the Huainan mining group was searching for customers for its coal, while Shanghai Electric Power was searching for suitable sites to build power plants to address the severe shortage of electricity at the time. The two companies easily reached a deal to build up a power plant at the site of the Huainan Mining Group's Tiji coal mine. The two-phase, 4 × 600 MW Tiji power project was approved in March 2005. (Huainan Mining Group reached a similar deal with Zhejiang Energy Ltd. around the same time to set up a power plant with a capacity of 2 × 600 MW.) The first phase of the Tiji project started generating power on July 26, 2007, and the second on October 15, 2007.[18] Electricity is transmitted via super high-voltage transmission lines and feeds into the East China Grid.

The full development of the coal-power base in Anhui was jointly led by the Huainan and Huaibei Mining Group corporations. The Huainan and Huaibei Coal Power Base, which is regarded as the first completed base among the thirteen planned coal-electricity bases (*Huainan Daily*, December 6, 2008), was officially commissioned on December 5, 2008, with representatives in attendance from Huainan and Huaibei, NDRC, Ministry of Railways, China Power Investment, and the National Power Grid. The base has rich water supplies and a total of 25.2 Bt of coal reserves, or 45% of the total reserves in East China. Consistent with the coal-power base policy, twelve pairs of new mines (twenty-four in total) and eight power-generating plants have been built or expanded at the Huaibei and Huainan base. Coal output reached 120 Mt in 2008, from less than 50 Mt in 2000. The installed electric capacity reached 13,000 MW in 2008, from 4,000 MW in 2000. Three transmission lines have been installed under the policy of "transmit electricity of Anhui to East China."

Although the Huainan-Huaibei project represents China's first coal-power base, there is no single legal entity that represents this base or any formal structure coordinating the activities of the numerous companies and organizations operating at the base. Apart from the two joint-venture power plants at Huainan previously described, this coal-power base, like all the other bases in China, is mostly a geographical rather than legal or institutional entity.

The equity sharing mode of coal-power base development, whether formal or informal, offers several advantages for both coal and power firms.

First, like all viable coal-power base approaches, it can reduce transportation costs. Traditionally, coal from Huainan needed to be transported first by a coal company's own railways and then by the national rail system. There are dozens of sub-national rail networks all over China, and each uses its own rail equipment, facilities, employees, schedules, and timetables. Although Anhui to Shanghai is only five hours by train, the rail between them belongs to two different networks. Bypassing the rail network in the coal supply chain – as the Huainan-Huaibei base has by means of local power plants and "coal by wire" – can therefore generate substantial cost savings.

Second, equity sharing joint ventures benefit both sides through risk sharing. Tianji is a 50–50 joint venture, with both power and coal companies sharing the profits as well as the risks. Power companies gain security of supply and cheaper coal procurement by eliminating transport costs and exposure to value added tax (VAT). By purchasing coal from a joint venture (JV) partner instead of on the open market, the power plant avoids VAT, which is 17%. In part as a result of this, the overall cost of each kWh of electricity generated in the Tianji power plant is 1 cent (RMB) lower than that in Shanghai. For their part, coal companies gain better access to capital as well as long-term customers in a traditionally volatile market where many mines may have to open and close in response to market conditions.

Third, this form of cooperation is believed by Chinese policy makers to facilitate the improvement of management systems in the coal company through the close cooperation with power companies. (The degree to which this benefit is actually achieved – and the mechanisms through which such transfer of expertise occurs – is a subject worthy of further study.)

Beyond these general advantages to cooperation, it should be noted that several specific factors also facilitated the base's development.

First, all of the Big Five power companies were expanding rapidly at the time and were racing to claim market share (*pao ma quan di*). Achieving scale and scope was seen as vital by these companies for increasing their competitiveness and influence in China's power market. The Big Five as well as other power-generating firms were making aggressive efforts to build power plants. Locations near coal thus became very attractive.

Second, it is a relatively short distance between Shanghai Power and Huainan Mining Group, which made logistical coordination between the two companies relatively easy. The availability of high-quality highways and trains linking the base to Shanghai made it easy for employees of the Tianji power plant to get back to the parent company in Shanghai.

Third, access to high-quality, high-voltage transmission lines made transporting electricity much more efficient and cost effective. According to Lu Yanshun, the former Deputy Minister of the then Ministry of the Machinery Industry, building coal-power bases focused on coal by wire would be impossible or impractical without high-quality transmission lines, as costs due to line losses over long distances would be prohibitive.

Fourth, the leadership vision and strategies of both cooperating companies were consistent. Many companies had expressed interest in similar joint ventures, but Wang Yuan, the head of Huainan Group, searched carefully and persistently for suitable collaborators. Coal prices change all the time in China, and some coal companies consider cooperating with power companies only when coal prices are low and change their minds when coal prices are high. Successful cooperation requires a long-term vision that involves sharing both profit and risk.

Fifth, the relationship between the two provincial governments was supportive and seen to provide mutual benefit. The Anhui provincial government supported this project as part of its larger initiative called "Transmit Anhui's Electricity to East China" (*wan dian dong song*), which enabled the base to gain priority access to the power grid. The Shanghai city government for its part considered Anhui to be a "rare garden" that could recruit industries that are necessary to the Shanghai economy but for multiple reasons are not optimal to site in Shanghai. Shanghai's focus on high technology and advanced sectors requires a series of supporting industries, including raw materials, power generation, construction, and other inputs. At the same time, Shanghai in recent years has grown extremely short of land and has suffered from increasing pollution. Shanghai seeks to maintain growth of its advanced industries without increasing pollution and is therefore enthusiastic about allowing its companies to locate polluting businesses outside Shanghai. Spurred by the demand for cheap labor and raw materials and the desire to avoid high taxes and other fees associated charged to polluters, many Shanghai firms have moved to Anhui in recent years. Meanwhile, Anhui is keen on gaining advanced technologies through cooperation with Shanghai.

3.6 The impact of the coal-power bases on the future of China's coal market

This section assesses how coal-power policy and integration of the Chinese coal and power sectors will impact China's coal market, focusing on the obstacles to base establishment that could determine the pace of these reforms.

As we have discussed, the broad direction of reform and industrial restructuring policy in the Chinese coal and power sectors is toward consolidation of coal production, integration of the coal and power sectors, and greater concentration of control over the coal sector by major coal and power SOEs (and thus indirectly the central government). Yet the early integration efforts described earlier are only the initial steps in the wholesale restructuring of coal and power that is envisioned. The coal-power base policy ultimately aims to produce a substantial fraction of China's coal by means of coal-power bases controlled by key SOEs. Whether coal-power base creation can be accomplished at this scale – and what impacts such reforms might have on China's coal markets – is still an open question. Building on our review of coal-power integration frameworks and progress to date, we make the following main arguments about how coal-power bases are likely to develop and what impacts this development may have on the coal market.

First, due to the many obstacles faced by coal-power base reforms beyond Shenhua, these reforms are more likely to have a significant impact on the structure of the Chinese coal industry in the medium to long term (after 2020) rather than in the next few years. While some integration is already underway, a wholesale restructuring of the industry along the lines of the Shenhua model – as envisioned by the 11th Five Year Plan of China's Coal Industry – faces a set of major barriers that must be overcome. We discuss these obstacles in Section 3.6.1 and identify the enabling conditions that are necessary for the coal-power base policies to succeed. The larger macroeconomic climate is one key variable that will affect the pace of integration. When either the coal or the power sector is experiencing significant financial risks, they will have incentives to push for integration.

Second, development of the coal resources in Xinjiang may be a crucial proving ground for the coal-power base initiative. As we will discuss in Section 3.6.2, successful base development in Xinjiang could turn the province into a major contributor to China's future coal supply.

3.6.1 Near-term obstacles to coal-power base development

In the near term, some major obstacles are likely to prevent coal-power base policy from achieving a wholesale restructuring of the coal and power sectors (which we define as a shift in the industrial structure of coal production such that more than one-third of China's coal is produced at integrated bases modeled after Shenhua). Efforts to build large coal-power bases and concentrate a large share of China's coal production at government-controlled bases may prove successful over time.

However, rather than exactly following the Shenhua template, the evolution of the bases is likely to be less uniform and will have to adapt to the particular conditions of local coal geology, politics, and industrial organization in order to succeed. Other coal-power bases will face challenges as they seek to follow in the footsteps of Shenhua.

Most such challenges stem back in some way to the difficulty of aligning interests between the different players – which can be local, provincial, or national in orientation – that must come together to form coal-power bases. For example, local governments may benefit from blocking cross-province mergers and acquisitions that could diffuse benefits and control away from them. The Huainan and Huaibei example discussed earlier showed how the interests of different provinces and the national interest could all align in favor of coal-power base formation, but it is just as possible that a coal-power base which has net benefits for China could make one locality worse off unless it is somehow compensated. A particularly thorny problem arises from the fact that coal-base formation often implies closure of small mines that provide local economic and employment benefits, or at the very least their integration into larger corporations.

While government policy has supported this kind of mine consolidation independently of coal-power base policy, most dramatically and successfully in Shanxi, it cannot be assumed that the process will occur quickly and seamlessly elsewhere. History demonstrates how difficult mining consolidation policy can be to implement. From 1998 onward the government consistently appealed for the closure of such small coal mines. Although the number of small mines in operation reportedly did decrease year by year, the share of national coal output produced by small mines remained at around 40% as of 2009, compared with its previous peak share of about 50%. This reflects the political reality that closure of small mines cannot happen by central government fiat. At first glance, the difficulty in closing the mines appears mainly due to local government resistance, as small mines are an important contributor to local government revenue in many areas. But more fundamentally, small mines often provide a crucial source of income for the local population in underdeveloped areas, which neither the central nor the local government is eager to cut off (Rui 2005). In addition, while acquisition and consolidation of small mines is feasible for large corporations, it may not be realistic for small companies that lack the necessary capital or political support.

Fundamental differences in the nature of the players in the coal-to-power value chain – and the levels of government to which they are most connected – may also retard coal base formation. As discussed earlier in this chapter, the distinct but interrelated development trajectories of the Chinese coal and power sectors have resulted in a highly concentrated power sector, with close ties to the central government, and a relatively fragmented coal sector, with closer connections to local governments. The big SOEs in power tend to have a technocratic approach to management – perhaps one reason the government, with its own technocratic leanings, considers their management to be superior – while the coal industry is less technologically advanced on average but also more used to competing within a market framework. (That most power companies are already closer to the government's vision of what constitutes an ideal corporation – notwithstanding the fact that the original shortcomings of large SOEs in expanding coal production drove coal sector liberalization in the first place – also probably explains in part why the power-to-coal mode has predominated over the coal-to-power mode.) The cultural gulf between power companies and coal companies can complicate the task of melding them into successful joint ventures or equity sharing arrangements.

Making the task of coal-power base formation even more difficult is the usual need for a rail linkage (except where coal-by-wire is feasible and economically attractive), as the state railway has every incentive to contest further erosion of its monopoly. The ability of the company or companies forming a base to control rail capacity and capacity expansion will be critical to the development of coal-power bases. Although some companies have already started to build up railways or take shares of railway construction, captive development of railways is still limited and has not kept pace with coal production and demand. The MOR was a monopoly operating China's railways for more than half a century and in that time managed to build up and defend entrenched interests around the sector, as exemplified by the inside plan and outside plan transportation system and various other fees collected over the years. SOEs favored by the central government – which are more often power companies than coal companies – may be most capable of carving out the political space to challenge the state railway and develop more captive rail through the coal-power base process. Although coal-power base construction could accelerate liberalization, the process will be gradual. Ultimately, this is one of the most unpredictable parts of creating China's coal-power bases, because reform of China's railways industry may continue to be slow notwithstanding some of the recent changes.

On top of all of these political economy obstacles, the simple fact remains that creating an organization as large, complex, and expensive as a coal-power base requires plentiful capital and project management skill even under the best of conditions.

Unique aspects of Shenhua's situation helped it sidestep most of these challenges. The government's gift of massive coal assets to Shenhua meant that the company for the most part did not have to acquire or merge with smaller mining companies in order to develop a base. The quality of these reserves meant that Shenhua could produce coal cheaply and be competitive from the outset. Large government loans helped it develop both its mines and, just as importantly, its integrated rail and port projects.

Shenhua's formation from scratch in a new coal province allowed it to sidestep many of the challenges of getting legacy players to work together, and its special connectedness to the central government helped facilitate resolution of any problems that did arise. As discussed in Section 3.4, it acquired power generation assets directly from the state and was given special dispensation to create a captive rail network. Shenhua CEO Ye Qing's background as deputy director of NPC and connections in many government agencies made the company's integration process much easier than it might have been for traditional SOEs.

For all of these reasons, no other coal-power bases have yet achieved the success and scale of Shenhua; most joint ventures to date have been on a small scale. The Huainan and Huaibei case shows the positive outcomes that can be achieved when incentives naturally align between players but does not say much about the prospects for coal-power bases to be formed on a more "top-down" basis where incentives are more divergent. A lesson of the Huainan-Huaibei case is that coal bases beyond Shenhua are most likely to flourish when coal and power firms pursue modes of integration adapted to local circumstances instead of trying to slavishly replicate Shenhua. (Huainan-Huaibei also highlights the value of "coal by wire" as a way to circumvent the rail monopoly.)

This is not to say that central government support does not matter. Aligning players with such diverse interests and cultures is challenging in the best of conditions; without strong support from the central government as a facilitator it is virtually impossible. Coal-power base policy has certainly signaled to relevant players that the central government supports horizontal and vertical integration as a general principle – although the extent of its support and involvement has varied by coal base. In addition, the overall level of government investment in coal, power, and transportation serves as an important background factor affecting the rate of base development. The rapid growth of the coal industry during the 11th Five Year Plan (2006–2010) was enabled by the government's ambitious plan of fixed capital investment in the Chinese coal sector. For the first three years of the 11th Plan, the investment was 538 billion yuan – 297 billion yuan more than the total investment in the 10th Five Year Plan (Pu 2009).

The broader macroeconomic environment may also have an influence over the pace of coal-power base development. Rapid macroeconomic growth accompanied by high coal prices tends to exacerbate the coal-power conflict and create losses for power SOEs. This scenario, which occurred in 2008 and 2010, gives the central government (which owns the Big Five) incentives to push harder and faster for integration. Alternatively, weak economic growth could create drivers for integration from the coal end of the value chain. (This kind of scenario could be triggered by a slowdown of government investment in the economy.) The most significant recent example has been the massive consolidation of the coal industry in Shanxi province since 2009, with small coal mines merging with larger companies. The total number of coal mines in the province was reduced from 2598 in 2009 to 1000 in 2011 and the number of coal-mining companies from 2200 in 2009 to 100 in 2010 (*Time Weekly* 2009). Unlike numerous previous examples of small mine closure in China, Shanxi's restructuring demonstrated that the central and provincial governments could move fairly quickly to close small mines and consolidate the sector. The central government aims for further consolidation across other provinces in the coming years; the recently renewed policy is intended to work in tandem with the coal-power base policy to increase industrial concentration in coal.

3.6.2 Developing Xinjiang under the coal-power base model

The biggest wild card with the potential to dramatically change China's coal industry and market in the near future is the development of the largely untapped coal resources in the Xinjiang Uygur Autonomous Region, likely utilizing a coal-power base model. In 2010, the NDRC indicated that it was considering developing Xinjiang as the fourteenth coal-power base.

Xinjiang has great potential to become a large coal-power base due to its extremely attractive coal resources. Its coal resources are estimated at 2,350 Bt, accounting for 40% of China's total, with 1,300 Bt of this considered "reliable" (likely to be convertible into proven reserves upon further exploration). By 2006 Xinjiang's proven reserves stood at 161 Bt, behind only Inner Mongolia and Shanxi. Xinjiang's coal is also of excellent quality. Most of the coal discovered in Xinjiang is low in ash (some is even classified as "super-low" in ash, with an ash content of 3% to 5%), super-low in sulphur and phosphorus, and would serve well as a medium to high calorific value steam coal. It is also suitable for liquefaction, gasification, and making chemical products.

Xinjiang's coal is also amenable to low-cost mining. Deposits are relatively shallow and suitable for extraction using very large scale, high productivity, high efficiency, open cast mines. The coal resources are concentrated, mainly in the three large coalfields of Zhungeer (estimated resources of 710 Bt), Turpan-Hami (660 Bt), and Yili (470 Bt), which together make up 79% of the total coal reserves in Xinjiang (internal document 2008). Reserves continue to be fortified with large new discoveries, especially since 2004 as soaring energy prices in China have driven intensified exploration in Xinjiang.[19]

With eastern China's coal reserves appearing to be in sharp decline, Xinjiang is poised to become the next important base for coal development in China. In keeping with coal-power base policy and the government aim of consolidating coal mines, the government of Xinjiang Autonomous Region plans to develop its coal resources through large corporations.[20] It has attracted dozens of promising enterprises from all over China to enter Xinjiang.[21] The coal mines under construction in Xinjiang are all very large, high-output, efficient mines.

According to the development plan of the Xinjiang government, coal output capacity was targeted for 130 Mt, 1.04 Bt, and 1.4 Bt for 2010, 2020, and 2030, respectively. By 2020 coal liquefaction capacity is slated to reach 31 Mt, coal-to-methanol capacity 23 Mt, coal-to-gas capacity 24 billion cubic meters, coal-to-ammonia capacity 3.9 Mt, and coal-to-coke capacity 10.50 Mt (internal document 2008). This is an ambitious plan, although the coal resources to support this ambition certainly do exist as previously described.

Two principal constraints will likely set the pace of Xinjiang's coal development: water and railway capacity. Among the three major coalfields only Yili has access to adequate water resources. And the cost of long-distance transportation of Xinjiang's coal remains high. Currently, the Lanzhou-Xinjiang railway is the only rail line connecting Xinjiang to the inland cities. The recently discovered Sha'Er coalfield is close to the Lanzhou-Xinjiang railway. Other coalfields such as the massive Zhundong deposit (the east portion of the larger Zhungeer coalfield) are also at least 800 km farther from China's inland provinces than Sha'Er, probably adding around 40 yuan per ton to the transportation cost.

The Xinjiang government has been investing itself and finding outside investors to fund water and railway projects. Newly built infrastructure transporting water from water-rich areas has been completed or is close to completion. It is expected that available water for the two coalfields could reach 2.5 billion cubic meters/year. In recent years, Xinjiang has also sped up railway construction inside its territory. Newly built railways include the Jinhe – Yinin – Huerguosi, Wuxi – Jinhe II, and

Table 3.5 *Projected coal output of Xinjiang (tonne)*

Coalfield	2010	2020	2030
Zhungeer	45	290	590
Turpan-Hami	15	150	300
Yili	10	40	85
Kubai – Nanjiang	10	20	25
TOTAL	80	500	1,000

Source: Internal document 2008.

Kuitun – Beitun lines among others. Other railway lines under construction include Urumchi – Zhungeer, Kashgar – Hetian, and Korle – Aksu.

However, these are all railways inside Xinjiang; transporting coal – or power – to the rest of China is a much bigger challenge. Estimates by Chinese experts have tended to find that new railways from Xinjiang are less expensive than very long distance power transmission lines. A large-capacity, high-efficiency railway was planned from Kumul in Xinjiang to Xi'an in Shaanxi, with an expected investment requirement of 63.8 billion yuan (internal document 2008).[22] According to the International Energy Agency, local governments estimated that rail capacity would be sufficient to export 500 Mt of coal eastward in 2015 and 1,000 Mt in 2020. As a result, Chinese government experts predicted Xinjiang coal output of 80 Mt, 500 Mt, and 1 Bt by 2010, 2020, and 2030, respectively (internal document 2008) – significantly lower than Xinjiang's estimates of coal output capacity as previously indicated. Table 3.5 breaks down these projections of coal output under rail constraints.

Considering the deteriorating output capacity in coal mines in eastern and central China, some Chinese firms and policy makers have appealed to the government to speed up the development of Xinjiang's coal resources and designate Zhundong, Tulufan-Hami, and Yili coalfields separately as China's fourteenth, fifteenth, and sixteenth large coal-power bases (internal document 2008). However, it is understood in Xinjiang that the exploration of the coalfields in the region and the rate of further development depend on the country's pace of economic development and its overall energy demand. While there is a wide consensus that Xinjiang will become a strategic energy base for China in the long term, the high costs associated with transporting its coal to the rest of the country create significant uncertainty around how quickly the region will develop into the massive energy producer that it is capable of being.

3.7 Conclusion

Driven by the coal-power conflict, coal-power integration is now a focal point of the Chinese government's energy policy. If successful, coal power-base policy will establish upward of thirteen major coal-power bases, each producing more than 100 Mt of coal with consuming industries on-site. The richness of Xinjiang's coal resources has sparked plans to develop this critical new region through a coal-power base approach as well. China's government hopes that up to half of production could eventually be produced by such bases. In the government's ideal scenario, the bases would help achieve the following three main objectives.

First, successful development of bases could help to alleviate the transportation bottlenecks that have long plagued the Chinese coal sector. Creating captive rail networks, siting coal-consuming applications near mines, and delivering "coal by wire" in the form of coal-generated electricity over long transmission lines could all help sidestep the high costs and constraints on capacity associated with the state rail network. With lower transportation costs, coal prices could be significantly lower. Among other things, this would serve the government's goal of keeping power prices low.

Second, coal-power bases might give the central government enhanced control over the coal sector by reorganizing it around large SOEs that are more closely tied to the central government. A large share of coal production has historically been regulated and controlled at the provincial level (both TVE production and local SOE production), and central government efforts to reign in small mines have had mixed success. The large power companies that consume most of China's coal, on the other hand, are controlled by the central government, and these companies suffered heavy losses when coal prices rose in 2008 and 2010.[23] The government found that it could not effectively control the coal sector by means of price caps. It hopes that it can do so by directly controlling a larger share of coal production through its control of SOEs running coal-power bases. Power sector losses could be mitigated by mandating lower coal transaction prices within integrated SOEs. Using this kind of internal transfer pricing at below market prices would represent a significant shift in how coal is priced in China. If a large share of China's coal were transacted in this manner, it might create (again) an unofficial two-tiered pricing structure in the coal market.[24] While SOE power generators would obtain large volumes of coal at internal transfer prices, likely at below market rates, the rest of the coal market might be permitted unfettered market pricing.

Third, coal-power base policy is expected by the government to help modernize and mechanize a larger share of China's coal production and to bring economies of scale to the sector. While up-front capital investment per tonne produced would increase compared to a TVE-dominated production structure, the idea is that marginal cost of coal production should decrease, all else equal. (It should be noted, however, that as China's coal production moves further from load centers in the east, as in the case of Xinjiang, cost reductions in production could be counterbalanced by increased coal transport costs.) The government also envisions improved efficiency simply through the better management capability that it attributes to large SOEs, particularly in the power sector, in comparison with smaller coal companies.

While there is legitimate hope that all three of these goals may be accomplished to some degree by the coal-power base strategy, the history of China's coal sector urges some caution as well. We suggest close observation of the following dynamics going forward. First, to what degree is the Ministry of Railways able to contest further erosion of its position? Captive rail may have been relatively feasible for a new and well-connected state player such as Shenhua to develop, but other coal-power base managers may find it more politically difficult to build out and control comparable private networks. On the other hand, perhaps the success of the "Shenhua model" and other nascent coal bases will make it increasingly difficult for the Ministry of Railways to stand in the way of further movement in this direction.

Second, will the "power-to-coal" and "equity sharing" modes of coal-power base creation prove successful on a larger scale? The "coal-to-power" case of Shenhua, where a new coal SOE was created and granted massive resource endowments, is probably unlikely to be repeated. Because the central government is closer to the power SOEs – both in terms of direct connectedness and in cultural match – than to most coal companies, it seems likely that a majority of future coal-power bases will be led by power companies. However, many of these companies have struggled thus far to successfully absorb existing entities in coal. They will need to find a way to do so – probably with strong central government support – if coal-power bases are to have a wider impact.

Third, will the SOEs running coal-power bases be as efficient as the central government hopes? The government seems to equate efficiency with use of modern equipment and technocratic management. These can indeed be of great value, as demonstrated by some of the lower production costs and more efficient use of manpower of modern integrated firms such as Shenhua. On the other hand, the fact that SOEs do not face the same threat of going out of business as non-state-owned entities can make them less responsive to market signals. As discussed previously in this chapter as well as in Chapter 2, coal sector reforms originally moved away from exclusive control by large state mines precisely because they could not produce enough coal to feed a rapidly growing Chinese economy.

To the extent that China's coal-power bases do prove widely successful, they have the potential to significantly impact both China's coal market and the broader global trade. Since 2009, China's import behavior has become a dominant factor determining the price of globally traded coal (see Chapter 9, this volume). In simple terms, when Chinese domestic prices are higher than global prices, the country imports. When domestic prices are lower, it tends to export. In 2009, China imported nearly 15% of all globally traded coal, and the Chinese arbitrage between domestic and international coal supply has had the effect of linking the global price of coal to the Chinese domestic price. The development of coal-power bases could radically alter coal price formation in China, and therefore has the potential to alter coal price formation globally. If in the long term coal-power base reforms put downward pressure on Chinese domestic prices and allow large SOEs to acquire coal at below-market prices, Chinese coal imports and by proxy global coal prices might be reduced.

Whether or not coal-power bases will overcome key obstacles to their development and bring about a wholesale restructuring of the Chinese coal and power sectors in the form of profound horizontal and vertical integration remains to be seen. But given the pressure to solve the coal-power conflict through reform of the relationship between the coal and power industries in China, further change in the industrial organization of the Chinese coal and power sectors appears imminent.

Notes

1. The rest of the players in the power industry are foreign and private power generation firms, which hold an insignificant share of the market.
2. It is worth noting that the coal-power conflict has deep roots in the era of the planned economy. Although closely connected to each other, the two industries have significantly different levels of income and profit. The salaries of the employees within the two industries were among the highest (power) and the lowest (coal) compared with the national level (Rui 2005). Only during the past decade, when coal was in high demand and coal prices were high in both domestic and global markets, were coal companies relatively better off. Even now, the income level of the coal industry, in terms of average salary for employees and profit made by companies, is still far behind that of the power industry. Because of these social inequalities, coal companies were more determined to increase the price of electricity coal.
3. As witnessed in 2009, small coal mines were forced to close or merge with large coal mines in Shanxi and other provinces.
4. It is believed that the Annual Coal Conference will "die" once the transportation constraint is removed. Naturally, the rail constraints could be removed if coal

and power companies cooperate on transporting electricity rather than coal – or if all of the three entities are shareholders and work on a cooperative project and share the benefit of the outcome.

5. There is, of course, no official data available on how much coal is moved inside versus outside plan, but interviewees believed the quantity of outside plan coal transported by the railways was a significant figure.
6. Of course, if possible, small coal mines hire trucks to transport coal. The problem is that the various fees and charges by road authorities are also quite high.
7. China's coal output was already at 2.7 Bt in 2008. The 2.6 Bt projection for 2010 might have indicated that the government expected a slowdown in coal production.
8. Chinese companies have not taken into account opportunity cost of not selling the coal, as they are still not so advanced in risk management.
9. This long-term point should be emphasized because we have argued that despite the failure of the conference in 2008, power companies were all still able to get enough coal and keep the electricity running, but both coal and power firms do not consider that is ideal, given the unstable short-term feature of that demand and supply fit.
10. It is reported that 200 Mt of annual rail traffic is equivalent to twenty to thirty-four 1,000 kV UHV AC transmission lines (i.e., they provide an equivalent amount of energy to end-user customers). However, the former's required investment is less than 1,000 billion yuan, whereas the latter's investment is between 2,400 and 4,080 billion yuan – a huge gap (Chen 2008).
11. A coal-to-olefins project in Baotou has been approved by relevant governmental authorities. Phase one of Wuhai coal tar processing project, with a capacity of 300,000 tons per year, has commenced trial operation. Meanwhile, Shenhua Group is actively working with SASOL on the feasibility study of indirect coal-to-liquid (ICL) projects in Ningxia and Shaanxi. It is also working with Dow Chemical on a pre-feasibility study of the Yulin coal chemicals project. More preparatory work on coal chemicals projects has also been carried out in Xinjiang, Hulunbeier, and Tongliao.
12. These are Baotou to Shenmu, with a capacity of 22 Mtpa; Shenmu to Shuozhou, and a linked line from Shouzhou to Shenhua's Huanghua port, with a 110 Mtpa capacity; Datong to Zhungeer, with a capacity of 50 Mtpa; and Huanghua port to Wanjia terminal at 40 Mtpa.
13. These include a new railway from Datong to Zhungeer; an upgrade to the existing line from Shenmu to Huanghua via Shuozhou; a new railway to Zhungeer with Shenchi, just east of Shuozhou on the line to Huanghua; and a new line from Huanghua to Wanjia terminal at the port of Tianjin.
14. Power companies have expressed particular concern about China's coal mines having such poor safety records and that managing the mines will be risky.
15. Traditional state-owned coal companies were called "coal bureaus" under the planned economic system in China. After reforms, many of them have been changed into limited companies and so on, but some "bureaus" remain in existence.

16. It is not yet known if they will be part of the thirteen bases.
17. The equity sharing mode comes in two main flavors: (1) two companies take direct equity stakes in each other, and (2) two companies take equity stakes in a third, separate asset or joint venture.
18. The design of Tianji power plant won China's Lu Ban Prize for High Quality Engineering.
19. Dananhu Coalfield was discovered in 2004 with 7.2 Bt of proven reserves. Fuyuan Coalfield was discovered in 2006 with a much larger proven reserve base of 46.6 Bt. Furthermore, Zhundong Coalfield was discovered in 2006, covering an area of 21,800 sq. km, with proven reserves of 68.5 Bt. With the estimated 390 Bt of total estimated reserves, it will become the largest coalfield in China. In 2008, Xinjiang revealed another surprise when the local Department of Land and Resources announced that a coalfield with 23 Bt in reserves was discovered. This was the Sha'Er Lake coalfield, about 800 m underground, covering an area of more than 300 sq. km in Shanshan county in the Turpan Basin (*China Daily* 2008).
20. Xinjiang has followed the central government policy to increase the region's coal production concentration. More than 1,500 small mines were closed before 2008. Currently about 300 mines remain, with an average coal output of 200,000 tons. These small mines account for 25% of total coal output in Xinjiang, while the remaining 75% is produced by the Autonomous Region–owned key coal mines and local state-owned mines.
21. Xuzhou Mining Group was the first large coal company to enter Xinjiang. The development mode was to restructure existing small mines by fully accepting small mines' debt and employees and by injecting new capital, technology, and management systems. Other major coal consumers have also entered Xinjiang to develop coal mines or transform coal to other products such as electricity: Shanghai Baosteel, Shandong Luneng Group, China Power Investment, and Hubei Yihua are all investing in the region. This has resulted in great improvement in technology and equipment in local coal mines. For example, Shenhua established Xinjiang Energy Corporation through the acquisition of the Urumqi Coal Bureau.
22. The estimated cost of railways from Hami to Xian will be 180 yuan/ton. In 2008, coal costs in Xinjiang were 90 to 100 yuan/ton and the selling price was 100 to 105 yuan/ton. This was far below the domestic market price due to Xinjiang's isolation from the majority of buyers in the national market. Compared with a 370 yuan/ton coal price in Xian, Xinjiang's coal would still be competitive (internal document 2008).
23. See the China Electricity Council for estimates of power sector losses during high coal price environments: http://www.cec.org.cn/cec-en/index.htm.
24. Two-tiered pricing already exists to some extent, as NDRC "guides" annual coal contract prices for some key power generators.

References

CCIA (various years). *China Coal Operation Report*. Beijing: China Coal Industry Association. Available at http://www.chinacoal.gov.cn.

CCIA (2010, March 4). *The Development Status of the Coal Industry in 2009*. Beijing: China Coal Industry Association.

Chen, Wanxiang (2008, April 30). "The comprehensive comparison of investment cost in transporting coal via railways and transporting electricity via transmission lines." *China Energy Network*. Available at http://zsyz.sei.gov.cn/ShowArticle.asp?ArticleID=146540. Accessed on May 2, 2008.

ChinaCoal (2009, April). "Shenhua builds up new railways and improves its own rail network." 63(10): 5.

China Daily (2008, December 2). "Coal's the goal." Available at http://www.chinamining.org/News/2008-12-02/1228198094d19758.html.

China Daily (2009, April 21). "Shenhua eyes Mongolian mine." Available at http://www.chinadaily.com.cn/business/2009-04/21/content_7698017.htm.

China Mining Federation (2010, January 12). "China's power plants sign 2010 coal supply agreements." Available at http://www.chinamining.org/News/2010-01-12/1263276794d33298.html.

Fan, Bi (2009). "Thoughts on comprehensive solution of coal and power disputes" (*guanyu xitong jiejue meidian maodun de silu*). *Macroeconomic Management*. Available at http://blog.sina.com.cn/s/blog_5937caf80100rdaf.html?tj=1 2011.

IEA (2013a). *Electricity Information 2013*. Paris: International Energy Agency.

IEA (2013b). *World Energy Outlook 2013*. Paris: International Energy Agency.

Internal document (2008). *Prospect of Xinjiang Coal Development*. Beijing: State Development Bank.

Lin, Wen (2009). "Development Road for China's Coal Industry Under the Global Financial Crisis." Coaltrans China, Beijing. April 20.

Ministry of Railways (2009, June 1). *2008 Statistical Bulletin of Railways*. Beijing: Statistics Center, Ministry of Railways.

Morse, Richard K. and Gang He (2010). "The World's Greatest Coal Arbitrage: Chinese Coal Import Behavior and Implications for the Global Coal Market." Program on Energy and Sustainable Development, Stanford University, Stanford, CA, USA.

NDRC (2007). *The 11th Five Year Plan of the Coal Industrial Development*, Beijing: National Development and Reform Commission.

NDRC (2008). "Industrial News Express." Available at http://www.ndrc.gov.cn/xxfw/hyyw/t20090414_272399.htm.

NDRC (2008, April 22). "Strengthen Exchanges and Co-operation; Promote Sustainable Development in the Global Energy." Keynote speech, Sun Qing of National Energy Authority at the Eleventh International Energy Forum, Rome.

Nolan, P. (2001). *China and the Global Business Revolution*. Houndsmill: Palgrave.

Nolan, P. and H. Rui (2004). "Industrial policy and global big business revolution: The case of the Chinese coal industry." *Journal of Chinese Economic and Business Studies* 2(2): 97–113.

Nolan, P., A. Shipman, and H. Rui (2004). "Coal liquefaction, Shenhua Group, and China's energy security." *European Management Journal* 22(2): 150–164.

Pu, Hongjiu (2009). "Experts advise the 12th Five Year Plan of China's coal industry." *Economic Consulting Report* (*jinji cankao bao*), August 26, p. 2.

Ren, Runhou [Chairman of Lu An Group] (2009). "Building the Larger and Stronger New International Lu An Group Fast and Well." Coaltrans China, Beijing, April 20.

Rui, H. (2004). "The rise and fall of China's TVE coalmines." *Mineral & Energy* 19(1): 2–15.

Rui, H. (2005). *Globalization, Transition and Development in China*. London and New York: RoutledgeCurzon.

Sagawa, A. and K. Koizumi (2007, December). "Present State and Outlook of China's Coal Industry." The Institute of Energy Economics, Japan. Available at http://eneken.ieej.or.jp/data/en/data/pdf/410.pdf.

Shenhua (1998). *Shenhua Group Annual Report*. Beijing: Shenhua Group.

Southern Weekend (2009, November 5). "Coal bosses question Shanxi coal restructuring." Available at http://www.infzm.com/content/36878.

State Power Information, China (2010, January 26). "Electricity price increase is certain. Will coal-power prices connection system success?" Available at http://www.powerplants.com.cn/fddt/dlzc/201001/t20100126_146146.htm.

Thomson, E. (2003). *The Chinese Coal Industry: An Economic History*. London and New York: RoutledgeCurzon.

Time Weekly (2009). "Shanxi Coal: restructuring is now under way." Available at http://money.163.com/09/0910/05/5IQUTD2000253B0H.html. Accessed on February 28, 2015.

Wang, An (2009). "China Coal Group's Reform and Development Status and Strategies." Coaltrans China, Beijing, April 20.

Wu, Yongpin (2009). "Strategic Vision of Coalmine Group against Financial Crisis." Coaltrans China, Beijing, April 20.

Xie, Juchen [CEO of the China Energy Fuel Ltd.] (2009). Interviewed, April 21, Beijing.

Xinhua News (2005, July 22).

Xinhua News (2009, January 12). "The enemy of coal and power work towards co-operation." *Xinhua News*. Available at http://news.xinhuanet.com/energy/2009-01/12/content_10644566_1.htm. Accessed on February 28, 2015.

Ye, Qing and Baoming Zhang, eds. (1998). *China's TVE Coalmines*. Beijing: Coal Industrial Publishing House.

Zhang, Guobao (2008, December 29). "The current situation: opportunities among the danger." *People's Daily*. Available at http://scitech.people.com.cn/GB/8591346.html. Accessed on February 28, 2015.

4 The causes and implications of India's coal production shortfall

Jeremy Carl

4.1 Introduction

India's coal market presents a conundrum. India has one of the world's richest endowments of coal and depends on coal-fired power plants for over 70% of its electricity. Steady economic growth has driven up the demand for electricity and thus for coal, even as the current electricity supply remains well short of actual needs (in part due to underpricing, theft, and infrastructure challenges). And yet India's coal production has failed to keep pace with the need, contributing to serious power shortages and putting the economy at risk.[1] Scheduled coal mining projects have failed to be approved in timely fashion, and a significant portion of the new production that has been brought on line has come from expansions of existing projects rather than greenfield operations.[2] Why is this so? Oft-used explanations such as "corruption" or "lack of political will" are inadequate, offering a mere description of the symptoms rather than an analysis of the underlying disease.

The problem has not been a lack of awareness of the problem at the highest levels of government. Former Prime Minister Manmohan Singh viewed the coal ministry as important enough that his office took over the coal ministry as part of the prime ministerial after the coal minister, Shibu Soren, was indicted on a murder change during Singh's first term. Current Prime Minister Narendra Modi also appears very focused on improving the performance of the coal sector (Das, 2014).

Likewise, contrary to popular perception, the difficulty is not a simple matter of the inefficiency of Coal India Limited (CIL), the state-owned enterprise (SOE) that is the world's largest coal company and controls the vast majority of coal production in India. CIL does have its problems, which include staff bloat (400,000 employees), difficult bureaucratic politics, and administrative encumbrances that go with being state-owned. It is inherently difficult to incentivize an SOE to respond to supply shortfalls as efficiently as a private company that can capture more of the profits from doing so. In Section 4.3, we will compare a private captive miner to CIL and discuss how industrial organization factors – especially related to corporate structure and flexibility – contribute to the coal production deficit. But CIL's production costs remain fairly competitive at the mine mouth, and simple privatization of the mining behemoth will not make the coal production deficit disappear.[3] There are a number of other factors across the coal-to-power value chain that contribute to India's poor performance in coal, including excessive regulation of coal distribution and price, the reliance of CIL on a government-owned rail monopoly for transport, and the poor interface between the coal and power sectors.

Arguably the most serious constraint on India's coal output is its unsettled land rights regime, which prevents the development of many greenfield coal projects. The extent of the uncertainty around land rights was dramatically illustrated by the Supreme Court's ruling in August 2014 that all coal mining rights issued since 1993 were in fact illegal (Chaturvedi and Zhong, 2014). Land acquisition issues are further complicated by the need to obtain environmental clearances in forested areas as well as the fact that several coal-rich regions still grapple with a Maoist insurgency known as the Naxalite rebellion. Section 4.4 discusses in detail these various problems around acquiring land for coal mining.

How India's coal market responds to these mounting supply and demand pressures has significant implications not only for Indian economic development but also for the world coal trade. The Planning Commission of the Indian government has projected that India will import 200 Mt of coal in 2017. But Indian planners have consistently been over-optimistic about the possibilities for domestic coal production (PTI [Press Trust of India], 2011a), and the actual figure could turn out to be higher. If the coal supply deficit continues to grow, India could soon become the world's largest coal importer, a shift that would have major ramifications for the global coal market. It is not beyond the realm of possibility that India could swallow up much of the export capacity from South Africa (typically the swing supplier for the Atlantic and Pacific basins) and Indonesia, and also that it could become a dominant market for Australian coal and potentially even a customer for American or Colombian supplies. Understanding the complex dynamics of India's coal production is not just relevant to the economic development of India – it is crucial to understanding the future of the global coal trade.

The future direction of the Indian market is perhaps the key uncertainty affecting the development of global coal markets over the next decade. If India is able to overcome the problems described in this chapter and develop its domestic coal supply more effectively, it will import less and greatly ease concerns about global coal supply. If, on the other hand, it needs to import hundreds of million tonnes annually, this could shake the global coal market to its foundations.

4.2 An overview of India's coal sector

4.2.1 Origins of coal mining in India

The modern Indian coal sector was born with the introduction of the steam locomotive to India in 1853. Coal was used to power the railways, and it was produced by both the Indian government and British private companies. By the time of the Partition of India in 1947, production had reached 30 million tonnes per annum (Mtpa) (Ministry of Coal, 2010). The industry continued gradual development as a private sector entity under Prime Minister Nehru and his immediate successors before Prime Minister Indira Gandhi dramatically altered the industry by nationalizing all privately held coal assets during the early 1970s and forming state-owned enterprise Coal India Limited (CIL) to manage almost all of India's coal mining. CIL stayed largely beyond the reach of the dramatic liberalization that swept the Indian economy in the early 1990s and is still reshaping it today.

Coal India Limited is the world's largest coal company by both production and market value metrics. While other state-owned coal miners and a limited number of private sector miners also produce coal in India in significant quantities, CIL mines 80% of India's domestically produced coal. Table 4.1 shows the growth in coal production at CIL and its precursor companies since the 1950s.

Table 4.1 *Production at Coal India Limited (and its precursor companies)*

Terminal year (financial year)	Five-Year Plan	Production in India (Mt)
1955	First	38.13
1960	Second	55.72
1965	Third	67.74
1973	Fourth	78.17
1978	Fifth	101.97
1984	Sixth	147.43
1989	Seventh	200.91
1996	Eighth	285.66
2001	Ninth	327.79
2006	Tenth	379.00
2011	Eleventh	440.00

Source: Coal India Limited (2010); PTI (2011b).

It was not until the mid-2000s that reform really began to take hold in the coal sector, beginning with the introduction of limited mining outside of CIL's reach performed by so-called "captive" miners. These mainly private sector entities mine coal for their own use in applications such as steel production or power generation. The reform trend continued with the initial public offering in 2010 of 10% of the shares of CIL. These reforms were meant to open up the sector and significantly expand production, but captive mining has grown more slowly than planners hoped and the limited privatization of CIL has not yet led to the kind of improvement in efficiency that is needed.

Along with the inherent inefficiencies of a state-run monopoly, the original nationalization of the coal industry and the strict price controls that were established brought with them increased corruption, which remains deeply rooted to this day. Mine nationalization led to the closure of small and supposedly inefficient privately run mines. Whenever prices would spike, "coal mafias" would re-open the mines illegally, paying local peasants paltry wages to extract coal in extremely unsafe conditions for sale on the black market. This dynamic continues to be a problem.[4] (In China as well, small mines have had significant safety problems while also functioning as a sort of swing capacity that can come on line when major SOE mines cannot produce enough coal – see Chapters 2 and 3 and the Appendix.)

4.2.2 Coal's Importance in India

India has an estimated 7% to 12%[5] of global coal resources. Coal constitutes more than 95% of India's domestic fossil fuel reserves, even using conservative estimates. As can be seen in Figures 4.1 and 4.2, coal-fired power correspondingly dominates India's electricity mix.

The coal sector is a major employer in India – and a major source of political patronage. CIL alone employs almost 400,000 workers, making it India's second largest employer, with the largest being Indian Railways. Coal mining also generates huge numbers of jobs outside the coal sector itself, most notably with Indian Railways. Coal represents 40% of rail freight traffic in India, and so a substantial number of the 1.6 million employees at Indian Railways owe their jobs to the coal industry (*Indian Railways Yearbook* 2006–2007). Including the dependents of workers, millions of Indians are directly dependent on the coal sector for their livelihoods, while hundreds of millions more are dependent on the power generated from coal. Given the massive scale of coal-dependent employment in India – often in regions that lack formal employment opportunities of other types – coal's importance in India's political economy is

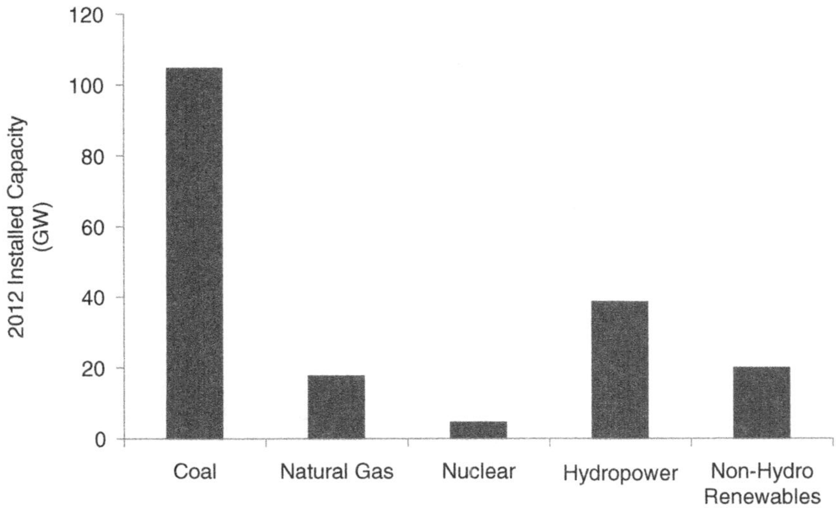

Figure 4.1 Installed power capacity in India by fuel source, 2012. Data source: Central Electricity Authority (2012).

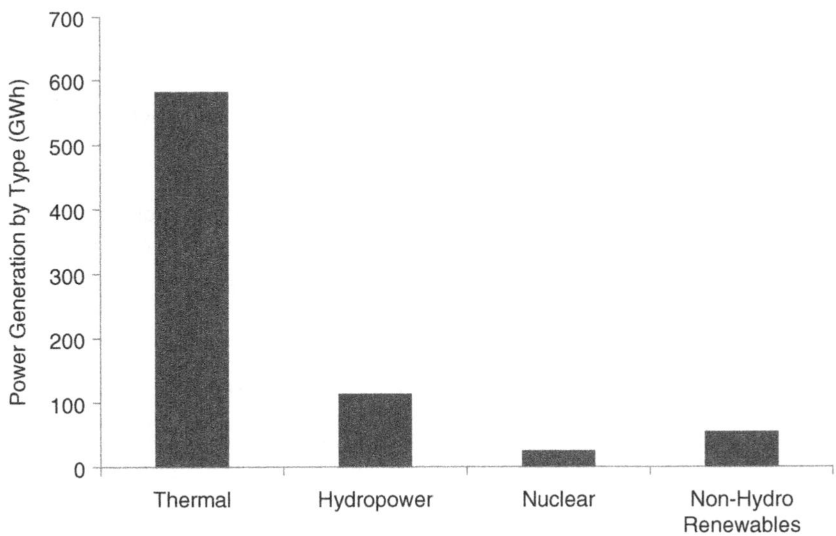

Figure 4.2 Power generation in India by type, January 2012. Data sources: Central Electricity Authority (2012); Agneya (2011).

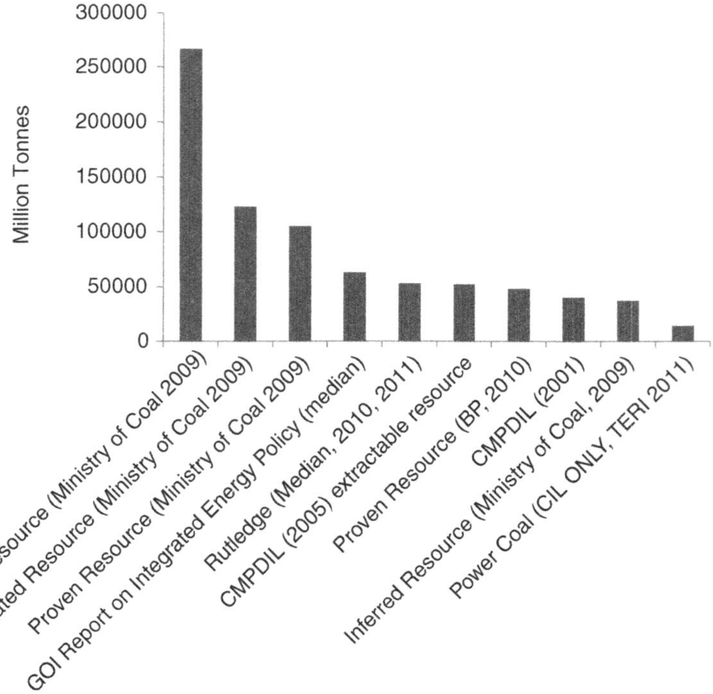

Figure 4.3 Estimates of India's coal resources and reserves by data source.

obvious. The vast and entrenched character of India's dependence on coal as a source of both employment and energy does much to explain the resistance of the sector to change and the political power of those who defend the status quo.

4.2.3 *Understanding India's coal reserves data*

Depending on the source consulted, one can get widely divergent views of the scale of India's coal resources. The importance of calculation methodology in determining reserves figures can be seen in Figure 4.3, which shows the size of the Indian coal resource as calculated by a variety of methods. The high figure of 267 billion tonnes (Bt) from the Indian government and the low figure of less than 15 Bt estimated by TERI, an Indian energy think tank,[6] bookend the range of estimates. Those who cite the former number as evidence of India's overall coal security in the power sector should understand the methodological justifications for arriving at the latter estimate. Estimates vary depending on whether one is talking about reserves, technically extractable resources, or economically extractable resources. Resources are often poorly mapped, and their extent is highly subject to geological assumptions. Furthermore, much of India's coal is of such low quality as to only be suitable for low-end industrial uses. As we shall further examine in this chapter, translating this nebulous concept of "reserves" (in principle, the amount of a resource that is economic to extract at current prices using current technology) into future production is even more difficult once "above ground" factors (such as industrial organization and land rights issues) are taken into consideration.

4.2.4 Coal quality

Part of the difficulty in measuring coal reserves is India's notoriously low coal quality. India's coal contains very high levels of ash, often four times those typical in basins outside of India. This has several important effects on the Indian coal market:

First, the high ash content of India's coal makes it unexportable. Even if miners could somehow produce more coal than India needed, they would not be able to sell the surplus on the international market.

Second, Indian power plants, which are by far the dominant consumer of Indian coal, are designed specifically for high-ash Indian coal, meaning that imported coal is rarely available as a "drop-in" solution.[7] Because the entire Indian coal value chain is tailored to these poor quality coals, all imported coal must be blended with domestic coal before use, which requires technical expertise and limits operational flexibility.

Third, coal in India was traditionally priced and graded on a unique-to-India system that made it difficult to compare Indian coals with foreign coals. This led to India's coal being underpriced relative to what it might have been expected to fetch on the open market.

Indian coal pricing regulations are slowly changing in an effort to address this last problem. In January 2012, India undertook a major price revision, jettisoning its previous reliance on an antiquated UHV (Useful Heat Value) measure used only in India and moving toward the Gross Calorific Value (GCV) method of analyzing coal quality in line with international standards.[8] The recategorization from UHV to GCV allowed for a substantial rise in price without going through the formal mechanism of a price increase. UHV forced coal producers to discount their coals to a degree based on moisture and ash content, something that is not necessary under a GCV standard, which looks only at the underlying energy content of the coal (PTI, 2012; Ranjan Bose, 2011). This reform also allowed Indian companies to charge more for their coal by creating a more precise measurement of coal quality and energy content.

Given the low quality of Indian coal, it is unsurprising that beneficiation, in the form of coal washeries, is often needed to utilize Indian coal effectively. Yet the availability of beneficiation in India falls far short of the need. CIL has recently increased the fraction of its low quality coals that sees beneficiation, but due to poor market incentive structures, beneficiation capacity is still a small fraction of what is needed. CIL plans to increase its washery capacity from 110 Mtpa to 250 Mtpa in five years, but given CIL's track record on past such goals, this target may not be achieved. Most current washeries are operated by the private sector (91 Mtpa), while only 19 Mpta are run by CIL.

4.2.5 Cost structure of Indian coal production

Despite difficulties with beneficiation, India does have a durable advantage in coal extraction costs, which are competitive with efficient producers worldwide even on a calorie-adjusted basis. Table 4.2 shows the average cost of production for some of CIL's subsidiary companies.

4.2.6 The linkage system, coal pricing, and coal distribution policy in India

Historically, India has not had a coal "market" in any conventional sense, where sellers are allowed to negotiate prices and volumes with independent buyers. Instead, all coal in India was distributed through a linkage system by a national coal linkage board linking individual coal consumers with particular mines. (This linkage system plays a role somewhat analogous to that which China's annual Coal Conference has historically played in that country – see Chapter 3.) India's coal linkage policy is set by the Ministry of Coal, which is responsible for regulating which production from which mines is provided to which consumers. The traditional Indian linkage system was more of a command-and-control type "quota" system than a market in any meaningful sense. India has never had market pricing in coal, a legacy of the colonial system, the planned Nehruvian economy, and the coal nationalizations of the early 1970s. Under such central planning, coal is distributed in ways that can be arbitrary, inefficient, or corrupt; well-connected companies and industries get supplies while players without connections often languish.

The linkage system has evolved over the years. In the days of full central planning, India's coal linkage system was relatively straightforward. One needed a linkage to a specific mine to get a power plant built, and that fuel

Table 4.2 *CIL cost of production (INR/tonne*) by subsidiary*

Company		2002–03	2003–04	2004–05	2005–06	2006–07	2007–08	2008–09	2008–09	2009–10
ECL	Opencast	562.29	493.25	525.92	478.47	505.08	742.99			
	Underground	1,966.14	2,259.82	2,593.40	2,517.27	3,059.96	3,608.88			
	Overall	**1,113.37**	**1,098.94**	**1,220.08**	**1,069.56**	**1,172.48**	**1,702.81**	**1,233.25**	**1,674.77**	**1545.47**
BCCL	Opencast	553.40	555.84	637.67	606.88	621.84	659.65			
	Underground	2026.05	2,245.44	2,527.50	2,887.17	3,284.55	4,059.40			
	Overall	**983.68**	**1,043.16**	**1,161.33**	**1,123.26**	**1,140.92**	**1,238.82**	**1,096.47**	**1,569.63**	**1459.61**
CCL	Opencast	474.60	503.85	509.55	500.62	514.58	549.86			
	Underground	1,805.99	1,888.83	2,348.71	2,552.79	2,881.52	3,464.67			
	Overall	**571.29**	**603.31**	**638.28**	**616.55**	**625.69**	**670.28**	**643.71**	**788.09**	**750.60**
NCL	Opencast	430.42	445.58	492.06	508.07	544.78	527.93			
	Underground									
	Overall	**430.42**	**445.58**	**492.06**	**508.07**	**544.78**	**527.93**	**579.46**	**619.46**	**595.47**
WCL	Opencast	487.83	507.50	513.93	546.53	619.62	706.75			
	Underground	1,162.95	1,218.77	1,321.64	1,360.61	1,455.44	1,618.05			
	Overall	**654.36**	**678.11**	**701.58**	**734.85**	**811.10**	**915.47**	**898.35**	**1124.77**	**1095.37**
SECL	Opencast	275.36	281.17	302.64	330.82	346.17	353.15			
	Underground	1,141.02	1,143.73	1,162.47	1,317.17	1,424.22	1,612.58			
	Overall	**484.23**	**479.41**	**483.07**	**526.06**	**543.19**	**576.95**	**542.34**	**671.42**	**660.58**
MCL	Opencast	187.56	167.72	194.60	197.93	214.46	229.56			
	Underground	1,122.46	1,080.71	1,161.49	1,112.82	1,292.05	1,395.00			
	Overall	**218.89**	**198.77**	**226.37**	**224.38**	**240.97**	**257.59**	**244.08**	**337.94**	**316.07**
CIL Overall	*Opencast*	*379.13*	*375.05*	*397.75*	*406.81*	*426.76*	*452.29*			
	Underground	*1,510.50*	*1,595.07*	*1,741.90*	*1,819.31*	*2,017.61*	*2,318.71*			
	Overall	***564.00***	***560.78***	***589.77***	***592.04***	***614.88***	***663.42***	***620.99***	***784.28***	***736.64***

* The exchange rate at the end of 2010 was 45 rupees to the dollar.
Note: Actual data is given for 2002–2008. For 2008–2010, the data is a best estimate.
Sources: Coal India Limited (2010) and private research data.

supply was guaranteed for the life of the plant (Chandra, 2011). Transport was planned and allocated, if imperfectly, with Indian Railways. Price was considered largely irrelevant as most players involved were government owned and managed entities. Transport price was determined by Indian Railways' governing board in a political process in which freight, and especially coal, subsidized passenger fares. Production pricing for linkages was not a matter of public record, but the overall price for coal was set by the Ministry of Coal at a fixed nationwide price depending on coal grade. Pricing was determined in theory using a cost plus mechanism, although different subsidiaries of CIL had vastly different cost structures, leading to big losses at uncompetitive subsidiaries of CIL.

The linkage system is bureaucratic and not well adapted to a market that includes private players and prices based on supply and demand. For long-term linkages in the power sector, requests must be routed through the Central Electricity Authority and the Ministry of Power. The linkages committees are composed of stakeholders from a variety of areas, including the railways, CIL, and the planning commission. This commission determines whether a long-term linkage will be granted and what amount of coal will be provided. A separate committee on short-term linkages, composed of similar groups, has the ability to change the linkage patterns from quarter to quarter. An additional linkage process, developed annually, determines coking coal linkages for steel production. Yet another linkage process serves so-called non-core (usage that is not power, cement, or steel) industries that consume coal; these linkages are granted for two years, in contrast to power sector linkages, which are typically granted for the expected life of the underlying power asset (Ministry of Coal, 2010).

A major coal-consuming project, particularly in the power sector, requires formal linkage with a mine. Those wishing to develop a coal-fired power plant must work with the Ministry of Coal, the Ministry of Power, and the Central Electricity Authority to get approval. Such approval will dictate the precise source, amount, and even mode of transport of coal to the project for the life of the plant.[9] Needless to say, flexibility has not been the strong suit of the system. While various reforms to the system have been made in recent years,[10] it still is maintained in its basic state, dramatically limiting the flexibility of the Indian coal market. For smaller consumers, the black market has often been an attractive place to buy coal due to the insufficient supply available through official channels.

The linkage system has become particularly limiting as coal demand has outpaced domestic supply. This has started to lead to changes. CIL's newest fuel supply agreements with power plants dictate that just 75% of coal will be supplied through linkages. The rest of a power plant's coal must be procured on either domestic or international private markets – often at a much more expensive price than the official linked price. However, since 2009, CIL has been reluctant to sign supply agreements for any amount of coal, since such agreements are legally binding and CIL is contractually obligated to supply at least 90% of the agreed upon quantity (Bhaskar and Singh, 2011).

Starting in the early 2000s, Indian officials, seeking alternative means to get coal into the hands of consumers who needed it most, instituted an e-auction system for coal purchases. The system involved an open auction for a limited supply of Indian coal (just a small percent, though continually rising) that would be sold at market price rather than distributed through the linkage system. The e-auction system went through a series of ups and downs in the ensuing years, including Supreme Court challenges, allegations of corrupt distributions, and frustrations about low availability. Nevertheless, the system has survived and had an impact. E-auctions allocated about 10% of India's coal production at the time of writing, with the coal selling at a price premium of approximately 60% to the current linkage price.[11] (International markets are only accessible to the most established Indian firms due to past problems in which buyers from India failed to come through on their commitments to foreign sellers.)

Overall, there has been a significant – though not sufficient – increase in the volume of coal that is being procured at the market-linked price. Before the mid-2000s, virtually all of CIL's coal was sold at regulated prices, but the advent of e-auctions as well as a shift to align grade A and B coal prices more closely with market prices has created more alternatives. In 2010, 15% of coal in India was sold at market-linked linked prices; in FY 2011, the figure was more than 21% (Bhuta and Mehta, 2011).

Linkage reform and the growth in the e-auction mechanism are leading, slowly, to a restructuring of coal allocation and pricing mechanisms in the Indian coal market. But such steps remain halting and insufficient. E-auctions allow price discovery for smaller buyers (at a substantial premium to the linkage price) and if expanded sufficiently could allow price discovery for the power sector as well. But too many factors – institutional, political, and economic – stand in the way of a more serious reform of the sector. What we will likely continue to see is increasing market procurement of electricity (and thus coal) by reformed state electricity boards (SEBs) in more dynamic areas of the country such as Gujarat or Maharashtra. However, the political power of incumbents in other areas who benefit from cheap coal seems likely to cause the linkage system to persist for some time.

4.2.7 Coal transport and delivery infrastructure

India's coal shipments are heavily reliant on rail. Bhattacharya and Murthy (2006) broke down the transport modes of India coal as of 2006: railways handled almost 60% of India's coal shipments; 20% of coal was transported by Merry-Go-Round train (a particular type of specialized, semi-automated train that moves coal short distances from pithead to local power stations); 17% went by road, also to nearby destinations; and a remaining 5% was transported by other means including coastal shipping. But rail is still the mainstay of coal transport, and a significant challenge is that areas of high coal production also have low rail density. Fundamentally, long-term growth of coal supply to key end users such as power plants will have to be sustained by a dramatically improved rail infrastructure.

4.3 The effect of firm ownership type on coal production

While issues around land acquisition are arguably the most severe impediment to coal production in India (see Section 4.4), the industrial structure of India's coal sector certainly affects how mining is conducted. As discussed, Coal India Limited is by far the dominant player in mining, but private captive miners play an important role, and there are also coal firms owned at the state rather than national government level.

Captive mining is attractive in principle because it bypasses the linkage system and may allow vertically integrated private players to harmonize coal production, transport, and end use. But the allocation of captive mining blocks – coal-mining areas allocated by the government to private or government companies – has itself become a locus of corruption. Recipients of captive mining blocks have often been politically favored constituencies and others who were not serious about mining but who wished to speculate in coal mine licenses. As one extreme example, an ayurvedic medicine producer was granted a coal mining license.

Table 4.3 highlights some of the advantages and disadvantages of different ownership types. CIL's connectedness to government actually makes land acquisition somewhat easier than for other players, but at the same time it reduces the company's financial flexibility and tends to encourage bloat. At the other extreme, captive miners are able to move more nimbly to assemble a coal supply chain, but they may lack the connections needed to overcome certain bureaucratic hurdles – so they have to finesse or work around these whenever possible (see Section 4.3.2). Mining companies owned by state governments tend to be intermediate in their characteristics between CIL and captive miners.

Table 4.3 *Strategic ownership advantages and disadvantages by firm type*

	Owned by national government (CIL)	Owned by state government	Captive
Relative ease of land acquisition	Easiest*	Easier	Most difficult
Relations with national government	Good	Fair	Variable (depends on owners)
Relations with state and local government	Fair	Good	Variable (depends on owners)
Financial flexibility	Low	Low	High

* This assessment is relative – even CIL has significant challenges with land acquisition.

In the remainder of Section 4.3 we compare CIL (Section 4.3.1) to a particular captive miner, a subsidiary of the Jindal Group (Section 4.3.2). These two contrasting examples help illustrate how degree of connectedness to the government can affect the character and efficacy of coal mining operations.

4.3.1 The functioning of Coal India Limited

Any analysis of the forces shaping India's coal industry must begin with a serious look at Coal India Limited. CIL contributes 80% of all Indian coal production. It is composed of ten different geographic subsidiaries (see Figure 4.4), all of which are wholly owned by the parent company. The subsidiaries have independent histories and different profiles. Some older subsidiaries, such as BCCL and ECL, have precursors that date back to the earliest days of coal mining in India. These tend to have large numbers of excess employees, primitive mining methods, and poor financial performance. Others, such as SECL and CCL, have more modern equipment, leaner workforces due to extensive outsourcing, and far better financial performance.

As a whole, CIL produces coal at minemouth prices that would be considered competitive in world markets (even when adjusting for Indian coal's lower calorific value) – when it can actually bring coal to market (Coal India Limited 2010). And CIL's average price increases over the last decade, which have grown at 4.9% per year on average, have lagged behind India's rate of inflation. However, as discussed previously, Indian coal does not trade internationally, both because high transport costs raise market prices and because India's coal has such high ash content that there is no export market for it.

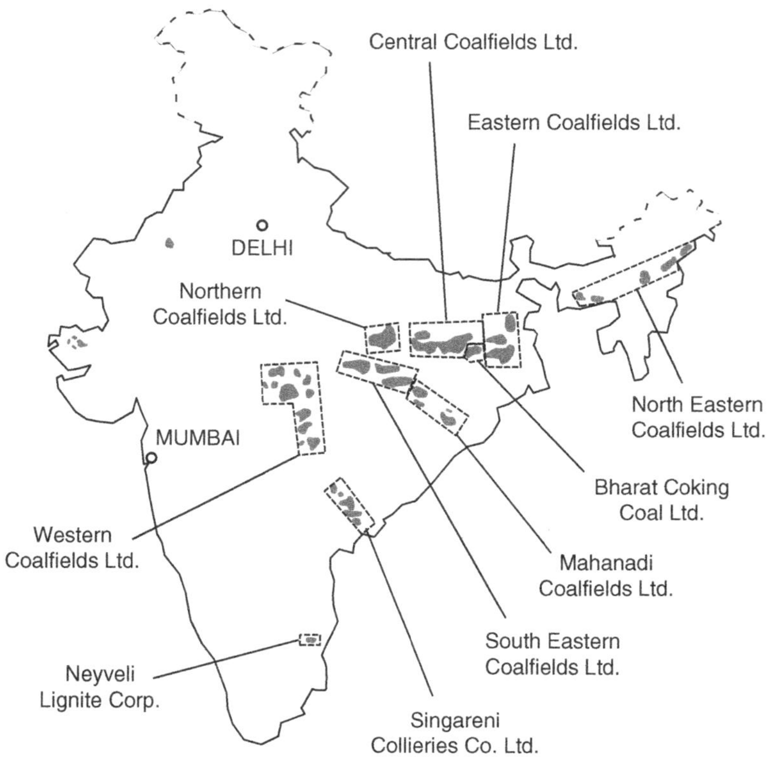

Figure 4.4 Coal India Limited subsidiaries and their coal resources. (This map shows India's land claim but should not be interpreted as taking any position on national borders.) Data source: Jha (2011).

Because domestic coal production is not able to meet all of India's needs, imports have become a necessity for the country. Because of the quality differences, it is not straightforward to compare the price of CIL's domestic production to the price of coal imports.[12] Nevertheless, rough comparisons on an energy-adjusted basis suggest that the prices of CIL coal are very competitive against imports (Table 4.4).

As compared to other domestic and international mining companies CIL has a much higher fraction of its revenue than is typical devoted to salaries and wages. Employee remuneration and benefits were INR 126.35 billion, INR 197.4 billion, and INR 166.5 billion in fiscal years 2008, 2009, and 2010, respectively, representing approximately 44% of company revenue (Coal India Limited, 2010). (The decrease between 2009 and 2010 was due to a large retroactive pay increase given to

Table 4.4 *CIL coal prices compared to imported coal prices.*

	COAL PRICES (INR/kcal)		
	Indonesian Coal (McCloskey Marker, 5,200 kcal/kg)	South African (globalCOAL RB Index,™ 6,200 kcal/kg)	Grade E Non-coking Coal from NCL (4,502 kcal/kg)
January 2006	262.73	314.25	184.11
January 2007	356.76	356.14	184.11
January 2008	465.85	628.91	206.41
January 2009	690.52	613.62	206.41
January 2010	582.30	643.35	225.93

1 Calculated based on the price of Indonesian coal of calorific value of 4,900 kcal/kg (assumed as net calorific value) as indicated by McCloskey Marker. GCV has been calculated on the basis of the formula net calorific value = GCV – (10.02 × moisture percentage). GCV of 5,160.52 has been rounded up to 5,200 kcal/kg
2 Calculated based on the price of South African coal of net calorific value of 6,000 kcal/kg as indicated by globalCOAL® RB Index™. GCV has been calculated on the basis of the formula net calorific value = GCV – (10.02 × moisture percentage). GCV of 6,120.24 kcal/kg has been rounded up to 6,200 kcal/kg.
3 Price of Grade E non-coking coal from NCL has been taken as representative of CIL's raw coal price, as the weighted average GCV of CIL's total raw coal production is 4,704 kcal/kg, which corresponds to Grade E non-coking coal, and NCL contributes the largest percentage of CIL's Grade E non-coking coal.
4 Calculated on the basis of IMF exchange rates as on the last working day in January 2006, 2007, 2008, 2009, and 2010 of INR 44.07, INR 44.17, INR 39.39, INR 49.02, and INR 46.37, respectively, per USD 1.00.
Source: Coal India Limited (2010).

workers in 2009, a common practice for CIL during labor negotiations.) This excessive employee compensation, a classic attribute of public sector companies, has been a drag on CIL's performance.

CIL's financial performance is also hurt by its having to maintain a substantial non-core infrastructure that would not be needed by a private company. As of 2010, CIL supported 590 educational institutions, 85 hospitals with 5835 beds, and 424 dispensaries, as well as water supply infrastructure for communities in which it operated. Overall, CIL spent 5% of its total profits on such activities (Coal India Limited, 2010).

The introduction of private shareholders through CIL's IPO in 2010 might reasonably have been expected to improve the performance of the company.[13] The company did appear to focus on its competitiveness in the run-up to the IPO. It hired new engineers and spent dramatically more money on exploration and assessment, almost tripling expenses for these activities relative to previous years in the three year period leading up to the IPO (Coal India Limited, 2010).

Early analysis of the IPO's impacts on CIL suggests that it has had some positive effects on the company, introducing an unprecedented degree of transparency while providing the enterprise with a welcome infusion of working capital, which had been in short supply. The publication of CIL's IPO prospectus provided the first public window into coal sector functioning in India and brought about greater pressure for performance improvements. The repeated updates on CIL performance that are now required have also focused more attention on governance issues in India. And the fact that 10% of CIL equity is now in the hands of private investors may deter the government from following through on its 2012 threat to compel CIL to procure coal on international markets if it cannot meet its customers' needs with domestic coal.

Yet despite the positive changes brought about by the IPO, it is as yet unclear whether these initial reforms will translate into the necessary production growth due to the constraints that are discussed later in this chapter. Production growth in the three years from financial year 2009 to financial year 2011 rose 36 Mt to a total of 440 Mt, a rise of just 9%. This is far less than what was needed and what was planned, indicating that, while the IPO was perhaps a necessary step toward increasing production, it is far from a sufficient one.

Though the IPO has given it much needed flexibility, CIL still faces substantial headwinds. Its excess labor, facilities expenses, and legacy culture (which even post-IPO remains that of a state-run corporation) will continue to present challenges as it seeks to modernize. Compounding the difficulty, the nature of these challenges varies a great deal by CIL subsidiary.

4.3.2 The functioning of private, captive miners: a case study of the O. P. Jindal mine in Raigarh

While CIL struggles with excessive bureaucracy and government regulation, private miners face different problems, notably poor quality coal blocks and the absence of critical infrastructure from rail and road connectivity to reliable power. One way of coping with the latter problem is to create a modern version of a "company town" complete with company-supplied infrastructure that the local government may be incapable of providing. (This is a common model for extractive industries in poor countries.)

In India, one can get a sense of the unique challenges in captive mining – and the ways captive miners deal with them – through an examination of the O. P. Jindal mine in Raigarh. The Jindal mine is run by the Jindal Steel and Power Limited (JSPL) unit of the Jindal Group, India's fourth largest conglomerate by sales. It was founded by self-made Indian entrepreneur O.P. Jindal and continued by his son, Naveen Jindal, who is also a member of India's Parliament. Because of the younger Jindal's prominent political position,[14] the Jindal group is ideally positioned to operate in the Indian market as a "dual firm" – a private sector entity that nevertheless benefits from deep political connections to key public sector entities necessary for successful operation in a highly politicized environment (Victor and Heller, 2009). The Jindal mine – the most impressive mine witnessed by the author while in India – provides a fascinating look into how a large, politically connected firm is able to combine skilled management to international standards with deep connections to the government to gain strategic advantage (*Ahimsa Times*, 2008).

The most notable aspect of Jindal's operation to an outsider is the way in which the corporation had to make up for the lack of adequate infrastructure. Chhattisgarh, the Indian state in which Jindal is located, has just one airport located in the capital city of Raipur, approximately 150 miles from Jindal's operations in Raigarh. Due to the poor state of India's road network, the trip between the two cities takes more than five hours – over bone-jarring roads with broken pavement.[15] Yet upon approaching within a few miles of the town of Raigarh and its associated mining areas, the quality of the roads improves dramatically. And after travelling hours without a discernible hotel or commercial restaurant, one enters a city with modern infrastructure, hotels, restaurants, and other amenities – many of which are supplied directly by Jindal.[16] Jindal is able to operate effectively by essentially taking over the primary functions that would ordinarily be provided by the state, including road and rail infrastructure, power, hotels and even hospitals and educational institutions for the families of workers – a virtual necessity in the far reaches of Chhattisgarh, one of India's poorest states. A Jindal official explained to the author, "We solve our problems directly [and] don't get involved with Babus [bureaucrats] and Netas [politicians]" (author interviews, 2008).

The same Jindal official said that the company makes extra efforts to employ locals (often more than one per family) as contractors or employees in low-skill areas of the company's operations. There was a particular emphasis at Jindal on land swaps and ensuring that at least one adult in each affected community would have a job with the mining company. This sometimes required training locals in heavy equipment operation. SOEs, with their more burdensome procedural constraints, would not have such flexibility in hiring (or infrastructure development) decisions. The ability to build local support in this way can prove a political advantage that helps expedite bureaucratic approvals. Jindal received its license in 1999 and just two years later was able to commence mining, which is light speed by Indian standards and compares quite favorably even to international permit application timelines.

The Jindal mine shows substantial value created through technological and engineering developments of which state-owned firms in India's coal sector are not capable. A mark of the technical sophistication of Jindal's operation is that it is able to operate in a coal block that CIL had considered unsuitable for production because of its extraordinarily high ash content. While there is high variability in coal quality around the world, commercially mined coal generally averages approximately 10% overall ash content, while Indian coal averages approximately 35% to 38% ash content. The Jindal block's ash content ranges from 37% to 62%; the higher-ash material is not even considered coal by international standards.

And yet, with a sophisticated on-site coal washery using advanced technology developed in South Africa, a quick construction timeframe (again, something that is difficult for a state-owned company to achieve), and the use of sophisticated blending methods not utilized by CIL, Jindal was able to create effective blends of high-ash coal for power consumption, direct utilization, and even steel-making. Jindal was able to produce coal for only 250 rupees (approximately USD 5) a ton at the minemouth and provide it at 1,050 rupees (USD 21) per ton at Raipur, Chhattisgarh's capital, where much of the coal was utilized. Using sophisticated production methods, Jindal was able to commercially utilize coal with a calorific value of less than 3,000 kcal/kg, approximately half that of a typical imported coal. CIL has lacked both financial incentives and regulatory flexibility to build sufficient washery capacity and more generally to pursue these kinds of innovative approaches. It remains to be seen whether these constraints on CIL will change in the post-IPO era.

Jindal has also utilized its flexibility as a private entity to deliver welfare benefits with speed and quality that would be difficult to replicate in a state-controlled company. With approximately fifteen villages and 13,000 people affected by Jindal's local mining operations, including one village that was relocated in its entirety, the company has developed community drinking water facilities, education facilities, a key bridge, and a zero-cost hospital for community members.[17] The ability to deliver increased welfare benefits to workers leads to higher worker productivity, fewer workers being poached, and a generally more productive workforce (as measured by tons produced per man shift).

Nevertheless, despite all of these ways in which Jindal could take advantage of its status as a private company, Jindal is still at a structural disadvantage from not being a state-owned entity. As one mine executive noted:

> Coal India still gets the best blocks and the rest of the blocks that are offered – these tend to be the rejected blocks from Coal India. We have tried to show that we can utilize these and hopefully that will serve as a motivator to others in the space. Nonetheless, it is a challenge as a private company. Coal India has the right to coal. We have to fight for it. (Author interviews, 2008)

On the other hand, it is conceivable that having to access more difficult resources is part of what helps keep private miners technically sharp in comparison with CIL.[18]

4.4 The effect of land issues on coal production

While reforms of CIL and other changes in the industrial structure of the coal sector can make a difference, the most significant factor behind India's coal deficit is conflicts over land. In this section we consider the various dimensions of the land problem. The most serious issue is the basic ambiguity over land rights that impede land acquisition for coal mining. However, obtaining environmental approvals to mine on forested land has also proved troublesome. In some cases, land is actively contested by insurgent groups.

4.4.1 Land rights and land acquisition

Coal India officials have made it clear on numerous occasions that land acquisition is the top impediment to their goals. In the first two years of the 11th Five-Year Plan (2007–2012), Coal India Limited acquired scarcely more than 5% of the total 62,000 hectares of land that it had planned to acquire for new mining permits by the end of the plan (Jog, 2010). In fact, it is even difficult to get land in India to build thermal power stations, a far less invasive form of land use than coal mining.[19] A presentation from a leading Indian industrial house at a private conference inquired as to whether land acquisition might be "an issue without a solution" (Pradeep Lenka, 2010).

There are several main factors restricting access to land for mining (and other industrial activities) in India. First, title to land and entitlement to compensation is often unclear. Second, the compensation that is offered may not be sufficient to encourage a voluntary sale. Third, legal procedures involving the transfer of land are often lengthy even when there is a willing buyer and seller, or, alternatively, a clearly evictable tenant. All of these concerns are discussed in greater detail in the following paragraphs.

There are several different types of land by Indian governmental classifications, and rules and regulations vary for acquiring each. The major types of land acquired for mining are as follows: homestead land, land under cultivation, derelict land or wasteland, communal and religious land, grazing land, and forest land. The most difficult forms of land to acquire are homestead and forest lands. For these land types, landholders have substantial legal rights and protections. Agricultural land tends to be easier to acquire, as new agricultural parcels in different locations can be substituted for the acquired parcels. (However, such substitutions cannot be made in a cavalier fashion, as moving a landholder even a short distance may place him in a dramatically different community, caste, or other community grouping.) Wasteland or derelict land can also be quite easy to acquire. In the case of the Gujarat mines visited by the author, which were located in the forbidding Rann of Kutch desert, company officials claimed land acquisition was easy; they said they were able to obtain acreage from the nominal landholders, who were not using the land, for an extremely low price. However, this is atypical. The vast majority of Indian coal lies in the heavily forested, agriculturally fertile area of East-Central India, and land acquisition becomes increasingly difficult as desirability of the land rises.

As emphasized by the typology of land rights proposed by Means (1999) and adapted here in Table 4.5, both strict legality and social

Table 4.5 *Typology of Indian land rights*

		Socially seen as legitimate?	
		Yes	No
Strictly legal?	Yes	Ownership rights: Obtained through inheritance or market purchase Customary use rights over village commons Legally protected tenancy	Women's independent land ownership Some legally protected tenancies
	No	Concealed contract tenancy: Common in India (rent exceeding legal maximum or actual tenancy time would confer ownership rights) "Illegalized" customary use: Informal economy cultivating tribal forest or other legally forbidden land	Encroachment on commons: Legitimacy will not be granted unless encroacher is of high social status. Alienation of tribal land for indebtedness etc. (forbidden by law in India)

Source: Adapted from Means (1999).

legitimacy of land acquisition may be called into question. In the case of the Gujarat lignite mines discussed earlier, strict legality was fairly easy to ascertain: land title was with the few nomadic groups who had obtained it through governmental recognition. Social legitimacy was also easy. The land truly was largely wasteland – barren desert largely unsuitable for grazing. And the Gujaratis had effective control. Through cooperation with a very pro-business government, they were able to lease their land effectively and quickly.

In other cases, legal title may be murky[20] and the social legitimacy of an acquisition may be in question.[21] For example, in areas like West Bengal or tribal regions of Jharkhand and Chhattisgarh with many mineral resources, land title may be poorly organized, and a single family may have used land for generations ("customary use") without any formal title having been recorded. Authenticating title claims of owners, legitimate rentership by tenants, and general ownership disputes have been at the heart of many recent delays in coal block acquisition (Jog, 2010).

Moreover, despite the wide variance of land tenure laws by state (some with Zamindari origins, some with colonial origins), actual administration of Indian land law takes place at the federal level, often leading to conflict between local, state, and federal authorities in land-related cases. Key Indian states for coal mining such as Chhattisgarh and Orissa (Zamindari legal foundation) and West Bengal (colonial legal foundation) have very different legal regimes and outcomes when it comes to land acquisition.[22]

Irrespective of the legalities involved and the formal rights of government companies to acquire land for mining under what is effectively eminent domain, what is written on paper may ultimately not be relevant to the process of land acquisition. Local governments may decide not to acquire land because of "law and order" issues and political unpalatability, regardless of what is written in the formal statute. Rival industries competing for similar land may aid strikes or further disruptions, and an overly active judiciary may make arbitrary and capricious rulings. And in no case is the state incentivized to expedite private sector claims.

Even if there is a legitimate legal case for land acquisition, the process for compensating those who own the land (or just live or work on it) is often fraught. (People living on land that miners attempt to use are officially classed by the Indian government as project-affected persons; landless individuals working on agricultural land can also be categorized as project-affected persons[23] and in theory must be compensated and "rehabilitated" in the parlance of the Indian government.) Rural

Indians are extremely reluctant to part with lands that may have been in the family from time immemorial, and often there are few available jobs for those who lose their land. When selling to state companies in particular, citizens may only be able to get an "official" price for their land rather than a true market price. Indian land law requires compensation by public entities at a so-called "benchmark" value, a value that invariably is based on tax records for actual usage and often far undervalues the land for the project. This disparity makes it hard to encourage voluntary sales to public entities like CIL, especially given that surrounding landowners who do not sell may see the prices of their land rise due to economic development.

Some have suggested changing the formula to compensate landowners for inconvenience and the extra value of land foregone through development. The following two mechanisms have been proposed for dealing with the reluctance of landowners to sell:

(1) Unequal compensation – give more money to those who sell earlier in a land purchase process to ensure more expedited property sales. Have a cutoff of 70% of a mining area being voluntary sales before going to the government to fix a notified price.
(2) Provide a meaningful cash bonus to landowners for on-time project completion to discourage strikes and other delays

However, as of this writing there had been no major recent changes in Indian land law to incorporate these and other incentives for landowners.

In practice, compensation to landowners and project-affected persons varies widely. Sometimes companies are to blame, but in other cases the root problem is that it is very difficult to determine who is legitimately affected. According to mining company officials and independent reports, pseudo-landless laborers sometimes attempt to get easy compensation from companies whether or not they even live in the area or are affected. One executive described a case of landowners who lost only part of their land receiving a financial "bonanza," but he said the bigger challenge is not excessive compensation to actual landholders but dealing with "squatters, migrants, and forcible occupants." In most cases, rehabilitation is not a major part of the costs of land acquisition, and companies are, in theory, more than willing to rehabilitate those who are legitimately affected by mining operations. But establishing actual ownership can involve assessing voter ID and food subsidy cards, census records, and a number of other documents. Many may be unclear, and others may be inauthentic.

Even dislodging those who are legally evictable can prove quite challenging. The notoriously slow Indian judiciary often delays cases for years, and Public Interest Litigation (PIL) with both sincere and rent-seeking intent often further delays land acquisition. The ambiguities in the law can be used by a corrupt nexus of local officials and corporate officials for the purposes of mutual enrichment.

In most cases, companies are allowed to negotiate directly with legitimate landowners upon receiving permission from the government. But such regulations vary substantially in each of India's states, and in some states, there is no provision for this at all. Even when there is, the process of getting permission from the government to negotiate with landowners can provide yet another opportunity for rent capture for government officials. The Land Acquisition Act is also heavily influenced by federalism, with various Indian state court decisions having a profound effect on its local implementation.

There are differences in the legal regimes governing land acquisition by state-owned and private companies. In theory, the Land Acquisition Act (1894) enables the government (and thus state-owned companies) to acquire any land for public purpose, while the Coal-bearing Areas Act (1957) is the only act that private companies can use to acquire land. Its criteria for acquisition are more restrictive – there is no ability to use eminent domain, for example – which creates some additional challenges for private companies. Private miners are sometimes forced to pay much more in R&R (resettlement and rehabilitation) than government projects would pay to acquire the same land. On the other hand, as already discussed, private companies may have more flexibility to offer a market price to landowners.

4.4.2 *Forest and Environmental Clearances*

4.4.2.1 *Environmental Issues From Coal Use in India*

The heavy reliance of India on coal use has substantial environmental implications. It contributes to the global climate change problem discussed in Chapters 1 and 13, but in general Indian policymakers – even those most concerned about climate change – see no real alternative but for India to continue to use substantial amounts of coal if it is to improve the standards of living of its massive population (see Box 4.1).

Coal also causes more localized environmental issues. Local air quality in Indian cities has been notoriously poor – Delhi has been listed as having the world's worst air quality in the world by the WHO (McIntyre, 2010) – and coal use is a major culprit. Given India's energy needs, coal use is

Box 4.1 Climate change and coal in India

In 2011, CO_2 emissions from burning coal in India, at 1,205 Mt, represented almost 4% of total global greenhouse gas emissions – and that share could grow if India's economy continues to grow and its coal demand along with it (IEA, 2013). Nonetheless, India's total CO_2 emissions are still less than a quarter those of the US and China (Watts, 2010; International Energy Agency, 2012). India's climate stance in international negotiations has been focused around the notion of "common but differentiated" responsibilities, a concept that originated in the Rio Earth Summit in 2002 (Rogers, 2009). India continued its tough stance in the Durban talks of 2010, refusing to agree to meaningful binding emissions cuts (Dhalmia, 2011). India continues to be extremely resistant to foreign pressure to reduce emissions, and even forward-looking former Environment Minister Jairam Ramesh publicly dismissed the notion of a binding carbon cap when Hillary Clinton visited India, an action that created a minor international incident (Carl, 2011). Or, as R. K. Pachauri, an Indian national and the former Director-General of the Intergovernmental Panel on Climate Change (IPCC) and an Indian national, observed

Can you imagine 400 million people who do not have a light bulb in their homes? . . . You cannot, in a democracy, ignore some of these realities and as it happens with the resources of coal that India has, we really don't have any choice but to use coal in the immediate short term.[24]

That even India's most eager climate campaigners and high-profile environmentalists are aware that ultimately India must use coal as sensitively as possible given environmental constraints, and that eliminating – or even substantially reducing – coal's role in India's energy system is, for the foreseeable future, totally impractical, says a lot about the position of coal in India's current energy system. As Chandra Bhushan, director of the mining project at the Center for Science and the Environment (India's best known environmental activist group), observed during an interview with the author:

Whether we like it or not, with the energy scenario we have in the country in the medium term we can't wish away coal, it's as simple as that. In the medium term coal will remain. I expect coal to be 50% of commercial energy until 2030. I don't expect that to go down. (Bhushan, 2010)

likely to continue, but pressure on coal-fired power may increase in more affluent metropolitan areas (Delhi, Mumbai, Ahmedabad, and others), much as has occurred in Shanghai, Beijing, and other affluent cities in China.

Subsidence (sinking) from poorly designed Indian coal-mines has devastated various regions of India, even in some cases leading to the evacuation of entire villages, particularly in the old coking coal region of Jharia (*American Chronicle*, 2008). The prospect of local environmental impacts has in some cases discouraged the sale of land to coal companies for mining.

4.4.2.2 The effect of forest and environmental clearances on land acquisition

Discussing the detailed environmental issues around coal in India is beyond the scope of this chapter. However, local environmental concerns are the basis for forest and environmental rules that can further complicate land acquisition for coal mining, and we discuss these challenges to the coal sector here.

Specifically, India has developed a set of government clearances that are necessary to obtain before opening up new areas for mining. Clearance from the Ministry of Environment and Forests is needed for conducting mining operations on forest lands (as is a clearance from the director general of mine safety attesting to the safety of operating conditions) (Bhuta and Mehta, 2011). Environmental clearances are also required to renew mining leases. While clearances are sometimes used for the purposes of rent extraction by officials, they do also represent a genuine effort by environmental officials in India to make sure that coal development is not uncontrolled in environmentally sensitive areas.[25] The many tribal residents of forest lands also act as a powerful force to make sure that clearances are granted legitimately. Because of the substantial political problems that could be caused by tribal displacement, officials are unlikely to simply give the green light for forest development of mines in the name of expediency. Getting the proper forest and environmental clearances is a time-consuming process at the local, state, and federal levels, escalating from the forest sheriff to the Minister of Forests.

Adding to the difficulties of obtaining the environmental clearance, the implementation of the CEPI regulation, which assesses the pollution of each industrial cluster, has created a new administrative hurdle. Four Indian coalfields representing 48% of production apparently did not meet the required specification and were at least nominally the reason for CIL to announce a substantial decrease in planned production of 39 Mt in 2011 (Bhuta and Mehta, 2011). The Bureau of Environment and Forests demanded that these units reduce their pollution, as required for any units with a failing score, before more units would be allowed to be put in operation (Bhuta and Mehta, 2011).

Environmental clearance problems are not limited to mines. Railways are increasingly becoming an infrastructure bottleneck, and environmental clearances can slow development in that area as well (Bhuta and Mehta, 2011). For example, in October 2010, the proposed new Tori-Shivpur-Hazaribagh rail line in Jharkhand from the North Karanpura coalfields was denied clearance by the Ministry of Environment and Forests, thus stranding coal assets not only for state-owned CIL but also potentially for several captive mines. This action was taken on the eve of the CIL IPO, highlighting the limitations of CIL's power. It stranded 24 Mt of production from CIL alone, and it was no coincidence that this production was from Central Coalfields Limited (CCL), the most Naxal-infested subsidiary of CIL (Singh, 2010) – see Section 4.4.3.

4.4.3 Naxalism and land rights

One of the primary manifestations of India's contentious relations over land – and one that has had significant effects on India's coal development – has been the Naxalite rebellion, a complex, coordinated insurgency against the Indian government concentrated heavily in resource-rich East-Central India. Naxalism has its origins in Bengali student radicalism of the 1960s, but over the ensuing decades has morphed into a powerful peasant-led rebellion in the countryside against the Indian state. Described by former Indian Prime Minister Manmohan Singh as the "single biggest internal security challenge ever faced by our country"[26] (a rather remarkable statement given the large number of historical Indian insurgencies, most notably its troubles in Kashmir), Naxalism is a multifaceted phenomenon, but one that many observers believe is best understood as a contest over resources (Spacek, 2010; PTI, 2010b; PTI, 2010d). Coal has been one of the most critical resources that Naxalites have contested, and Naxal activities in coal country have substantially limited production in several areas.

Figure 4.5 helps illustrate the relationship between Naxalism and resource contestation. India's coal-bearing areas are overwhelmingly in the poor, remote, east-central portion of the country. These correspond in large degree to areas that are also forested and tribal, with Jharkhand and Chhattisgarh, two of India's most coal-rich states, being specially created states for tribal people.

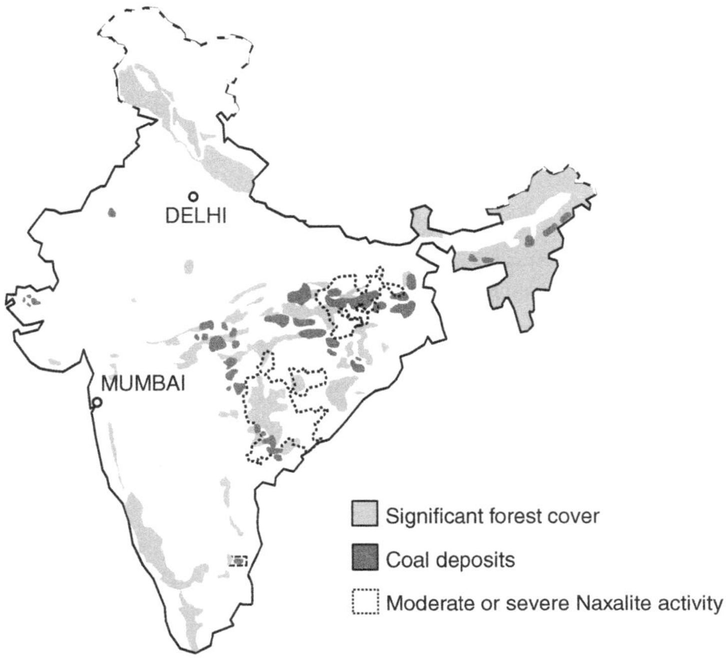

Figure 4.5 Overlap between forests, coalfields, and insurgents. (This map shows India's land claim but should not be interpreted as taking any position on national borders.) Data sources: *Forest cover*: Forest Survey of India (2013); *Coal deposits*: Bhattacharya and Murthy (2006); *Naxalite/Maoist impacts*: Institute for Conflict Management (2014).

Naxalism has had an important effect on resource extractive industries. Several companies contacted by the author freely confessed they would be reluctant to open mines in heavily Naxal-affected areas even if they could obtain the necessary rights. Another, which boasted of its success in dealing with Naxalism and its unconcern about a facility located in the heart of Naxal country, ultimately had that facility destroyed by Naxals little more than a year after speaking with the author (Naxal Terror Watch, 2008).[27] In other cases, Naxalism operated at the level of the transport Naxalites, who have a strong presence in Andhra Pradesh, Orissa, Jharkhand, and West Bengal – five states that possess approximately 85% of India's coal resources. Naxalites have killed coal mine security officers and burned vehicles from a coal survey team from the Mineral Exploration Corporation of India. In addition, Naxals have become heavily involved in illegal coal mining, which they use to fund

Table 4.6 *Loss of coal production to Naxalism at Central Coalfields Ltd.*

Year	Loss of coal production (Mt)
2007–2008	6.1
2008–2009	8.2
2009–2010	11.5
2010–2011 (through June)	11.2

Source: Press Information Bureau (2011).

their insurgency operations (PTI, 2010d). In Andhra Pradesh, Naxalites have destroyed vast quantities of mining equipment in raids, while mining explosives are frequently captured for military purposes by Naxalite leaders, often in raids on poorly armed and trained security police who suffer significant fatalities (Carl, 2006).

After years of relative denial, the government has become increasingly open about the challenges of Naxalism to minerals production and extraction. Shriprakash Jaiswal, the Minister of Coal in 2010, told India's Lok Sabha (lower house) that CIL subsidiaries were losing substantial production because of strikes and threats of Naxalite violence (Table 4.6). The problem became particularly severe in heavily tribal Jharkhand and Chhattisgarh (Bhaskar, 2009), and the government expressed concern that the violence could seriously dent India's overall coal production (Jog, 2010).

While bureaucratic hurdles have prevented many legitimate companies from mining, Naxals have used illegal coal mining to fund their insurgency.[28] In this way, the instability that the Naxals cause in India's coal mining areas flows directly from revenues that Naxalites themselves obtain from coal. This dynamic has been particularly potent in Jharkhand (Bhaskar, 2010). Combating Naxalism presents problems of federalism because, as India's coal minister has noted, military action against Naxals must be coordinated at the state level, whereas the ministry operates at the federal level.

A key question is whether the losses due to Naxalism will grow or shrink over time. In the last several years, the government has seemingly managed to curb the number of states across the country that suffer from severe Naxal activity, but the insurgency remains a serious problem in key coal-bearing regions. Naxalism joins environmental clearances and general land rights problems to form a powerful trio of factors impeding the growth of India's coal sector.

Table 4.7 *Sector-wise actual (2012–2013) and projected (2016–2017) coal demand in India (Mt)*

Sector	2012–13	2016–17
Power (utilities)	512.00	682.08
Power (captive)	43.00	56.36
Cement	30.24	47.31
Sponge iron and others	35.30	50.33
Coking coal steel	52.30	67.20
Others	100.00	77.22
TOTAL	772.84	980.50

Source: IndiaStat (2014).

4.5 Demand for domestic and imported coal

As India's domestic coal production is hampered by the industrial organization and land-related problems discussed in Sections 4.3 and 4.4, respectively, demand continues to grow. Barring major setbacks for India's economy, power and industry demand for coal is likely to continue growing at rates near 8%, while coal production will be lucky to grow at rates approaching 5%. While any attempt to quantify the domestic production shortfall going forward is necessarily speculative, both official Indian documents and informal interviews by the author, as well as CIL's recent performance, suggest that a gap of multiple hundreds of millions of tons on an annual basis over the next decade is not out of the question. This section considers the demand side of the equation, focusing on the organization of the power sector, which accounts for the large majority of demand (Table 4.7). Power producers seek the lowest-priced coal available, but in a number of cases they may need to turn to higher-priced coal from e-auctions as well as imported coal. This section discusses the advantages, disadvantages, and obstacles to procuring coal from these other sources.

4.5.1 The nature of coal demand from the power sector

As of November 2011, the electricity sector in India had an installed capacity of 185.5 GW, making it the world's fifth largest (Central Electricity Authority, 2012). Thermal power plants constituted 65% of the installed capacity, hydroelectric about 21%, and the balance of capacity was from a combination of wind, small hydro, biomass, waste-to-electricity, and nuclear. Coal-fired plants account for 55% of India's installed electricity capacity, compared to 92% in South Africa, 77% in China, and 76% in Australia, but in terms of actual generation coal is still well over 70% due to the higher utilization of coal-fired power plants.

The Indian electricity sector was fundamentally restructured by the Electricity Act of 2003, which has had implications for coal demand that we will discuss further in this section.[29] Before the restructuring, State Electricity Boards (SEBs) were in control of licensing all companies that wished to sell power within a particular state, and they were often the only buyers of this electricity.[30] This allowed significant scope for politicization and corruption. Extensive cost subsidies, late payments, and rampant theft were characteristic of India's power industry (and continue to be a problem even post-reform).

The Electricity Act of 2003 deregulated who could build generation assets. It also, in theory, removed licensing requirements in rural areas, although from a practical perspective, difficulties in defining "rural" in various states mean that a third of India's rural areas still have licensing regimes.

Another key contribution of the act was to dramatically expand the scope of captive power that a company could generate for its own use, which paved the way for captive coal mines and their associated power plants. As discussed previously, this has been an important reform for the coal sector. At the same time, the impact of captive power (and captive mining) is limited by the fact that it can only really be exploited by major players with sufficient access to capital. Smaller private companies remain reliant on e-auction and other similar mechanisms for procuring coal and thus are vulnerable to shortages or high prices.

The Electricity Act allowed the central government to play a much bigger role in facilitating state to state interconnection of the power grid. For the first time in India, the act opened up the possibility of interstate electricity trading and allowed states with more advanced and transparent regimes to leapfrog ahead in building out their electricity supply. By providing for open access in transmission and, theoretically, phasing out cross-subsidies between industries – so that businesses would not be subsidizing residential users, and especially, farmers – the Electricity Act helped develop the first truly national Indian electricity market. SEBs were unbundled and metering of electricity was made mandatory (although this requirement was not always observed, as in the "Free Power to Farmers" scheme that swept through various areas of India in the last decade).

Despite its many imperfections in design and implementation, the Electricity Act of 2003 produced competition in Indian electricity markets, with ripple effects in the coal market. Competitive private generators selling to commercial customers were able to pay more for coal than the notified price used for standard customers, leading to an increased appetite for coal imports and grand plans such as the Ultra Mega Power Projects (UMPPs) that would run on imported coal.

Even with their somewhat diminished role post-restructuring, SEBs remain critical in determining the future of Indian coal demand. The financial problems of SEBs – which stem from billing inefficiencies, theft,[31] fraud, stagnant and/or artificially low tariffs,[32] and the need to buy expensive spot power to avoid blackouts – make them uninterested in power purchase agreements with more expensive coal-fired power plants. This in turn has an impact on fuel purchase agreements and what power plants get built. Independent power producers (IPPs) are pushed by the SEBs to reduce exposure to more expensive fuel sources such as imported coal or even "market rate" coal from e-auction and other emergent market mechanisms domestically. This reduces demand for these kinds of coal. It also means that IPPs that do not have long-term PPAs or dedicated linkages internally – and are therefore forced to use more expensive coal, perhaps imported – will see substantially lower plant load factors (PLFs) and thus lower profitability. Given that there is little practical way to pass on the price of more expensive coal in the form of higher power prices, and coal imports will increase their losses relative to not generating power at all, Indian buyers are likely to continue to resist importing coal. (See Section 4.5.2 for more discussion about factors deterring coal imports.)

More forward-looking SEBs have worked to improve their situations in part by making themselves more attractive places for IPPs and captive power producers to set up shop – and thus more attractive markets for coal producers to sell into over the long term (Asthana, 2011). As a result of actions such as increasing power tariffs, states like Gujarat and Maharashtra come much closer to balancing their books in the power sector than most states in India (*Power Today*, 2011). Their relatively healthier SEBs are able to procure power from generators using more expensive coal, which in turn increases coal demand.

4.5.2 Coal imports and other market-priced coal sources

As previously discussed, one basic obstacle to coal imports is that the price of imported coal is significantly higher than that of domestic coal even on an energy-adjusted basis (see Table 4.4). Analysts have shown that the disparity between CIL official prices and import prices has been relatively consistent over time (although CIL prices are not adjusted to track international price fluctuations[33]). This represents a large subsidy to coal-consuming industries, particularly the power sector (smaller scale buyers as well as those in less politically connected industries receive less benefit). As private sector miners are not allowed to sell any of their coal on the open market, they are an unrealistic alternative to imports. While this situation could be changed by governmental regulatory changes, there is tremendous institutional resistance to such change, both from industries that benefit and from those who profit from corruption tied to the existing system.

In addition to financial impediments, there are logistical obstacles to increased imports – specifically, constraints on how much coal can be imported due to limited rail infrastructure and, especially, port capacity. India's state-owned ports have traditionally lagged international rivals in virtually all key metrics, from cost to unloading time to number of berths available (Acharya, 2003). Partly for this reason, there has been a recent boom in private ports to accommodate new coal imports, as privately run ports have been able to respond more nimbly to demand than their state-owned counterparts. While existing, mostly publicly owned ports were clogged and able to accept only smaller Handymax and Panamax vessels (50,000–75,000 DWT), newer private ports emerged unexpectedly which were able to deal with the 150,000 DWT capesize vessels that have increasingly become the industry standard for moving coal. In particular, the launch and huge expansion of ports such as Mundra in Gujarat, which is projected to handle up to 60Mt of coal imports annually in the western state of Gujarat, far from India's coalfields and closer to India's industrial centers, could remake the coal market in those areas (Cowhig and Singh, 2010).

There could be some concerns about import supply, with key swing exporter South Africa suffering from production and transport issues (see Chapter 5) and Indonesia, India's main supplier, enacting policies that could negatively affect its exports (see Chapter 7). CIL has previously made noise about acquiring foreign coal assets, but, as with many similar SOE ventures in India, these appear unlikely to bear fruit (Coal Guru, 2010; Shanker, 2011).

Difficult Indian weather in the monsoon season (which also complicates domestic transport) provides a further barrier to imports. And finally, many importers with whom the author spoke expressed reluctance to do business with Indian power companies, whom they considered to be unreliable purchasers.

For all the challenges that coal imports face, they have nevertheless grown, as shown for steam coal in Figure 1.5 in Chapter 1, and they are likely to remain an important part of the coal supply picture.

4.6 Conclusions

The Indian coal market has confounded experts, both foreign and domestic, over the last decade of halting reforms. Most analysts thought that in the wake of the tremendous growth of the Indian economy, the Indian coal market would have to enact fundamental reform, as the entrenched interests holding back reform would prove less powerful than the forces demanding it.

What actually happened was more complicated. While there have been substantial reforms in India's coal market over the last decade – most notably the partial privatization of CIL but also including e-auctioning of coal, the growth of captive mining, the loosening of the linkage system, and other structural changes – what these changes revealed is that the challenges facing India's coal market were more fundamental and less situational.

Partial privatization of CIL did introduce some accountability and transparency, but the government was still willing to give excessive CIL labor contracts for political reasons and in other ways to interfere with operations. E-auctions were useful in getting more coal to market at true market prices, but the inherent corruption within India's political system invariably ensured that those who benefitted most were the most politically connected. Coal sector reform also did not address the fundamental problems with railways and ports, neither of which was under the control of India's coal sector. Likewise, the fundamentally unreformed power sector, dominated by political rent-seeking and bankrupt SEBs, was not willing or able to pay the sorts of rates that would have driven the fundamental innovations necessary to bring new coal supplies to market.

Most importantly, these structural reforms could do nothing to ameliorate the biggest problem facing India's coal sector: the lack of available land for mining, the result mainly of institutional difficulties around land rights and environmental protection. The difficulty in obtaining land for mining purposes has been the number one challenge to India's coal sector – and it looks likely to remain so going forward.

Notes

1. Seventy-six percent of the growth in generation capacity that was scheduled for the 11th Five-Year Plan (2007–2012) was to be in the thermal sector. The thermal sector includes coal supply, which was intended to account for almost 85% of that newly installed thermal capacity. As it turned out, India's hydroelectric and nuclear power generation was above plan in the financial year of 2011, but the thermal power that represents the vast majority of Indian output was 19% below the plan, representing a substantial deficit (Central

Electricity Authority, 2012). These fuel shortages have been directly responsible for power shortages, including the stunning blackouts that gripped India in July 2012 and were the largest in Indian history (Mail Today Bureau, 2012).

2. For example, 60% of the way through India's 11th Five-Year Plan (2007–2012), only 77 of 142 scheduled coal mining projects had been approved, and of those only 32 were operational. Incremental production over India's previous five-year plan was just 57 million tonnes per annum (Mtpa) out of 104 scheduled for these projects, meaning that Coal India Limited (a state-owned enterprise responsible for the majority of Indian coal production) was producing at a mere 15% of the scheduled 381 Mt. Of the incremental production achieved, 45% came through expansions of existing projects and not greenfield operations (Ministry of Coal, 2010).
3. Detailed information on costs can be found in Coal India Limited (2010).
4. Rampant illegal mining is increasingly becoming an important political issue in India, particularly given the growing political power of the Reddy family of Andhra Pradesh, which has dominated illegal mining in that area. For example, see the *New York Times* article "Despite Scandals, Indian Mining Bosses Thrive" at http://www.nytimes.com/2010/08/19/world/asia/19india.html?pagewanted=1.
5. The amount of India's overall coal resource is subject to significant dispute and was downgraded by the World Energy Council (WEC) due to insufficient exploration data and the fact that many reserves were not economically recoverable under current conditions. The 7% figure should be seen as a lower bound. Indian authorities continue to maintain that the higher reserve numbers are accurate, and other independent estimates have ranged between the Indian government's numbers and the WEC estimates (Ians, 2008).
6. TERI comes to these estimates by assessing India's actual proven (by internationally accepted standards) coal reserves currently accessible for meeting power needs.
7. Some plants now under construction will be designed for imported coal.
8. Under the old "useful heat value" (UHV) system, coal was graded in quality "bands," and CIL often provided coal that was at the higher end of the band range. By moving to a GCV standard, pricing became more precise and CIL was able to charge more exactly for its quantity of delivered coal, which had the net effect of raising prices.
9. A full discussion of India's coal linkage policy can be found on the Indian Ministry of Coal website.
10. Reforms have included greater flexibility on coal linkages, the development of e-auctions for coal, and the growth of captive coal mining.
11. Almost 80% is allocated through the Ministry of Coal's administrative committees to public and private sector applicants from various sectors. An additional 10% is sold through online e-auctions, which have resulted in considerable revenues. In 2009–10 e-auction prices were, on average, almost 60% above notified prices.
12. CIL's coal is mostly of inferior quality to anything that is traded internationally and energy/sulfur content–based adjustments cannot be made on a

straightforward basis as each power plant is designed around a very particular quality of coal and must often blend coals together if that quality is not available.

13. See, for example, Wolf and Pollitt (2008, 2009), which sought to quantify the benefits in firm performance associated with the privatization of state-owned enterprises in the oil and gas sector.
14. Jindal's mother was also a minister in the Haryana state government at the time of writing.
15. Author trip to Raigarh from Raipur.
16. Author visit to Raigarh site.
17. Several of these facilities were viewed by the author during his site visit. They compared quite favorably to government-sponsored facilities that were either inadequate or non-existent. They also compared favorably to facilities in nearby villages not affected by the mine.
18. In the oil sector, for example, the fact that private companies often have access to less geologically favored blocks helps drive them to improve their technical and risk management capabilities (Nolan and Thurber, 2012).
19. http://www.bloomberg.com/news/2012–03–07/tata-favors-wind-solar-as-coal-fed-plants-impossible-to-build.html.
20. In the prospectus for its 2010 IPO, Coal India Limited listed the following possible reasons for unclear land title:
    - Conveyance or sale deeds have not been executed.
    - The conveyance deeds have not been registered in the land records maintained by the relevant authorities and are insufficiently stamped.
    - The process by which changes in beneficial ownership are formally recorded in the land registries in India have not yet been carried out in the records of the local land registries.
    - The lease deeds have not been executed.
    - The lease deeds have not been registered in the land records maintained by the relevant authorities and are insufficiently stamped.
    - Eviction suits have been filed.
    - The lease deeds have expired and have not yet been renewed.
21. As Rodrik, Subramanian, and Trebbi (2002) note, formal property rights are frequently not the best institutional mechanism for coordinating complex social and economic interests in developing countries.
22. As Means (1999) notes, "Since land is a State subject in India, it is not possible to suggest specific policy options except with reference to particular states" (p. 3). Land became a state subject in 1935, thirteen years before Indian independence. So, in many ways the land issues plaguing the industry today are a legacy of British rule that was designed to increase local control of what was then a very fragmented Indian state. While the central government can provide guidance, legally its role is restricted in doing anything more than that.
23. Forty-three percent of rural Indians are landless. While not all of these are technically landless agricultural laborers, there is no question that landless labor is a major problem in resettlement and rehabilitation.
24. "Cleaning up Coal," Richard K. Morse, Foreign Affairs July/August 2012. Available at http://www.foreignaffairs.com/articles/137685/richard-k-morse/cleaning-up-coal.

25. The author interviewed several Indian forest and environmental officials in the context of his work, several of whom were quite knowledgeable and took their role seriously.
26. "Naxalism Biggest Challenge: PM" *Hindustan Times*, April 13, 2006. Available at http://www.hindustantimes.com/nm9/naxalism-biggest-challenge-pm/article1-86531.aspx.
27. The author had spoken to an Essar executive in 2007 who had dismissed concerns about the security of a coal-consuming steel plant in a part of Chhattisgarh State with a very heavy Naxalite presence.
28. A leading trade union representative and former Member of India's Parliament claimed in 2010 that more than 10,000 smaller mafia groups stole approximately 6 Mt per year of coal, leading to approximately USD 240 million in losses (PTI, 2010b).
29. The electricity sector restructuring in India was prompted in part by the disastrous Dahbol power plant investment of the 1990s by Enron, a textbook example of political and regulatory failure that gave a black eye to the entire Indian power system.
30. State Electricity Boards (SEBs) in India are government-owned corporations chartered in each state that are in charge of the distribution of power within that state. They work in conjunction with the federal ministry of power on issues related to the generation and distribution of electricity. SEBs have historically tended to serve the interests of politically connected individuals and groups – representing constituencies ranging from favored neighborhoods in Mumbai represented by a powerful politician to specially protected industries and agriculture that receive free or reduced tariffs.

    SEBs across India face losses into the billions of dollars due to power theft ranging from simply tapping in illegally and informally to lines (an epidemic in urban India) to more sophisticated schemes by major customers with political clout. Currently the SEBs' collective debt is almost USD 20 billion – an enormous amount for what is still a moderately sized economy relative to the global giants. As independent analysts have noted, without the explicit guarantee of state governments these SEBs would universally be considered bankrupt entities incapable of getting a loan.
31. Power losses (often a euphemism for theft) can be as high as 28% in some regions.
32. For example, while costs rose 16.5% from 2007 to 2009, tariffs rose just 6.5%. Free power for farmers and artificially low tariffs demanded by the states, particularly for consumers, can prove a major burden. Some outside estimates have suggested that as much as a 25% tariff rise will be needed for the SEBs to become financially healthy (TNN, 2012).
33. While they remain significantly below international prices, CIL official prices have been adjusted up over time.

References

Acharya, Vijayendra, et al. (2003). "India Port Report: 10 Years of Reforms and Challenges Ahead." *I-Maritime Consulting.*

Agneya (2011, February 14). "Renewable Projects in India: What to Adopt – CDM or REC?" Available at http://agneyablog.wordpress.com/2011/02/14/renewable-projects-in-india-what-to-adopt-cdm-or-rec/ (accessed June 13, 2014).

Ahimsa Times (2008, December). "Savitri Jindal a Jain Devotee from Terapanth Sect Declared the Richest Woman of India." *Ahimsa Times*. Available at http://jainsamaj.org/magazines/ahimsatimesshow.php?id=266 (accessed June 16, 2014).

American Chronicle (2008, August 8). "Underground Mine Fire Engulfing Jharia Town of Jharkhand State, India." *Americanchronicle.com*. Available at http://www.americanchronicle.com/articles/view/70817 (accessed June 27, 2014).

Asthana, Shishir (2011, July 11). "Rs 1,00,000 Cr SEB Debt Bomb Is Waiting to Explode – Firstbiz." *Firstbiz*. Available at http://firstbiz.firstpost.com/economy/rs-100000-cr-seb-debt-bomb-is-waiting-to-explode-22341.html (accessed June 27, 2014).

Author interview with Jindal Coal Mining Official (2008, November 10).

Bhaskar, Utpal and Ruchira Singh (2011, October 27). "Coal India Keeps Away from Fuel Supply Pacts." *Livemint*. Available at http://www.livemint.com/Home-Page/zqNOplAEPQCs1gIz3Ihe0K/Coal-India-keeps-away-from-fuel-supply-pacts.html (accessed June 13, 2014).

Bhaskar, Utpal (2010, September 29). "Illegal Coal Mining Revenue Going to Naxals." *Livemint*. Available at http://www.livemint.com/Companies/hbYVbMuknQEwoHBYrCwsKJ/Illegal-coal-mining-revenue-going-to-Naxals.html (accessed January 17, 2014).

Bhaskar, Utpal (2009, December 1). "Naxals Put the Squeeze on Transport of Jharkhand Coal." *Livemint*. Available at http://www.livemint.com/Companies/jk8rau3k9IbIOLed5jbTLI/Naxals-put-the-squeeze-on-transport-of-Jharkhand-coal.html (accessed June 27, 2014).

Bhattacharya, Anjan and P. K. Murthy (2006). "India's Coal Situation and Tata's Role." Proceedings of International Pittsburgh Coal, Pennyslvania. Available at http://www.engineering.pitt.edu/Coal_Conference/2006_Plenary_Presentations.aspx (accessed September 14, 2014).

Bhushan, Chandra (2010, March 19). Interview by author.

Bhuta, Jonas and Saurabh Mehta (2011). "Coal India." March 21, 2011.

BP (2010). *BP Statistical Review of World Energy, June 2010*. Available at www.bp.com/statisticalreview (accessed March 4, 2011).

Carl, Jeremy (2011). "Ramesh Was Right." Unpublished manuscript.

Carl, Jeremy (2010). "Captive Mining in India." Proceedings of Coaltrans India 2010, Mumbai, India.

Carl, Jeremy (2006). "India's Hidden Civil War: Consequences for Energy Security." *Journal of Energy Security*. Available at http://www.iags.org/n0731061.htm.

Central Electricity Authority (2012). "Highlights of Power Sector." *Government of India, Ministry of Coal*. Available at http://www.cea.nic.in/reports/monthly/executive_rep/jan12/1–2.pdf (accessed September 5, 2014).

Chandra, Rohit (2011, November 21). "India's Coal-ed Streak." *Center for the Advanced Study of India*. Available at http://casi.sas.upenn.edu/iit/rohitchandra (accessed June 13, 2014).

Chaturvedi, Saurabh and Raymond Zhong (2014). "All Coal-Mining Rights Issued in India since 1993 Ruled Illegal." *The Wall Street Journal*, August 25.

Coal Guru (2010, June 15). "CIL Has Put on Hold Its Initiative to Acquire Foreign Assets." *Coal Guru*. Available at http://coal.steelguru.com/india/506/cil_has_put_on_hold_its_initiative_to_acquire_foreign_assets (accessed September 5, 2014).

Coal India Limited (2010, August 9). *Coal India Limited Prospectus*. Available at http://www.sebi.gov.in/dp/coaldrhp.pdf (accessed June 13, 2014).

Cowhig, Jackie, and Ruchira Singh (2010, March 24). "ANALYSIS – India Surprises with Big New Coal Ports." *Reuters*. Available at http://in.reuters.com/article/2010/03/24/idINIndia-47181020100324?loomia_ow=t0:s0:a49:g43:r1:c1.000000:b32535222:z0 (June 27, 2014).

Dalmia, Shikha (2011, December 13). "India Proudly Sinks the Durban Climate Change Talks." *Reason.com*. Available at http://reason.com/blog/2011/12/13/india-proudly-sinks-the-durban-climate-c (June 27, 2014).

Das, Krishna N (2014, May 21). "Modi Exploring Breakup of Coal India, Opening Up Sector – Sources." *Reuters*. Available at http://in.reuters.com/article/2014/05/21/india-coal-india-idINKBN0E10KB20140521 (accessed September 4, 2014).

Dun & Bradstreet India (2010). "Regulatory and Policy Environment." *Dun & Bradstreet India*. Available at http://www.dnb.co.in/IndiasEnergySector/Regu_Power.asp (accessed June 27, 2014).

Energy Information Administration (2013, May 30). "India's Oil Reserves." *EIA*. Available at http://www.eia.gov/countries/country-data.cfm?fips=IN (accessed October 21, 2010).

EPPO (2014). "Oil Industry Conversions." Available at http://www.eppo.go.th/link_ref.html (accessed September 5, 2014).

Forest Service of India (2013). *India State of Forest Report 2013*. Dehradun, India. Available at http://fsi.nic.in/details.php?pgID=mn_93 (accessed September 4, 2014).

Ians (2008, September 26). "Railways to Levy 5–7 Percent Surcharge on Freight." *Thaindian News*. Available at http://www.thaindian.com/newsportal/business/railways-to-levy-5–7-percent-surcharge-on-freight_100100543.html (accessed June 13, 2014).

IEA (2013). *World Energy Outlook 2013*. Paris: International Energy Agency.

Index Mundi (2014). "India – GDP – Real Growth Rate – Historical Data Graphs per Year." India – GDP – Real Growth Rate – Historical Data Graphs per Year. Available at http://www.indexmundi.com/g/g.aspx?c=in&v=66 (accessed June 13, 2014).

Indian Railways Yearbook (2006–2007). Ministry of Railways, Government of India. Available at http://www.indianrailways.gov.in/railwayboard/view_section.jsp?lang=0&id=0,1,304,366,554,941,947 (accessed March 4, 2015).

IndiaStat (2014). "IndiaStat.Com – India's Comprehensive Statistical Analysis, Data Information & Facts About India." Available at indiastat.com (accessed September 12, 2014).

Infocus (2011, December 7). "Power Distribution Losses Trip SEBs." *Infocus*. Available at http://www.dhanbank.com/pdf/reports/IF-7-Dec-2010.pdf (accessed June 27, 2014).

Institute for Conflict Management (2014). "India: Chronic Conflict Map 2014" (Institute for Conflict Management, South Asia Terrorism Portal). Available at http://satp.org/satporgtp/countries/india/database/conflictmapindia_2014.html (accessed September 4, 2014).

International Energy Agency (2012). *CO_2 Emissions from Fuel Combustion: Highlights 2012*. Paris: International Energy Agency. Available at http://www.indiaenvironmentportal.org.in/files/file/CO2emissionfromfuelcombustionHIGHLIGHTS.pdf, Accessed March 3, 2015

Jha, NC (2011, March 24). "Meeting of the Indo-US coal working group." Presentation in New Delhi. Available at http://coal.nic.in/India's%20coal%20industry%20&%20CIL.pdf (accessed September 4, 2014).

Jog, Sanjay (2010, June 23). "Naxals, Green Issues Hit Coal Projects." *Business Standard*. Available at http://www.business-standard.com/article/markets/naxals-green-issues-hit-coal-projects-110062300059_1.html (accessed June 26, 2014).

Lenka, Pradeep (2010). *Land Acquisition in India*. Proceedings of Coaltrans India 2010, Mumbai, India.

Mail Today Bureau (2012, June 7). "Coal Shortage to Hit Power Generation, Brace up for More Blackouts." *Indiatoday.in*. Available at http://indiatoday.intoday.in/story/coal-shortage-to-hit-power-generation-coal-india-ntpc/1/199551.html (accessed June 06, 2014).

McDonald, Alan and Holger Rogner (2011, October 24). "IAEA Projects Slower Nuclear Growth after Fukushima." *IAEA*. Available at http://www.iaea.org/newscenter/news/2011/nuclgrowth.html (accessed June 27, 2014).

McIntyre, Douglas (2010). "The 10 Cities with the World's Worst Air." *DailyFinance*, November 29. Available at http://www.dailyfinance.com/2010/11/29/10-cities-with-worlds-worst-air/ (accessed March 1, 2012.).

Means, Robin (1999). "Access to Land in Rural India: Policy Issues and Options." Working Paper, South Asia Regional Office, World Bank. Available at http://www-wds.worldbank.org/external/default/WDSContentServer/IW3P/IB/1999/09/14/000094946_99060905321228/Rendered/PDF/multi_page.pdf.

Ministry of Coal (2010) "Government of India, Ministry of Coal." Government of India, Ministry of Coal. Available at http://www.coal.nic.in/ (accessed September 5, 2014).

Ministry of Railways CMS Team (2011, February 19). "Year Book (2006–07)." *Indian Railways*. Available at http://www.indianrailways.gov.in/railwayboard/view_section.jsp?lang=0&id=0,1,304,366,554,941,947 (accessed June 13, 2014).

Naxal Watch (2008, April 25). "Maoists Attack Essar Steel Plant in Chhattisgarh." *Naxal Terror Watch*. Available at http://naxalwatch.blogspot.com/2008/04/maoists-attack-essar-steel-plant-in.html (accessed March 7, 2011).

Nolan, Peter A. and Mark C. Thurber (2012). "On the state's choice of oil company: Risk management and the frontier of the petroleum industry." In D. G. Victor, D. R. Hults, and M. C. Thurber, eds., *Oil and Governance: State-owned Enterprises and the World Energy Supply*. Cambridge: Cambridge University Press.

Power Today (2011, September). "What Ails SEBs?" *Power Today*. Available at http://www.powertoday.in/News.aspx?nId=qvhelsDh8X92zsK2IlhhzA== (accessed June 27, 2014).

Press Information Bureau, Government of India (2011). "Loss of Coal Production in Naxalite Areas." *Press Information Bureau English Releases*. Available at http://pib.nic.in/newsite/erelease.aspx?relid=64116 (accessed March 20, 2011).

Press Trust of India (2012). "Industry Bodies Oppose Coal India's Gross Calorific Value (GCV) Grading System." *The Economic Times*, January 10. Available at http://articles.economictimes.indiatimes.com/2012–01–10/news/30611752_1_coal-prices-coal-controller-coal-india-limited (accessed June 13, 2014).

Press Trust of India (2011a). "Coal Deficit May Widen to 200 Mn Tonne by End of 12th Plan." *Business Standard*, September 14. Available at http://www.business-standard.com/article/economy-policy/coal-deficit-may-widen-to-200-mn-tonne-by-end-of-12th-plan-111091400156_1.html (accessed September 5, 2014).

Press Trust of India (2011b, December 07). "Coal India Lowers Its Production Target to at Least 440 MT – NDTV." *NDTV Profit*. Available at http://profit.ndtv.com/news/corporates/article-coal-india-lowers-its-production-target-to-at-least-440-mt-293633 (accessed June 13, 2014).

Press Trust of India (2011c). "Coal India to Set up 20 New Washeries." *The Hindu Business Line*, February 14. Available at http://www.thehindubusinessline.com/companies/article1454514.ece (accessed June 16, 2014).

Press Trust of India (2010a, June 3). "Arrest Me for Speaking for Naxals: Arundhati." *IBNLive*. Available at http://ibnlive.in.com/news/arrest-me-for-speaking-for-naxals-arundhati/117722–3.html?from=justinhome (accessed April 5, 2011).

Press Trust of India (2010b). "Have No Say on Naxalism; Will Target Mafia: Coal Minister." *Dna*, May 25. Available at http://www.dnaindia.com/india/report-have-no-say-on-naxalism-will-target-mafia-coal-minister-1387636 (accessed March 7, 2011).

Press Trust of India (2010c, October 18). "India to Offer Shale Gas Blocks by End of 2011." *Livemint*. Available at http://www.livemint.com/Money/ppVNJocZscmsWFmA7bS2xJ/India-to-offer-shale-gas-blocks-by-end-of-2011.html (accessed June 27, 2014).

Press Trust of India (2010d). "Naxals Lending Money and Investing in Illegal Coal Mining." *Business Standard*, June 11. Available at http://www.business-standard.com/article/economy-policy/-naxals-lending-money-and-investing-in-illegal-coal-mining-110061100074_1.html (accessed June 27, 2014).

Ranjan Bose, Pratim (2011). "CIL May Soon Price Coal Using Gross Calorific Value Metric." *The Hindu Business Line*, November 7. Available at http://www.thehindubusinessline.com/companies/article2607124.ece (accessed June 13, 2014).

Rodrik, Dani, Arvind Subramanian, and Francesco Trebbi (2002). "Institutions Rule: The Primacy of Institutions Over Geography and Integration in Economic Development." *Journal of Economic Growth* 9(2): 131–165.

Rogers, Kelly (2009, September 21). "History and Relevancy of "Common But Differentiated Responsibilities." From Kyoto to Copenhagen (blog). Available at http://blogs.dickinson.edu/copenhagen/2009/09/history-and-relevancy-of-common-but-differentiated-responsibilities/ (accessed June 26, 2014).

Shanker, S (2011). "Indonesia Set to Tighten Coal Export Norms Further." *The Hindu Business Line*, October 10. Available at http://www.thehindubusinessline.com/companies/article2526105.ece (accessed June 27, 2014).

Singh, Manmohan (2006). Speech, Chief Minister's Meeting on Naxalism, New Delhi, April 13. Available at http://www.hindustantimes.com/news-feed/nm9/naxalism-biggest-challenge-pm/article1–86531.aspx (accessed on September 5, 2014).

Singh, Sundeer Pal (2010). "Jharkhand Refuses Forest Clearance for CIL Rail Link." *Business Standard*, October 6. Available at http://www.business-standard.com/article/companies/jharkhand-refuses-forest-clearance-for-cil-s-rail-link-110100600010_1.html (accessed September 5, 2014).

Spacek, Michael (2010, May 20). "Economics of War." India's Forgotten War – Blogging Naxalism (blog). Available at http://naxalwar.wordpress.com/tag/resources/ (accessed February 7, 2011).

TNN (2012). "Coal Price Revision to Hike Power Cost." *The Times of India*, February 2. Available at http://timesofindia.indiatimes.com/city/kolkata/Coal-price-revision-to-hike-power-cost/articleshow/11721088.cms?referral=PM (accessed June 27, 2014).

Trading Economics (2014). "India GDP Growth Rate." *Trading Economics*. Available at http://www.tradingeconomics.com/india/gdp-growth (accessed August 2010).

Victor, David G. and Thomas C. Heller (2009). *The Political Economy of Power Sector Reform*. Cambridge: Cambridge University Press.

Watts, Jonathan (2010). "India Discloses Carbon Emissions for First Time in More than Decade." *The Guardian*, May 25. Available at http://www.theguardian.com/environment/2010/may/25/india-carbon-emissions (accessed June 26, 2014).

Wolf, Christian and Michael G. Pollitt (2009, March). "The Welfare Implications of Oil Privatisation: A Cost-Benefit Analysis of Norway's Statoil." EPRG Working Paper 0905. Available at http://www.econ.cam.ac.uk/dae/repec/cam/pdf/cwpe0912.pdf (accessed September 8, 2014).

Wolf, Christian and Michael G. Pollitt (2008, March). "Privatising National Oil Companies: Assessing the Impact on Firm Performance." EPRG Working Paper 0805. Available at http://www.eprg.group.cam.ac.uk/wp-content/uploads/2014/01/eprg0805.pdf (accessed September 8, 2014).

World Coal Association (2010). "Coal Statistics." Available at http://www.worldcoal.org/resources/coal-statistics/ (accessed July 7, 2010).

5 Market, investment, and policy challenges for South African coal

Anton Eberhard

5.1 Introduction

South Africa is a key participant in global coal markets. It is not the biggest: China, the United States, and India are much larger producers and consumers of coal; Indonesia, Australia, Russia, and Colombia are larger exporters. Yet South Africa's coal industry is noteworthy in a number of respects: it is a relatively low-cost producer (along with Indonesia and Colombia), has the world's largest coal export terminal, and is positioned conveniently between Atlantic and Pacific coal markets. As a result, it can play an important role as a potential swing producer, able to export competitively to either Europe or Asia.

South Africa has substantial coal reserves that give it scope for expanding its coal exports, thereby generating much needed export earnings and reducing the country's negative trade balance and current account deficit. However, increased coal exports face serious barriers and obstacles, including inadequate rail capacity to the coast. There is a lack of planning and investment coordination between privately-owned mines, state-owned rail infrastructure and ports.

South Africa is also a major consumer of coal, mainly for electricity production but also to feed the world's only commercial coal synfuel (coal-to-liquids, or CTL) plants. It is among the twenty-five most carbon-intensive economies in the world but does not yet face any binding international treaty obligations to reduce its greenhouse gas emissions. Nevertheless, global warming and other environmental concerns are beginning to factor in to coal-based investment decisions in the country, even if they do not halt them. In the Copenhagen Accord, South Africa made a voluntary commitment to reduce its greenhouse gas emissions below a business-as-usual scenario.

This chapter explores the interplay between South Africa's domestic and export markets for thermal coal and what might shape their development in the future. It first examines the industrial organization and political economy of the coal sector in South Africa. An overview is provided of coal mining companies, how the current market structure emerged historically, the development of rail and port facilities, and coal costs and prices. Policy and legislative developments are also described. Finally scenarios are presented for how local and export coal markets might evolve.

The coal sector in South Africa offers both challenges and opportunities. Its case highlights the policy tensions that can be associated with coal production and use. South Africa is a relatively poor country with major energy needs, and yet it also has significant industrial capacity, outsized greenhouse gas emissions, and a sense of concern about climate change. It is a major exporter in world coal markets, and yet it also faces rail and port constraints and at times pressure to reserve its coal for domestic use. Like some of the other cases discussed in this volume – for example, Australia (Chapter 6) and Indonesia (Chapter 7) – the South African example highlights the importance of strong government facilitation in building out the value chain from coal mine through transport to end use. It remains to be seen whether South Africa will develop integrated policy and regulatory frameworks, and more purposeful investment strategies and programs, that will set the country on a sustainable development path while maximizing potential export benefits.

5.2 Overview of the South African coal sector

South Africa's global position in hard coal production has slipped; in 2012 it ranked sixth, behind China, the United States, India, Australia, and Indonesia. Total coal output in 2012 was 258 million tonnes (Mt). South Africa ranked fifth as a hard coal exporter, behind Indonesia, Australia, Russia, and Colombia.[1] Its exports totaled 73.6 Mt in 2012 (IEA, 2013).

South Africa's economically recoverable coal reserves are estimated at between 15 and 67 billion tonnes (Bt); 96% of reserves are bituminous coal, metallurgical coal accounts for approximately 2% and anthracite another 2%.[2] Production is mainly steam coal of bituminous quality. The majority of South Africa's reserves and mines are in the Central Basin, which includes the Witbank, Highveld, and Ermelo coalfields (see Figure 5.1). Coal production in the Central Basin is likely to peak in the next decade. The Waterberg coalfield is the focus of recent exploration efforts and could become a major coal mining center in the future, subject to infrastructure and water constraints. Production in this area will double in the next five years. Other coalfields in the Limpopo Province are also being explored, with a focus on coking coal.

Coal seams are relatively thick and close to the surface, which allows for low-cost mining: a quarter of South Africa's bituminous coal is to be

Table 5.1 *2012 global steam coal production and exports in million tonnes coal equivalent (Mtce)*

Producers	Mtce/annum
China	2,895.1
United States	769.2
India	498.6
Indonesia	439.9
South Africa	258.5
Russia	201.5
Australia	200.3
Kazakhstan	107.6
Columbia	85.5
Poland	67.5
Rest of world	248.3
World	**5,772.0**

Exporters	Mtce/annum
Indonesia	379.8
Australia	159.2
Russia	115.9
Colombia	81.6
South Africa	**73.6**
United States	50.6
Kazakhstan	31.3
Vietnam	19.1
DPR Korea	12.0
China	8.7
Rest of world	30.9
World	**962.7**

Source: IEA (2013).

Figure 5.1 Major coalfields of South Africa. Coal deposits are shown in gray. Data source: Birtles and Soden (2010).

found between 15 and 50 m below the surface and much of the remainder between 50 and 200 m. Half of reserves are in seams that are 4–6 m thick and a further third in 2–4 m seams (Petrick Commission, 1975). Approximately half of production comes from opencast mines, and the balance from underground mines. Ash contents vary, but are high and can range up to 65% in the Waterberg field. (Export grade coals generally require washing so that their ash content does not exceed 15%, or more recently 20% for exports to India.) Heating values of export coals of around 6,200 kcal/kg (26 MJ/kg) net as received were common, but average values are declining and some export coals are now around 5,500 kcal/kg (23 MJ/kg). Sulfur contents are generally between 0.6% and 0.8%. Thermal coals used for domestic power and synfuel production have much lower calorific, and higher ash, values and are supplied mostly from screened run-of-mine production (although about a third of the coal supplied for electricity production derives from the middlings fraction from coal washing).

A further by-product of coal washing or beneficiation is large quantities of discard or "duff" coal, which now amount to more than 1 BT – a total that is being added to at the rate of about 60 million tonnes per annum (Mtpa). This presents both an environmental challenge as well as a potential resource for power production using fluidized bed combustors, although, as Lloyd (2000) notes, the coal is referred to as discard for a reason: it is of poor quality (high ash and sulfur), and its volatiles diminish further as stockpiles age.

A large proportion of the steam coal for export and domestic markets is produced by eight mega-mines, with a production of more than 10 Mtpa each. Seven of these mines are in the Central Basin. The remaining large mine is in the Waterberg. Five companies account for over 70% of coal production in South Africa: Anglo American, Exxaro, Sasol, BHP Billiton, and Glencore Xstrata.

Just over two-thirds (by mass) of domestic coal consumption is for electricity generation by Eskom, the national power utility. Coal-to-liquid-fuel (CTL) plants, operated by Sasol, account for just under a quarter of coal consumption. Small merchants, who supply mainly residential users, and small businesses account for about 2%, metallurgical industries about 3%, and cement, chemical, and other industries the remaining 5% (DMR, 2009).

Coal thus plays a vital role in South Africa's energy economy: it accounts for 70% of primary energy consumption, 93% of electricity generation and 30% of petroleum liquid fuels. However, coal production and use also result in a number of serious environmental impacts. While greenhouse gas emissions and climate change loom large in the

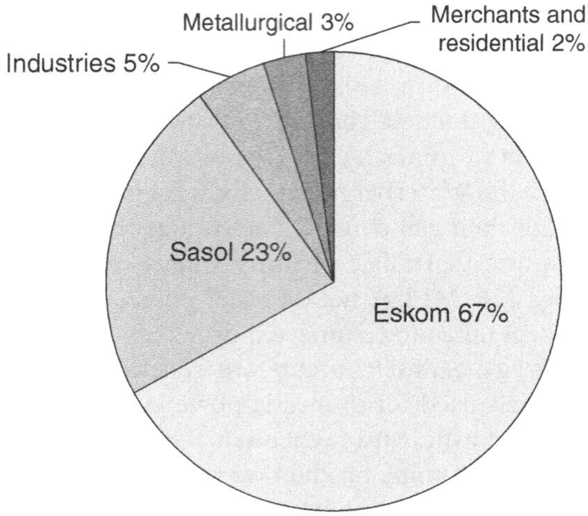

Figure 5.2 Coal use in South Africa (excludes exports). Source: Prevost (2010).

future, a number of other coal mining related environmental problems are emerging. Perhaps the most serious immediate environmental problem is that of acid mine drainage. Mining breaks up the rock mass, allowing free access of water, which reacts with iron sulphide naturally present in the coal and its host rocks to form sulfuric acid. (Different mining techniques create different exposure to water, with the opencast mining that is common in South Africa resulting in the most acid production.[3]) Acidic water dissolves aluminum and heavy metals, including iron and manganese, creating runoff that is toxic to animal and most plant life. Mine drainage also acidifies soil and rivers. Destruction of biodiversity, reduced agricultural production, soil erosion, and serious water pollution are now evident in the Olifants River and are beginning to emerge in the Vaal River catchment. Environmental studies indicate that further large scale development of coalfields, especially in the central basin, could result in irreversible and costly environmental damage and could prejudice the quality of scarce and essential water resources in South Africa's most productive economic region (McCarthy and Pretorius 2009).

Acid mine drainage is less of a problem in the Waterberg because of low rainfall and consequently lesser water ingress; in fact, the opposite problem arises – one of water shortage. Insufficient water could be a significant environmental constraint limiting large-scale development of the Waterberg coalfields.

5.3 Historical development of the sector

The current market structure of South Africa's coal industry, with its five dominant production companies, has its roots in the historical development of mining and industrial policy over the past 120 years. Coal, and later coal-fired power stations, provided the energy for diamond and gold mining and associated industry and infrastructure, including the railways. For the first half of the twentieth century, nearly all coal mines were owned by gold-mining houses.

Anglo American, for many years South Africa's largest company, has its origins in gold mining and remains a major coal producer. Another multinational company currently involved in the coal sector in South Africa, Glencore Xstrata, has its origins not only in Swiss capital (Glencore) but also in Lonrho's mining interests in Southern Africa (Duiker Mining).

The history of the third major current coal company, BHP Billiton, is partially intertwined with General Mining, or Gencor (which absorbed TransNatal collieries and Rand Mines). After the Afrikaner nationalists came into power in 1948, there was a conscious attempt to bolster Afrikaner capital as a counterweight to English "colonial" mining capital. Gencor was one of the products of this initiative. In 1994, Gencor absorbed Billiton's mining assets from Billiton's then-owner Royal Dutch/Shell Group. Gencor then divested all of its base metals, minerals, and coal mining operations three years later as Billiton Plc, which in 2001 merged with BHP.

The other current South African coal major, Exxaro, also owes its origins to overt empowerment politics, but of a different kind. After the democratic revolution of 1994, policies were initiated to promote black economic ownership. Exxaro came about from a merger of Eyesizwe and certain Iscor/Kumba interests. The former was created from divested coal assets from Anglo and BHP, and the latter from the coal division of Iscor, the national iron and steel company, which was originally established by the old nationalist government.

A series of cartels initially controlled the industry, including the Transvaal Coal Owners Association (TCOA), which was formed in 1908. The purpose of the TCOA, in the words of the President of the Transvaal Chamber of Mines, was "to put an end to what was considered in some quarters a ruinous competition" (Marquard 2007: 79). The TCOA, along with the Natal Associated Collieries (NAC), was highly influential in helping to shape state policy for the coal industry in its early period.

Until the 1940s, between a sixth and a third of annual coal tonnage was exported. However, doubts as to the extent of reserves and the imperative

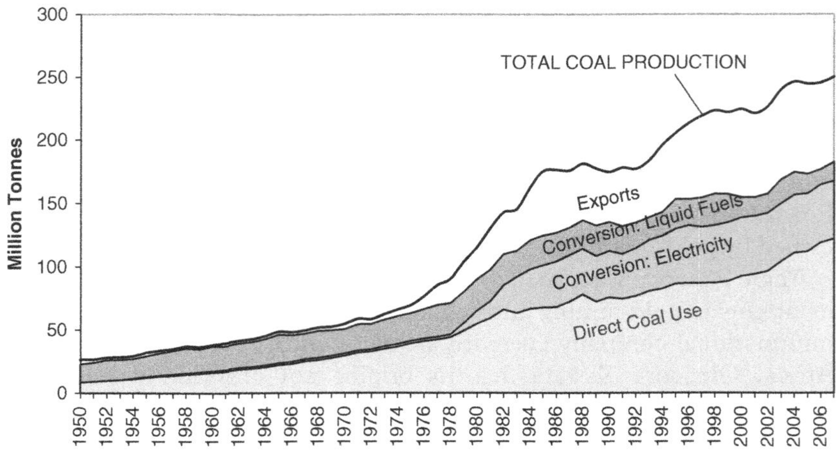

Figure 5.3 South African coal production, consumption, and exports: 1950–2007. Source: Marquard (2007).

of supplying local demand led the state to impose restrictions on exports, which decreased as a result to around 2% of production between 1950 and 1970. During this period, the state also imposed price controls on the domestic market with the aim of promoting industrialization through cheap energy inputs. The regulated price was set using a rate-of-return methodology which did not recognize depreciation costs, resulting in coal prices that were among the lowest in the world (Marquard 2007).

The consequences of this pricing policy were low profitability, limited investment, and inefficient mining techniques. Extraction rates and levels of mechanization were low, the wasteful bord and pillar mining method was common, and more than half of the coal resource was left in the ground. Low wages and low domestic prices provided few incentives for investment and technology development.

The 1970s saw a new era in the development of South Africa's coal industry, with expanded investments in electricity generation, synfuels, and coal exports. These investments were spurred by higher labor costs, following increased unionization, and changes in government policy that promoted mechanization, domestic energy security, and increased exports.

The replacement of hand loading with mechanization facilitated the introduction of open-cast mining. The first mechanized dragline was introduced in Gencor's Optimum mine in 1970. Labor productivity and coal extraction rates increased substantially. "Continuous miners" and long-walling were introduced later. Coal output more than doubled between 1970 and 1980, and quadrupled by the end of the century.

The expansion of South African coal exports was catalyzed by a contract signed in 1971 between the TCOA and seven Japanese steel mills for 27 Mt of blended coking coal over a ten-year period. Exports started at 100,000 tonnes per annum in late 1972 and increased to 2.7 Mtpa from 1976 to 1986.

Expanded exports were also facilitated by improvements in coal washing technology. Engineers at Anglo American developed a two-stage washing process that economically produced both a low-ash coking coal and large quantities of high-grade steam coal. Not all of the resulting high-quality coal could be absorbed on the domestic market.

The 1973–74 oil crisis created an opening for South African coal exports into Europe (see Chapter 1). Given the higher prices and profits in coal exports, industry began to lobby for increased export allocations, which were subsequently granted. However, existing rail and port facilities at the time were a major bottleneck. A condition of the Japanese coking coal contract was the development of a new port that could handle larger bulk carriers. Agreement was reached between the TCOA and the government to expand rail capacity and to build a new port at Richards Bay. Both were planned for greater capacities than the Japanese coking coal contract. The TCOA exporters were required to guarantee the financing of the railway line and ensure that its capacity would be adequately utilized. Shareholders also financed, built, and continued to manage the coal handling facilities at Richards Bay, which was opened in 1976 with a capacity of 12 Mtpa. Extensive cooperation between the state and key industry players made this development possible. Exports expanded significantly as step changes were made in rail and port capacity (Figure 5.4). Export permits were eventually abolished in 1991.

In the face of controlled domestic prices, the potential value of exports was a critical driver of the development of the South African coal mining industry. It motivated a push for more efficient extraction rates and opened up the possibility of cross-subsidizing delivery of cheaper, lower grade coal to the domestic market. (Following years of intensive lobbying by the coal industry, domestic wholesale and retail coal prices were progressively increased and finally deregulated in 1986; pithead prices were deregulated in 1987.)

The domestic coal market grew substantially as the national electricity utility, Eskom, embarked on an ambitious capacity expansion program in the 1970s and 1980s. At the same time, Sasol invested in new CTL, or synthetic fuel, plants.

The end of apartheid in 1994 accelerated the internationalization of South Africa's coal industry. Collieries that had previously functioned as divisions within gold mining houses were restructured into separate

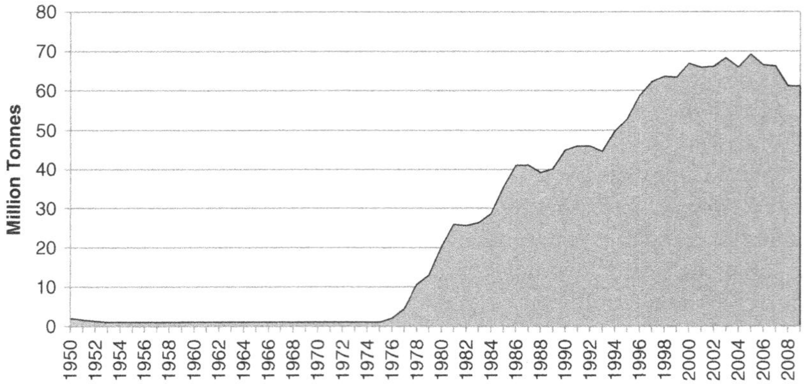

Figure 5.4 South African coal exports: 1950–2008. Source: Marquard (2007).

subsidiaries with their own local and overseas marketing operations. As global commodity markets expanded, there were clear strategic and efficiency advantages for multinational companies to create and manage global coal (and other commodity) businesses that stretched across countries, rather than country specific multi-product businesses.

The most recent regulatory development of significant importance in the coal mining industry in South Africa has been the advent of black economic empowerment deals which have seen not only the creation of a new coal major, Exxaro, but also of a host of smaller black-owned companies which are competing for share in domestic and export coal markets. Together, black-owned coal companies now control more than 30% of South African coal production (Prevost, 2010).

5.4 Industrial organization and economics

All coal mining in South Africa is undertaken by private companies. The main coal port at Richards Bay is also controlled by private shareholders. However, the railways from the mines to the ports are controlled by a state-owned monopoly, Transnet.

As we have seen, local coal use is dominated by Eskom, the state-owned power company, and by Sasol, the large synthetic fuels company which was started by the state but was later privatized. Sasol owns its own coal mines, while Eskom relies mainly on long-term contracts with adjacent private mines owned by the four coal majors. Eskom's favorable supply arrangements with dedicated coal mines are increasingly under threat as mines divert higher quality coals for export and as higher electricity

Table 5.2 *Coal production in South Africa, 2009*

Company	Mtpa
Anglo Coal	57
Exxaro	40
Sasol	40
BHP Billiton	33
Xstrata	19
Other	69

Source: Company annual reports.

demand (and the associated requirement for coal) has forced it to turn to supplementary short-term contracts – not only with the coal majors but also with a variety of small mining companies.

5.4.1 Coal producers

Anglo American, Exxaro, Sasol, BHP Billiton, and Glencore Xstrata account for over 70% of coal production in South Africa. We review the characteristics of their production and exports in this section. The remaining production is undertaken by a plethora of smaller private companies, many of them new entrants that have benefited from black-empowerment policies.

5.4.1.1 Anglo American

Anglo American is South Africa's largest coal producer. It is also one of the world's largest diversified mining groups. Its Anglo Coal Division has operations in South Africa, Australia, South America, and Canada. It owns and operates eight mines in South Africa. In 2012, Anglo Coal produced 57.1 Mt, of which 40 Mt went to the domestic thermal market (mostly Eskom but also 5 Mt to Sasol). It has an export entitlement of 19.8 Mtpa through the Richards Bay Coal Terminal, although actual exports have been slightly lower at around 17 Mtpa.

5.4.1.2 Exxaro

Exxaro is a South African-based, majority black-owned mining group with a coal production capacity in 2012 of around 40 Mta which was mostly for Eskom with about 3 Mt for export markets (it has an export entitlement at Richards Bay of 6.3 Mtpa). In addition to its mines in the Witbank field, Exxaro is currently the one large operator in the Waterberg. Its Grootgeluk mine has the world's largest coal beneficiation complex, where 7600 tonnes per hour of run-of-mine coal is upgraded in six different plants. A 14.6 Mt expansion of the Grootgeluk mine will supply the new Eskom 4,800 MW mega coal power station from 2014.

5.4.1.3 Sasol

Sasol Mining (Pty) Ltd is South Africa's third largest coal producer, with production capacity of 43–46 Mtpa, although total production has been lower – about 40 Mt in 2012. The vast majority of Sasol Mining's coal production is from its mines in the Highveld coalfield which supply its Secunda CTL plants. It exports up to 3.6 Mtpa.

5.4.1.4 BHP Billiton

BHP Billiton is the world's largest diversified resources group and a major producer and marketer of thermal coal for export. It has coal mining operations in New Mexico in the United States, Australia, and South Africa. It also has coal-exporting interests in Colombia and Indonesia. BHP Billiton Energy Coal South Africa (BECSA) is its largest coal operation, which produced 33.3 Mt in 2012. BHP Billiton used to be one of the largest coal producers in South Africa but now ranks fourth after Anglo, Exxaro, and Sasol. It retains a 17.95 Mtpa export entitlement.

5.4.1.5 Glencore Xstrata

Glencore Xstrata Coal is the world's largest producer and exporter of thermal coal and has operations in Australia, South Africa, Colombia, and Canada. Glencore Xstrata's South African coal production in 2012 was 18.7 Mt, of which 13.1 Mt were exports. Glencore Xstrata Coal South Africa holds a 20.9% interest in the Richards Bay Coal Terminal Company Ltd (RBCT).

5.4.2 Major domestic coal consumers

In this section we explore in more detail the structure and dynamics of local coal demand, primarily for Eskom's power stations and Sasol's CTL plants.

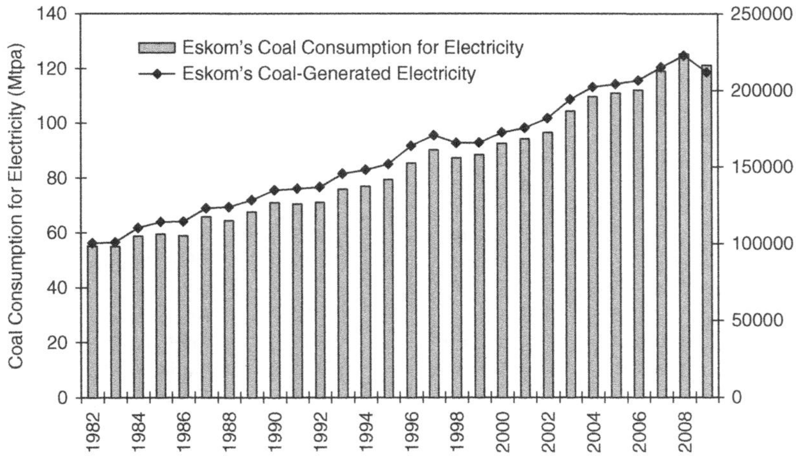

Figure 5.5 Eskom's coal-generated electricity in South Africa (GWh, right axis) and coal burned to generate this electricity (Mtpa, left axis). Source: Eskom annual reports.

5.4.2.1 Eskom

5.4.2.1.1 Eskom's coal-fired power stations and coal contracts South Africa's power industry is dominated by the state-owned national utility Eskom, which generates 96% of the country's electricity. Its electricity prices are regulated by the National Energy Regulator of South Africa. In 2012, Eskom's total generating capacity was 41.6 GW, of which 35.7 GW were coal-fired power stations. In 2012, Eskom generated 238 TWh of electricity, of which 218 TWh was produced by coal-fired stations. It burnt 125 Mt of coal, amounting to about 67% of the total used in South Africa (Eskom, 2012).

Eskom's coal-fired power stations use conventional pulverised coal technology, with average thermal efficiencies of 32%. Coal quality is poor with average calorific values of 4,500 kcal/kg (19 MJ/kg), ash content of 29%, and sulfur content of 0.8% (Eskom, 2012 Annual Report). Coal quality has been deteriorating in recent years as coal suppliers reserve higher grades for more lucrative export markets. Electrostatic precipitators are employed to reduce particulate emissions, but none of the power stations have flue-gas desulfurization. Eskom currently emits 232 Mt of CO_2 per annum.

Eskom's coal-fired power stations are mine-mouth, mostly supplied by conveyor belts. Nine of these stations have long-term coal contracts: six of these are "cost-plus" and three are "fixed price." In the cost-plus

Table 5.3 *Eskom's coal-fired power stations in 2012*

Power station	Location	MW	Principal coal supplier
Arnot	Middelburg	2,040	Exxaro
Camden	Ermelo	1,440	(various)
Duvha	Witbank	3,450	BHP Billiton
Grootvlei	Balfour	1,180	(various)
Hendrina	Hendrina	1,895	Optimum
Kendal	Witbank	3,840	BHP Billiton
Komati	Middelburg	1,000	(various)
Kriel	Bethal	2,850	Anglo Coal
Lethabo	Sasolburg	3,558	Anglo Coal
Majuba	Volksrust	3,843	(various)
Matimba	Lephalale	3,690	Exxaro
Matla	Bethal	3,450	Exxaro
Tutuka	Standerton	3,510	Anglo Coal

Source: Eskom Annual Report, 2012.

contracts, Eskom and the coal supplier jointly provided capital for the establishment of the colliery. Eskom pays all the costs of operation of the colliery and the supplier is paid a net income by Eskom on the basis of a return on the capital invested (ROI) by the coal supplier in the colliery. The ROI is divided into two components. The fixed portion is a set ROI, payable irrespective of coal production, and the variable portion is based upon tonnages supplied to Eskom. The ROI is generally escalated for half of the duration of the contract and is typically between 15% and 25%. In the fixed price contract, coal is supplied at a predetermined price, that is, a base price which is escalated by means of an agreed formula. There are no early termination provisions in the contracts. Cost-plus contracts have no real incentives for cost minimisation. Future contracts are all likely to be fixed (indexed) priced.

In the past, nearly all of Eskom's coal was supplied on long-term contracts. However, Eskom is now also exposed to short- or medium-term contracts on about 30% of its coal supply. Small coal contractors supplying via road transport are an increasing phenomenon. In part this is due to Eskom's power stations being run at capacity factors higher than those envisaged in the long-term supply contracts. A further driver has been Eskom's commitment to increase its procurement expenditure with black-owned firms as part of government's overall black economic empowerment drive. The growth in short-term contracting was also spurred by the fact that the coal mines that were originally developed for two of Eskom's power stations, Majuba and Tutuka, were unable to meet production expectations, so these are now also supplied mainly by road and, in the case of Majuba, some rail. Increased use of road transport has seriously impacted road infrastructure, and Eskom has had to invest directly in road repairs. A feasibility study for a larger capacity railway line from Ermelo to Amersfoort for the Majuba power station was undertaken in the 1990s but was never acted upon; similarly, a railway connection to Tutuka power station was never built. The return to service of Eskom's old Camden, Grootvlei, and Komati power stations will mean even more road haulage of coal. In 2012, 23% of Eskom's coal was delivered by road, 7% by rail, and 70% by conveyor belts.

The average price Eskom pays for coal has increased quite substantially as its reliance on short-term contracts has increased. Nevertheless, Eskom's average coal costs are still very low compared to international prices. Eskom's long-term coal contract price is an average of R135 (US$13.5) per tonne compared with about R400 (US$40) per tonne for short-term suppliers. Its average coal cost in 2012 was R270 (US$27) per tonne and is expected to increase to an average of R438 (US$44) by 2017 (Eskom, 2012).

5.4.2.1.2 Eskom's performance In recent years, Eskom has not been able to meet fully the demand for electricity in South Africa, and there have been a number of blackouts. In the period 1999–2004, the government was considering unbundling Eskom and introducing a competitive electricity market: private investment in generation was encouraged, and Eskom was prohibited from building new power stations. The reforms were part of a wider effort to improve efficiency in state-owned enterprises and boost the overall competitiveness of the economy. However, by 2004 the design of the new market had not yet been approved – partly because of insufficient political buy-in and union resistance and partly because of tacit resistance by Eskom. Nor had contracting arrangements for independent power producers (IPPs) been put in place. With no new private investment, the government backtracked on its market reforms and once again charged Eskom with the responsibility of generation expansion planning and new investments, although it is still government policy that IPPs may provide up to 30% of electricity generation.

One consequence of this policy detour was that Eskom fell four years behind in terms of power station investments. Growing demand soon outstripped supply and blackouts began to be experienced from 2006, first in the Western Cape and later nationally. The situation was exacerbated by poor operating performance of Eskom's existing plants and the run-down of its coal stockpiles, in part because of higher burn rates, unseasonal rain, and supply problems but also because Eskom's financial managers were pushing for lower inventory costs. In January 2008, Eskom load-shed large energy users such as mines and minerals beneficiation industries in order to avert a total system collapse. Although these users resumed operations a week later, they were subjected to electricity consumption quotas which lasted until late 2008, when demand fell away as the global recession impacted the South African economy.

A memo prepared by Wingfield consultant Susan Olsen for Eskom, and subsequently leaked to the press, paints a damning picture of Eskom's coal contracting in the period leading up to the blackouts. She questions why Eskom failed to secure long-term contracts for the Camden and Majuba power stations in the seven years after Ingwe was released from its obligations. She questions further why existing long-term contracts at the other power stations were not renegotiated to meet higher burn requirements and criticizes Eskom's failure to enforce even minimum performance in the existing contracts. As export prices escalated, some of the tied collieries (such as Optimum, Grootgeluk, and Douglas/Middleburg) focussed on meeting export qualities and tonnages at the expense of quality and tonnage for Eskom. Apart from Majuba, coal supply problems have been most acute at Tutuka's new Denmark colliery, which is incapable of producing at its contractual level. Olsen concluded: "[Eskom's] Generation Primary Energy (GPE) division as it is structured and managed cannot meet current needs much less future requirements. GPE lacks leadership, knowledge, and direction. It is haemorrhaging staff and is left with those who have delivered it to its current condition" (Olsen, 2007: 4).

Eskom's CEO failed to act on these warnings and the 2008 blackouts ensued, caused in the end by a lack of sufficient coal for Eskom's power stations. Ultimately, Eskom's CEO was sacked, and a new executive team is seeking to restore supply security. Coal stockpiles were rebuilt to an average of more than forty-five days. The level of coal stockpiles differs from power station to power station: Matimba, which is supplied reliably by an adjacent mine, has a stockpile of around twenty days; the more vulnerable Majuba station has a stockpile of about seventy-five days.

However, supply and quality problems remain. Only three of Eskom's power stations – Kriel, Matla, and Matimba – are currently supplied entirely by conveyor systems from adjacent mines. Conveyor supplies to the Lethabo, Hendrina, Kendal, Duvha, Tutuka, and Arnot power stations have to be supplemented by road deliveries of coal. The Camden power station is entirely supplied by road and, as previously mentioned, Majuba is supplied by both road and rail. The World Bank loan to Eskom, announced in 2010, includes funding for the development of a heavy haul coal line to Majuba. Deteriorating coal quality at Duvha, Matla, and

Tututka, in particular, have led to these power stations operating well below their rated capacity.

5.4.2.1.3 New investment The electricity supply/demand situation remains tight, and Eskom is now playing catch-up. It has embarked on a massive investment program which will see its capacity increased by about 12 GW over the next seven years. The planned capex program over the next five years amounts to close to R400 billion (US$40 billion).

Much of Eskom's new generation capacity will be coal-fired. It has brought back into service three old coal stations (Camden, Grootvlei, and Komati) and at the time of writing was constructing two new coal fired power stations: the 4,764 MW Medupi plant in the Waterberg (to be supplied by Exxaro) and the 4,800 MW Kusile plant in the Witbank coalfield (to be supplied mainly by Anglo Coal's New Largo colliery). These new power stations will use supercritical technology. Medupi will not initially have flue-gas desulfurization (FGD), while Kusile, located in the more densely settled Mpumalanga Province, will have FGD fitted upfront. Both will evidently be "carbon-capture ready" – i.e. they could be retrofitted with this technology at a later stage.

Investment decisions in new base-load capacity have not yet been made beyond the Kusile plant. Earlier plans envisaged the construction of two further mega coal power stations by Eskom but, as we shall see later, environmental concerns now make this unlikely. Nuclear and gas options are being explored and a new renewable energy independent power producers (IPP) programme is resulting in a number of wind and solar projects coming online. In addition to Eskom's coal power stations, it is possible that some coal-fired Independent Power Producers could enter the market. In November 2012, the Minister of Energy indicated that 2.5 GW of coal power would be procured from IPPs in the period to 2025.

5.4.2.2 Sasol

Sasol operates the only commercial coal-to-liquids fuel production facilities in the world. Its Secunda plant produces around 160,000 barrels per day of petroleum (just over a quarter of South African consumption) as well as a range of petrochemical products. It is also one of the largest single sources of carbon emissions in the world. The facility consumes around 44 Mtpa of high-ash (35%) and low-calorific value (less than 5,000 kcal/kg) coal, amounting to about a fifth of domestic coal use. Sasol's greenhouse gas emissions total 72.7 Mt CO_2 equivalent per annum.[4] While synfuel production at Sasol's existing Secunda plant will increase by up to 20% by 2015, the expansion will not result in much incremental coal production; additional feedstock will be natural gas from Mozambique and coal fines from stockpiles.

Sasol is currently undertaking a pre-feasibility study for a new US$5–7 billion, 80,000 barrels per day CTL plant called Mafutha in the Waterberg coalfield. It has signed a joint venture agreement with Exxaro to explore the feasibility of a coal mine that would supply 25 Mtpa to the new Sasol plant. However, government departments are wary, both of the tax and fiscal support measures being sought by Sasol and of the impact Mafutha will have on South Africa's greenhouse gas emissions.

As previously mentioned, Sasol's coal is supplied almost entirely by its subsidiary Sasol Mining from adjacent mines, through an internal transfer price which is not publicly disclosed. According to Sasol's 2008 annual report, Sasol Mining earned R5 billion from "inter-segment" turnover, which would imply a transfer price for coal of less than R120/tonne (US$12/tonne). DMR data for 2008 indicates an average price of R127/tonne for coal for synfuels production. Data derived from Sasol's 2009 Annual report indicate that internal coal prices for its synfuel operations for that year were also less than R150/tonne (US$15/tonne). (Sasol Mining also exports small amounts of high value coal, which contributes about a third of its coal revenue.)

5.4.2.3 Industry and residential coal consumption

Small merchants trade around 2% of domestic coal consumption, mostly to small businesses and households. About 1 Mt is sold into a retail market for domestic space heating during the cold winters of the interior. This application generates high levels of particulate air pollution. Larger industries, including cement and chemicals, consume a further 8 Mtpa. Finally, metallurgical coal accounts for about 5.4 Mtpa.

5.4.3 Coal rail infrastructure

A major current constraint in moving coal to markets, particularly export markets, is South Africa's aging and inefficient rail infrastructure. Nearly all of South Africa's export coal is transported via rail from the central coal basin to Richards Bay on the East Coast. The state-owned rail monopoly, Transnet, owns and operates a dedicated coal track from Witbank to the sea, a distance of 580 km. The double line is bi-directionally signaled and fully electrified. Two 100-wagon trains are coupled to form one 200-wagon train at Ermelo, in Mpumalanga, typically using CCL-type wagons. There are approximately twenty-eight 200-wagon set trains in operation on the coal line on any given day.

A ten-year contract between Transnet Freight Rail (formerly called Spoornet) and the coal mining companies came to an end in 2004/5, and new fixed contracts are being negotiated. Some of the pricing elements in the contracts are the distance to the Richards Bay Coal Terminal and the time taken to load wagons. This means that if a coal-mining company invested in efficient loading facilities and could load wagons quicker, a lower tariff would be paid. The average time to load a 100-wagon train is about 4.6 hours. With efficient loading facilities, such as those found at BHP Billiton's Douglas colliery in Mpumalanga, a 100-wagon train could be loaded in 2.3 hours. Smaller coal miners generally have longer loading times.

The Waterberg field is 1,050 km from the coast and does not yet have a dedicated coal rail link to Richards Bay. Currently, small amounts of coal (less than 1 Mtpa) are transported by rail from the Waterberg via the existing Transnet Freight Rail network.

There is an apparent lack of planning and investment coordination between Transnet, coal producers, and ports. While port capacity at Richards Bay has increased, coal exports actually decreased for a while, before picking up slightly. Rail capacity would have needed to increase to 76 Mtpa by 2008 and 91 Mtpa by 2010 to match expansions at the Richards Bay Coal Terminal; however, rail capacity in these years remained less than 68 Mtpa, and actual performance was even lower, with some costly derailments. Shortages of locomotives, wagons, and skills are also cited as problems. Transnet counters that coal train schedules have outpaced the availability of coal from the mines. Only 61 Mt were railed to Richards Bay in 2008, and a similar amount in 2009. Since 2010 there have been improvements and 68.5 Mt were railed in 2012.

Transnet's tardiness in responding to expected growth in coal markets (especially for export) needs to be seen within the context of its overall business. Top-line income from Transnet's rail freight business (of which its dedicated export coal line to Richards Bay contributes a quarter) accounts for more than half of its total annual revenue. But profit margins from rail are small. Transnet's other businesses, namely its ownership and operation of South Africa's ports (other than Richards Bay) and petroleum pipelines, account for the lion's share of its profits. Historically, its general rail freight business has generated insufficient income for reinvestment in much needed infrastructure upgrades. For example, in 2009 rail contributed R241 million to its profits, pipelines R590 million, and ports R5262 million (Transnet, 2010). Transnet is probably the only integrated rail-ports-pipeline state-owned monopoly in the world. Its port and pipeline operations currently subsidize its rail operations. This situation may not persist as pressure builds for more effective regulation of the rail sector and also for port and pipeline operations to provide more competitively priced and better quality services to customers.

Transnet is undertaking feasibility studies on expanding the coal link to Richards Bay, initially to 81 Mtpa by 2014/5 and thereafter to 91 Mtpa. It estimates that about R40 billion (US$4 billion) would need to be spent over a decade to accomplish this, including investment in additional rolling stock, new bridges and tunnels, and more double-track. Options are also being considered for strengthening the railway line via Swaziland to Richards Bay. Transnet appears unwilling to embark on this upgrade without long-term guaranteed usage agreements with coal exporters.

While Transnet is wholly government-owned, with a Board appointed by the Minister of Public Enterprises, it is fully corporatized and subject to the Companies Act. In the absence of strong policy directives from either the Minister of Public Enterprises or the Minister of Transport, the Board and management, in the past, have appeared to be more focused on turning around Transnet Freight Rail's poor financial performance than on supporting South Africa's long-term strategic investment priorities. Recently there is evidence of a more ambitious investment strategy to support the National Development Plan.

There remains a high degree of dissatisfaction by private coal miners and exporters with the rail bottlenecks. Derailments are not uncommon, blocking the line and reducing export potential. Alternative business models have been proposed by coal mining companies, including possible private ownership and operation of rolling stock. However, these options are dependent on the government agreeing to a new policy on public-private partnerships in the rail sector and providing more purposeful directives to the Board of Transnet.

5.4.4 Coal ports and exports

As mentioned previously, the Richards Bay Coal Terminal was opened in 1976 with an export capacity of 12 Mtpa. A phase 2 expansion doubled capacity to 24 Mtpa in 1979. Phase 3 in 1984 increased capacity to 44 Mtpa. Further upgrades increased capacity to 63 Mtpa in 1991, 72 Mtpa in 1999, and 76 Mtpa in 2008. The phase 5 expansion in 2010, at a cost of R1.2 billion (US$120 million), has taken capacity to 91 Mtpa. Feasibility studies are being done to increase RBCT's capacity to 105 Mtpa in five years, although this next phase is likely to be delayed until rail capacity catches up. RBCT is the largest single coal terminal in the world.

Coal exports through Richards Bay declined from 2005 to 2009, mainly because of the rail constraints mentioned earlier. Recent years

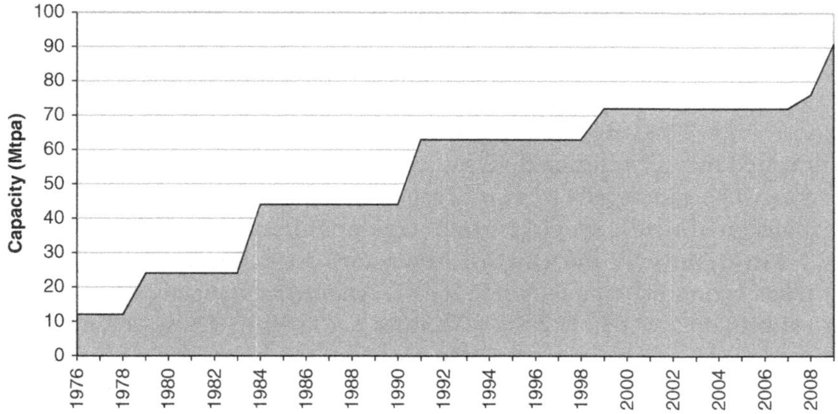

Figure 5.6 Richards Bay Coal Terminal capacity. Source: RBCT website and presentations.

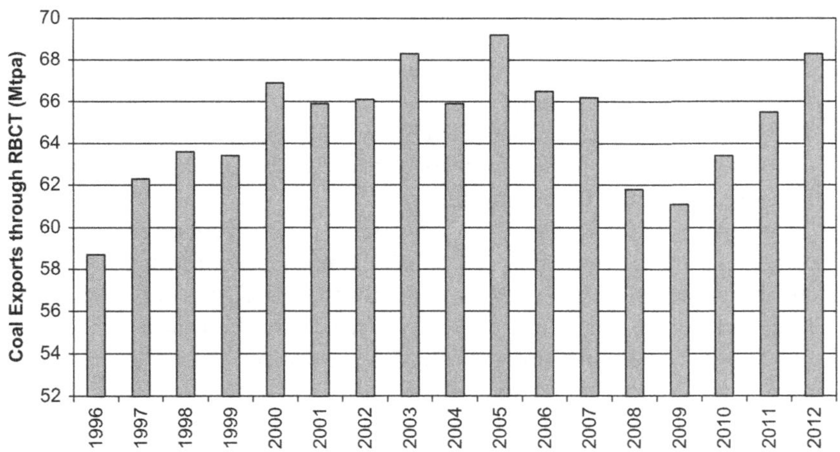

Figure 5.7 Coal exports through Richards Bay Coal Terminal (Mtpa). Source: RBCT website and presentations.

have seen volumes pick up, reaching 68 Mt in 2012. About 25% of the port's export capacity remains unused. South Africa's global ranking as a coal exporter has declined. In 2000 it was second behind Australia, but by 2002, Indonesia had overtaken South Africa as well. Russia surpassed South Africa in exports by 2004, with Colombia doing the same by 2008.

Until recently, Anglo, BHP, and Xstrata were the major shareholders in RBCT, with export allocations of 19.8, 17.95, and 15.05 Mtpa, respectively, accounting for three-quarters of the terminal's capacity. The balance of the export allocations were held by Exxaro, Optimum, Total, Sasol, Kangra, and Koornfontein.

RBCT has been proactive in expanding the terminal despite the rail constraints. The RBCT shareholders commissioned a market survey that indicated that an additional 40 Mtpa of coal was potentially available for export. The additional 19 Mtpa of capacity that became available with the expansion of RBCT from 72 to 91 Mtpa was allocated as follows:

- 4 Mtpa (through the Quattro expansion) was allocated to emerging black economic empowerment (BEE) exporters, typically with annual export volumes below 250,000 tonnes per annum. These volumes are made available on a commercial basis and do not come with entitlements to shareholdings in RBCT. The Department of Minerals and Energy–led Coal Industry Task Team (CITT) allocated the Quattro export entitlement to eighteen BEE emerging companies for a minimum three-year period to facilitate mine development plans, structured financing, and the development of long-term relationships with international customers. At the end of each year past performance is reviewed and the next three-year period considered. Umhlathuze Coal coordinates and facilitates the exports of the companies involved (Quattro Common Users).
- 6 Mpta to the South Dunes Coal Terminal Consortium which represents a number of BEE exporters and includes Kumba (now part of Exxaro), Eskom, and Goland Coal.
- The remaining 9 Mtpa was made available through a tender process to larger exporters (in excess of 250,000 tonnes per annum), including those exporters who wanted a shareholding in RBCT or could sign long-term contracts. The tender was oversubscribed, and allocations were made by ranking bidders according to several criteria. BEE exporters were given priority, and bidders also had to demonstrate export quality coal reserves and capacity support from Transnet Freight Rail.

Black-owned coal companies now have about a third of export allocations from RBCT, more or less in proportion to their overall contribution to coal production. In 2004, less than 1% of the terminal was owned by black-controlled companies.

Rail constraints mean that it will be some time before the full 91 Mtpa export allocation can be transported to the port. RBCT has to decide how the export allocations will be realized over the next few years while rail constraints persist. Export allocations may be cut proportionally according to shareholding and Transnet's rail capacity, with the possibility of individual shareholders being able to expand their exports if others are not able to fulfil their allocations.

Small amounts of coal are also exported via the Richards Bay dry bulk and multi-purpose terminals (3 Mtpa) as well as via Durban (0.9 Mtpa) and Maputo's Matola port (4 Mtpa). Due to environmental constraints, coal exports via Durban may have to be shifted to Richards Bay. Grindrod Freight Services, which operates the Maputo Port concession as well as the coal terminal sub-concession, is exploring plans to expand capacity to between 16 and 25 Mtpa. However, these plans will only be meaningful if rail capacity is expanded accordingly. In the meantime 23 Mtpa of export capacity remains unused at Richards Bay.

5.4.5 Coal export destinations

South Africa's coal exports go to Europe (including Belgium, Denmark, France, Germany, Italy, Netherlands, Portugal, Spain, Switzerland, and the UK), Asia (mainly India but also China, South Korea, Malaysia, Pakistan, and Taiwan), the Middle East (Israel, Turkey, United Arab Emirates), Africa (mainly Morocco but also Kenya, Mauritius, Reunion, and Senegal) and in small amounts to Latin America (Argentina, Brazil, Chile, and Mexico).

McCloskey coal consultancy forecasts indicate that demand for coal imports in Europe will slowly decline but that import demand from India and China and the Pacific Basin more generally will increase substantially. This swing to the east is already evident in South Africa's coal exports: sales of coal to India have increased in recent years while sales to Europe have fallen, from about three-quarters of South African exports in 2005 to about a fifth in 2011.

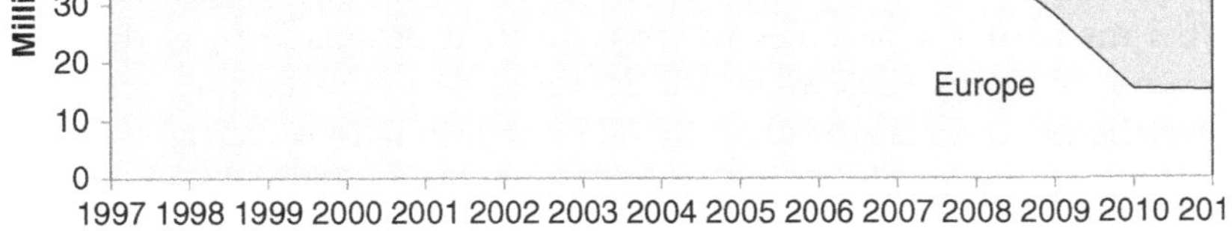

Figure 5.8 Destinations of South Africa's coal exports. Source: IHS McCloskey.

South Africa is effectively a swing producer between the Atlantic and Pacific coal markets. South Africa's main competitors for European markets are Russia and Colombia. In the Pacific coal market, Australia and Indonesia remain the main exporters to China, India, Japan, Korea, and other Southeast Asian countries, but South African coals are increasingly competitive in India and other Asian countries. Even Colombian coal has found its way to Asia.

5.4.6 Coal costs and prices

Although coal mining productivity in South Africa, at 6,500 tonnes/man-year on average (or 3,700 tonnes/man-year if outside contractors are included), is comparatively poor relative to other world exporters, costs per tonne are relatively low, mainly due to low wages. However, per capita wages in the sector have been increasing while employment has declined from around 123,000 in 1980 to around 37,800 (or 65,400 including outside contractors) in 2008 (Baruya, 2007; Prevost, 2009).

Export coals are washed or beneficiated to reduce ash content. All costs up to the washing stage obviously have to be divided by the washing yield (typically around 60%, but lower in the Waterberg) to determine the actual costs of transported or exported coal. South African coal generally has a good reputation on the market because washing guarantees a homogenous product. Quality control at RBCT is also tight. In the past, two major coal qualities were generally used by traders, which differ mainly in their volatile matter: RB1 = min 22% as received volatiles and RB2 = min 25% as received volatiles. The calorific value for both grades is 6,000 kcal/kg. In recent years a third grade – RB3 – has been added which has a lower calorific value (5,500 kcal/k) and a higher ash value (23%) as opposed to 15% for RB1 and RB2. However, most mining companies have a number of coal brands, and a typical physical coal contract may have up to 20 quality elements.

There is evidence that the quality of export coal is declining and an increasing volume of coal with a calorific value of below 6,000 kcal/kg on a net as received basis is being marketed. This has implications for the customer mix for South African producers. The lower the calorific value, the harder it becomes to service distant destinations, as the cost per unit of energy rises when full delivered cost is taken into account. Nevertheless, there is growing demand from destinations such as India for coals with calorific values ranging down to 5,500 kcal/kg.

Domestic coal costs are obviously much lower than FOB export costs due to the shorter transport distances and lack of beneficiation/washing. The new, smaller miners tend to be higher cost producers whereas the larger coal majors benefit from economies of scale and better efficiencies. The long-term trend in costs is upward as labor costs increase, smaller mining operations multiply, lower quality deposits are mined, and the development of the more remote Waterberg field accelerates.

Estimates made in 2007 of FOB *costs* for South African coal ranged from US$27 to US$48/tonne. Ex-mine, washed costs account for about two-thirds of the total, inland transport around a quarter and port charges about 10% (Baruya, 2007; Schernikau, 2009). Mining costs increased substantially in 2007/8, in part because of the boom in commodity prices, and although prices of iron and steel and other commodities have since fallen, mining costs are still greater than 2006 levels. Rail freight charges on the Witbank-Richards Bay line were estimated in 2007 to lie between US$6/tonne for large producers and US$11/tonne for smaller producers (who often have additional truck haulage costs to link with the railway). Transnet's Annual Report (2009) indicates that average coal freight charges are around US$9/tonne or US 1.5 cents/tonne kilometre. Richards Bay port charges are relatively low and have been estimated to be between US$1.5–3.5/tonne (Baruya, 2007). RBCT reports that average user charges in 2008 were slightly over US$3/tonne.

Seaborne freight rates are a key determinant of CIF prices for coal. Standard routes to Europe typically use capesize (120–200 kt of bulk product) or panamax vessels (60–80 kt). Indian cargoes and other less standard routes often use handy-size vessels (20–45 kt). Historical sea freight charges are shown below for capesize vessels from Richards Bay to Europe. Charges peaked at US$60/tonne in June 2008 before plunging back to US$8/tonne in October 2008 and then fluctuating between US$10–20/tonne since.

Internationally, we have seen a shift in recent years from longer-term, annually priced export contracts toward index-priced and shorter-term deals. The recent price spike is only one reason for this change in behavior; another reason is the emergence of coal price indices (see Chapter 1). Export prices are captured in the API4 index, which is a weekly assessment published by Argus/McCloskey of spot physical prices for FOB Richards Bay coal cargoes. API4 index prices are now also used as the basis of derivative contracts. Buying or selling coal against the index allows companies to manage their price risk by using forward contracts or swaps. Forward contracts are traded over the counter (OTC) bilaterally between parties and settled against the index. Coal derivatives started appearing only in 1998 and took until 2002 to reach two times physical volumes. But the market is growing steadily, and by 2007 paper volumes had reached a multiple of ten in the Atlantic market.

5.5 Coal policy, governance, and regulation

5.5.1 Coal policy

5.5.1.1 Policy challenges

It is a surprising fact that although South Africa's economy is overwhelmingly dependent on coal, especially for power production, and coal is the third largest commodity export earner (after platinum and gold), the country has no explicit coal policy. In part, this reflects the preoccupation of the democratic government since 1994 with policies and programs to increase household access to energy (especially electricity). Little attention was given to security of supply or broader industrial energy issues, which were assumed to be largely unproblematic. In the absence of government action, the largest coal user in the country, Eskom, has in recent years initiated a forum for government departments and enterprises to discuss coal supply challenges for the electricity sector. And, as we shall elaborate, private coal companies have recently brought all relevant stakeholders together to develop a "South African Coal Roadmap."

The latest official energy policy statement is the Energy Policy White Paper published in 1998, much of which is out-of-date. The coal section in the document amounts to only three pages out of a total of 102. It advocates continued deregulation, maintaining a coal resource database, promotion of low-smoke coals for households, the use of discard coal, promotion of end-use efficiency and clean-coal technologies, and investigating the use of coal-bed methane.

A number of key policy questions remain unanswered. Chief among these is the issue of South Africa's carbon intensive economy and its potential future exposure to international climate change treaty commitments aimed at reducing greenhouse gas emissions. South Africa's commodity exports, whose production currently depends on coal-based electricity, might in the future face carbon tariff barriers from its major trading partners.

In the meantime, two new coal-fired power stations are currently being built, ensuring that CO_2 emissions will continue to increase until at least 2020. At that time, some of Eskom's 1970s-vintage coal power stations are due for decommissioning. A crucial question will be whether South Africa should replace them with further coal-fired power stations or instead pursue lower carbon alternatives. Assuming that coal remains the mainstay of power production as it is now, the declining coal reserves in the Central Basin will create a need for new mines and transport infrastructure to ensure continuity of coal supply to existing coal power stations. If the plan is to start moving away from a carbon-intensive economy, the question is whether alternative energy resources can be developed in time and at a scale that can replace the country's coal generation without overburdening its developing economy.

Another open question is whether South Africa will be able to respond more purposefully and successfully than it has so far to opportunities to export coal to Asia. If coal exports are to expand, a number of interventions will be necessary to ensure more efficient allocation of prospecting and mineral resource rights and to mobilize capital and public-private partnerships that would support major expansion of coal mining and rail infrastructure.

If South Africa does have a default coal policy, it could be summarized crudely as "export the best and burn the rest!" But even this default-operating mode has run into problems as investments in rail capacity have lagged behind those in coal mine and port capacity.

5.5.1.2 Eskom initiatives

As mentioned earlier, this policy void is being tackled by two initiatives. First, Eskom is mounting an effort to address its concerns around the quality, security, and price of its coal supplies. Already announced coal mining investments will not fully meet expected power station coal demand. Furthermore, coal quality has been deteriorating, negatively affecting the performance of its power stations. And Eskom's coal costs are escalating, placing pressure on electricity prices and the company's financial performance.

During Eskom's power crisis in 2008, the utility assembled a team of CEOs and senior managers in the coal sector and charged it with developing an emergency plan to restore Eskom's coal stockpiles. Eskom has since taken much more seriously the whole area of primary energy procurement and management. During the course of 2008, the major coal companies made commitments to contribute additional allocations – even diverting a portion of planned exports – in order to respond to Eskom's need. A senior coal executive was seconded from Anglo/Kumba to Eskom.

Eskom is pursuing a number of strategic actions: the development of an optimal portfolio of long-, medium-, and short-term coal contracts; movement to coal contract prices based on costs of efficient operations and fair returns on capital invested by the mining industry; investment in rail and conveyor coal transport options;[5] coal quality management and selective beneficiation, especially for Matla and Duvha power stations; risk-based coal stock management; improved cooperation with major stakeholders; and strengthening of organizational capabilities in Eskom's primary energy division.

Eskom has helped convene a wider Coal Energy Forum, which includes key government ministries, the mining industry, unions, and Transnet, in order to develop approaches for establishing scenarios of future coal needs, facilitating approval of coal prospecting and mineral rights, planning new coal mining investments, accelerating required investments in rail and road infrastructure, and sourcing key geological, engineering and management skills. However, industry participants complain of a lack of government leadership and progress in these areas.

5.5.1.3 South African Coal Roadmap

A second initiative is the South African Coal Roadmap, which aims to identify the factors that will affect the sector going forward as well as the best options for future development of the coal industry. The initiative seeks in a general sense to facilitate policy, planning, and strategy development that can help achieve optimal outcomes. Participants include all relevant stakeholders in the coal sector. The process commenced in 2007 at the instigation of a handful of concerned coal industry leaders, but progress has been slow. Initial concerns around competition (anti-trust) law were raised but now appear to have been resolved. The roadmap was coordinated by the Fossil Fuel Foundation, an association of stakeholders involved in the coal sector. Focal areas were established in geology (resources, reserves, exploration); coal characterization and analysis; mining (handling, storage, environment); beneficiation; technology; economic, social, labor, and skill development issues; infrastructure and logistics; current and new markets; safety, health, and environment; climate change; and local and international policy and regulation. The South African Coal Map was published in 2013 and outlined a number of scenarios and policy actions.

The roadmap process faced many challenges, and this initiative still has to demonstrate that it will lead to integrated policy proposals that will shape investment decisions and enable the development and expansion of the coal industry in a way that meets local demand and realizes export growth opportunities. The government's participation has also been complicated by the split of the Department of Minerals and Energy into two separate government departments: the Department of Mineral Resources and the Department of Energy, each of which has its own

Minister. Coal prospecting and mining rights are controlled by the Department of Mineral Resources without any explicit linkage to the broader energy policies and plans being developed by the Department of Energy.

5.5.2 Legislative framework

Developments in the coal sector have been affected by key legislative interventions over the past decade. First, there has been a concerted effort to increase ownership and participation by black South Africans in the economy and in the mining sector. Second, and related to these objectives, there has been a fundamental overhaul of the licensing regime for prospecting and mining. The impact of these two initiatives has been a dramatic increase in black economic empowerment: currently over 30% of coal mines, as well as the Richards Bay Coal Terminal, are owned by companies controlled by black South Africans. A third legislative development has been the establishment of a new royalty payment regime. Collectively, these major changes in the legal framework, which have brought with them bureaucratic delays and uncertainties around prospecting and mining rights, have had the effect of slowing investment in the sector.

South Africa's Constitution provides for redressing historical socio-economic inequality and discrimination. Black economic empowerment is central to the government's economic transformation strategy of empowering historically disadvantaged South Africans (HDSAs). The Broad-Based Socio-Economic Empowerment Charter for the South African Mining Industry (signed in 2002) and the Broad-Based Black Economic Empowerment Act (2003) have established objectives of increasing HDSA representation in management to 40% within five years and transferring ownership of 26% of mining assets to HDSAs within ten years. Black empowerment goals were also incorporated in the Minerals and Petroleum Resources Development (MPRD) Act No. 28 of 2002.

The MPRD Act also heralded a profound change in mineral resource rights. These rights were previously held by the landowner; the legislation transferred resource ownership to the state, which is now empowered to grant, control, administer, and refuse prospecting or mining rights, subject to various conditions. Since 2004, when the legislation became effective, there has been an extended and laborious process to convert "old order" to "new order" rights.

The new legislation creates a "use it or lose it" principle. The holder of *unused* old order rights has exclusive rights to apply for new order prospecting or mining rights within one year. The holders of *active* old order prospecting or mining rights have to apply for new order rights within two and five years respectively, failing which the rights will lapse.

Codes of Good Practice for the Minerals Industry were set forth by the government in April 2009 that further define objectives for black ownership, management control, employment equity, human resource development, preferential procurement, community and rural development, beneficiation, and housing and living conditions. The codes reaffirm the objective of attaining 26% ownership (in terms of voting rights, economic interest, and net value) by HDSAs by 2014. Controversially, the codes also require that the full ownership transitions embedded in BEE deals be completed within two years of transactions, implying that debt incurred for acquiring shareholdings must be fully redeemed within that time period.

Conversion or granting of mining rights may be refused if these empowerment objectives are not met. There are also specified periods for prospectors or miners to commence their operations. Prospecting or mining rights may only be transferred with the permission of the Minister. A change of control over prospecting or mining companies also requires ministerial permission.

The Mineral and Petroleum Resources Royalties Act of 2008 introduces new royalty taxes on minerals. The Act took many years to finalize and was implemented in 2010. Royalties are capped at 5% for refined minerals and 7% for unrefined minerals. Effective rates can be a lot lower. Coal is classified as an unrefined mineral.[6]

While these policy and legislative initiatives have had a positive effect on black economic empowerment, the overall impact on the mining sector in South Africa has been more mixed. Investment and production of most commodities in South Africa has stagnated or even declined over the past decade, and South Africa mostly missed the global commodity boom in the period up to 2008. In part this was because some of the ore bodies are becoming more difficult and expensive to mine. Inadequate infrastructure has been a constraint as well. But it also seems likely that delays and inefficiencies in implementing new legislation have dampened investment in the sector.

5.6 Issues that will shape future development of the coal sector

Looking to the future, it seems clear that the development of South Africa's coal industry will be driven, first, by domestic electricity demand and, second, by export demand from countries such as India and China.

Coal consumption in South Africa will continue to be dominated by the existing stock of coal-fired power stations, the first of which might only be decommissioned in 2021, and probably later. Whether any more power stations are built after the two that are currently under construction and whether another CTL plant is built will depend on precautionary steps South Africa might take to reduce its carbon-intensity and greenhouse gas emissions in the face of global concerns around climate change.

Demand from Europe for South African coal exports is expected to decline in the coming years as European countries decarbonize their power sectors, but Asian demand is likely to grow – especially from India but also from China and other countries in the East. South Africa's ability to respond to this growing demand will hinge on the development and implementation of a coordinated strategy for investing in new mines and rail capacity to get the coal to its ports.

5.6.1 Coal production and domestic power demand

The major driver in the development of domestic coal markets is undoubtedly Eskom's investments in additional coal-fired plants. In 2008, Eskom estimated that it would need around 200 Mtpa of coal by 2018 and that South Africa would need forty more coal mines at an investment of R100 billion. A number of old mines are nearing the end of their useful life and, according to Eskom estimates, new mines will have to contribute around 180 Mtpa within ten years to meet demand from Eskom, Sasol, other domestic consumers, and export customers.

Eskom's estimates were definitely on the high side. Electricity demand fell in 2008/9 as a result of the recession and has remained stagnant for six years. Eskom's forecasts have had to be revised downwards. The author's own modelling suggests that Eskom's coal demand will peak at a little over 155 Mtpa around 2021, *provided South Africa commits to CO_2 emission restrictions*, as will be discussed in the next section.

The large Medupi and Kusile coal power stations will come on line between 2014 and 2020 and will increase domestic coal demand by around 30 Mtpa. Long-term contracted coal supplies from Eskom's existing power stations will also need to be supplemented with medium to short-term contracts as plants run at full capacity and adjacent mines struggle to meet demand. Private power plants (IPPs and self-generation) could add another 15 Mtpa of demand by 2025. Overall coal consumption for power production will decline from the mid-2020s when some of Eskom's oldest coal-fired power stations will need to be decommissioned. Between 2021 and 2030, approximately 10 GW of coal-fired capacity is slated to be taken out of service. Alternative plans include investing in

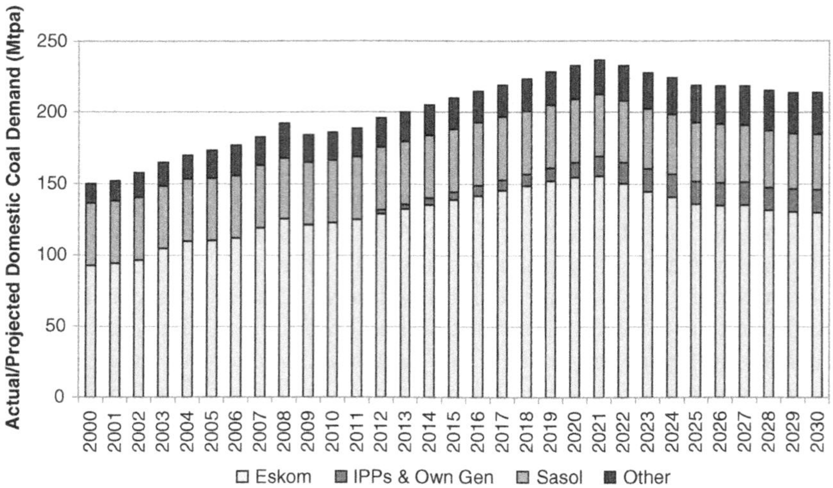

Figure 5.9 Scenario for domestic coal demand evolution in South Africa (Mtpa). Actual figures are shown for 2000–2008; 2009–2030 projections are from author's own modeling performed in 2010.

life-extensions of these plants, implying coal consumption could plateau before falling.

While Eskom will account for the majority of growth in domestic demand for coal, Sasol's earlier plans to build another CTL plant would have resulted in a step change in demand in the course of the next decade. If the proposed Mafuta plant were to be built, Sasol's existing demand of around 44 Mtpa could increase to around 70 Mtpa. However, there is still a great degree of uncertainty as to whether this investment will actually materialize; acceptance of CO_2 caps would make this CTL investment unlikely.

Figure 5.9 presents the author's modelling of total South African domestic coal demand through to 2030. Eskom and IPP demand is assumed to grow as previously indicated and Sasol's demand remains the same. Other domestic coal demand is projected to grow at a modest 2% per annum.

What is clear from this analysis is that Eskom's coal needs will continue to dominate the development of South Africa's coal industry in the next decade, accounting for about two thirds of domestic demand.

Exports could rise to 105 Mtpa in the mid 2020s provided that a number of dormant mining investment plans are revived, existing rail constraints are tackled, and RBCT proceeds with its next expansion. The need for more coherent policy action to relieve infrastructure bottlenecks is discussed further in Section 5.6.4.

5.6.2 The impact of climate change concerns

South Africa's CO_2 emissions from fuel combustion were around 347 Mta in 2012, making it the eighteenth largest emitter of CO_2 globally. Eskom's coal-fired power stations account for around 220 Mtpa and Sasol's CTL plants about 60 Mtpa. South Africa's per capita emissions as a result of fuel combustion are 6.94 t CO_2 per annum compared to a global average of 4.44 t CO_2 per annum; and South Africa emits 1.2 kg CO_2 per \$ GDP compared to a global average of 0.6 and the OECD (Organization for Economic Cooperation and Development) average of 0.44 kg CO_2/\$GDP (IEA, 2012).

While South Africa might justifiably argue that it is still a developing country or emerging economy and needs to foster rapid economic growth in order to create employment and overcome poverty, it also has to confront the reality that its carbon emissions are way above the global average and will need to fall in per capita and GDP terms, assuming movement to a level of resource usage which is equitable globally (i.e., equal per capita emission rights[7]).

Policy debates in South Africa are also raising the spectre of tariff barriers by trading partners on its commodity exports because their production depends almost entirely on coal-fired electricity inputs. Further arguments are being made that South Africa needs to explore the potential benefits of developing competitive "green" industries and jobs, including in the renewable energy and energy efficiency sectors. While the country has huge potential to improve the efficiency with which it uses energy and hence carbon, and is becoming more efficient as energy prices increase and the structure of its economy changes (relying less on mining and more on service industries), energy conservation alone will not solve South Africa's energy conundrum. Further energy supply options will be necessary, and within this context some hard decisions will need to be made on coal's future contribution to South Africa's energy economy.

South Africa is developing a climate change mitigation strategy. The government has overseen the development of long-term mitigation scenarios, and these were endorsed at Cabinet level in mid-2008, not as a specific policy but rather as a roadmap or set of strategic options. A National Climate Change Summit followed in March 2009 at which then-President Motlanthe announced that the country's greenhouse gas (GHG) emissions would peak in 2025, plateau, and then decline. At the

Copenhagen summit in late 2009, President Zuma made a voluntary offer that South Africa would reduce its greenhouse gas emissions below a business-as-usual scenario by 34% in 2020 and 42% in 2025, conditional on "the provision of financial resources, the transfer of technology and capacity building support." A subsequent National Climate Change Response White Paper, published in 2011, reaffirmed these commitments.

The modeling underpinning these numbers suggests that South Africa would need to limit its total greenhouse gas emissions (fossil fuel combustion plus other sources) below 614 Mtpa CO_2 equivalent. The electricity sector currently accounts for about 45% of South Africa's greenhouse gas emissions. If it were to take a proportional burden-sharing commitment, then CO_2 emissions from power generation would need to be capped at around 275 Mtpa. If we add to Eskom's current CO_2 emissions (220 Mt) those from the coal fired power stations currently being built (more than 60 Mt), then it is clear that the 275 Mtpa CO_2 target will be exceeded by 2020 and only re-attained in 2025 after old coal power stations are decommissioned.

These assumptions have been incorporated into the preferred scenario in the Integrated Resource Plan for the electricity sector published by the Department of Energy in October 2010. The plan suggests that these targets can be met through a relatively modest increase in costs (less than 10% above the least-cost, coal-rich scenario). However, deeper cuts in CO_2 emissions would result in significantly higher costs and are unlikely to gain sufficient industry and political support.

The modeling in Figure 5.9 assumes the preceding scenario. If the overall cap is taken seriously, then there will be no room for Sasol's planned Mafuta CTL plant.

The adoption of carbon capture and storage (CCS) technology (see Chapters 11 and 12), of course, would remove the above constraints on building further coal power stations or CTL plants. While South Africa has a modest CCS program, mainly aimed at mapping geological sites for potential storage, it is unlikely to be able to afford the investment costs of this technology in the near to medium term, and thus CCS is unlikely to impact the preceding scenarios before 2025.

5.6.3 Increasing exposure of domestic coal to short-term contracts and international prices

Domestic coal prices in South Africa, on average around US$20/tonne, are among the cheapest in the world, but domestic prices are rising and are likely to continue to do so in the immediate future. This chapter has documented Eskom's increasing exposure to short- and medium-term road haulage coal contracts, which are more expensive than its existing long-term contracts with mines adjacent to its power stations. The growth of small mining companies has meant average industry costs have increased. Recent applications by Eskom to the National Energy Regulator of South Africa (Nersa) for electricity tariff increases have underscored the increase in primary energy costs.

A number of mines with Eskom coal contracts have also diverted better quality coals to exports, which are more lucrative. Even though coal export prices are well below 2008 highs, they are still much higher than domestic prices, even after allowances have been made for transport and beneficiation costs. Higher export prices have had some effect in pushing up domestic prices in new local contracts. However, the linkage between domestic and export prices should not be overstated. Eskom and Sasol account for 90% of domestic coal use. 70% of Eskom's coal still comes from long-term contracts with tied collieries, and Sasol owns its own coal mines. And while it is true that the differential in quality between export and domestic coal has been diminishing, especially with lower quality coals being exported to India, these are still essentially two different brands of coal. Eskom and Sasol use high-ash, low-calorific value coal, while most of the export coals are washed to improve their quality. The optimal development of many new coal mines, especially in the Waterberg, will involve both local supply to Eskom and exports.

Because of the lack of investment in new coal mines and looming coal shortages for Eskom in the central basin post 2018, political pressure has been building to restrict exports and to control domestic coal prices. If implemented, these measures could have the unintended consequence of discouraging further investment in new mines. Government, Eskom, and industry need to come together to plan the development of new coal mines in a way that both domestic and export demand can be met optimally.

The clear trend is for domestic coal prices to increase, although both Eskom and Sasol will be able to blunt the impact of these cost and market pressures to some degree.

5.6.4 Policy and infrastructure constraints on export growth

This chapter has documented the potential for exports from South Africa to grow. South Africa is a relatively low-cost producer and sits on the cusp between the Atlantic and Pacific coal markets. Yet South Africa largely missed the commodity boom in the period up to 2008. Global trade in steam coal has doubled over the past decade. Indonesia has seen a five-fold increase of its steam coal exports over the same period, Colombia has increased exports 2.3 times, and Australia 1.6 times (IEA, 2012). Yet South Africa's exports have been stagnant. Legislative and regulatory risks have restricted investment, but policy and infrastructure planning failures have probably had an even more significant and negative impact.

As noted earlier, in the 1970s, the South African state, in partnership with private coal companies, embarked on an ambitious infrastructure program to grow coal exports. This effort was remarkably successful; as Figure 5.4 demonstrates, exports were lifted from negligible levels to quantities that now place the country fifth globally as an exporter.

In contrast, there is no evidence today of the state focussing on the potential for growth in coal exports. And there is insufficient coordination between the investment plans of coal miners, Transnet rail infrastructure planning, and RBCT port expansion. Richards Bay has bravely proceeded with port expansion, but Transnet remains a bottleneck. The resulting uncertainty around export potential complicates decisions by coal companies to invest in expanding mining production.

This situation is exacerbated by the progressive mining-out of the central basin (South Africa's most favorable export coal resource) and the shift to develop the more remote and challenging Waterberg field, with its high ash and lower yield coals. The Waterberg is remote from South Africa's industrial heartland. It has one urban center, a constrained water supply, and a single railway line with 3 Mtpa capacity for coal exports. Exports from coal deposits in this region will only be economic if further multi-product mines are developed which supply large quantities of low-grade coal for power production alongside smaller volumes of coal for export. Massive investment in expanded rail infrastructure for exports from the Waterberg will need to be committed and water constraints will need to be overcome, which entails integrated management of competing water demands for mining, power production, irrigation, and human settlements.

Failure to develop an integrated and rational coal policy, coupled with domestic energy security concerns, could result in short-term and ill-judged initiatives to re-introduce export quotas and domestic price regulation – the same policies which held back the development of the sector in the 1950s and 1960s as discussed previously. A well-thought-out roadmap is certainly the most needed and urgent priority for the development of South Africa's coal industry.

5.7 Summary and conclusion

In summary, growth in South African coal production will be driven primarily by domestic power demand. However, despite South Africa having a couple of hundred years' worth of coal reserves, and coal being the cheapest electricity generation option, environmental constraints (climate change, acid mine drainage, and water shortages) could cause total coal-fired generating capacity to peak in the mid 2020s. Under this scenario, domestic demand for coal would plateau and then fall as old coal-fired power stations are de-commissioned. Nonetheless, new coal mines would still have to be developed as existing mines exhaust their reserves (especially in the Witbank coalfield) and as Eskom and IPPs expand their production. Local coal demand will also be impacted by rises in coal prices as a result of increased exposure to higher-cost, short-term contracts, competition with low-grade exports, and possibly carbon taxes.

While South Africa clearly takes climate change concerns seriously, current decision-making still understandably prioritizes electricity sufficiency and development of the country. In the face of a serious electricity shortage, South Africa approved the massive Medupi and Kusile coal plants that are currently under construction. The immediate attention of Eskom is clearly focused on tackling serious problems that constrain the availability of coal for electricity generation. At the same time, planning for climate change mitigation has begun. This suggests that South Africa needs to begin seriously pursuing realistic baseload alternatives to coal, which might among other things entail a serious effort to develop the country's shale gas resources, which appear to be large.

The rise in domestic coal demand over the next decade will not necessarily constrain growth in exports. Indeed, in some instances there are good synergies in the development of both markets (for a comparison point, see Chapter 7 for a discussion of possible synergies in Indonesia between development of coal for export and for domestic use). As already explained, specific coal seams in the Central Basin can be economically beneficiated/washed for exports with the middlings fraction available as local steam coal. Other seams provide only low-quality local steam coal and are not suitable for exports in any case. And the economics of developing the Waterberg coalfield require multi-product mines. The superior profits earned on exports are an important source of investment income for expanding local mining.

Although some of Eskom's power stations have suffered from increasingly poor quality coals as better quality coals are diverted to exports, it would still be a mistake to re-impose restrictions on exports with the idea of expanding local supply. These are mostly separate markets. Any problems with deteriorating quality of supply to Eskom should be dealt with through tighter management of contracts.

Coal exports have huge macroeconomic benefits in terms of South Africa's balance of payments and current account. The potential for substantial expansion of coal exports (especially to India) is constrained by the absence of an integrated coal development policy, more distant and more difficult-to-mine coalfields, water shortages, and insufficient infrastructure, especially rail capacity to the ports.

The need for strong coordination between coal producers, Transnet, and RBCT to strengthen South Africa's reputation as a well-organized and reliable coal supplier would seem obvious. But we have yet to see purposeful policy and investment actions that would resolve the coordination problem between private mining companies and the state-owned railways. Transnet faces pressure to invest in expanded capacity from a large number of smaller producers, few if any of whom are prepared to sign long-term take or pay agreements on rail freight. The coal majors have been prepared to do this in the past, but they have experienced caps on their export allocations as black economic empowerment policies have focused on widening entry into the sector.

Growth in exports could come from both the Central Basin and the Waterberg. There will have to be consolidation and optimization of mines in the Central Basin, while in the Waterberg, the requirements of opencast mining plus multiple washing processes imply large-scale development and investment plus deployment of technology, favoring the involvement of coal majors.

The contrast between the purposeful and integrated development of coal export capacity in the 1970s and the current mismatch between mines, rail, and ports is striking. How do we explain this anomaly? The restructuring of the mining industry, with the entry of many smaller and less capitalized companies, has clearly disturbed the previous relationships between the main industry players and the government and thereby made coordination more difficult. The emphasis of government policy has been on changed ownership and increased access to mining and export rights rather than on growing and expanding the sector.

But clearly, state-owned Transnet itself has been a major problem. The period for at least fifteen years after 1990 saw widespread changes in Transnet's management, declining performance, and inadequate investment. Poor governance and inadequate incentives have contributed to these outcomes. While there has been a more recent attempt through new management appointments to improve efficiencies and financial performance within Transnet, rail capacity continues to lag port capacity. Recent engagements with the coal industry have been encouraging but have not yet translated into long-term contracts that would facilitate new investment and expanded exports. These difficulties have been exacerbated by capacity challenges within government and an absence of cross-sector planning.

Coal is fundamental to South Africa's energy economy. It is also a valuable export earner. With more coherent government management of the sector, the potential is there for maximizing the economic benefits from the coal industry within the constraints of environmental sustainability.

Notes

1. The United States also ranks as an exporter, although more than half of its exports are coking coal. Australia, Indonesia, and Russia also export significant quantities of coking coal. Nearly all of South Africa's exports are steam coal.
2. The Department of Minerals and Energy, in its South Africa's Mineral Industry 2001/2 Report, estimated economically recoverable coal reserves at 55.3 Bt. This figure is based on the Bredell report (1987). The earlier Petrick Commission (1975) estimated reserves at 53.9 Bt and De Jager (1982) at 58.4 Bt (Alberts, 1987). Prevost of the Minerals Bureau downgraded the reserves to 39.1 Bt in 2000, but that was during a period of declining coal prices in real terms, and one of largest coal deposits was uneconomic at the then ruling prices. The BP Statistical Review of World Energy put South African coal reserves at 30.4 Bt. Hartnady (2010) calculates remaining reserves to be 15 Bt. A new coal resources study by the Council of Geosciences estimates coal resources and reserves to be 67 Bt, two-thirds of which are in the Waterberg (report still unpublished in 2013).
3. In bord and pillar mining, only the pillars come into contact with water. Longwall mining, where the roof is allowed to collapse into the mined-out void, increases the contact area and also facilitates the ingress of rainwater. In opencast mining, the rock mass is completely fragmented, maximizing the contact between water and rock; it is, therefore, the most acid producing mining method.
4. From synfuels as well as production of chemicals.
5. Eskom has established a working group with Transnet to facilitate rail solutions; the Majuba Rail project and combined road/rail container system for Camden are being implemented; and a feasibility study on the Waterberg rail link has begun.
6. For an unrefined mineral, the royalty is calculated as gross sales multiplied by a percentage equal to 0.5 + [earnings before interest and taxes / (gross sales × 9)] × 100.
7. As of this writing, it is far from a foregone conclusion that international climate negotiations will converge on this principle.

References

Alberts, B. C. (1987). Planning the Utilization of South Africa's Coal Reserves, *Journal of the South African Institute of Mining and Metallurgy* 87(11), pp. 371–395.

Baruya, P. (2007). *Supply Costs for Internationally Traded Coal.* Paris: IEA Clean Coal Centre. IEA/OECD.

Birtles, A., and Soden, G. (2010, July 2). The Easy Pickings: Mostly Gone in South Africa's Coal Fields. *Mining Review.com*. Available at http://www.miningreview.com/the-easy-pickings-mostly-gone-in-southern-africa-s-coalfields/ (accessed September 19, 2014).
BP (2010). *BP Statistical Review of World Energy*. London: BP plc.
Bredell, J. H. (1987). South African Coal Resources Explained and Analysed. *Geological Survey of South Africa*. Report number 1987–0154.
Chamber of Mines of South Africa (2008). Facts and Figures. Available at http://www.bullion.org.za/.
Department of Energy (2011). *Integrated Resource Plan 2010–2030*. Pretoria.
Department of Environment (2011). National Climate Change Response. White Paper, Pretoria.
Department of Mineral Resources (2009). Unpublished Coal Statistics. Pretoria.
De Jager, F. S. J. (1982). Coal Reserves of the Republic of South Africa, Geological Survey. *Bull* 74, pp. 289–330.
Eskom (2012). *Revenue Application: Multi-Year Price Determination 2013/4 to 2017/18 (MYPD3)*. Johannesburg.
Hartnady, C. (2010). South Africa's Diminishing Coal Reserves. *South African Journal of Science* 106(9/10), pp. 18–22.
IEA [International Energy Agency] (2013). *Coal Information*. Paris: OCED.
IEA (2012). *Key World Energy Statistics*. Paris: OCED.
IEA (2008). *World Energy Outlook*. Paris: OCED.
Jefrey, L. S. (2005). Characterization of the Coal Resources of South Africa. *Journal of the South African Institute of Mining and Metallurgy* (February), pp. 95–102.
Leger, J. (1991). Coal Mining: Past Profits and Current Crisis. In S. Gelb, ed. *South Africa's Economic Crisis*. Washington, DC: Zed Books, pp. 129–155.
Lloyd, P. (2000). The Potential of Coal Wastes in South Africa. *Journal of the South Africa Institute of Mining and Metallurgy* (January/February), pp. 69–72.
Marquard, A. (2007). The Development of Energy Policy in South Africa. PhD thesis, University of Cape Town.
McCarthy, T. S., and Pretorius, J. P. (2009). Coal mining on the Highveld and its implications for future water quality in the Vaal River System. Working Paper, School of Geosciences, University of Witwatersrand, Johannesburg.
Olsen, S (2007). Generation Primary Energy. Confidential Memo to Jacob Maroga, Eskom CEO. Wingfield Consultancy, New Hampshire, USA.
Petrick Commission (1975). *Report of the Commission of Enquiry into the Coal Resources of the Republic of South Africa*. Pretoria: Government Printer.
Prevost, X. (2009/10). Unpublished SA coal statistics. DME / XMP Consulting, Pretoria.
Prevost, X. M. (2002). Coal. *South Africa's Mineral Industry 2001/2002*. Pretoria: Minerals Bureau.
Schernikau, L (2009). The Renaissance of Steam Coal. PhD dissertation, Technical University of Berlin.

Annual Reports

Anglo American (2012). Available at http://www.angloamerican.com/~/media/Files/A/Anglo-American-Plc/reports/annual-report2012.pdf.

BHP Billiton (2012). Available at http://www.bhpbilliton.com/home/investors/reports/Documents/2012/BHPBillitonAnnualReport2012.pdf.

Eskom (2012). Available at http://financialresults.co.za/2012/eskom_ar2012/divisional-report/statistical-overview.php.

Exxaro (2012). Available at http://www.exxaro-reports.co.za/reports/ar_2012/.

Richards Bay Coal Terminal (2012). Available at www.rbct.co.za.

Sasol (2012). Available at http://www.sasol.com/sites/default/files/publications/integrated_reports/downloads/Sasol_Integrated_Annual_Report2012_1351837531724%20-%2030%20June%202012.pdf.

Transnet (various years). Available at www.transnet.co.za.

Xstrata (2010). Available at http://www.glencorexstrata.com/assets/Uploads/2012-Annual-Report-FINAL.pdf.

6 Australia's black coal industry

Past achievements and future challenges

Bart Lucarelli

6.1 Introduction

6.1.1 Background

When one discusses the black coal industry of Australia, one is really referring to the coal industries of Queensland and New South Wales (NSW). Ninety-nine percent of Australia's black coal reserves (19.2 billion tones [Bt]) and 96% of its demonstrated economically recoverable black coal resources (57.5 Bt) are located in Queensland and NSW, and almost all of Australia's black coal exports also come from those two states (JORC, 2009;[1] Geoscience Australia, 2012). For fiscal year 2012–13, Australia's production of raw black coal was 526.7 million tonnes (Mt), with NSW and Queensland accounting for 517.2 Mt, or 98% of FY 2012/13 production (Bureau of Resources and Energy Economics, 2013).[2] This concentration of resources, production, and exports of black coal in Queensland and NSW is expected to be a permanent fixture of Australia's black coal industry.

Australia's state governments are given wide control over the planning, development, extraction, and sale of coal and other mineral resources. They also have regulatory control over the rail and port industries that provide the inland transport services for moving coal to both domestic and export customers. In short, Australia's state governments are the

The author wishes to express his sincere thanks to Mark Thurber, associate director of PESD, and Richard Morse, affiliated researcher at PESD, for their expert reviews of this chapter and its earlier versions and their many helpful suggestions on how to better organize the chapter, present data as tables and charts, and frame the arguments and conclusions expressed in this chapter. The author also wishes to express his deep gratitude to Michael Friederich, coal exploration consultant with Coal Discovery and located in Brisbane, Australia, for sharing his many insights into the early (1960–70s) history of Queensland's coal industry and technical, economic and regulatory issues that are shaping Australia's black coal industry today and to Stuart Murray, previously of London Commodity Brokers and now an independent coal consultant. Finally, the author wishes to thank the staff of Coal Services Pty Ltd for making available to him reports of the Joint Coal Board from 1949 to 2000.

nexus where public-private interfaces occur, and the states of Newcastle and Queensland are where all the black coal is. This chapter, therefore, focuses on the black coal industries of Queensland and NSW, which are collectively referred to here as "Australia's black coal industry."[3]

While the focus of this overall volume is steam coal, the current structure and past development history of Australia's black coal industry do not allow one to separate the steam coal segment of the industry from its coking coal sibling (see Box 6.1 for definitions of coal types covered in this chapter). The interrelationships between these two industry segments are simply too strong to be ignored. This chapter therefore discusses the role of both industry segments in shaping the historical expansion and current structure of the black coal industry of Australia. In particular, development of the large coking coal deposits of the Bowen Basin in Queensland and the Hunter Valley in NSW was the driving force behind the industry's export-led expansion during the period 1950–1985. Since 1985, steam coal has captured a larger share of Australia's production and exports of saleable black coal. Over the next twenty years, steam coal is expected to play an even greater role within Australia's black coal industry – partly due to its greater availability and partly due to the larger demand for steam coal in Asia.

Box 6.1 Coal definitions and terminology

Black coal resources and reserves are classified in this chapter largely according to definitions adopted by the Australia's coal industry and its Commonwealth and state governments, which classify bituminous, anthracite, and sub-bituminous coals as "black coals," with lignite grade coals classified as "brown coals."[4] Black coals are used for a large variety of applications. Power generation is by far the largest of those end uses, accounting for 75% of total domestic consumption of black coal in 2011 (Bureau of Resources and Energy Economics, 2013).

Black coals used in the steel making industry and for making coke are referred to as "coking coals" (sometimes also called metallurgical coals), of which there are three primary types: hard coking coals, soft coking coals, and semi-soft coking coals. A fourth type of coal used in the steel making industry is known as PCI (pulverized coal injection) coal. PCI coals are low volatile matter steam coals without coking properties. As such they have qualities that make them very similar to steam coals. However, PCI coals are used primarily for steel making (for which they earn a premium), and due to their low volatile contents, they are not

Box 6.1 (cont.)

preferred coals for generating power. The production and sales of PCI coal tend to be classified under the coking coal category.

Coals used in the power and other industries for raising steam are referred to as steam coals (synonymous with "thermal" or "steaming" coal, which are the names used by BREE and Coal Services Pty Ltd, respectively), consistent with International Energy Agency (IEA) terminology.

Black coal statistics for Australia are expressed in a number of ways. The first figure usually cited is raw coal or run-of-the-mine production. Raw coal refers to coal as it is produced at the mine site before washing and other methods of beneficiation. After washing and beneficiation, coal is classified as "saleable black coal," which in any particular year can be exported, sold into the domestic market, or added to end of the year stocks at the mine site, export ports, or the sites of domestic customers. In this chapter, all references to coal production are to saleable coal production unless otherwise noted as raw coal production.

6.1.2 Main themes and organization

As discussed above, Australia is distinctive among the case studies in our book in the degree to which subnational regulation and intervention have shaped the coal industry. Section 6.2 of the chapter therefore focuses on how the evolving regulatory framework in NSW and Queensland – in conjunction with geographical, geological, technological, and demand factors – shaped the historical development of the coal industry in these respective states.

Another unique aspect of the Australian case has been its passage in 2012 at the federal level of a law that put a price on carbon. Although this law was subsequently rescinded by the Abbott administration in July 2014, its passage in 2012 showed that, as the one developed country among the major coal producers we study in this book, Australia has a population with perhaps more widespread interest in tackling climate change issues than countries where basic needs are not yet met. On the other hand, coal exports are a major contributor to Australia's economy, and coal still provides most of the country's electricity. The tension between coal's contribution to climate change and its contribution to the economy have led to significant political contention around carbon policy, which we will discuss at length in Section 6.3 of this chapter. (It also makes Australia a logical innovator in technologies to burn coal with fewer CO_2 emissions, a topic that we discuss in some detail in Chapter 11.)

Section 6.4 considers which resource, regulatory, and market factors may have the most influence on the black coal industries of Queensland and NSW over the next twenty years. It documents particular challenges facing the black coal industries of Queensland and NSW – such as the difficulty in developing new transportation infrastructure in the face of volatile prices and uncertain domestic regulations – and describes the likely futures that may emerge from the resolution of these regulatory challenges over time. One key question is whether environmental regulations may reshape Australia's black coal industry over the next two decades in much the same way that concerns around the impacts of large scale wood fuel use during the industrial revolution of the eighteenth and nineteenth centuries did in the UK, Europe, and the US.

6.2 Post-WWII history of Australia's black coal industry (1948–2013)

The history of Australia's black coal industry can be conveniently broken into (a) the "early years" period, which starts in 1797 and ends in 1947, and (b) the post-WWII period, which extends from 1948 to the present. This chapter focuses on the post-WWII period and analyses the post-war history of Australia's black coal industry as four discrete development phases:

- The recovery phase (1948–1959)
- The rapid growth phase (1960–1986)
- The competition phase (1987–2003)
- The volatile price phase (2004–2014)

Figure 6.1 presents the annual saleable black coal production levels for Queensland, NSW, and other states over the post-WWII period for each of the four development phases.[5]

6.2.1 The recovery phase (1948–1959)

Prior to WWII, Australia's black coal industry was thoroughly dominated by production from New South Wales. Regulations were few and those that did exist were either weakly enforced or ignored by industry and labor unions. Frequent strikes led to disruptions in NSW's black coal supplies and strangled its budding coal export business (Shaw and Bruns, 1948, pp. 156–169; Joint Coal Board, 1948, pp. 6–7). Labor disputes and other problems persisted through the war, and many analysts remained

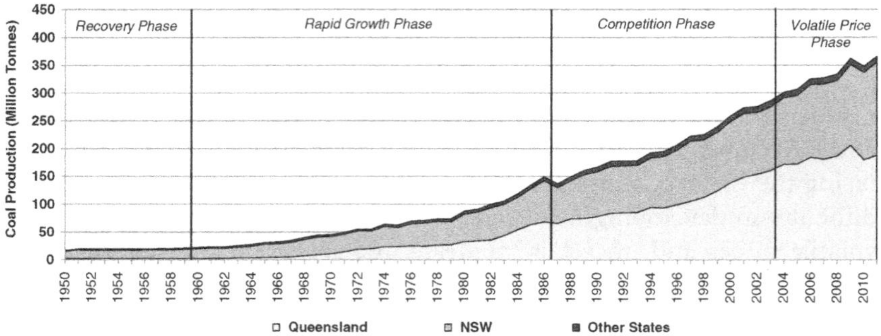

Figure 6.1 Australian saleable black coal production by state, 1950–2011. Data sources: Joint Coal Board, "Black Coal in Australia," 1980–81, p. 12, table 5 (for period 1950–80); Joint Coal Board–Queensland Coal Board, "1994 Australian Black Coal Statistics," p. 9, table 9 (for period 1981–92); Coal Services Pty Ltd. & Queensland Dept. of Natural Resources & Mines, "2002 Australian Black Coal Statistics," p. 8, table 9, (1993–2001); and Coal Services Pty Ltd., "Australian Black Coal Mining Summary, 2011–12," May 13, 2013, p. 12, table 5 (for 2002–08).

pessimistic about the future of Australia's black coal industry even into the early 1950s (Shaw and Bruns, 1948, pp. 156–159).

However, a combination of government interventions initiated after the war and the application of improved mining technology helped significantly improve the health of the industry. The 1950s turned out to be a time when miners and mine owners, with prodding from the government, reached a point of peaceful co-existence, which supported the industry's recovery and subsequent expansion. This post-WWII recovery phase saw the Australian black coal industry transformed from an inefficient and unreliable supplier of coal into the important market player that would eventually become the world's largest and most efficient exporter of high-quality coking and steam coal.

6.2.1.1 *Regulatory environment: labor issues and the government's push for enhanced control*

During WWII, the Commonwealth government passed a number of laws aimed at ensuring a reliable supply of coal for the war effort. However, these legislative initiatives were largely ineffective. The Commonwealth government, in February 1941, created the Commonwealth Control Board to resolve labor-management disputes at an early stage, but it was soon replaced by the Conciliation Commission in November 1941 (Productivity Commission, 1998, vol. 2, p. E3). At a special conference in Canberra convened by the Commonwealth government in May 1942, mine owners and labor unions agreed in what is known as the Canberra Code to protocols for resolving labor-management differences without strikes (Productivity Commission, 1998, vol. 2, p. E3). But new labor disputes cropped up shortly after the agreement was signed, leading to significant disruptions of coal supply to industry. The growing number of management-labor disputes was partly due to the fragmentation of the powerful Miner's Federation into communist and non-communist factions, which attempted to outdo each other with respect to their salary and other demands (Productivity Commission, 1998, vol. 2, p. E2).

The Commonwealth and state governments did not help matters by largely acceding to the demands of the unions in the interest of ending strikes and keeping the coal supply flowing (Productivity Commission, 1998, vol. 2, p. E3). Reasonable government concessions between 1940 and 1942, such as the establishment of a miner's pension fund and the lifting of prior penalties imposed on striking miners, only encouraged the unions to engage in more work stoppages (Shaw and Bruns, 1948, pp. 156–159).

The overall result was that the WWII period was "characterized by near critical shortages of coal for Australia's war effort and near disastrous levels of industrial disputation in the black coal industry," with an estimated 23% of potential production lost due to labor strife (Productivity Commission, 1998, vol. 2, p. E3).

Eventually, the Commonwealth government's ad hoc and largely ineffective war-time powers over the coal industry were subsumed into the Coal Production (War-Time) Act of 1944, which established a "Commonwealth Coal Commissioner with wide powers to take control of any coal mine and distribution networks in the interest of supporting the nation's war effort. The Coal Commissioner also had broad authority to set coal prices throughout Australia" (Joint Coal Board, 1948–49, p. 4).

When the war ended, the Commonwealth and NSW governments decided that tighter regulatory control over the NSW coal industry was necessary (Commonwealth Bureau of Census and Statistics, 1951, p. 877). The industry was in a precarious state at the close of WWII, with exports practically nonexistent and the domestic power, town gas, rail transport, steel making, and cement industries all clamoring for a more reliable and lower cost supply of coal. Coal shortages after WWII were expected to have serious adverse effects on Australia's post-war reconstruction, which depended heavily on adequate supplies of iron and steel products, building materials, and transport (Joint Coal Board, 1947–48, pp. 6–7).

To address these concerns, the Commonwealth government passed the Coal Industry Act No. 40 and the NSW government passed the Coal Industry Act No. 44 (Joint Coal Board, 1948–49, p. 4). Both laws took effect in 1946 and are commonly referred to as the joint acts. These acts established the Joint Coal Board (JCB) and the Coal Industry Tribunal (CIT), with the latter given broad powers to determine wage levels and working conditions.[6]

The passage of the joint acts transformed the regulatory setting for the NSW black coal industry from one based almost totally on a laissez-faire approach to one in which the Commonwealth and NSW governments played a strong role in regulating and rehabilitating the industry (Productivity Commission, 1998, vol. 2, p. E4).[7] Before the JCB and the CIT could exercise their powers, however, the Communist Party of Australia and its supporters within the more radical labor unions launched a rear guard action, now referred to as the 1949 Coal Strike, which shut down the NSW's coal industry for seven weeks. Eventually, the Labor governments of NSW and the Commonwealth crushed this strike and the radical unions by passing and implementing emergency legislation that led to the jailing of many of the strike's leaders.[8]

The JCB and the CIT ended up playing pivotal roles in neutralizing the more radical unions and removing labor unrest as a development issue for the black coal industry during the 1950s. A key contributor to their success was the active role they took in improving the working conditions in underground mines and the infrastructure of mining communities. This is not to say that work stoppages and labor-management disputes did not flare up again during the postwar period. During the 1970s and 1980s, disputes again caused lost output (Productivity Commission, 1998, vol. 2, pp. E4–E5). However, the losses, while significant, never threatened the overall reliability of coal supply as had events prior to 1950.

Although the joint acts paid lip service to the notion of "private ownership and operation of coal mines," the early emphasis of the JCB was on increased state control and intervention. In particular, the joint acts authorized the JCB to take over the operations of inefficient coal mines, as well as supporting enterprises, to ensure that "coal produced in NSW was available in such quantities and with such regularity as will meet requirements throughout Australia and in trade with other countries" (Joint Coal Board, 1948, p. 5).[9]

Pursuant to this mandate, the JCB itself owned and operated collieries (coal mines) in NSW from 1947 (its first year of operations) until March 1957. Its share of total raw coal production ranged from 5% in 1947 to a peak of 16% during 1951. Between 1950 and 1954, JCB-owned mines supplied more than 10% of NSW's raw coal production (Joint Coal Board, various years). Eventually, after much prodding from the Commonwealth government, the JCB, in March 1957, sold off its ownership in three NSW coal mining companies and, with those divestments, the JCB stopped playing the joint role of black coal producer and regulator (Joint Coal Board, 1957–58, pp. 32–33).[10]

6.2.1.2 Technology: increased mechanization

Technology was another significant contributor to production growth and improved labor relations in the Recovery Period. Until 1935, coal mining in Australia was largely a manual operation, based on two-man teams equipped with picks, shovels, and hand-push trolleys and, in some cases, wheelbarrows, to remove coal from coal seams and transport it to the surface. In deeper mines, the coal was dumped from the trolleys and wheelbarrows at the bottom of a vertical shaft and then hoisted to the surface in buckets by winch (Australian Academy of Technological Sciences and Engineering, 2000, p. 758).

At the start of the recovery phase, the Australian black coal industry remained woefully under-mechanized when compared with the coal industries of other coal-producing nations (Joint Coal Board, 1950–51, p. 15). Prior to the 1950s, mechanization was limited to the use of hand-held machine cutters and picks for cutting the coal from the rock face and for drilling bore holes to place explosives. In certain instances, winches powered by electricity and steam engines were used to transport coal out of the mines (Australian Academy of Technological Sciences and Engineering, 2000, p. 758). The few new technologies that were adopted were used to partially mechanize underground coal mines, which, during most of the 1950s, accounted for more than 80% of Australia's black coal production.[11]

Most coal mining companies in those early days did not have strong enough balance sheets and technical capabilities to mechanize their mines. The JCB recognized this industry deficiency from the outset and implemented programs to encourage companies to mechanize their operations and regulations to punish those that did not. These incentives and supports included free technical assistance to mine operators in need of advice on how to plan the future development of their mine sites and financial assistance in the form of loans and even grants to those mine operators who chose to purchase equipment for mechanizing their mines (Joint Coal Board, 1950–51, p. 14).

The JCB also established "an equipment pool from which colliery companies could either purchase or hire the most modern types of mining machinery" and provided financial and technical assistance to domestic manufacturers of mining equipment who wanted to diversify into types of mining equipment only available offshore (Joint Coal Board, 1950–51, p. 14). The purchase-hire program had an important demonstration effect, encouraging many other companies to adopt the same equipment as early adopters in order to remain competitive in the industry.

The coal industry also benefited in the 1950s from significant improvements in exploration methods, which included the development of improved seismic techniques, data logging equipment, drilling rigs, and core lifters; improved electro-mechanical instruments and tools; and advanced instrumentation for monitoring the performance of drills.[12] These technological advances in minerals exploration contributed to the discovery within NSW's Hunter Valley and Queensland's Bowen Basin of high-quality, coking coal deposits with low stripping ratios that were amenable to open-cut mining methods. However, the new mining technologies drew more attention than these exploration improvements because they directly replaced labor, which resulted in strong union opposition. Mechanization of the mines also had an immediate and visible impact on coal production.

Major improvements in productivity were achieved through mechanization of both bord and pillar mining underground and open-cut mining. (See Box 6.2 for a description of bord and pillar mining.) For underground mines, the introduction of continuous miners – starting with the first demonstration in the Huntley Colliery in NSW – was a key innovation that allowed the cutting, boring, and loading phases of bord and pillar mining to be completed as one fully automated operation (Illawara Coal Port and Australian Academy of Technological Sciences and Engineering, 2000, p. 759). Continuous miners also allowed mechanical extraction of coal from the pillars without the use of explosives and with a substantial reduction in the number of coal miners required per tonne of coal extracted (Australian Academy of Technological Sciences and Engineering, 2000, p. 759).[13]

The successful demonstration of the continuous miner at the Huntley Colliery eventually led to the lifting of a ban (originally prompted by the concerns of labor) on the mechanical extraction of coal from pillars and the start of multiple shift operations in underground mines, which previously operated on a single shift (Australian Academy of Technological Sciences and Engineering, 2000, p. 759). The continuous miners would first cut the coal from the work face of the mined area using a revolving drum head containing concentrically positioned teeth (Productivity Commission, 1998, vol 2. p. B7). The coal would then be moved toward the center of the mine and transported out of the mine.

Box 6.2 Bord and pillar mining

During the recovery phase and for most of the rapid growth phase, mining companies developed their underground mines according to the bord and pillar mining system, which consisted of a series of parallel main tunnels that intersected secondary tunnels known as "cut-throughs," which were driven at right angles to the main tunnels (Joint Coal Board, "Fifth Annual Report for the Financial Year 1951–1952," appendix II, p. 48). The roof of the mine was supported by the pillars of coal as shown in Figure 6.2.

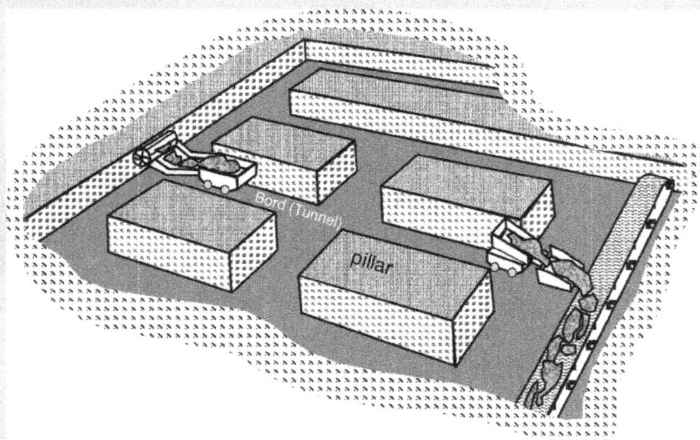

Figure 6.2 Diagram of a bord and pillar mine.

The bord and pillar system involved removal of the in situ coal in two "workings." The first working involved the extraction of coal from the tunnels followed by the removal of coal in the pillars, which was known as a second working. The percentage of coal left in pillars after completion of the second working was always quite high, at least 50% of the total recoverable coal and oftentimes much higher (Joint Coal Board, "Fourth Annual Report of the Joint Coal Board for the Financial Year 1950–51," p. 21). Once most of the pillar coal had been extracted from a specific section of the mine, the mine operator allowed the roof to collapse under "controlled conditions," which involved removing the temporary roof supports.

The coal extraction process in both workings needed to happen in four distinct phases: (i) coal cutting; (ii) boring (drilling holes in the coal seam for placing explosives); (iii) shooting or placing and detonating the explosives; and (iv) loading the coal onto wagons for transport out of the mine. Manual cutting, boring, and shooting of coal seams during the first working was very time-consuming, dangerous, and an unproductive use of labor. The manual

Box 6.2 (cont.)

removal of the pillar coal during the second working was also very dangerous and inefficient (Joint Coal Board, "Fourth Annual Report of the Joint Coal Board for the Financial Year 1950–51," p. 49). Although a maximum of 60% of the in situ coal could be extracted using bord and pillar methods, the extracted amount was typically much lower when manual methods for removing pillar coal were used. (Longwall mining systems, which were not widely applied in NSW mines until the late 1980s, allow for 60% to 80% of the in situ coal to be extracted – see Box 6.3.)

The use of continuous miners also facilitated mechanization of coal transport from the coal face to the surface of the mine. At first, the extracted coal was loaded from the continuous miner onto a rubber-tired shuttle car, which would transport and then discharge the coal onto a belt conveyor that would carry the coal to the surface of the mine (Australian Academy of Technological Sciences and Engineering, 2000, p. 762). Eventually, shuttle cars were replaced by electric conveyors equipped with steel-cored, non-flammable belts. These conveyors, which were connected to the back of the continuous miners, allowed the extracted coal to be conveyed from the mine face to either the surface of the mine or a central collection area within the underground mine. This transport arrangement allowed continuous miners to be more fully utilized and to achieve the lowest cost of coal extraction and loading.

By the end of the recovery period in 1959, seventy-one continuous miners were in operation in NSW, covering 39% of NSW's coal production (Joint Coal Board, 1959–60, pp, 29–30). The use of continuous miners and other improvements to the methods of underground bord and pillar mining led to very large productivity gains for Australia's underground coal mines during the recovery period.[14] Output per man-shift worked (OMS), a standard measure at the time of coal mine worker productivity in Australia, increased from 2.93 tonnes per man-shift in 1950 to 5.13 tonnes per man-shift in 1959 for NSW underground mines (Joint Coal Board, 1960–61, p. 146, table 22).

Worker productivity gains of even greater magnitude were achieved through increased mechanization of open-cut mines operating in NSW. In 1950, NSW's open-cut mines had a worker productivity of 8.55 tonnes per man-shift. By 1959, NSW's open-cut mines produced 19.12

tonnes per man-shift (Joint Coal Board, 1960–61, p. 146, table 22). These open-cut productivity gains were achieved through the application of massive electric-powered draglines, large diesel-fuelled trucks, and other pieces of large earth-moving equipment, which collectively resulted in more economical overburden removal and coal extraction. As it had for underground mines, the JCB played an important role in promoting the adoption of advanced machinery for open-cut mines.

For all the positive role it played in promoting advancing technologies for open-cut mining, the JCB also held back the productivity of open-cut mining for a time with well-intentioned but misguided policy restrictions. Open-cut mining was first applied in Australia in 1937 at the Blair Athol mine in Queensland and in 1940 it was introduced into NSW. The first open-cut mines to operate in NSW were prolific producers of coal, but due to poor advance planning the open pits were rapidly depleting by the early 1950s, leading to disruptive swings in production. In response, the JCB in 1954 attempted to limit production from open-cut mines by requiring coal producers to first try to meet their coal orders from existing underground mines. Only after the underground capacity was exhausted were coal producers allowed to rely on open-cut mines for any unmet demand. As a result of the JCB's policy of using open-cut mines as producers of last resort, NSW production from open-cut mines dropped from 1.8 Mt and 14% of total NSW production in 1950 to 0.6 Mt and 3% of total NSW production in 1959 (Joint Coal Board, 1960–61, p. 146, table 22). The JCB eventually realized the folly of this restriction and dropped it by the late 1950s. Open-cut mining during the subsequent rapid growth phase (1960–1986) became the primary vehicle for growing the black coal industries of NSW and Queensland.[15]

These technological advances had important indirect impacts beyond the sheer increases in productivity that they yielded. Namely, they contributed to the decline of labor union power at the coalmines of NSW and Queensland by reducing the need for unskilled and semi-skilled labor in both underground and open-cut mines. The advanced machines and methods required improved technical skills on the part of the remaining labor force, and these skilled workers were paid higher salaries and provided with improved working conditions and benefits.

The improved pay and working conditions along with the decline in union power led to reduced labor-management tensions and a huge reduction in lost output due to industrial work stoppages in NSW (Figure 6.3). Production losses due to industrial work stoppages, which had stood at 1.7 Mt and 11% of possible production in 1950,

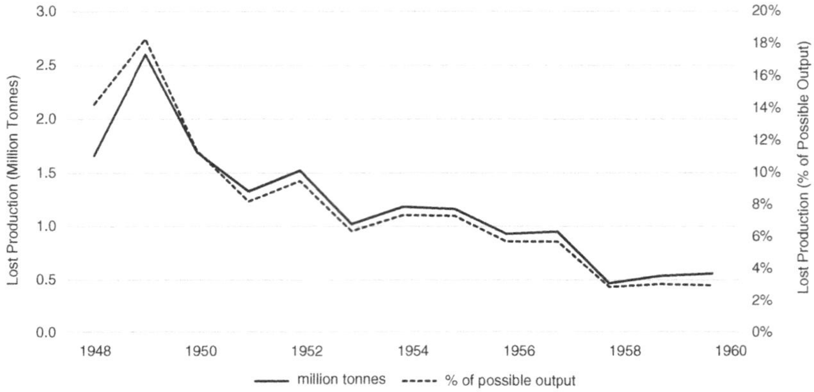

Figure 6.3 Lost production due to work stoppages at NSW coal mines, 1948–1960. Source: Joint Coal Board, "Fifteenth Annual Report of the Joint Coal Board for Financial Year 1961–62," pp. 174 and 175, tables 27 and 28.

had by 1959 been reduced to a relatively small amount – less than 600,000 tonnes and 3% of possible production. By the close of the recovery phase, industrial work disputes were no longer seen as a threat to the survival of NSW's black coal industry (Joint Coal Board, 1960–61, p. 150, table 26.).

6.2.1.3 *Market trends: development of exports and shifts in the domestic market*

The application of advanced mining technologies also resulted in substantial reductions in the cost of production that allowed Australian black coals to compete in international markets. This in turn justified the expansion of rail[16] and port infrastructure in both NSW and Queensland. During the 1950s, the Port of Gladstone was transformed from a port that catered to declining primary industries, such as cattle and other agricultural products, into a coal export center.[17] Starting from an extremely low base of 68,000 tonnes in 1950, exports grew by 38% per annum over the 1950s, reaching 1.2 Mt by 1959. Because export markets were growing simultaneously with the expansion in mine capacity, the industry was able to grow its output without causing a price collapse.

The domestic market remained the dominant source of demand through the 1950s, with domestic uses of coal accounting for 99% of coal sales in 1950 and 94% in 1959.[18] During the recovery phase, the

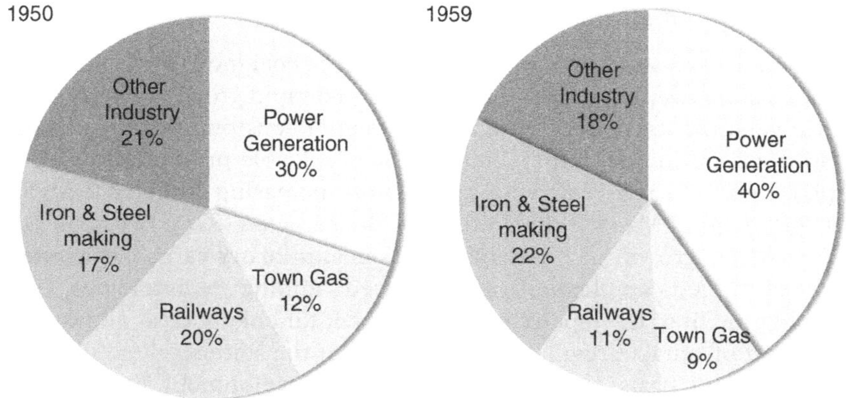

Figure 6.4 Share of black coal consumption by industry, 1950 and 1959. Sources: Australian Official Yearbook, #42, 1956, p. 1012; Joint Coal Board, 16th Annual Report, 1962–63, p. 68.

domestic markets for coal started to transition away from the shipping, railway, and town gas industries to the electric power and steel making industries (Figure 6.4).[19] The demise of the coal bunker fuel market was already a fait accompli by 1950, with the small remaining amount of coal used to fuel ships limited to those plying interstate trade in Australia. The beginning of the end for coal in the rail and town gas industries was also noticeable by 1959 (Figure 6.4). The growth in the domestic power and steel making markets for black coal largely offset the declines in coal usage in the rail and town gas industries prior to 1970.

Growth in saleable production overall remained modest during the recovery period. In 1950, saleable production of black coal for all of Australia stood at only 16.8 Mt (see Figure 6.1). By 1959, saleable production increased to 19.8 Mt, representing a compound annual growth rate (CAGR) of only 3.2%. Most of the production was from NSW, which accounted for 77% of Australia's saleable black coal production over the period 1950–1959, with Queensland accounting for 14% of saleable production and other states for the remaining 9% (Joint Coal Board, 1980–81, p. 12, table 5). The coal production mix began to shift during the recovery phase. The primary coal product in 1950 was steam coal, accounting for 82% of total coal sales. By 1959, steam coal sales had dropped to 72% of total black coal sales as coking coal emerged as the Australian black coal industry's fastest growing market.

6.2.2 *Rapid growth phase (1960–1986)*

Between 1960 and 1986, the Australian black coal industry achieved an astonishing twenty-six years of uninterrupted rapid growth (Figure 6.1). It diversified its sources of supply, its export customers, and its coal products during this time. The CAGR of saleable production for this phase was 8% per year, with production increasing from 22.2 Mt in 1960 to 148.7 Mt in 1986 (Joint Coal Board, 1986–87, p. 14, table 8). The strong growth of Australia's black coal industry can be attributed to continued application of advanced mining technologies, the rapid growth of the Japanese export market for coking coal, and development of suitable regulatory approaches at the state level (sometimes on an ad hoc basis) that supported industry development. Queensland emerged as a key coal province during the rapid growth phase, becoming Australia's largest exporter of black coal in 1972 and its largest producer of "raw" black coal in 1985 (Joint Coal Board, 1986–87, p. 14, table 8).[20]

6.2.2.1 *Regulatory environment*

6.2.2.1.1 Finding workable regulatory approaches at the state level The development at the state level of sometimes inelegant but sufficiently effective regulatory approaches played an important role in facilitating the production expansion observed in the rapid growth period. Queensland and New South Wales maintained their own separate laws and regulations for the awarding of coal exploration and mining licenses, known in Australia as tenements.[21] At the start of the rapid growth period, these mining laws were still attuned to the production methods and technologies in use prior to WWII, when coal mining was a small-scale, labor-intensive industry.[22]

During the 1960s and 1970s, especially in the state of Queensland, large mines were developed in outlying areas lacking rail and basic community infrastructure and amenities. Developing these new mining areas required the private companies to make massive investments in new rail and port facilities and to support the establishment of new communities and towns for housing the miners and their families. The existing laws and supporting regulations did not provide an adequate framework for developing these large-scale, capital-intensive, mechanized mining operations (Galligan, 1989, pp. 44–76).

To get around the limitations of outdated mining laws, mining companies were forced to engage in negotiations with the minister responsible for mining in each state in order to obtain licenses over large areas.

The terms and conditions of the agreements reached between each company and the relevant minister were then embodied in an act of that state's Parliament (Official Yearbook of Australia, 1976, p. 932).[23] These special laws were in turn amended to cover subsequent projects. In Queensland, one such special law, the Central Queensland Coal Associates (CQCA) Agreement Act, No. 58 of 1968, provided most of the regulatory support for open-cut coal projects implemented during the 1970s and 1980s (Galligan, 1989, pp. 60–66). As discussed further in Chapter 7 of this volume, Indonesia copied this approach of tailored government-to-business agreements during the 1980s.[24]

However, establishing mining licenses by passing and amending special laws was expensive for individual companies and politically risky for both companies and governments. New, more comprehensive state laws sought to improve this situation. In 1972, the Queensland government passed the Queensland Mining Act of 1968 to 1971 to resolve the issues related to ad hoc licensing agreements (Galligan, 1989, pp. 60–66). NSW followed with the Mining Act of 1973, which become effective as a law in 1974. South Australia also passed a new mining law in 1971. But even these so-called "modern mining acts" were found to be wanting, and the ad hoc approach of government-to-business (G-T-B) agreements continued to be used by the governments of Queensland and NSW for issuing mining licenses for large mining projects (Official Yearbook of Australia, 1990, p. 932).

The role of the JCB also changed during the rapid growth phase. When the JCB was established in 1946, the Commonwealth and NSW governments authorized it to take any actions necessary in order to ensure that:

(a) Coal is produced in the State (NSW) in such quantities and with such regularity as will meet requirements throughout Australia and in trade with other countries;
(b) The coal resources of the State are conserved, developed, worked and used to the best advantage in the public interest; and
(c) Coal produced in the State is distributed and used in such manner, quantities, classes and grades and at such prices as are calculated best to serve the public interest and secure the economical use of coal and the maintenance of essential services and industrial activities. (Joint Coal Board, 1947–48, p. 5)

The JCB was also mandated to promote the welfare of workers engaged in the coal industry in the state.[25] By 1975, the JCB had been so broadly successful in achieving its four objectives that its role shifted from organizing the mechanization of mines, setting prices, and controlling production to publishing a detailed statistical report titled "Black Coal in Australia" and overseeing safety programs.[26] (Another key step in the maturation of the sector was the establishment in 1971 of the Australasian Joint Ore Reserves Committee (JORC), which has played an extremely valuable role ever since in developing world-class standards for classifying resources and reserves.[27])

6.2.2.1.2 The re-emergence of labor issues as a concern Even in the more mature regulatory environment of the rapid growth period, some labor issues returned, although coal producers were able to keep them from causing serious disruptions in coal supply. In all years in the 1960s except 1969, lost production due to industrial strife was less than 3% of possible production. However, from 1970 on, labor disputes once again caused significant lost production, between 7% and 12% of possible production in 1971, 1975, 1979, 1981, and 1985 (Figure 6.5) (Joint Coal Board, 1986–87, p. 91, table 95).

The strikes were largely over wages. It soon became apparent to miners that coal companies were not only earning banner profits due to the strong growth in their export markets but that inflation, which was running in double digits during most of the 1970s, was taking a serious bite out of their rising nominal wages. Figure 6.6 shows the impact of inflation on the real annual salary of the average NSW coal mine employee over the period 1968–1986. Although nominal wages increased smartly over most of the rapid growth period, the real wage rate for the average NSW coal mine employee increased by 44% between 1968 and

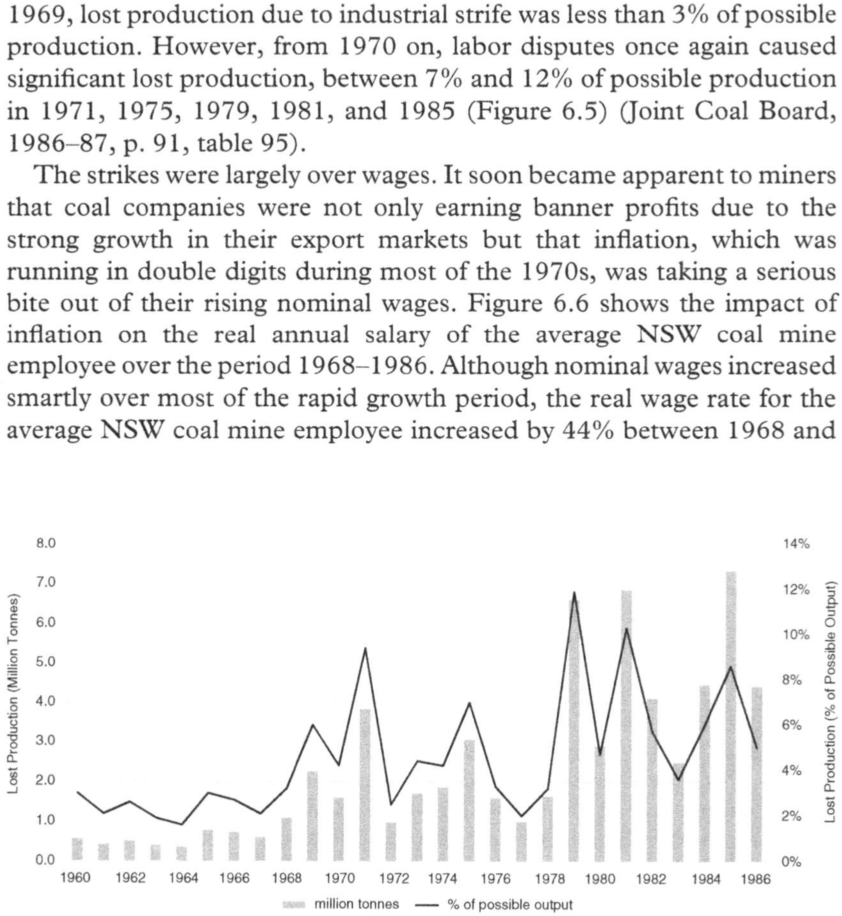

Figure 6.5 Lost production due to industrial strife in Australian coal mines, 1960–1986. Source: Joint Coal Board, "Black Coal in Australia, 1986–87," p. 91, table 95.

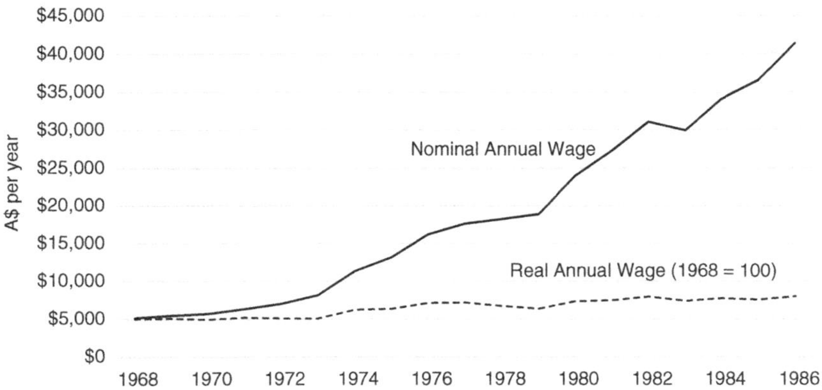

Figure 6.6 Annual nominal and real wages for NSW coal miners, 1968–1986. Sources: ABARE, "Australian Commodity Statistics 2009," p. 9, table 9, for CPI data. Wage data taken from JCB Annual Report 1972–73, p. 163, table 86, for years 1968–69; JCB Black Coal in Australia (BCIA) 1983–84, p. 60, table 62, for years 1970–82; from JCB BCIA, 1985–86, p. 62, table 66, for weekly salary data and multiplied by 52 to give annual salary estimates for years 1983–85; and JCB BCIA 1986–87, p. 66, table 65, for weekly salary data and multiplied by 52 to give annual salary estimates for 1986.

1977 (almost 5% per year) but only by 10% over the next nine years (1977–1986) or slightly more than 1% per year. Facing this slowdown in the growth of their earning power, it is hardly any wonder that miners returned to their strike-prone ways of the 1940s.

6.2.2.2 *Technology: longwall mining and further development of open-cut mining*

During the rapid growth phase, the continued mechanization of mines and the growth of open-cut mining relative to underground mining drastically improved efficiency and boosted production. The penetration of continuous miners in NSW shot up from 37% of underground coal production in 1960 to 87% by 1966 (Joint Coal Board, 1960–61, p. 160, table 34 and 1965–66, p. 30).[28] Following on the heels of this very rapid mechanization, new longwall mining techniques provided another big boost in productivity for underground mines (see Box 6.3). In parallel, new high-capacity equipment enabled big productivity gains in open-cut mining.

Longwall mining was designed to address several shortcomings of continuous miners. First, "continuous" miners actually only operated about 50% of the available time at the coal face (Australian Academy of Technological Sciences and Engineering, 2000, p. 760). A continuous miner has to stop operating to allow an empty shuttle car to "berth" next to it after a loaded shuttle car departs as well as when roof supports have to be put in place when moving from tunnel to tunnel (Australian Academy of Technological Sciences and Engineering, 2000, p. 760). Second, a continuous miner could work only around 50 m along the coal face before it needed to be moved to a new location (Australian Academy of Technological Sciences and Engineering, 2000, p. 760). Third, removing the pillars with a continuous miner, although safer than doing the job manually, still posed risks to the eight-man crew working the continuous miner, which became a growing safety concern over time (Joint Coal Board, 1951–52, pp. 48–49). Mine equipment manufacturers in the UK designed a new underground mining approach (later branded the longwall mining system) that could achieve almost 100% continuous operation with an improved level of worker safety through the use of hydraulic roof support assemblies to which a conveyor and coal cutter, known as a shearer, were attached (Australian Academy of Technological Sciences and Engineering, 2000, p. 760).

Box 6.3 Longwall mining

The longwall mining system improves productivity and safety through the use of hydraulic roof support assemblies to which a conveyor and coal cutter ("shearer") are attached. Prior to installing the hydraulic roof supports and putting the longwall mining machine in place, continuous miners were used to cut two parallel tunnels into the coal panel. These tunnels, which were typically placed 200 m apart, were used to place pillars to support the mine walls and roof while allowing the mining area to be properly ventilated. At the front of the coal panel, the continuous miner was used to cut another tunnel, known as the long wall, perpendicular to the two parallel roads. Finally, the continuous miner was used to cut ventilation shafts and escape tunnels as needed throughout the longwall coal mine. (Note that the descriptions in this box are based on Productivity Commission, 1998, "The Australian Black Coal Industry" Inquiry Report, AusInfo, Canberra, vol. 2, appendix B).

At that point in the development of any longwall mine, the hydraulic roof supports were put in place and the longwall

Box 6.3 (cont.)

mining machine, consisting of a shearer and conveyor system, was moved against the long wall. The shearer was used to extract the coal and place it on the conveyor belt in one continuous operation. After the coal was removed from a particular area of the mine, the hydraulic assemblies, which are self-advancing, were moved forward into the coal panel, allowing the process of coal cutting and removal to occur on an almost continuous basis. The roof area directly behind the mined-out area was then allowed to collapse as the longwall miner proceeded into the coal panel. (The roof material, known in Australia as "goaf," contains coal and needs to be properly ventilated and then sealed underground to avoid risk of dust explosions and spontaneous combustion.)

Once the coal was cut from the panel, it was loaded onto a conveyor, which either carried the coal directly to the surface or to a central storage area for later transfer to the surface by rubber-tired loader or second conveyor. The use of a flexible conveyor to remove the cut coal away from the coal face while moving the shearer and roof assemblies forward allowed the process to be very close to continuous. The only instances where the system needed to be shut down were (a) for maintenance, either routine or forced, and (b) if the position of the shearer needed to be reversed.

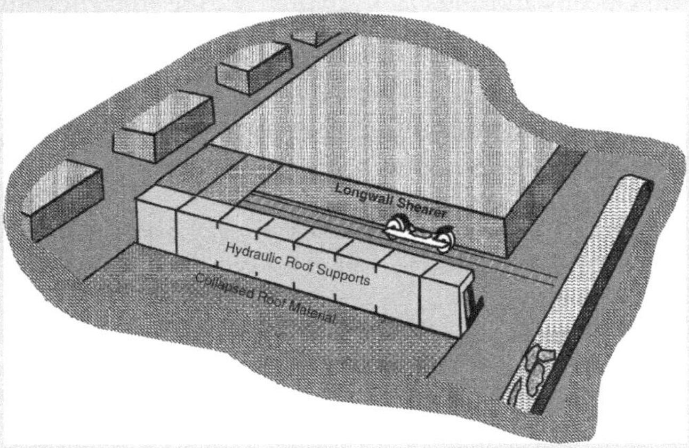

Figure 6.7 Diagram of a longwall mining operation.

The longwall mining system was first used in the UK in 1963 (Productivity Commission, 1998, vol. 2, p. B8). The Joint Coal Board attempted to apply the UK system in NSW between 1967 and 1970, but operating problems emerged as the UK design was not suited for NSW mining conditions. By 1972, through trial and error, the JCB developed a modified design that worked for NSW conditions (Joint Coal Board, 1972–73, pp. 48–49).[29] In 1978, the longwall mining system accounted for only 4.3% of NSW underground production (1.63 Mt), but by the end of the rapid growth period, its share of underground coal produced in NSW had increased to 32% (16.6 Mt longwall, versus 51.8 Mt total underground production) (Joint Coal Board, 1986–87, p. 22).[30]

Toward the end of the rapid growth period (1980–1986), longwall mining methods started to displace bord and pillar mining methods in underground mines throughout Australia. The shift to the longwall mining system and other improvements in underground mining technology allowed worker productivity in Australia's underground mines to increase from 5.31 tonnes per operating man-shift (man-shift) in 1960 to 13.98 tonnes per man-shift in 1986, a productivity gain of 4% per year (CAGR basis) (Joint Coal Board, 1961–62, p. 20, table II, and 1986–87, p. 78, table 83).

Advanced equipment and technology were also applied to open-cut mines in Queensland and NSW. In this case the efficiency advance came from the economies of scale enabled by new equipment. Table 6.1 shows the massive increase in handling capacities of the overburden removal and coal extraction equipment used in Australia's open-cut mines. Worker productivity in Australia's open-cut mines on a raw

Table 6.1 *Handling capacities of major pieces of equipment used in Australian open-cut mines for overburden removal and coal extraction*

	Typical capacities in cubic meters	
Type of equipment	1950	2001
Draglines	2	46–60
Shovels	2	14–46
Heavy excavators/bulldozers	1.5–2	10.0

Source: Australian Academy of Technological Sciences and Engineering, "Technology in Australia 1788–1988 (updated November 21, 2001)," p. 763.

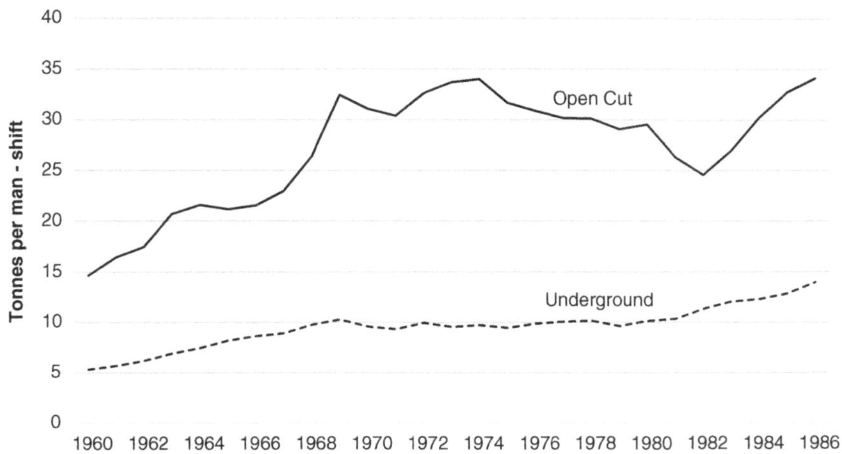

Figure 6.8 Raw coal output per man shift for Australia's underground and open-cut black coal mines, 1960–1986. Source: Joint Coal Board "Black Coal in Australia, 1986–87," p. 75, table 80.

coal basis climbed from 14.64 tonnes per man-shift in 1960 to 34.10 tonnes per man-shift in 1986, an improvement of 3% per year (CAGR) (Figure 6.8) (Joint Coal Board, 1960–61, p. 150, table 2 and 1986–87, p. 75, table 80).

Open-cut mines were more productive than underground mines to begin with (see Figure 6.8), so the shift to more open-cut mining was itself a huge benefit for productivity. At the start of the rapid growth phase, open-cut mines accounted for only 3% of NSW raw coal production and 2% of total Australian black coal production. By the close of this period, open-cut mines accounted for 40% of NSW raw coal production and 67% of all Australian black coal production (Joint Coal Board, 1985–86, p. 75, table 80). Open-cut mines accounted for 67% of Queensland's raw coal production in 1970 and more than 93% of its raw coal production in 1986 (Joint Coal Board, 1960–61, p. 150, table 2, and 1986–87, p. 22, table 18).

6.2.2.3 Market trends: increasing exports to Asia

Production growth during the industry's rapid growth phase was largely export driven, with the previously dominant domestic market declining from 91% of total saleable production of black coal in 1960 to 30% of total saleable production by 1986 (Figure 6.9). Between 1960 and 1986,

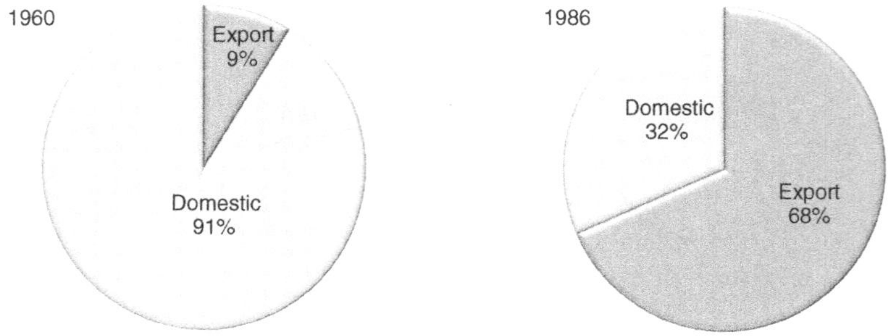

Figure 6.9 Exports and domestic coal sales, 1960 and 1986. Sources: Joint Coal Board, "23rd Annual Report, 1969–70," p. 249, table 9; Joint Coal Board, "Black Coal in Australia, 1986–87 (January 1988)," p. 97, table 100 (export), and p. 126, table 133 (domestic).

Australia's black coal exports increased from 1.9 Mt to 95.7 Mt as demand for steam and coking coal in Asia grew at an impressive 28% annual growth rate while Australia's domestic consumption grew at a much slower rate of 3% per year (Joint Coal Board, 1986–87, p. 97, Table 100).

The market for coal in Australia's railways and in sea vessels disappeared completely during the 1950s. Town gas took a bit longer – until 1967 – to fade away completely.[31] From 1960 onward domestic coal consumption became increasingly dominated by the power industry, followed by steel and iron production, as shown in Figure 6.8. The oil shocks of the 1970s created strong demand for steam coal to replace oil in power generation, which helped lead to a shift in Australia's export mix from almost entirely coking coal in 1960 to a nearly even mix of coking and steam coal in 1986 (Figure 6.11).[32]

Japan, with its strong focus on assuring energy security, was the largest export customer for Australia's black coal producers throughout the rapid growth phase of development (Joint Coal Board, 1960–61, p. 75).[33] Australia's reliance on the Japanese export market, especially in the initial part of the rapid growth phase, created predictable and long-term revenue streams that allowed mining companies to obtain long-term financing for both mining and transport infrastructure projects in NSW and Queensland. For the Japanese buyers, the top concern was security of coal supply, with price appearing to be of secondary concern. In this context, the interests of the buyer (Japanese steel mill or power utility) and seller (Australian coal producer) were aligned. The Commonwealth government, however, did not initially view the relationship in such symbiotic terms. Instead, it worried that Japan would exert market

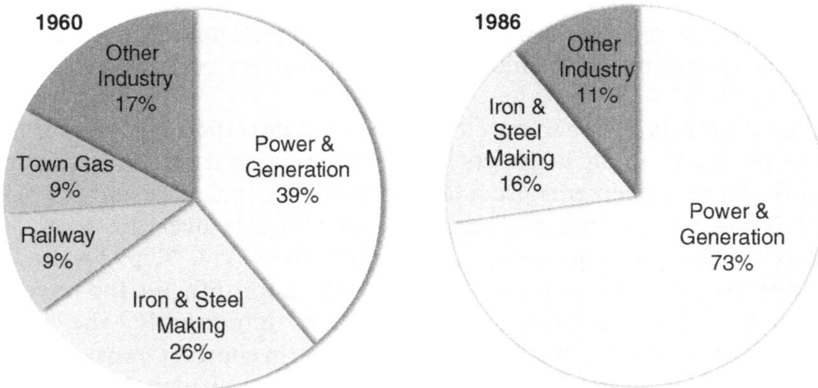

Figure 6.10 Shares of domestic coal consumption by industry, 1960 and 1986. Sources: Australian Official Year Book, 1962, p. 1090; Joint Coal Board, "Black Coal in Australia, 1986–87," p. 126, table 133.

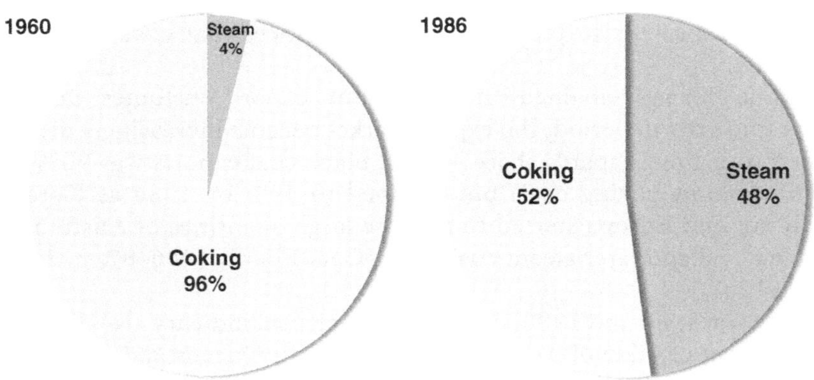

Figure 6.11 Steam coal and coking coal shares of total exports, 1960 and 1986. Source: Joint Coal Board, "Black Coal in Australia, 1986–87," January 1988, p. 97, table 100.

power over Australia's coal producers, forcing coal producers to accept below-market prices for their exports and thereby reducing royalties and tax revenues for the Commonwealth and state governments. As a result of this concern, the Australian government in 1973 imposed price controls on all coal exports. These price controls were not removed until 1993 but were only weakly enforced after 1983 (Bowen and Goodray, 1993). (A 2001 analysis by the Australian government found that Japanese buyers actually paid a premium for Australian steam coal between 1990 and 2000.[34])

Coal exports to Japanese customers were governed by long-term, cost-plus coal supply contracts with terms ranging from ten to fifteen years. These agreements provided for a starting coal price and pre-set escalation factors that allowed price adjustments over the term of each contract (Productivity Commission, 1998, vol. 2, p. D-12).[35] However, this arrangement became much more difficult for buyers to police than originally expected. More importantly, the true market price, as measured by spot and later term contract transactions, diverged significantly year to year from the prices calculated under the previously agreed cost-plus mechanisms (Productivity Commission, 1998, vol. 2, p. D-12). This divergence from market prices became especially problematic during the 1970s, when the price of oil and other commodities increased dramatically as a result of the twin oil price shocks of 1973 and 1978 (Productivity Commission, 1998, vol. 2, p. D-12). At the same time, these spikes in oil prices, which highlighted energy security concerns, played a significant role in supporting exports, especially to Japan, and boosting prices that coal exporters received.

While Japan remained the dominant export customer during the rapid growth period, the export market became increasingly diversified over time. Japan's share of total black coal exports was 90% in 1960 (mainly coking coal) but dropped to 49% by 1986 as Korea, Taiwan, and Europe started to procure large quantities of Australian coking coal and later steam coal (Joint Coal Board, 1986–87, p. 117, table 126).

Between 1960 and 1986, Australia's black coal industry also diversified the supply side of its export business, with Queensland replacing NSW as Australia's largest black coal exporter in 1975 and its largest producer of saleable black coal in 1983.[36] The shift in supply-side dominance from NSW to Queensland was not due to lack of growth in NSW's exports of black coal, which increased from 1.9 Mt in 1960 to 42.2 Mt in 1986 (Joint Coal Board, 1986–87, p. 117, Table 126). It was just that Queensland's export growth was so much faster. Starting from a very small export tonnage of 48,000 tonnes in 1960 (representing just 2% of Australia's black coal exports), Queensland's black coal exports grew to 53.5 Mt in 1986, which translated into a 56% share of total black coal exports (Figure 6.12) (Joint Coal Board, 1986–87, p. 97, table 100).

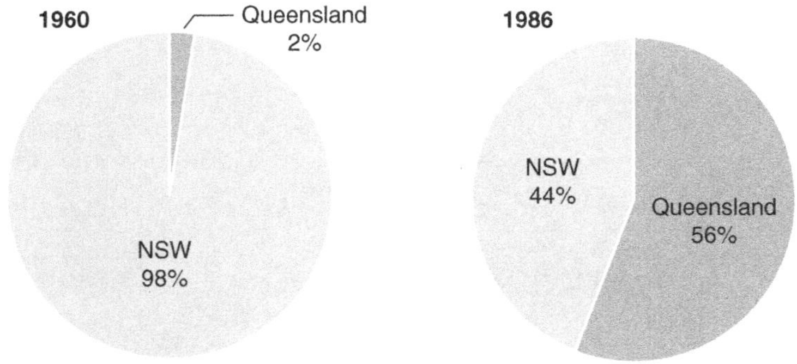

Figure 6.12 Queensland and NSW shares of black coal exports, 1960 and 1986. Source: Joint Coal Board, "Black Coal in Australia, 1986–87," p. 97, table 100.

In addition to having vast coking coal deposits that were amenable to open-cut mining, Queensland had better topography, with multiple locations suitable for construction of deep-sea coal ports, such as Gladstone, Hay Point, and Abbot Point. Exports from NSW's coal industry, on the other hand, were largely limited to the port of Newcastle due to draft limitations along most of the coast and mountain ranges between the inland Hunter Valley coalfields and the coast. Queensland's coal export terminals, especially the Abbot Point Coal Terminal, also had the advantage of being located closer to the key markets of North Asia. Abbot Point has a roundtrip sea vessel journey to Japan that is at least two days shorter than the equivalent trip from Newcastle.

6.2.2.4 *Transportation infrastructure: port and rail expansion*

The major coal handling ports of NSW (Newcastle) and Queensland (Gladstone, Hay Point, and Abbot Point) were either built or expanded during the rapid growth phase. The port of Newcastle is one of the world's largest coal export ports and has the longest history of all of Australia's world-scale coal export ports. (Its first commercial shipment of coal – 50 tonnes – occurred in 1799.) The port of Newcastle became a significant player in the coal export market when it commissioned the Carrington Coal Terminal (CCT), with a ship loading capacity of 16 Mtpa, in 1976. It subsequently added the Kooragang Coal Terminal (KCT), with an initial ship loading capacity of 25 Mtpa, in 1984 (Newcastle Port

Figure 6.13 Coal piles being kept moist at Port of Newcastle. Source: Reuters.

Corporation, 2014). Figure 6.13 shows a photo of the Kooragang Coal Terminal.

In Queensland, the port of Gladstone (see photo in Figure 6.14) brought on line the Barney Point Coal Terminal (BPCT) in the mid 1960s[37] and the RG Tanna Coal Terminal (RGTCT) in 1980 (Gladstone Ports Corporation Ltd., 2011).[38] The port of Hay Point, which is located north of Gladstone, is Australia's second largest coal export port and consists of two coal terminals: Hay Point, which has been operating since 1971[39] (and which currently serves as a dedicated terminal for coal exports from the BHP Billiton Mitsubishi Alliance), and Dalrymple Bay, which was commissioned in 1983 as a common user facility with a ship loading capacity of 15 Mtpa (North Queensland Port, 2012). The port of Abbot Point contains a coal terminal that was commissioned in the early 1980s and is currently under lease to Adani Abbot Point Bulk Coal Pty Ltd, a subsidiary of Adani Pty Ltd. Smaller coal terminals exist at Brisbane in Queensland and Port Kembla in NSW (Fish, 2010).

The states of Queensland and NSW also constructed extensive rail networks during the rapid growth phase to haul coal from inland mines to coastal ports for export. These rail networks included spurs that delivered coal to specific power plants throughout each state. For the

Figure 6.14 Ships waiting to be loaded at Port of Gladstone. Source: Reuters.

entire rapid growth phase, the railway systems of NSW and Queensland were owned and operated by the governments of the states within which they operated.[40]

The development of Queensland's rail system started in 1884, when the Griffith government secured a 10 million pound loan for railroad expansion. These early rail lines were intended primarily to haul passengers and agricultural products to market, which is not surprising since Queensland's coal production amounted to less than 100,000 tonnes in 1881 and remained below 1 Mtpa until 1913 (Official Australia Yearbook, 1916, "Non-Metallic Minerals: Coal," p. 426). It was not until the late 1960s that the government of Queensland made a serious effort to develop a rail network for hauling coal from the Bowen Basin to the ports of Gladstone and Hay Point. The first major coal haul line was the Moura Short Line, which was completed in 1968 with funding from the Thiess Brothers and other mining companies (Galligan, 1989, pp. 202–203). It was built and operated by government-owned Queensland Railways (QR) as part of its network (Galligan, 1989, pp. 202–203). Between 1960 and 1984, QR constructed additional coal haul lines to the port of Gladstone from coal mines at Moura, Koorilgah, Laleham, Curragh, Gregory, and Gordonstone. Spur lines and other expansions were built in the

Blackwater area as well as on the Goonyella system from the port of Hay Point to Goonyella and Blair Athol. Many of these lines were funded by Utah International, Thiess Brothers, and other mining companies that were developing mines in the Bowen Basin (Galligan, 1989, pp. 43–76). In the mid-1980s the main coal haul rail lines in central Queensland were electrified and more than 1,600 kilometers of railway lines were upgraded to allow increased amounts of coal to be moved to export ports in central Queensland (Queensland Coal).

6.2.3 The competition phase (1987–2003)

By the late 1980s, Australia's coal industry had matured to a significant extent, and the government was able to pull back from the more intense regulatory involvement that had marked previous stages. At the same time, competitive threats to the export industry began to develop, while dropping global energy prices created a more adverse environment for Australia's coal exports. Indonesia emerged as the most serious competitor with the development of its sub-bituminous and bituminous steam coal resources on Kalimantan (see Chapter 7). The greatest concern of both the Australian Commonwealth Government and the state governments of Queensland and NSW was whether its black coal industry could compete on price against Indonesia in Asian markets. Indonesia had cost advantages due to its favorable market location and the low stripping ratios of its Kalimantan coal deposits. The author labels the sixteen-year period from 1987 through 2003 as the competition phase, in recognition of the growing level of export competition that emerged from Indonesia as well as to a lesser extent South Africa and, for a limited period, China.

While technological changes were perhaps less revolutionary in this period, incremental technological development and the shift to open-cut mining continued as a way to reduce costs and make Australia's coal more competitive. A commission assembled by Australia's government in 1997 suggested that the industry's competiveness could be improved by eliminating the bord and pillar production still used in some underground mines and reducing fees levied by government-owned port and rail networks (Productivity Commission, 1998). Underground mines in NSW increased their production of raw coal "from long wall faces" from 16.6 Mt in 1986 (32% of total underground coal production) to 36.4 Mt (77% of underground production) by 2003 (Joint Coal Board, 1986–87, p. 23, table 20) and Coal Services Pty., Ltd/ Queensland Department of Mines and Energy, 2006, p. 14, table 15).[41]

Due to its huge cost advantage, open-cut mining remained the method of choice for those deposits amenable to it. Between 1986 and 2003, saleable coal produced from all Australian open-cut mines increased from 98 Mt to 219 Mt respectively (Joint Coal Board, 1986–87, p. 18, table 13 and Coal Services Pty Ltd/Queensland Department of Mines and Energy, 2006, p. 13, table 14).

6.2.3.1 *Regulatory environment: decreasing government involvement in a mature industry*

During the competition phase, the regulatory frameworks of NSW and Queensland changed in ways that both supported and reflected the maturation of the coal industry, in part by moving toward less direct government involvement. Queensland's Parliament signed the Mineral Resources Act of 1989 into law in 1992. The Parliament of NSW passed similar legislation, the Mining Act of 1992. These acts, which were passed mostly in response to the move to large open-cut mining projects in Queensland and NSW during the late 1960s and through the 1970s, allowed for the allocation of larger coal exploration blocks and mining areas and for the provision of community services, taking away most of the need for government-to-business (G-T-B) agreements.[42]

By the late 1980s, state-owned rail and port companies in NSW and Queensland were financially strong enough to support infrastructure expansion on their own. Mining companies could now meet their requirements for participating in the development of new infrastructure via business-to-business (B-T-B) agreements with the government-owned corporations that operated the rail and port networks, without the need for the state guarantees they had required previously.

In the later part of the competition phase, the governments of NSW and Queensland successfully privatized assets of their port and rail systems. In 2002, the NSW government allowed private companies to own and operate rolling stock on its fixed rail lines. The bulk of NSW's freight rail network is now being operated by Asciano, a private company traded on the ASX, under a long-term lease with the government of NSW.

By the start of the competition phase, Australia's black coal industry had matured to a point where there was no longer felt to be a need for industry-specific regulatory bodies, such as the JCB, the Coal Industry Tribunal (CIT), the Queensland Coal Board (QCB), and the Mines Rescue Board – entities that had previously held broad mandates to regulate production, pricing, and technology in Australia's black coal industry. The JCB was refocused in 1992 on "concern for the occupational health and welfare of mine workers" and then dissolved in 2002, with its remaining functions (along with mine rescue and statistical reporting) taken on by Coal Services Pty Ltd, a newly formed agency owned by the NSW coal industry (National Archives of Australia). The CIT and QCB were dissolved in the late 1990s, with the functions of the QCB taken over by Queensland's Department of Mines and Energy and those of the CIT folded into the appropriate labor relations agencies of the Commonwealth and NSW governments.

With basic industrial organization settled, regulators could turn to other issues as well. The 1993 Commonwealth Native Title Act required holders of mining tenements to conduct negotiations with native landholders (aboriginal persons) on just compensation and other matters related to native land rights (Department of Foreign Affairs and Trade). Other laws related to the allocation of water rights and environmental impact assessments were either passed or amended during the competition phase. These required amendments to the Queensland Mineral Resources Act of 1989 and the NSW Mining Act of 1992.

6.2.3.2 Market trends: decreasing real prices and increasing competition

As oversupply caused oil prices to begin falling from their late 1970s highs, downward pressures on real coal prices started to emerge. Between 1980 and 1986, Australian black coal producers experienced a 17% decline in the average real price for their black coal exports (Table 6.2 and Figure 6.15). This decrease in real coal prices was largely due to the general collapse in the price of energy worldwide between 1982 and 1986. The worst of the real price decrease, however, was yet to come. Mirroring plummeting oil prices, real prices for Australian black coal collapsed by an additional 33% from 1986 to 1987, and in 1996 they were still 29% lower than in 1986. The Asian Financial Crisis started in 1997 and further impacted Asian demand, suppressing prices for Australia's black coal through 2003. Exchange rate movements, which had a positive impact on the A$/t price during the last six years of the rapid growth phase had, on average, a negative impact on Australia's black coal industry between 1987 and 1996 (see Figure 6.16).

The emergence in 1995 of Indonesia as a major supplier of steam coal into the Asia Pacific region (see Chapter 7) posed a major new challenge to Australian suppliers. From 1995 through 2003, Indonesian coal suppliers sold their coals to power companies located in East and Southeast Asia at a significant energy-adjusted discount to the delivered prices

Table 6.2 *Nominal and real (1980) prices (A$/t) for Australian black coal, 1980–2013*

				(% Change)	
Year	Nominal	Real		Nominal	Real
1980	$41.93	$41.93	1980–1986	31%	−17%
1986	$54.87	$34.68	1986–1987	−13%	−33%
1987	$47.66	$28.16	1986–1996	0%	−29%
1996	$54.93	$24.58	1986–2003	−8%	−46%
2003	$50.36	$18.91	2003–2004	48%	43%
2004	$74.53	$27.09	2003–2013	45%	8%
2013	$108.31	$29.16	1986–2013	97%	−16%

Source: BREE Resources and Energy Statistics 2013, p. 27, table 26, and p. 43, table 42, for nominal price data, which were converted into 1980 real prices using Australian GDP deflators obtained from Australian Bureau of Statistics.

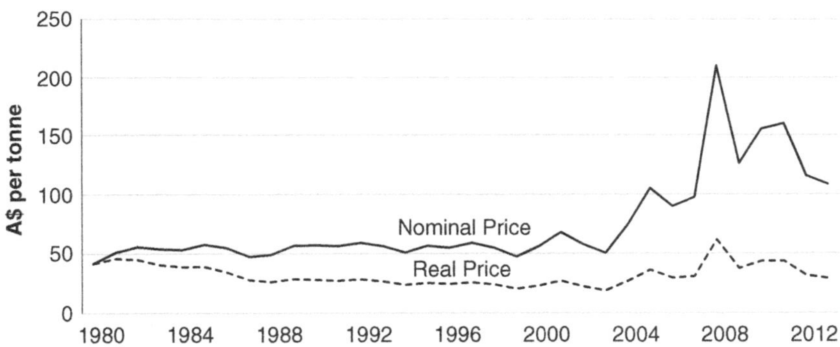

Figure 6.15 Nominal and real prices (1980 A$) of Australian black coal in A$ per tonne, 1980–2013. Source: BREE Resources and Energy Statistics 2013, p. 27, table 26, and p. 43, table 42, for nominal price data, which were converted into 1980 real prices using Australian GDP deflators obtained from Australian Bureau of Statistics.

charged by Australian producers, which enabled them to take market share from Australian producers. They were able to do so because of the low stripping ratios of their bituminous and sub-bituminous coal deposits as well as government-subsidized diesel prices that kept Indonesian coal production costs very low. By offering their coals at a

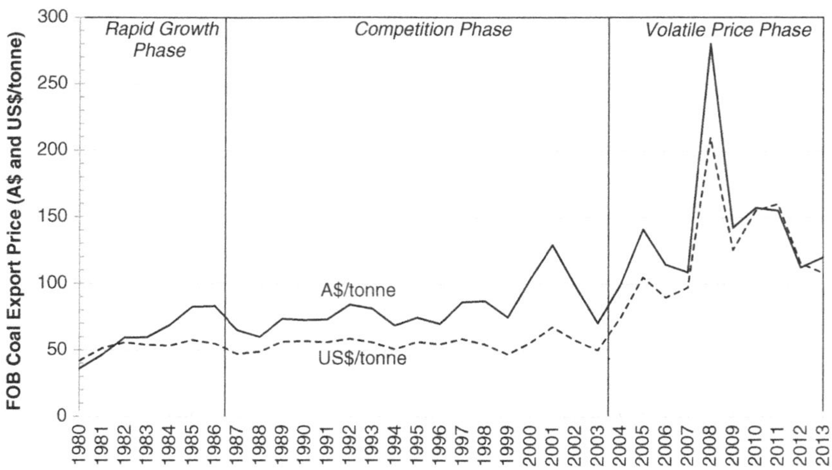

Figure 6.16 FOB price of Australian black coal exports in A$ and US$ per tonne, 1980 – 2013. Prices are annual weighted averages for black coal (steam + coking coal exports). Source: BREE Resources and Energy Statistics, 2013, for data from 1980 to 2012; BREE, *Resources and Energy Quarterly*, March 2014.

significant discount to Newcastle grade coals on an energy-adjusted basis and also allowing their customers to "capture the benefits" of any savings on transportation costs, Indonesia's coal producers were able to gain an increasing share of the Asian steam coal market over time. By 2006, Indonesia was the world's largest steam coal exporter in raw tonnage terms.

Improvements of 40% in output of Australian mine workers between 1987 and 1996 (Joint Coal Board, 1996, p. 23, table 24) helped the competitiveness of Australian coal, but a significant portion of these productivity gains were offset by higher wages and costs of other mining inputs as well as monopoly rents charged by government-owned ports and rail companies.[43] Moreover, during the competition phase, the governments of NSW and Queensland were either unable or unwilling to expand their rail and port infrastructure quickly enough to allow them to fully compete for Asian markets. Indonesia, on the other hand, had a more flexible inland transportation system based on easily expandable components like river barges, trucks, and floating transshipment facilities, which allowed Indonesian suppliers to quickly ramp up production (see Chapter 7).

6.2.4 *Volatile price phase (2004–2014)*

Despite a tightening of the Asian coal market in 2004, most analysts believed that the nominal price of coal would remain stable into the foreseeable future and that real black coal prices would continue to decline. Instead, nominal coal prices rose steeply and real prices rose moderately over the volatile price phase (Figure 6.15), with a significant increase in volatility. For both coking and steam coal, the largest price increases occurred between January 2007 and July 2008. They were due to (a) sudden increases in steam coal imports by China; (b) the shutdown of nuclear power plants in Japan for safety reasons, which created an unexpected increase in that country's demand for steam coal; and (c) coal supply shortfalls in China, Australia, Indonesia, and South Africa due to exceptionally bad weather plus chronic rail and port infrastructure constraints in Australia.[44] The transient nature of these and other factors has also contributed to price volatility. Other causes of increased volatility since 2004 have included global economic conditions (especially the onset in 2008 of recession in the United States and Europe[45]) and shifts in whether PCI-type coals are sold into coking or steam coal markets as a function of the price differential between these markets.[46]

Australia's regulatory environment during this period continued to shift away from direct government intervention, notably through the privatization of rail networks. New South Wales privatized its railway company in 2002, and Queensland followed later, selling off 60% Queensland Rail in 2010 and all but 5% of its remaining holdings by the end of 2013. For the most part, the regulatory framework for coal at the end of 2013 was mature and, especially compared with those in some other coal exporting countries like Indonesia and South Africa, effective at supporting necessary investment (although major new federal taxes on carbon and minerals extraction, which will be discussed in Section 6.4, injected significant new investment uncertainty). State-level regulation continues to be a defining feature of Australia's coal sector, with Queensland and New South Wales each having some relative strengths and weaknesses in their regulatory regimes.

Despite the generally favorable regulatory environment, Australia has wrestled in recent years with rail and port constraints. Coal price volatility, weather and geographical factors, and a lack of government follow-through at times have all contributed to these issues. The volatile price movements for Australia's black coal exports between 2007 and 2013 have at times led to "stop-and-go" decisions in which mining companies and state governments make "firm" decisions to proceed with large mining projects and related transport infrastructure projects one month, and then a few months later reverse those decisions due to changes in market conditions and financial constraints.

6.2.4.1 Regulatory environment

6.2.4.1.1 Rail privatization In 1998, the Commonwealth government's Productivity Commission challenged the state governments of NSW and Queensland and their state-owned railway corporations to improve their productivity and lower their costs of service to the coal industry. The Productivity Commission made two main recommendations with respect to rail freight:

(a) State governments should discontinue the practice of setting rail freight rates for hauling coal at a level that allows the state railways to earn monopoly profits or the state government to collect implicit royalties above the legally mandated royalty rate.
(b) State governments should encourage third-party access to fixed rail infrastructure and allow coal producers and other investors to own and operate rolling stock on the fixed rail network.

The government of NSW implemented most of the Productivity Commission's recommendations in 2002 when it reorganized its state-owned company FreightCorp in preparation for divestment to private investors. In that same year, the Commonwealth government and the government of NSW combined the Australian government-owned National Rail Corporation's freight operations and rolling stock with the NSW government's reorganized FreightCorp and in 2002 sold the combined assets to Toll Holdings[47] and Patrick Corporation, which formed Pacific National to operate and manage those assets.[48]

In 2012, Pacific National's rail division (Pacific National Rail, or PNR) transported approximately 90 Mt of coal from the Hunter Valley, Southern, and Western coalfields of NSW to domestic customers and export customers through the Port of Newcastle and Port Kembla. PNR claimed that, in 2009, it delivered 93% of New South Wales's export coal and was Australia's second largest coal hauler behind Queensland Rail.[49]

The Queensland government vacillated for a number of years before fully implementing the recommendations of the Productivity Commission. Its delay was most likely motivated by the importance of the monopoly rents earned by Queensland's state government from Queensland Rail. However, on November 22, 2010, the government of Queensland took an historic step when it successfully conducted an IPO for 60% of Queensland Rail National (QR National).[50] A few months prior to the IPO, a consortium of coal producers representing 98% of

Queensland's coal export business had proposed to the Queensland government that it be allowed to buy only QR National's Central Queensland fixed coal rail network for A$4.5 billion, but it withdrew the proposal after a few months of fruitless negotiations with the Queensland government (Energy Publishing Inc., "Coal Chain Australia" Nov. 2010, pp. 1–2). Between November 2011 and December 2013, the Queensland government proceeded to sell off its shares in QR National until its ownership stake was reduced to only 5% of total shares of QR National.[51]

QR National was divested as an integrated railroad company with ownership of both the fixed rail networks and rolling stock. However, as part of the sales arrangements, third parties such as major coal producers were allowed to own and operate coal haul trains on the QR National fixed rail network. Large coal producers, such as BHP Billiton, Xstrata, and Rio Tinto, considered buying their own rolling stock and hauling their coals to export ports but eventually dropped the idea (Energy Publishing Inc. "Coal Chain Australia" Dec 2010, pp. 2–3). Nonetheless, the Queensland government's privatization of its railroad business may eventually lead to a more vertically integrated black coal industry if coal producers do move in this direction of owning the rolling stock and port facilities used to transport and export their coals.

In 2010, QR National's coal division started a service to transport coal from mines in NSW to the Port of Newcastle (QR National). In 2013, it delivered 194 Mt of black coal to export ports in Queensland and NSW as well as to customers in both states (Energy Publishing Inc. "Coal Chain Australia Weekly" June 15, 2010, p. 1).[52] It announced plans to double its coal haul tonnage in Queensland by 2015 (Energy Publishing Inc. "Coal Chain Australia" December 2010, pp. 2–3).

6.2.4.1.2 Overall assessment of the regulatory frameworks in NSW and Queensland Both NSW and Queensland now have in place comprehensive and proven legal frameworks for allocating areas for exploration and development of new coal mines. They also have clear regulatory procedures and frameworks for developing new rail and port infrastructure projects and managing labor relations, mine safety, and matters that impinge upon international relations.

These mature frameworks do not appear to have significantly constrained the expansion of steam coal production in either state. Complaints by coal producers tend to concentrate around two areas: (a) the slow pace of implementing requests for project approval, a common complaint in any jurisdiction where conflicting interests need to be considered before a public decision can be made, and (b) the reversal of key decisions at the first sign of economic adversity (author interviews).

Although the state of Queensland is described by a number of industry sources as a more attractive regulatory venue than NSW for developing new coal mines, its state bureaucracy is also criticized for not showing the political will and courage to make controversial decisions, with the two most frequently cited examples being the following (author interviews):

- Slowness in resolving overlapping claims between the separate industries of coal mining, coal bed methane (CBM) extraction, and underground coal gasification (UCG) production;[53] and
- Delays in approving new port and rail projects and then a failure to hold to those decisions in the face of short-term changes in the domestic economy.[54]

In the case of NSW, these sources claimed that the NSW government's bureaucracy is less friendly to miners. The regulatory process in NSW has been described by a number of sources interviewed as an adversarial process that creates a much higher level of investor uncertainty than coal miners have experienced in Queensland.[55]

But in general, the regulatory frameworks in both states provide a reasonable degree of transparency, stability, efficiency, and equity relative to the frameworks in Indonesia (Chapter 7) and South Africa (Chapter 5). Government agencies responsible for implementing the laws and regulations largely reach decisions that the industry views as fair and balanced. The rules and regulations are subject to public comment at reasonable intervals and the most important aspect of implementing either a new coal mine or a rail or port project – the review and approval of a project's environmental impact statement (EIS) – allows for the public to provide its comments over a six-week period from the date that the EIS is submitted to the Office of the Coordinator General (OCG), the government entity responsible for overseeing any significant infrastructure or mining project (Environmental Protection Agency, 2014).

Public hearings on any EIS are organized by the project proponent, who is required to provide a full and transparent response to any public comments (Environmental Protection Agency, 2014). At the end of the public review period, the OCG makes a final decision on whether a project should be allowed to proceed on environmental grounds and whether additional conditions are to be attached to that approval (Environmental Protection Agency, 2014). In most cases, the project proponent is required to issue a supplemental EIS that states the actions the developer will take to address concerns raised by the public and deemed by the OCG to be significant.

Once a coal mine is operating, issues related to worker safety are handled by the Coal Services Pty Ltd, the successor to the Joint Coal Board, while health-related issues are handled by the mines departments of both state governments. Over the past fifty years, the JCB/Coal Services and the Queensland and NSW mines departments have successfully resolved issues related to worker health and safety that previously led to strikes and other worker actions that caused significant losses in coal output prior to 1960.

As of December 2013, the regulatory issues of greatest concern to most coal mining companies were not the existing regulatory frameworks and the specific actions of state government civil servants. Rather, they were two new federal policies: the carbon pricing mechanism (CPM) that was passed as part of the Clean Energy Act (CEA) of 2012 and the 30% minerals resource rent tax (MRRT) on mining companies. Industry trade journals and Australian newspapers unanimously pointed to the CPM and the MRRT as creating the greatest level of investment uncertainty at that time.[56]

6.2.4.2 Market trends: impact of China and India on Australia's export markets

By 2012, eight years after the end of the competition phase, Queensland accounted for 53% of Australia's black coal exports and NSW for the remaining 47% of black coal exports.[57] The black coal export mix in 2012 was 54% steam coal (181.6 Mt) and 46% coking coal (154.2 Mt) (Bureau of Resources and Energy Economics, 2013, pp. 43–52).

Japan remained Australia's largest single export customer in 2012, accounting for 29% of 2012 black coal exports. Significant growth in demand from Korea and Taiwan resulted in these two countries accounting for 19% of Australia's 2012 black coal exports. However, starting in 2009, strong demand by China and India for Australia's hard coking coal and by China for its steam coal (Tables 6.3 and 6.4), contributed to a shift in market share to China and India and away from Japan, Taiwan, and Korea. By 2012, India and China in total imported 24% of Australia's black coal exports (Bureau of Resources and Energy Economics, 2013, pp. 43–52).

Figure 6.17 shows China's imports of coking coal and steam coal between 2000 and 2013. China's imports of steam coal from all countries increased by almost 300% between 2008 and 2009. As shown in Figure 6.18, China's exports of steam coal dropped in concert with the increase in imports. If not for China, FOB prices for Newcastle-grade steam coal (6,300 kcal/kg, GAR coal) would have collapsed far below

Table 6.3 *Chinese imports of Australian coking and steam coals and percent share of total Australian exports, 2004–2012*

	Australian coking coal			Australian steam coal		
Year	China's imports from Australia (Mt)	Australia's total exports (Mt)	China's share of Australia's total exports (%)	China's imports from Australia (Mt)	Australia's total exports (Mt)	China's share of Australia's total exports (%)
2004	4.24	117.18	4%	2.22	107.35	2%
2005	5.3	120.48	4%	3.99	110.82	4%
2006	4.3	131.97	3%	3.22	111.62	3%
2007	2.5	136.92	2%	1.48	115.07	1%
2008	3.5	125.24	3%	8.4	136.36	6%
2009	36.17	157.26	23%	13.92	134.98	10%
2010	17.39	140.46	12%	16.67	143.32	12%
2011	10.33	142.4	7%	28.46	158.44	18%
2012	20.42	154.24	13%	38.06	181.62	21%

Source: BREE, Resources and Energy Statistics, 2013.

Table 6.4 *India's imports of Australian coking and steam coals and percent share of total Australian exports, 2004–2012*

	Australian coking coal			Australian steam coal		
Year	India's imports from Australia (Mt)	Australia's total exports (Mt)	India's share of Australia's total exports (%)	India's imports from Australia (Mt)	Australia's total exports (Mt)	India's share of Australia's total exports (%)
2004	15.2	117.2	13%	1.3	107.4	1%
2005	16.4	120.5	14%	1.5	110.8	1%
2006	19.6	132.0	15%	1.1	111.6	1%
2007	24.2	136.9	18%	0.5	115.1	0%
2008	24.3	125.2	19%	0.9	136.4	1%
2009	31.4	157.3	20%	0.6	135	0%
2010	30.9	140.5	22%	0.5	143.3	0%
2011	29.3	142.4	21%		158.4	0%
2012	30.7	154.2	20%		181.6	0%

Source: IEA, 2010 Coal Information for Indian Imports 2004–2010; see BREE, *Resources and Energy Statistics, 2013*, for other data.

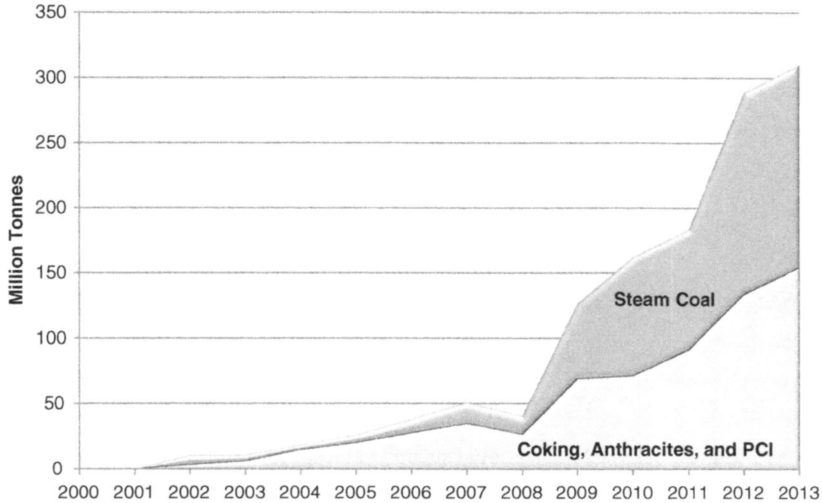

Figure 6.17 China's imports of steam and coking coal between 2000 and 2013. Source: China Coal Statistics, Export Information Administration.

the \$60–\$70/t price that prevailed for this grade of Australian steam coal during 2009. China's imports of steam coal continued to increase over the next four years, growing by an additional 190% from 126 Mt in 2009 to 324 Mt in 2013. Without the support provided by China's demand to the steam coal industries of both Australia and Indonesia, FOB prices in Asia for steam coals would likely have dropped far below the lows reached in 2013.

As described in Chapter 9, Chinese coal buyers are highly opportunistic, entering and exiting from the coal import markets depending on price. With the higher coal prices that prevailed for both coking and steam coals from January 2010 through October 2011, it was not surprising to find that China's imports of Australian steam coal plateaued between 2010 and 2011. Because China's domestic coal consumption is so large, even small percentage changes in its coal imports can have a very significant effect on international price, contributing to coal price volatility throughout Asia.

With respect to coking coals, the Chinese steel industry experienced robust growth between 2004 and 2008, and its demand for coking coal grew commensurately. However, the Chinese steel industry continued to procure most of its coking coal requirements from domestic suppliers until 2009 when, in response to very attractive FOB prices for Australian

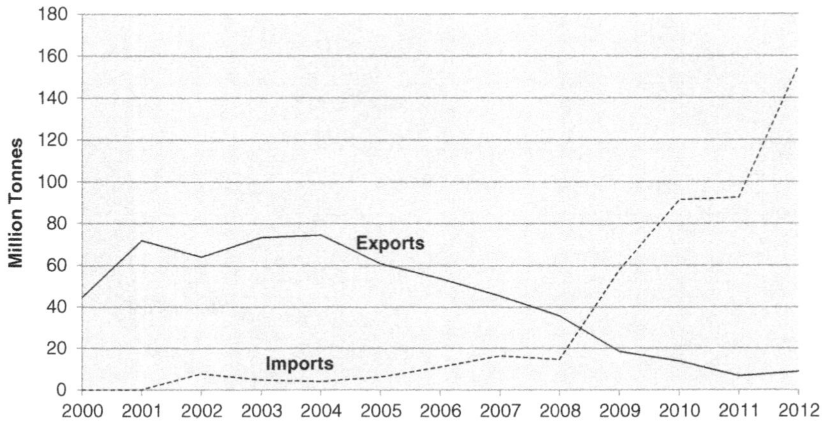

Figure 6.18 China's annual steam coal imports and exports, 2000–2012.
Source: China Customs Statistics, Export Information Administration.

coking coals and the very low costs of chartering Panamax and Capesize vessels, the Chinese mills increased their purchases of hard and soft coking coals from Australia from 4.2 Mt in 2004 to 36.2 Mt in 2009 (Table 6.3). In 2012, China imported 53.6 Mt of coking coal, of which at least 20.4 Mt was supplied by Australian exporters.

India has over the past decade accounted for a very large and growing share of Australia's hard and semi-soft coking coal exports. For example, between 2004 and 2010, India increased its share of Australia's coking coal exports from 13% to 22% (Table 6.4). India's steam coal imports from Australia have been small and declining since 2004 and are not expected to increase much due to high shipping costs and premium values that those coals can earn in other export markets. Also, Indonesian coals have qualities that make them more suitable as a blending stock for existing Indian power plants than Australian steam coals, plus they are significantly cheaper on a $/GJ basis.[58]

6.2.4.3 Transportation infrastructure: port and rail constraints

Between 1987 and 2004, the governments of NSW and Queensland accomplished timely expansions of their rail networks, which allowed substantial increases in their exports of both coking and steam coal up until 2004. From 2004 onward, however, the governments of NSW and Queensland were unable to keep up with the rapid growth in demand by the coal industry for additional rail infrastructure, contributing to long vessel queues at the Port of Newcastle and later at Hay Point and even

Gladstone. (Section 6.3.2 will further discuss port capacity issues and how the governments of NSW and Queensland are trying to deal with them.)

Between 2004 and 2011, the port of Newcastle dominated the news as Australia's most capacity-constrained port, with long lines of ships often queuing up for days at a time and incurring significant demurrage charges. However, from November 2009 through June 2010, DBCT and HPCT also suffered from problems that captured negative headlines. Vessel queues at Hay Point's DBCT hit seventy vessels and twenty-nine days' average turnaround in November 2009 due to mechanical problems with one of the terminal's coal in-loaders (Energy Publishing Inc. "Coal Chain Australia Weekly," Nov. 2009, pp. 5–9). In December 2009 and January 2010, bad weather kept vessel queues consistently above fifty days, which in turn led to vessels loading at DBCT experiencing higher than normal demurrage charges, ranging from $4/t to almost $10/t. In March 2010, cyclone Ului caused extensive damage to the Hay Point Coal Terminal, forcing its closure from March 11 to April 2 and pushing demurrage charges above $4/t (Energy Publishing Inc., "Coal Chain Australia Weekly," Feb. 2010, pp. 2–3, and March 2010, pp. 1, 8).

Gladstone and Newcastle, on the other hand, experienced much lower vessel queues and demurrage charges. Gladstone's RGTCT experienced rather light vessel queues of five to twenty-four vessels, with demurrage charges well under $4/t. Newcastle experienced moderately high vessel queues of eight to thirty-nine days but demurrage rates of under $2.50/t (Energy Publishing Inc. "Coal Chain Australia Weekly," Feb. 2010, pp. 2–3, and March 2010, pp. 1, 8).

Another natural occurrence that can affect security of supply from mines in Queensland is excessive rain during the cyclone season, which runs from December to March each year. In December 2010, excessive rains and flooding led to closures of most open-cut mines in Queensland and the exhaustion of mine and port stockpiles, leading most mine owners to declare force majeure under their coal supply contracts (Energy Publishing, Inc. "Australia Coal Report," Dec. 2010, pp. 2–3).

In short, Australia's coal exports can be significantly affected in the short term by weather-related conditions that dramatically limit rail speed and port loadings between December and May of each year. Worsening weather conditions are a factor that may increase in importance over the next two decades and should be viewed as a chronic issue, albeit one that is seasonal in nature. But the problems over the past six years are also partly attributable to a shortfall in rail and port capacity investment, which as discussed above stems from both price volatility and government and industry decision-making processes. The resulting shortage of investment in rail and port capacity has in turn affected the industry's level of investment in mines.

Both weather issues and underinvestment have limited Australia's coal exports over the past decade and seem likely to continue to do so into the foreseeable future.

6.3 Future challenges and issues

The coal industry is one of Australia's largest export earners and largest employers. In 2012, Australia's black coal industry exported steam and coking coals worth A$38.6 billion, runner up to iron ore, which in that same year generated A$57.1 billion in export revenues (Figure 6.19). The Australian Coal Association (ACA) claimed in 2008 that the industry directly employed about 50,000 people and supported another 150,000 indirect jobs (Australians for Coal, 2014).[59] The future trajectory of the industry is therefore a matter of great import for the country. Looking forward, the coal industry faces three broad challenges:

- Depletion of coal resources;
- Rail and port infrastructure constraints; and
- Regulatory uncertainty caused by recently enacted laws, since repealed, related to GHG reduction (the Carbon Pricing Mechanism, or CPM, in the Clean Energy Act) and new taxes on the mining sector (the Minerals Resource Rent Tax, or MRRT).[60]

In this section we will discuss each of these three broad challenges in turn and consider how the industry and regulators are approaching them.

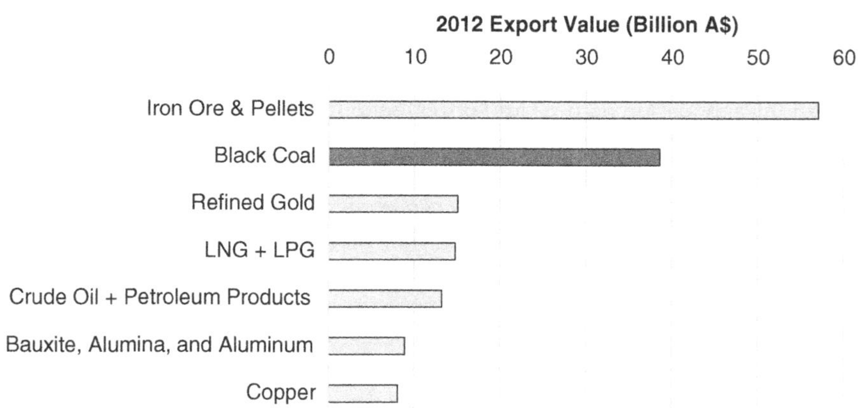

Figure 6.19 Value of different Australian exports in 2012 (billion A$).
Source: BREE, "Resources and Energy Statistics, 2013," table 28.

6.3.1 *Australia's coal resources and their depletion*

Depletion of Australia's coal resources is probably the least pressing of the three issues described above. Geoscience Australia[61] estimates that Australia had, as of December 2011, black coal Economic Demonstrated Resources (EDR) of 57.5 Bt (Table 6.5), up from 39.1 Bt in 2008, and enough to support 109 years of raw coal production at 2012 production levels (Geoscience Australia, 2009, p. 5, and 2012, p. 6, table 1). Almost all of those reserves are located in Queensland (61%) and NSW (36%).

Geoscience Australia estimates that NSW had, as of December 2011, recoverable black coal EDR of 20.6 Bt. The resources are contained in six basins: (1) Sydney-Gunnedah, (2) Hunter, (3) Gloucester, (4) Oaklands, (5) Newcastle, and (6) Western and Southern, with most of NSW's black coal EDR located in the 500 km long, 150 km wide Sydney-Gunnedah Basin. The Sydney Basin is estimated to contain 30% of Australia's black coal EDR (Geoscience Australia, 2013, p. 20). Coals from the NSW coalfields are typically exported out of the Port of Newcastle under the Newcastle brand, which has a gross calorific value of 6322 kcal/kg gross as received (gar) and 6000 kcal/kg net as received (nar).

As of December 2011, Queensland had in situ black coal EDR of 34.9 Bt (Geoscience Australia, 2013, p. 20). Almost all of Queensland's coal reserves are located in the Bowen (35% of total EDR), Surat (11% of

Table 6.5 *Australia's demonstrated coal resources and JORC reserves as of December 2011 (in Bt)*

	Demonstrated resources				Inferred resources and company estimates	
			Sub-economic			
Coal type	Economic (EDR)	Accessible EDR (AEDR)	Para-marginal	Sub-marginal	Inferred resources	JORC reserves (% of AEDR)
Black coal						
in situ	71.1	n/a	1.4	5.4	79.0	n/a
recoverable	57.5	50.7	0.8	4.0	57.3	19.2 (38%)
Brown coal						
in situ	49.1	n/a	37.5	16.9	121.2	n/a
recoverable	44.2	34.2	33.4	15.2	100.7	n/a

Source: Geoscience Australia, Canberra, "Australia's Identified Mineral Resources, 2012," p. 6, table 1.

EDR), and Galilee (8% of EDR) basins with small EDR quantities located in the Clarence-Moreton and Callide basins. Queensland's coal resources range from high-quality coking coals to low-rank brown coals, with its black coal industry primarily built around the high-quality coking coals that are located in the Bowen Basin and command premium prices in export markets. However, export-quality black coal, especially steam coal resources, can be found throughout the state and in particular in the Surat and Galilee basins located inland from the Bowen Basin. (Possible new mine developments in the Surat and Galilee basins are discussed below.) Queensland's coking coal resources are produced mostly from large open-cut mines, and approximately one-third of the remaining steam coal resources located in the Bowen Basin are amenable to open-cut mining.[62] (Any other steam coal mines developed in the Bowen Basin are likely to be underground longwall mines that will entail higher mining costs.)

Incremental expansions of existing steam coal mines will lead to significant increases in the aggregate amount of steam coal produced in NSW and Queensland over the next decade. Table 6.6 shows an estimate prepared in 2009 of potential production increases that would result from expansions of existing mines. The level of new output that may be generated from the sum of these coal mine expansion projects is estimated at 90 Mtpa by 2015 and an additional 26 Mtpa by 2020.[63]

New steam coalfields that may be developed in NSW after 2015 will either be located in the Gunnedah Basin or the Hunter Valley. Coal from these new developments will be more expensive to deliver to export customers due to the longer distances from the new mine sites to the

Table 6.6 *Forecast increases in steam coal mining capacity (Mt) due to expansion of existing mines in NSW and Queensland, 2010–2040*

State	Mine type	2010	2015	2020	2025	2030	2040
Queensland	Open-cut	80.7	106.0	119.8	108.3	95.1	66.4
	Underground	0.9	1.2	1.2	1.2	1.2	1.1
	Subtotal	81.7	107.3	121.1	109.6	96.4	67.5
Newcastle	Open-cut	111.4	145.5	159.1	156.2	146.2	88.7
	Underground	52.3	80.9	79.4	60.8	50.8	40.9
	Subtotal	163.7	226.5	238.5	217.0	197.0	129.6
Both States	Open-cut	192.1	251.6	279.0	264.6	241.4	155.1
	Underground	53.2	82.2	80.6	62.0	52.0	42.0
	TOTAL	245.4	333.7	359.6	326.6	293.4	197.1

Source: Summary of coal resource database prepared by Michael Friederich, consulting coal geologist, 2009.

Table 6.7 *Greenfield mining projects in Galilee and Surat basins that are at an advanced stage of development (as of May 2014)*

Developer	Basin	Deposit	2012 resource estimate* (Bt)	First shipment date	Export capacity (Mtpa)	Distance to port (km)	Capex (billion A$)
GVK Hancock	Galilee	Alpha	1.8 (open-cut)	2018	30	495	7.5
GVK Hancock	Galilee	Kevins Corner	6.1 (open-cut + underground)	2018	30	495	9.0
Waratah Coal	Galilee	Waratah	4.3	Unknown	40	495	6.0
Adani Group	Galilee	Carmichael	9.0	2018	60	495	16.5
Xstrata	Surat	Wandoan	7.6 (open-cut)	Unknown	22	380	6.0
Yancoal	Surat	Cameby Downs	1.4 (open-cut)	Unknown	12–15	430	n/a

* GVK Hancock and Xstrata resource estimates are for Measured, Indicated, and Inferred Resources; Xstrata also estimated that it has 730 Mt of Marketable Reserves out of 1 Bt of Proved and Probable Reserves.

Sources: GVK Hancock (http://gvkhancockcoal.com/hancockcoal.com.au/go/current-projects/kevin-s-corner); Waratah Coal:(www.dip.qld.gov.au/projects/mining-and-mineral-processing/coal/waratah-galilee-coal-project.html); Xstrata:(www.wandoancoalproject.com.au/mediacentre_protocols.cfm); and Cameby Downs (www.syntechresources.com.au).

Port of Newcastle and the need to either construct new rail lines or expand existing rail lines.

It will likely also be necessary to fully develop the Surat and Galilee basins in Queensland (see Table 6.7) if Australia is to sustain a national black coal production level of more than 300 Mtpa of saleable steam coal after 2025.[64] The Surat Basin lies to the south of the Bowen Basin and is reported to contain at least 4.0 Bt of undeveloped steam coal resources.[65] Most of the Surat Basin coals can be extracted using open-cut mining methods. The coals are also viewed as attractive plays because of their moderate ash (10–12%) and low sulfur content (0.6%). One downside of the Surat Basin coals is that they are relatively difficult to grind (low Hardgrove Grindability Index, or HGI), which could initially limit their uptake by Asian power plants designed for more easily grindable coals from NSW and Indonesia.[66] Some Surat Basin mines, such as Kogan Creek and Wilkie Creek, are already in production, but most of the Surat Basin coal deposits still remain undeveloped. Two companies, Xstrata and Syntech Resources, were planning to bring their very large Surat Basin coal deposits into production, which could have added 50 Mtpa of new steam coal production by 2015, but these developments were on hold at the time of writing – due primarily to market conditions but also secondarily to rail capacity concerns.[67]

The second area of interest within Queensland is the Galilee Basin, which is located in north-central Queensland, west of the Bowen Basin. The Galilee Basin is estimated to hold more than 10 Bt of in situ JORC-compliant steam coal deposits. The areas of primary interest are those located in the eastern part of the Galilee Basin and referred to in the industry as the "Alpha deposit." These coal resources are all steam coals. The seams are shallow and the overburden can be removed with little or no blasting. The coals are relatively high in ash and will need to be washed before railing to export ports. The yield of saleable coal after washing is expected to range from 60% to 85%, but the resulting coal quality should equal the quality of washed Surat Basin coals, except for typical HGI, which will be much higher for Galilee Basin coals (> 50 versus 32 to 35 for Surat Basin coals). Galilee Basin projects are being pursued by GVK Hancock Coal, Adani, and Waratah Coal. As with the Surat Basin projects, market conditions have pushed back their development, although the GVK Hancock developments (Alpha and Kevins Corner) still appear to be proceeding on a delayed timetable and remained the furthest advanced greenfield coal mine projects in Australia at the time of writing.[68]

Over the next two decades, steam coal will capture a larger share of Australia's black coal markets than it does today. The main reason for this shift will be the eventual depletion of the high-quality coking coal deposits of the Bowen Basin and their replacement with steam coals produced from within the Bowen Basin and Hunter Valley and from new open-cut mines located in the Surat and Galilee basins in Queensland and the Gunnedah Basin in NSW. New underground mines in the Bowen Basin could push up costs by replacing open-cut mines. More generally, it is expected that average industry-wide costs for coal production will increase significantly as new underground and open-cut mines are brought into production.

6.3.2 Managing rail and port infrastructure constraints

As discussed in Section 6.2, rail and port capacity shortages have constrained Australia's black coal exports off and on since 1980. They reemerged as a serious issue in 1996, with demurrage charges for the Port of Newcastle hitting around A$100 million in 1997, effectively doubling port charges from A$2.80/t to A$5.40/t. Despite being given prominent mention in the report of the Productivity Commission on Australia's black coal industry, infrastructure limitations persisted. Vessel queues at the Port of Newcastle ranged from 70 to 80 vessels over extended periods of time in 2007 and 2008, causing significant damage to Australia's reputation as a reliable coal supplier of black coal exports. In 2010, the "Port of Newcastle disease" spread to Hay Point's Dalrymple Bay Coal Terminal (DBCT), with vessel queues of thirty vessels and an average turnaround per vessel of twenty-five days. The Kooragang/Carrington Coal Terminals (KCCT) at the Port of Newcastle had a much longer vessel queue (forty-five vessels) but a much shorter turnaround time for each vessel (fourteen days) (Energy Publishing Inc. "Coal Chain Australia," December 2010, pp. 4–8).[69]

Despite these unresolved infrastructure constraints, Australia still managed to increase its black coal exports by a respectable 4.7% per year between 2006 and 2009. The real question is what could have been achieved if adequate rail and port infrastructure had been available during those years.[70]

The governments of Queensland and NSW are attempting to address both port and rail capacity concerns. Queensland put forward a plan in 2008 to expand ports and associated rail systems from 238 Mtpa in 2008 to 406 Mtpa by 2020 (Table 6.8).[71] The government of NSW similarly put forward a plan to expand the Port of Newcastle's coal handling capability from 102 Mtpa in 2008 to 208 Mtpa in 2020 (Table 6.8). If the new terminals at Abbot Point, Newcastle, and Gladstone are completed as scheduled, the total nameplate coal handling capacity of the ports of Queensland and Newcastle will increase from 263 Mtpa by the end of 2008 to 406 Mtpa, an increase of over 50%.

Table 6.8 *Nameplate capacity forecasts for major coal handling terminals at ports in NSW and Queensland*

		Nameplate capacity (Mtpa)					
Port/coal terminal	Location	2006	2008	2010	2012	2015	2020
Newcastle	NSW						
1. Kooragang Coal Terminal		64.0	77.0	91.0	101.0	101.0	101.0
2. Carrington Coal Terminal		25.0	25.0	25.0	25.0	25.0	25.0
3. NCIG Coal Terminal		0.0	0.0	30.0	45.0	66.0	66.0
Subtotal		**89.0**	**102.0**	**146.0**	**171.0**	**192.0**	**192.0**
Port Kembla	NSW	16.0	16.0	16.0	16.0	16.0	16.0
NSW total		**105.0**	**118.0**	**162.0**	**187.0**	**208.0**	**208.0**
Gladstone	Queensland						
1. RG Tanna Coal Terminal		51.0	72.0	72.0	72.0	72.0	72.0
2. Barney Point Coal Terminal		7.0	7.0	7.0	7.0	0.0	0.0
3. Wiggins Island (planned)		0.0	0.0	0.0	0.0	27.0	84.0
Subtotal		**58.0**	**79.0**	**79.0**	**79.0**	**99.0**	**156.0**
Hay Point	Queensland						
1. Dalrymple Bay Coal Terminal		55.7	85.0	85.0	85.0	85.0	85.0
2. Hay Point Coal Terminal		40.0	44.0	44.0	55.0	55.0	55.0
Subtotal		**95.7**	**129.0**	**129.0**	**140.0**	**140.0**	**140.0**
Abbot Point	Queensland	15.0	25.0	50.0	80.0	100.0	100.0
Brisbane	Queensland	5.0	5.0	5.0	5.0	8.0	10.0
Queensland total		**173.7**	**238.0**	**263.0**	**304.0**	**347.0**	**406**
TOTAL		**278.7**	**356.0**	**425.0**	**491.0**	**555.0**	**614.0**

Note: 2006–2008 figures are actual nameplate capacities; 2010–2020 figures are government forecasts.

Source: Queensland government, "Coal Transport Infrastructure in Queensland: Overview of Future Expansion," September 2008 (www.transport.qld.gov.au).

Some new capacity has already come online in the last few years, which has noticeably improved turnaround times for vessels at most Australian coal ports.[72] The first stage of the Newcastle Coal Infrastructure Group (NCIG) in NSW terminal achieved commercial operation in late 2010 with an initial coal handling capacity of 30 Mtpa.[73] The RG Tanna Coal Terminal in Queensland was expanded from 72 Mtpa to 100 Mtpa.

At the same time, the outlook for other planned expansions was more uncertain at the time of this writing. To meet industry needs for additional coal handling capacity, the Gladstone Ports Corporation (GPC) is supporting the development of a new coal terminal, known as the Wiggins Island Coal Export Terminal (WICET), within the Port of Gladstone. The first stage of WICET, with a ship-loading capacity of 27 Mtpa, was expected at the time of writing to be operational in November 2014, roughly two years later than the original completion schedule. The ultimate coal loading capacity for WICET is 84 Mtpa but due to uncertain market conditions and regulatory uncertainty, a schedule for completing later stages of WICET has not yet been announced (Wiggins Island Coal Export Terminal, 2014 and 2011). As a result of the delay in WICET, the associated Surat rail project was delayed (and then put on indefinite hold when Xstrata decided to suspend development of its Wandoan project due to weak market conditions and regulatory uncertainty).[74] These delays had spillover effects on the timing for the expansion of Yancoal's Cameby Downs project as well as the timing for implementing a number of new coal projects such as New Hope's Elimatta project and Stanmore's "Range" coal project.[75]

The coal ports located in Queensland and NSW had for all of 2010 a capacity utilization factor of 73%. If these ports could, on average, achieve an 80% capacity utilization factor, total black coal exports could reach 490 mt by 2020, up from 261 mt in 2008. Assuming that the current split between coking coal and steam coal exports continues through 2020, steam coal exports could increase from 120 Mtpa in 2008 to 221 Mtpa without running up against any port or rail constraints.[76]

However, the budget deficit problems of the Queensland government and the regulatory uncertainty caused by the Commonwealth governments led by Kevin Rudd and then Julia Gillard through September 2013 combined with uncertainty in the international market to create doubts about the timing for most new projects. Notably, the Surat Basin railway project and a number of the Galilee and Surat Basin coal projects have been affected, in turn affecting the schedule for the new Wiggins Island Coal Export Terminal that is planned for the Port of Gladstone. The following section will discuss two federal policies that created the greatest concern among investors: the Carbon Pricing Mechanism and the Minerals Resource Rent Tax that were passed under Gillard (and later repealed under Tony Abbott).

6.3.3 Federal policies on carbon pricing and mineral resources taxation

Between January 2008 and June 2010, the Commonwealth government, led by then Prime Minister Kevin Rudd, proposed sweeping changes to environmental rules and regulations and natural resource taxes – changes that generated political controversy and a fair amount of uncertainty in Australia's coal mining and power industries. Due to his advocacy of these controversial issues and, in the minds of many, his non-consultative style of governing, Rudd was removed from power by his own party in June 2010. In his wake, he left behind two proposals – a carbon pollution reduction scheme (CPRS) and a resources super profits tax (RSPT). Both proposals were restructured and renamed by the succeeding Australian Labor Party administration led by Julia Gillard, with the Carbon Pricing Mechanism (CPM) succeeding the CPRS and the Mineral Resources Rent Tax (MRRT) succeeding the RSPT.

6.3.3.1 Introduction of the carbon pollution reduction scheme (CPRS)

A 2008 study by Ross Garnaut, one of Australia's most distinguished economists, provided the rationales and principles upon which the government's carbon pollution reduction scheme (CPRS) proposal was later structured (Garnaut Review, 2011). In particular, Garnaut proposed that Australia play its part in mitigating the impacts of climate change by passing into law a cap-and-trade scheme to reduce emissions of greenhouse gases (GHGs). He recommended that the government target a 20% reduction in Australia's 2000 GHG emissions by 2020 and a 90% reduction by 2050, conditional on other nations agreeing to reduce their GHG emissions as well. The scheme would use the UNFCCC[77] definitions of greenhouse gases (CO_2 plus five others) and the methodology adopted by the UNFCC to convert those different gases into CO_2 equivalent values (CO_2e) for determining the GHG quantities emitted by a polluter.[78]

Relying heavily on Garnaut's report and follow-on advice, the newly elected Rudd government prepared a green paper in early 2008 that spelled out the principles and mechanics of the proposed CPRS.[79] The green paper was submitted to the public for its comments in July 2008. Based on extensive public inputs and government analysis of those inputs, a CPRS white paper was prepared and issued in December 2008.

The white paper was issued in two volumes and titled "Carbon Pollution Reduction Scheme: Australia's Low Pollution Future."[80] It was comprehensive, logically organized, professionally edited, and insightful. Most importantly, it addressed significant issues raised by the public during the green paper consultation period in a direct and clear manner. The white paper laid out the following structure for the CPRS:

- Baseline year of 2000 for carbon emissions reduction targets
- Start date of July 2010
- Agriculture sector exempted
- Plan covers 1,000 entities (accounting for 75% of CO_2e emissions)
- Targeted reduction by 2020 (in terms of year 2000 baseline emissions):
  (a) 5% as an unconditional commitment
  (b) 15% if a global agreement is reached in which all major economies commit to substantially restrain their CO_2e emissions and all developed countries accept binding CO_2e reduction targets equal to Australia's
- Long-term (2050) reduction target of 60% of year 2000 CO_2e emissions
- Starting CO_2e permit price of A$40/t

Once a white paper has been approved by the Cabinet, the next step in the Australian legislative process is to prepare "exposure draft legislation" and submit the draft bill to either the House of Representatives or the Senate to be considered for passage as an act of Parliament. The CPRS exposure legislation was put forward as a package of six related bills, of which the Carbon Pollution Reduction Scheme 2009 was the main bill.

Minister of Climate Change Peggie Wong, despite also being a senator, decided to have the bill originate in the House of Representatives, where there was strong support for it.[81] The draft CPRS legislation was introduced into the House and read for the first time on May 14, 2009.[82] The House of Representatives completed the necessary three readings of the revised CPRS bill on June 4, 2009.

Recognizing that the CPRS legislation would face stiff opposition in the Senate, Rudd decided to submit to the Senate a "gutted" version of the bill that had just passed the House. The revised CPRS legislation contained the following concessions:

- The start date for the CPRS was delayed until July 2012.
- Polluters were issued 85% of all required carbon permits for free.
- All other permits were priced at A$10/t during the first year of the CPRS.[83]

But the Senate remained unmoved by Rudd's concessions. The draft legislation was submitted to the Senate for its consideration on June 15, 2009, but it failed to pass the necessary three readings and was rejected in August 2009. The opposition senators focused the debate on the CPRS around one key point, which was that Australia should not unilaterally commit to a CO_2 cap-and-trade scheme before other far larger polluters such as the United States, Japan, Korea, Taiwan, and China agreed to a proportional reduction in their GHG emissions through a binding international agreement. In their view, consideration of the CPRS in the Australian Senate should be delayed until after the United Nations Climate Change Conference in Copenhagen.[84]

Australia's coal mining industry opposed the revised draft CPRS legislation for more specific reasons. First, industry representatives were upset at being ruled ineligible for transitional support under the emission-intensive trade-exposed sectors (EITES) program. They claimed that the government's rule for determining eligibility, which was based on CO_2e emissions per dollar of revenue and CO_2e emissions per dollar of value added over two years (2007 and 2008), arbitrarily resulted in the coal industry being declared ineligible. Coal industry representatives argued that the coal industry would have qualified for EITES's funding if a longer period of time, say four to five years, had been used.

Second, and more importantly, the government had decided to include fugitive methane emissions from both open-cut and underground coal mines as one of the coal industry's CO_2e emission obligations. The coal industry objected, arguing that methodologies for making such estimates were incomplete, if not fatally flawed, and that no other governments were including fugitive emissions from mines in a cap-and-trade program because such emissions were so difficult to measure for the case of open-cut mines. Industry respondents recommended that this provision of the CPRS be dropped until an international agreement could be reached on how to deal with fugitive methane emissions from all coal mines – both underground and open-cut mines – as a CPRS obligation. The coal industry was also frustrated by what it felt was favored treatment for the LNG and pipeline gas industries.

A revised CPRS bill known as the Carbon Pollution Reduction Scheme Bill 2009 (No. 2) was introduced in the Australian House of Representatives on October 22, 2009, which approved sending it to the Senate on November 16. The following day, November 17, CPRS No. 2 was introduced into the Senate, but it was rejected on December 2. In early 2010, the Rudd government tried a third time to pass the CPRS into law via a revised bill, known as the Carbon Pollution Reduction Scheme Bill 2010. This bill passed through the House on February 11, was introduced into the Senate on February 22, and was once again voted down.

As a result of the three failed attempts, Rudd decided to drop any further attempts to pass the CPRS into law until 2013. He blamed the failure of the COP15 Copenhagen Climate Change Conference and the lack of a broad international commitment to reduce emissions of GHGs. In fact, Rudd probably realized that he would never get the CPRS passed in the Senate and was simply looking for a face-saving way to put it on the back burner.

Once the Australian Senate had rejected the CPRS bill twice, the government could under the Australian Constitution have dissolved Parliament and called a general election. In March 2010, Rudd proceeded to call a general election for October 2010 but, ironically, based not on the Senate's multiple rejection of the CPRS legislation but rather on its two-time rejection of a health care reform bill.

6.3.3.2 Introduction of the resource super profits tax (RSPT)

Before the election could be held, however, Rudd attempted to impose a "resource super profits tax" (RSPT) on the mining industry. Rudd announced the RSPT to the nation and the mining and natural resource industries in May 2010. Except for its total opposition to the tax, the mining community had a hard time voicing specific complaints because so much was still unknown about the proposed tax scheme. A fact sheet issued by the Australian government on the RSPT was embarrassingly short on details (Australian Government). An article in Sydney's *The Daily Telegraph* provided a clearer picture of how the RSPT was expected to work than the government's own documents. It described the following elements (Carswell, 2010):

- A 40% tax would be imposed on all profits earned by mining as well as oil and gas extraction companies that earn a profit greater than 5.7%, which, at the time, was the government's long-term bond rate.
- Profit was to be calculated after the deduction of royalties, operating expenses, depreciation, federal company tax, and a capital allowance of around $5 million for each company.
- The federal company tax would be reduced from 30% to 28%.
- If a company failed to make a profit, the Commonwealth government would reimburse that company for 40% of its initial investment in a mine.
- The tax was expected to be applied retroactively.

The Commonwealth government expected to earn A$12 billion in extra tax revenues from the RSPT starting in 2012. The monies were to be used to "bail out" the government superannuation funds (using about one-third of the RSPT revenues) and to finance transport infrastructure projects in NSW and Queensland.

As opposition to the RSPT grew in June 2010, rumors circulated that the Rudd government would offer a compromise to the mining sector that would increase the threshold profit rate from 5.7% to 10% or 12%. It was also rumored that the Rudd government would drop the proposal to reimburse 40% of the cost of any failed mines. But these rumors were squashed on June 11 by Wayne Swan, the government treasurer, who announced that consultations with industry were ongoing and that a final design would only be announced after these consultations had been completed.

Industry reactions to the RSPT were, as could be expected, uniformly negative. Many of those opposed to the tax (and even some in favour) felt that it was a hypocritical attempt by Rudd to position his government for the upcoming federal election in November 2010. In response to the proposed RSPT, Xstrata announced that it had put its Wandoan coal project on hold, which would also delay the Surat rail project and WICET. Other coal producers including BHP threatened to do the same.

Whatever Rudd's rationale for proposing the RSPT, it turned out to be a serious political miscalculation. In late June 2010, the Australian Labor Party, under considerable pressure from the public and declining in the polls, decided to remove Rudd as its choice for prime minister. He was replaced by his deputy, Julia Gillard, who took immediate action to remove the biggest issue of the election – the RSPT – from the table. In consultation with the mining and oil and gas industries, Gillard on July 2, 2010 proposed a new scheme for "improved resource tax arrangements" that was developed in consultation with the chief executives of BHP, Rio Tinto, and Xstrata. She labeled this new mining tax as the Minerals Resource Rent Tax (MRRT). (The MRRT will be discussed further in Section 6.3.3.4.)

Gillard also moved up the date for the general election from October to August 2010. As one of her many campaign pledges, Gillard vowed that she would not impose a carbon tax prior to convening a citizens' assembly, which would determine whether there was sufficient public support for a carbon taxation scheme.

The August 2010 election resulted in Australia's first hung parliament in seventy years, with Labor winning seventy-two seats in the House of Representatives and the Liberal/National Party Coalition holding seventy-two seats. Labor, against all odds, managed to cobble together a government with Gillard as Prime Minister by joining with four independents and one member of the Greens Party, which gave it a majority of two members in Parliament. Both the independents and the Green member in the House of Representatives made strange bedfellows with Labor. Most Australian political analysts were predicting that the coalition between Labor, two independents, and the Greens Party would quickly collapse and a new election would be called sometime in early 2011.

6.3.3.3 *Resurrection of carbon policy via the carbon pricing mechanism (CPM)*

Gillard kept the CPRS on hold until December 21, 2010, when she convened an ad hoc committee known as the Multi-Party Climate Change Committee (MPCCC), which consisted of members of the ruling Labor Party, the Greens and two independent parliamentarians. The MPCCC released a proposal on February 24, 2011 to replace the CPRS with a scheme known as the Carbon Pricing Mechanism (CPM). The CPM would have the following broad features (Multi-party Climate Change Committee, 2011; Leslie, 2011):[85]

- A carbon tax would start on July 1, 2012.
- The carbon tax would be fixed for three to five years after which it would be replaced by a cap-and-trade system, presumably based on the CPRS.
- The following entities/activities would initially be subject to the carbon tax: (i) stationary energy sources such as power plants, (ii) transport, (iii) industrial processes, and (iv) fugitive emission sources. Agricultural and land use activities generating significant GHG emissions would not be taxed on these emissions.

The February 24 proposal was widely criticized by the opposition and local media as breaking Gillard's promise not to impose a carbon tax without first conducting broad public consultations (Morton and Arup, 2010).[86] The proposal was also notable for its lack of detail, especially given all of the research and debate that had informed the CPRS.

However, Gillard's options may have been limited by the political realities facing her government in early 2011. Tony Windsor, who at the time was one of the two independent parliamentarians supporting the Gillard/Labor Government, agreed to have the proposal released for public consultations but was decidedly against imposing a carbon tax by July 2012 (ABC News, 2011b). The Greens, on the other hand, were threatening to leave the government if Gillard did not take meaningful action to pass a tax on carbon. The February 24 proposal was likely the best that Gillard could achieve given the diverging views within her cabinet on how to proceed with the CPM.

Further complicating Gillard's situation were her very low public approval numbers. For example, a public opinion NewsPoll survey published on March 7, 2011 indicated that public approval ratings for the federal Labor party had fallen to "record lows" and that, within 9 months of taking office, Gillard had become less popular than previous Labor PM Kevin Rudd (ABC News, 2011a). By

February 2012, Gillard's hold on power was considered by most Australian political analysts to be so weak that a new election was very likely sometime during 2012. Despite her weak polling numbers, Gillard managed to obtain approval for the Clean Energy Act (CEA) containing the CPM from both the Parliament and the Senate in October and November 2011 respectively. Parliamentary and Senate approval of the MRRT followed shortly thereafter. Both bills officially became law on July 1, 2012.

Box 6.4 describes in detail the features that were included in the CPM.

Box 6.4 Provisions of the carbon pricing mechanism

This information comes from the Australian government's document "Securing a Clean Energy Future – the Australian Government's Climate Change Plan," July 2011.

- Start Date: July 1, 2012
- CO_2 emission reduction targets:
  - 2020 – 5% reduction in CO_2 emissions from 2000 levels irrespective of other country actions; 15–25% if other countries act
  - 2050 – 80% reduction in CO_2 emissions from 2000 levels
- The unit prices for CO_2 emissions will be set for three years as follows:
  - 2012: A$23/t
  - 2013: A$24.15 (est.)/t
  - 2014: A$25.40 (est.)/t

(Carbon prices for 2013 and 2014 were to be escalated by 2.5% per year plus the annual rate of inflation.)

- From July 1, 2015, the government will convert the "fixed carbon pricing mechanism" into a CO_2 emissions trading scheme (ETS), similar in most respects to what was proposed by Rudd under the CPRS. Under the ETS, the price of CO_2 will be partially determined by the market, with a transitional cap and floor price imposed through 2018.[87]
- The Australian Government's plan defines CO_2e as the CO_2 equivalent values for four of the six greenhouse gases (GHGs) covered under the Kyoto Protocol. These are: carbon dioxide (CO_2), methane (CH_4), nitrous oxide (N_2O), and perfluorocarbons (PFCs) generated by the aluminum refining industry. The two other GHGs covered under the Kyoto Protocol, sulphur hexafluoride (SF_6) and hydroflurocarbons (HFCs), "will face an equivalent carbon price, which will be applied through existing synthetic greenhouse gas legislation."

Box 6.4 (cont.)

- Sectors covered:
  - Stationary energy sources such as power plants and refineries;
  - Industrial processes (steel making, metal processing, cement making);
  - Fugitive methane emissions from gassy underground and open-cut coal mines, CBM and LNG production processes, natural gas wells, and waste dumps;
  - Emissions from non-legacy wastes; and
  - *Possibly* heavy on-road vehicles after July 2014.
- Sectors and activities exempted from the plan:
  - Agricultural and forestry sectors;
  - Light on-road vehicles; and
  - Households and small businesses, except through the indirect effects of higher input prices that might result from the carbon prices imposed on other liable entities.

Entities in covered sectors are liable for the tax if they have GHG emissions greater than 25,000 tonnes per year CO_2e (or 10,000 tonnes per year for the case of landfill dumps). The Australian Government estimated in 2011 that 500 business entities were liable under the CPM and that 50 of those entities were liable for around 75% of the CO_2 emissions covered under the CPM.

Commercial sectors that use liquid fuels for domestic transportation (rail, ship, and aircraft) and off-road vehicles such as earth moving equipment used at mining sites are not directly liable under the carbon pricing mechanism. Instead, their fuel use tax credits (for either VAT or excise tax) are reduced by an amount that would have an equivalent impact to applying the CO_2 price per tonne directly to their businesses as a tax.

The most noticeable differences between Rudd's CPRS and Gillard's "Climate Change Plan" (the portion of the CEA in which the CPM is addressed) were: (a) the more politically astute approach taken by the Gillard government when presenting its proposal to the general public and (b) the structuring of the CPM in a manner that would make it difficult for the opposition to rescind once it was passed into law. Under Rudd, the CPRS was advertised as affecting 1,000 businesses. Gillard argued that only 500 businesses would be liable under her carbon pricing scheme, at least for the initial three-year period when carbon prices were fixed.

Another important difference was in overall tone. Rudd and his cabinet were big on describing how the carbon price would apply and emphasizing its punitive nature and financial costs rather than other aspects of the CPRS package, such as transitional support. Even the title of the Rudd carbon pricing program, "The Carbon Pollution Reduction Scheme," had a punitive sound to it.

The CEA, for its part, did identify a few entities that would be severely affected by the CPM, including twenty-five gassy underground mines in NSW[88] and 2000 MW of heavy polluting brown coal power plants in the states of Victoria and South Australia,[89] but it also highlighted the potential for owners of coal-fired power plants with CO_2 emissions intensities above 1.2 tonnes CO_2 per MWh of electricity generation to receive substantial financial support. These payments, known as "payments to close," would be made to the owners of those plants on the condition that power sector reliability would not be adversely affected by the plant's closure, that worker entitlements had been fully paid, and that the power plant owner completed an appropriate environmental remediation of the site.

Nonetheless, both programs were similar in that they effectively put the black coal industry in their cross-hairs. As discussed previously, the Rudd government started the process of alienating the black coal industry when it declared the industry ineligible under the Emissions Intensive Trade Exposed Sectors (EITES) program and when it made open-cut and underground coal mines liable for fugitive methane emissions.[90] Gillard maintained these two aspects of the CPRS in her plan. The Exposure Draft Clean Energy Act legislation unequivocally stated that coal mining is not "an activity that, under the program, is taken to be an emissions-intensive trade-exposed activity" (House of Representatives "Clean Energy Bill," 2011). It still held the coal mining industry liable for its fugitive methane emissions. However, the Clean Energy Act did offer the coal mining industry a modicum of financial support, equal to A$1.3 billion, for the purpose of implementing technologies that would allow them to reduce their fugitive methane emissions. It also provided funds to support job retraining for miners who found themselves out of a job.

However, the coal industry was not mollified by these concessions. The Australian Coal Association (ACA) argued that the offer of financial support to the owners of gassy underground mines was a pittance and that the CPM would force the owners of gassy underground mines to close them down. The ACA estimated that this would cost 4,000 mining jobs and also cause many more indirect job losses.[91]

In August 2011, the NSW Government, which was controlled by the Liberal-National Party Coalition, announced the results of a modelling exercise that also predicted dire economic consequences from the CPM. The modelling results from the NSW Treasury directly contradicted the Commonwealth Treasury Department's claims that the CPM would have only minor and temporary impacts on employment and the national economy. Specifically, the NSW government argued that the CPM would cause the loss of 31,000 jobs in the state of NSW and annual losses in State GDP of A$3.7 billion per year.[92]

These criticisms of the CEA and CPM, while resonating with a certain segment of the electorate, were largely rear guard actions that did not stop passage of the Clean Energy Act. In addition to having a tenuous but sufficient base in the Parliament and Senate,[93] the Gillard government skillfully employed the following tactics to ensure passage of the Clean Energy Act into law:

(a) Cordoned off those entities that would lose big under the Clean Energy Act, such as the coal industry and power plants that rely on brown coal, and provided them with a modicum of transitional support;
(b) Redistributed over 50% of the funds generated by the CPM to lower and middle income voters in the form of tax breaks and uplifts in social benefits;[94]
(c) Funded projects and programs that would encourage the use of renewable energy and improvements in energy efficiency, expanding public support among progressives and environmentalists; and
(d) Allowed only a very short (six-week) public consultation period.

Once passed into law, the CPM had provisions that were "hard-wired" to set the initial fixed carbon prices and, thereafter, the caps and floors for the first three years of the flexible carbon price period. In other words, the CPM could not be derailed by merely changing ministry regulations. The law would need to be either amended or fully rescinded, which would require a majority vote in both the Parliament and the Senate.[95]

Most of the debate during 2011 was about the impacts that the CPM would have on the competitiveness of the Australian coal industry in Asian export markets due to changes in input prices paid by coal producers that would make Australian coal producers less competitive in international coal markets. However, the larger, unspoken risk was that the passage of a carbon pricing law in Australia might lead large coal consuming countries in Asia, such as Japan, Korea, Taiwan, and China, to adopt a similar regime for charging large emitters of CO_2. Such additional carbon pricing initiatives could create large, long term shrinkage of the highly profitable export markets for Australia's black coal. Fortunately for Australia's coal producers (if not for climate), such a scenario has not yet come to pass.

6.3.3.4 The mineral resources rent tax (MRRT)

As discussed previously, the mineral resources rent tax (MRRT) was the successor to Rudd's unpopular resource super profits tax (RSPT), crafted by the new Gillard government in consultation with the mining and oil and gas industries. The government announced the MRRT in September 2010 along with an amendment to the existing Petroleum Resource Rent Tax (PRRT), which extended the PRRT's coverage to include all offshore and onshore oil and gas projects, including projects that extract coal bed methane (CBM) and projects located in the Northwest Shelf and in the Timor Sea Joint Development Area.[96] Exposure draft (ED) legislation released in July 2011 proposed a commencement date of July 1, 2012 for the MRRT and the amended PRRT, with their application limited to four commodities: (i) black coal (steam and coking coals); (ii) iron ore; (iii) oil; and (iv) natural gas.[97] These four commodities together accounted for 66% of the value of Australia's mineral resource exports in 2012 (Bureau of Resources and Energy Economics, 2013, p. 27, table 26; p. 46, table 46; p. 92, table 99).

The MRRT is a complex tax system that was criticized as having considerable compliance costs for both large and small coal and iron ore mining companies. Tax analysts predicted that it would require numerous amendments over time in order to clarify and broaden the tax calculation methods to account for anomalous situations.

One of the main challenges that the Australian Tax Office (ATO) was expected to face in implementing the MRRT was defining consistent procedures and methods for "netting back" revenues and costs to the "taxing point," which for most coal mining operations is set at the point where the resource leaves the run of mine (ROM) stockpile. For those mining operations where an ROM stockpile does not exist or where the coal is beneficiated before being stockpiled, the taxing point would logically be the point prior to delivery of the coal to a beneficiation plant, such as a crusher or washing plant. In practice, most of Australia's black coal producers export their coals from one of the major export terminals in NSW and Queensland and their coals are priced at those load points. The process of netting back a coal's cost and revenues to one of the two taxing points mentioned above could become a complex and possibly litigious tax exercise.

The Gillard government made a number of concessions to coal miners with respect to the MRRT. In particular, the basic MRRT tax rate for all affected commodities, except coal, was set at 30% of net adjusted profits, defined as gross revenues at the taxing point less mining expenditures, royalties, and a mining loss allowance. In recognition that coal miners employ specialist skills to extract the coal and bring it to the taxing point, the Gillard government decided to reduce the effective MRRT to 22.5% through the application of an extraction allowance of 25% {[30% × (100% – 25%)] = 22.5%}. The MRRT also provided coal mining companies with three allowances that reduced MRRT revenues considerably.[98] Many of the allowances had uplift arrangements that preserved the present value of the portion of those allowances that might be carried forward to future MRRT years. In addition the MRRT payment in any year was made fully deductible from net income for purposes of calculating a taxpayer's income tax liability.[99]

Despite these significant concessions, the Association of Mining and Exploration Companies (AMEC), which represents small and medium sized Australian mining companies, remained strongly opposed to the MRRT. In its response to Treasury during the public consultation period on the MRRT, it noted that the ED Legislation lacks "over 17 topics – some of overriding importance to its members."[100] AMEC also charged that the MRRT was biased in favor of large, established companies such as BHP Billiton, Rio Tinto, and Xstrata and discriminated against emerging, small and medium sized mining companies.[101]

However, the MRRT ultimately attracted enough political support to pass the House in November 2011 and the Senate in March 2012.

6.3.3.5 *Fall of the Gillard government and implications for the CPM and MRRT*

In the end, Gillard's passage of the CEA and MRRT proved to be pyrrhic victories. During 2012 and the first half of 2013, Gillard faced many challenges to her continuation as PM from Kevin Rudd. Moreover, her political standing was weakened further by (i) a number of political scandals, which depleted her meager majority in Parliament, and (ii) worsening public opinion polls, which showed that Labor was headed for a massive election defeat against the opposition Coalition.[102] With little room to maneuver, Gillard on January 30, 2013 set new elections for Parliament and the Senate for September 14, 2013.

However, her expectation that she would remain Labor's candidate for PM in that election was not to be realized. On June 26, 2013, after two failed attempts, Rudd finally defeated Gillard in a Labor Party leadership vote, and on June 27 he was sworn in for a second time as PM of Australia. Rudd's second term in office proved to be an extremely short one, as the September 2013 election was won decisively by the Coalition. At the time of writing, Australia's Commonwealth government was controlled by the

Coalition, with Tony Abbott as PM. It set at the top of its legislative agenda the repeal of the laws that established the CPM and MRRT, and it achieved both objectives in 2014.

In summary, less than a year after her signal achievements of passing the CPM and MRRT into law, Gillard not only lost her position as PM but also saw the Coalition come to control Parliament and rescind both the CPM and the MRRT. Given the political challenges faced by Gillard during her brief term as PM, it is remarkable that the CPM and MRRT ever became law in the first place.[103]

The fraught political dynamics that developed around the CPM show what can happen when a government trying to tackle the climate change challenge, perhaps with an insufficiently deep popular mandate to do so, must confront domestic industries that could be disrupted by such action. The MRRT, while perhaps slightly less of a flashpoint, illustrates the political challenges for governments that seek to capture supposed "windfall profits" from domestic industries. The successive Labor governments of Kevin Rudd and Julia Gillard were particularly ambitious to pursue legislation in both areas at the same time.

Given all of the political and regulatory uncertainty that surrounded the passage of the CPM in particular, it is hardly any wonder that major investment decisions to open new mines and to construct new rail and port infrastructure in Queensland and NSW were constantly being second-guessed by private investors as well as the state governments of Queensland and NSW.

6.4 Conclusion: possible futures for Australia's coal industry

6.4.1 Australia's success story

At the beginning of the 1950s, there were many sceptics who felt that Australia's black coal industry would never amount to much due to the actions of radical trade unions and the competitive threat posed by cheap oil. However, the post-WWII history of Australia's black coal industry has turned out to be a Cinderella story come true. Its transformation from a threatened and declining industry in the early 1950s into the world's largest black coal exporter in 1984 is truly impressive.

One can cite many reasons for this successful transformation of the coal industry. The foremost reason is undoubtedly the proactive policy steps taken by the state governments of NSW and Queensland. The NSW government did a good job in the 1950s of correcting past deficiencies of the industry. Queensland implemented policies from 1960–1980 that supported the cost-effective and timely development of new mines and associated infrastructure. These industry and government actions, coupled with strong demand from Japan for high-quality coking coal and, after 1978, steam coal, propelled the rapid growth phase of the industry between 1960 and 1986.

The rebirth of Australia's black coal industry was also made possible by the following non-regulatory factors:

- The discovery of vast deposits of high-quality coking and steam coals in the Bowen Basin area of Queensland and the Hunter Valley area of NSW. It was helpful that many of these new deposits were amenable to open-cut mining and located within a reasonable distance of deep sea ports.
- The widespread commercial availability of advanced mining technologies for discovering these and other black coal deposits and then extracting them using either open-cut mining methods or advanced underground mining methods. The aggressive application of these advanced technologies led to huge increases in mine worker productivity over the period from 1950 to 1990 and led to low-cost development of Australia's black coal resources.
- The economic expansions of Korea and Taiwan (1980–1995) and the twin oil price shocks of the 1970s, both of which helped create strong export markets for Australia's black coal. Japan, with its focus on energy security, long played the role of "anchor customer" for Australia's export industry.

By the end of the twenty-six-year rapid growth phase (1960–1986), Australia was not only the world's largest black coal exporter but had also achieved substantial diversification of its customer base and product offerings, with the following specific achievements:[104]

- Exports in 1986 accounted for 69% of total sales (compared with only 9% in 1960).
- Steam coal supplied 48% of total exports in 1986 (up from less than 1% in 1960).
- Queensland emerged as Australia's largest exporter of black coal, jumping from 48,000 tonnes (less than 3% of total exports) in 1960 to 50.8 Mt (57% of total exports) in 1984.
- Japan's share of total exports was reduced from 90% of total exports in 1960 to 45% in 1986.

However, the growth trajectory for Australia's black coal industry started to plateau during the industry's competition phase (1987–2003). Australia's black coal industry experienced a dramatic slowing in its annual rate of growth, mainly due to factors such as (a) the entry of Indonesia as a significant exporter of steam coal and (b) the inability of the governments of NSW and Queensland to expand their port and railway systems quickly enough. This slowdown in the industry's rate of expansion affected both domestic consumption and exports. The competition phase was also a time when both Queensland and NSW made significant changes to their regulatory frameworks. Queensland passed into law its Mineral Resources Act of 1988, and NSW followed with the Mining Act of 1992. During the 1990s, the Commonwealth government lifted controls on export prices for coal (1993) and started to discuss greenhouse gas issues as a serious public policy issue (1996).

The volatile price phase (2004 to 2014) that followed brought on a troublesome trend towards start and stop project implementation in which governments and market participants would pursue the development of new mine, rail and port projects when prices were buoyant and then put these same projects into suspended animation when prices swooned.[105] The recent regulatory uncertainty associated with Australia's ill-fated carbon pricing and resource rent tax measures further complicated project development through 2014.

6.4.2 Factors that will shape Australia's coal industry in the future

The availability of sufficient coal resources in Australia to support the industry's continued prosperity is not in question. The states of NSW and Queensland are estimated to have sufficient black coal resources to allow Australia to sustain its 2012 level of raw coal production (526.7 Mt) for 109 years.[106] Resources under development in the Galilee, Gunnedah, and Surat basins by GVK Hancock, Waratah Coal, Xstrata, Yancoal, and others may add to these resource totals over time.

Coals from the three largely undeveloped basins of Galilee, Gunnedah, and Surat are mostly steam coals, and hence the share of steam coal in the overall mix of black coal exports is likely to grow significantly over the next two decades. If the two "advanced-stage" Surat Basin projects and the two Galilee Basin projects eventually enter commercial operation by 2018 or shortly thereafter, Queensland could be exporting an additional 110 Mtpa of steam coal by the end of this decade. (To put this figure in perspective, Australia exported 181.6 Mt of steam coal in 2012.)

Transportation infrastructure constraints and regulatory uncertainty could cause Australia's actual output to fall short of this potential. The governments of Queensland and NSW have implemented a number of rail and port projects since 2008 that have alleviated immediate infrastructure constraints. However, as demand for Australia's black coal exports increases over time, Australia's black coal industry will once again face logistical constraints similar to those experienced between

2004 and 2010 unless remaining proposed rail and port expansions are implemented according to their 2010 schedules. If NSW and Queensland adhere to those schedules, they could increase the coal loading capacities of their ports to 614 Mtpa and plausible levels of exports of steam and coking coal from 336 Mt in 2012 to close to 540 Mt by 2020.[107]

However, such schedule adherence is highly unlikely for either Queensland or NSW given today's complex planning environment and challenging fiscal conditions. Part of the reason that chronic infrastructure shortages are likely to re-emerge as a competitive issue has to do with the types of technology being implemented in Australia – namely, large rail and fixed land port systems. To maximize economies of scale, port and rail expansion projects need to be very large, which means very high first costs and long lead times. Given the volatile price environment and regulatory uncertainties that Australian coal producers are currently facing, it is unlikely that they will be in a position to implement greenfield mining projects or very large new rail and port projects on an expedited basis. Moreover, regulatory clearances have in general become increasingly complex and contentious in Australia due to heightened concerns about GHG emissions and land use impacts of new coal mine developments. The planned scale of new mine and transport projects adds to the sense of uncertainty and difficulty in implementing them.

A historical reason for shortages of transportation infrastructure was the reliance on state-owned entities to make the necessary investments in the rail and port systems. Now that Queensland and NSW have both privatized their port and rail systems, any lack of responsiveness of the rail and port operators cannot be blamed on government ownership. At the same time, private ownership does not necessarily mean that rail and port markets will be fully competitive, or that "for-profit operators" will be incentivized to invest in new infrastructure in the face of volatile global coal prices and regulatory uncertainty.

The most acute regulatory uncertainty had been caused by federal laws introducing the carbon pricing mechanism (CPM) and to a limited extent, the minerals resource rent tax (MRRT). These new measures, and uncertainty about their implementation and long-term survival, had, until they were repealed in 2014, slowed the decision-making processes of coal mining companies that were interested in developing new coal resources and expanding existing mines and related transport infrastructure in Australia. Notably, the massive new steam coal projects planned for the Galilee, Gunnedah, and Surat basins were put on hold, at least partly, due to regulatory uncertainty between 2012 and July 2014. By mid-July 2014, however, coal and other commodities were entering the second year of depressed prices, which are now cited as the principal reasons for the lack of new investments in new coal mines, rail networks, and coal terminals.

The CPM received an especially high level of political scrutiny in Australia. If it had not been repealed, domestic sales of steam coal, especially brown coals, would likely have suffered significant adverse impacts, including the shut-down of these low cost producers of electricity, especially those which were approaching their retirement age. However, a number of "young" brown coal plants, such as the Millmerran and Callide plants in Queensland, may have also been adversely affected by these GHG reduction measures, despite their use of higher efficiency supercritical boiler technology.

Early retirement of the brown coal plants would not have had an immediate adverse effect on the black coal industry. On the contrary, such an outcome might have led to higher dispatch for the remaining coal-fired power plants that are fired on higher rank steam coal. Over the long term, however, the impact of the CPM on domestic sales of black coal would have been substantial, particularly if the price of CO_2 emission permits was allowed to increase significantly over time. This would have discouraged the building of new coal-fired power plants and encouraged development of new power plants based on natural gas and renewable energy resources. As existing black coal power plants were retired, the domestic steam coal market would have declined over time in much the same way as the coal market in the shipping and railway industries did in the 1960s.

For now, the domestic power industry remains a very significant market for black coal producers in NSW and Queensland. In 2012, Australia's power industry consumed 47.8 Mt of black coal, which constituted 67% of all domestic black coal sales. Therefore, from the perspective of steam coal producers in NSW and Queensland, the CPM might have deprived them over time of a very significant and lucrative domestic market for which substitute export markets were simply not available.

But the bigger concern for Australia's black coal producers must have been and remains with their export markets. If the governments of East Asia (China, Japan, Korea, and Taiwan) were also to implement aggressive GHG emission reduction programs that favored alternatives to coal and thereby reduced demand for Australia's black coal exports, the impact on their core business would be devastating. NSW, with its higher proportion of steam to coking coal than Queensland, would experience greater near-term impacts from any regional carbon tax or cap-and-trade programs. However, over time, the effects would also be strongly felt in Queensland in the form of either reduced or cancelled investments in the Galilee and Surat basins.

6.4.3 *Alternative futures*

Given this overall outlook for the industry and in particular the uncertainties concerning CO_2 emissions reduction programs throughout the Asia Pacific region, it is difficult to know the exact path that Australia's black coal industry will take. It will likely face a future somewhere on a continuum between the following two scenarios:

1. A "business as usual" (BAU) scenario, which assumes (a) weak actions by Asian governments to reduce GHG emissions, (b) no major technological breakthroughs on carbon capture and storage (CCS), underground coal gasification (UCG), and other advanced energy conversion and GHG emissions reduction technologies that will be discussed in Chapter 11, and (c) the governments of NSW and Queensland continuing to play their traditional roles of supporting the expansion of transportation infrastructure for the black coal industry.
2. An aggressive GHG reduction scenario that assumes strong GHG emissions standards and implementation by Asian governments of cap-and-trade policies, perhaps coupled with significant technological breakthroughs for CCS and UCG.

Under the BAU scenario, one can expect coal producers to continue delivering energy from coal almost exclusively in solid form using existing rail and port networks expanded to meet their increased production. One might also expect power plant owners to build new, CBM-fired, combined cycle gas turbine (CCGT) plants rather than new coal-fired power plants under this scenario, especially given the increased natural gas supplies from CBM in Australia. However, the possibility depends for the most part on the relative prices of coal and natural gas, which at present are heavily skewed in favor of steam coal.

Over twenty to thirty years, there might be a significant shift to gas-fired CCGT plants as coal-fired plants are retired and new CCGT plants are put in place that run on either CBM or UCG-derived syngas. However, the lifetimes of coal-fired power plants are at a minimum forty years, with fifty- to seventy-year operating lives not uncommon. The potential for early retirement of black coal-fired power plants in Queensland and NSW is, therefore, not very strong under a BAU scenario. Victoria's brown coal power plants are a different matter. Given their age, they offer the strongest prospects for early retirement under a BAU scenario and replacement with gas-fired CCGT plants located either in Victoria or in other states that would transmit power to Victoria.

But the power sector transition would be very slow. It is difficult to imagine a case where a significant reduction in CO_2 emissions from the coal-fired power sector would ever occur under a BAU scenario by 2030.

Instead, the domestic use of steam coal in Australia would be expected to maintain its current use level (in million tonnes) for the foreseeable future.

Under the aggressive GHG reduction scenario, where East Asian governments (Japan, Korea, and Taiwan) implement aggressive programs to reduce GHG emissions in line with previous UN Framework Convention on Climate Change (UNFCCC) recommendations (25% to 30% reduction in 2020 CO_2e emissions from 1990 levels and even greater reductions by 2050), Australia's black coal industry could be facing a very stark future. China is also motivated to diversify its energy mix away from coal, although the extent to which it will be able to do so in the near term remains an open question.

The impacts on the steam coal portion of Australia's black coal industry will depend largely on technology. If "hoped-for" reductions in cost and improvements in efficiency of CCS systems actually occur over the next decade, power companies and other large coal consumers are expected to adopt these systems for new plants and possible existing plants. Under this case, one would expect some reduction in steam coal usage due to expedited retirement of aging coal-fired power plants and their replacement by new gas-fired power plants and to a small extent by power plants that rely on renewable energy resources. The outcome would be dependent on the magnitude of CCS cost reductions as well as efficiency improvements relative to the cost and efficiency improvements in competing technologies such as CBM, UCG, and renewable energy.

However, as this author will argue in Chapter 11, it is unlikely that cost reductions and improvements to CCS systems will be great enough under an aggressive GHG reduction scenario to maintain the current export market position of Australia's black coal over the next decade. If aggressive GHG reduction standards are adopted by governments throughout the Asia-Pacific region, it is likely that Australia's steam coal industry, in order to survive, will need to diversify from one that is based on the extraction and marketing of solid, high-carbon fuels to a two-tiered industry consisting of a declining or stagnant solid fuel segment and a fast-growing liquid and gaseous fuel segment, which will rely on the extraction of gaseous and liquid fuels from coal resources that are at depths that make them uneconomic to extract as solid fuels. Under this scenario, the CBM industry located in Queensland and to a much lesser extent in NSW would continue to grow and prosper. Coupled with a still-to-be-created UCG industry, these two new industry segments might offer an alternative pathway for the industry's growth and prosperity.

Under the aggressive GHG reduction scenario, there will be clear winners and losers. Those mining companies adhering to the old methods of extracting coal will see flagging growth over time and eventually become declining industries. This has happened in the past to the town gas industry as well as ships and railway locomotives running on coal. This scenario may appear stark, but Australia's black coal industry has been here before. What remains to be seen is whether it can repeat the development miracle of the past sixty years in a brave new world of carbon emission limits, with new technologies less proven than those adopted by the coal industry after WWII. We will turn to this question in Chapter 11.

Notes

1. According to data compiled by Geoscience Australia, an Australian government agency with responsibility for assessing the mineral resources of the country, Australia had, as of December 2011, 19.2 billion tonnes (Bt) of proved and probable JORC reserves of black coal and 57.5 Bt of Demonstrated Economic Resources remaining to be extracted (Geoscience Australia, 2013).
2. The Bureau of Resources and Energy Economics (BREE) has reported Australian black coal statistics on a fiscal year (FY) basis starting in FY 2011 with each FY starting in July of the starting calendar year and ending on June 30 of following calendar year (Bureau of Resource and Energy Economics, 2013). Prior to FY 2011, the Australian Bureau of Agricultural and Resource Economics (ABARE) provided periodic reports on Australia's black coal production, exports, and prices (ABARE, 2009). Coal Services Pty Ltd also reports black coal data on both a calendar year and FY basis (Coal Services Pty Ltd, April 2013). The author uses FY estimates throughout the chapter unless otherwise noted, but refers to each FY by the starting calendar year.
3. Australia also has a sizeable brown coal industry, which is located in the state of Victoria. This industry and its CO_2 emissions problem are not discussed in this chapter but are worthy of a separate chapter on the issue of Australia's brown coal utilization and its future in a carbon-constrained world.
4. Brown coals, which are not discussed in this chapter, have very low energy contents – less than 4,200 kcal/kg (gar) and either very high ash contents (> 30%) or very high moisture contents (> 35%), which have in the past made them unacceptable coals in most if not all export markets. Brown coals are used mostly for generating electricity in Australia with a small amount used for making briquettes. In addition to following the above-mentioned coal rank terminology, coal producers in Australia report their resources and reserves according to JORC 2004, which is short for "The Australasian Code for Reporting of Exploration Results, Mineral Resources and Ore Reserves: The JORC Code, (2004 edition)." The organization that issues the JORC Code also provides updated guidelines on applying the JORC Code to specific cases. These updates, as well as the original 2004 JORC Code, are available online from www.jorc.org.

5. The production and export data referred to in this chapter have been obtained from official reports published by (i) Coal Services Pty Ltd and its predecessor, the Joint Coal Board, and (ii) the Bureau of Resources and Energy Economics (BREE) and its predecessor, ABARE. The data provided are mostly by FY, which is the reporting period adopted for this chapter. For the sake of brevity, each FY is expressed as the starting year of that fiscal year. As an example, FY 1960–61 is referred to as 1960. The latest Coal Services Pty Ltd coal statistics report, as of October 2014, covered FY 2011–12, while BREE has published aggregated coal statistics covering FY 2012–13 with forecasts covering later fiscal years.
6. The CIT also assumed the authority over coal industry worker issues, which was previously held by the Commonwealth government's Industrial Relations Commission and the NSW government's Industrial Commission of New South Wales. The joint acts had almost identical provisions, except for powers allowing the JCB to control coal collieries and take over their assets and operations on a compulsory basis, which were contained in the NSW Coal Industry Act only.
7. In 1948, the state of Queensland established the Queensland Coal Board with powers roughly similar to those of the Joint Coal Board. However, Queensland did not seek Commonwealth involvement in the form of joint coal industry acts. Victoria and South Australia also determined that they did not need to involve the Commonwealth government. Instead they asked that they be exempted from the still active Coal Production (War-Time) Act of 1944, which the Commonwealth government did. Surprisingly, the Coal Production Act of 1944 appears to have remained in force until the 1960s, albeit without the coal commission, which discontinued its operations in 1947.
8. See Bullimore, Kim. "1949 Coal Strike: Labour's 'boots and all' sell-out" (www.greenleft.org.au) for a less than flattering analysis of the Australian government's response to the 1949 coal miner strike and the events that eventually led to the decline of union power over the coal industry after WWII. Also see Deery, Phillip (ed.), 1978. *Labour in Conflict: The 1949 Coal Strike*, Sydney: Hale and Iremonger.
9. The JCB, which had the power to take over the operations of inefficient coal mining companies, took control of a number of underperforming collieries during its first twenty years of existence, some of which were transferred to it by the Commonwealth Coal Commissioner's Office in 1947.
10. The divestment decision was certainly helped along by a major financial scandal that occurred within the JCB's accounting department in 1949. The JCB board of directors uncovered matters of gross negligence and managerial incompetence within its accounting department related to the operations of the JCB-owned collieries. The JCB board described the situation as "a complete breakdown in the accounting section which was no longer able to present an authentic statement of the Division's activities." (Joint Coal Board, 1948–49, p. 13). The JCB also noted that the accounting department's "reports and cost statements were not being presented on the due dates and information in reports that had been received was subsequently found to be unreliable and, in some cases, completely misleading" (Ibid.). As

a result of these financial irregularities, the JCB suffered what was considered a significant loss at that time: £350,441 (Ibid.).

11. Coal produced from Australia's underground mines peaked at 91% of total production in 1958 and 1959 (Joint Coal Board, 1961–62).
12. Tenements are those areas subject to mining licenses that are "leased" to some individual or entity for a fixed period of time on commercial terms. A person holding a mining license for a particular area is said to have "tenure" over the land. In Indonesia, tenements are referred to as concession areas.
13. Continuous miners also allowed labor reductions by reducing the number of coal faces that needed to be operated, maintained, and ventilated and by allowing a more continuous completion of the cutting, boring, and loading steps of coal extraction.
14. Longwall mining methods, which were not widely adopted in Australia until the 1980s, relied on the use of a specialized mining machine known as a longwall miner, which sheared the coal from a 100 m × 200 m wide work face and deposited it onto a conveyor for transport to the surface. This method of mining allowed for (a) coal recovery rates of up to 80%; (b) very high output levels per worker; and (c) improved worker safety relative to bord and pillar mining methods. Sensors used in this process of mining helped detect the amount of coal remaining in the seam, while robotic controls enhanced the efficiency of the process (Warner, 1979).
15. In Queensland, where such restrictions did not apply, open-cut mining methods were fully applied by Utah International and Thiess Brothers, which in the mid-1960s discovered large deposits of high-quality coking coal in the Bowen Basin areas of Queensland that were amenable to open-cut mining (Galligan, 1989, pp. 124–125). By the mid-1970s, Queensland produced 88% of its coal using open-cut methods versus 12% for NSW. The role played by these advanced open-cut technologies as well as longwall underground mining methods is discussed in more detail under the heading "rapid growth phase."
16. Diesel- and electric-powered locomotives started to displace coal-fired locomotives in the 1950s.
17. Today, Gladstone is the second largest coal export port in Australia (Newcastle remains the largest). But it was not until the 1970s that Gladstone and Newcastle were expanded into world-class coal shipping ports. These developments are discussed in Section 6.2.2.
18. Domestic consumption in 1950–151 was greater than 100% of total saleable production for that same year, indicating that coal stockpiles were drawn down to meet the strong domestic demand in that fiscal year.
19. Between 1941 and 1946, railway locomotives were the biggest domestic consumers of black coal. Over this period, railways accounted for 23% of domestic black coal consumption versus 20% for the power sector, 13% for the town gas industry, and 2% as bunkers used by the shipping industry (Official Year Book of Australia, 1951, p. 878). From 1949 onward, the power industry was the largest single consumer of black coal in Australia (Official Year Book of Australia, 1951, p. 878, and 1956, p. 1012).
20. In 1987, Queensland surpassed NSW as a producer of "saleable" black coal.

21. Limited information is available from government sources concerning the specific regulations and laws that were in force during the rapid growth phase.
22. For example, during the 1960s, Queensland's coal mining companies were regulated under a mining law that was passed in 1925 (Queensland Government, 1925).
23. This ad hoc method of awarding mining licenses was very similar to the approach followed by Indonesia between 1970 and 2000. In Indonesia, government-to-business (G-T-B) agreements, known as coal contracts of work, effectively launched Indonesia's steam coal industry, which to this day is dominated by companies operating under G-T-B CCOWs (see Chapter 7). This method of awarding mining licenses has now been officially replaced by a system of mining licenses similar to the one being followed in Australia.
24. Utah International, which produced more than 80% of Queensland coal exports during the 1970s as a party to the CQCA, also became a leading developer of mining areas on Kalimantan during the early 1980s (Galligan, 1989).
25. The JCB arguably viewed promoting the welfare of workers as "subsidiary" to the other three objectives (Joint Coal Board, 1948–49, pp. 5–36).
26. In 1983, the Commonwealth and NSW governments conducted a review of the powers, structures, and membership of the JCB. The two governments decided to keep the JCB's existing structure and powers but changed the members of the JCB board to include union and mining company representatives (Parliament of Australia, "Bills Digest No. 42 2000–01: Coal Industry Repeal Bill 2000"). They also required the JCB to appoint an independent chairman of the board. Although the JCB was allowed to continue its more limited operations, its continued role was now starting to be questioned within the government and industry.
27. JORC was established in 1971 by the Australian Mining Industry Council, now known as the Minerals Council of Australia, to address "unacceptable reporting associated with the Poseidon nickel boom and bust in Western Australia," an event that occurred during the late 1960s and destroyed investor confidence in the declarations of reserves by many publicly traded mining companies. JORC issued a number of reports between 1972 and 1989, which recommended improved procedures for publicly reporting and classifying coal deposits. These reports formed the basis for the first JORC Code, which was issued in 1989. It was immediately adopted by the Australian Stock Exchange (ASX) and imposed by the ASX as a mandatory requirement for all publicly listed coal mining companies to use when reporting their ore reserves. The JORC Code is now in its fourth edition, with the 2012 JORC Report representing the latest edition. It is widely accepted in Australia and throughout Asia as the highest standard for reporting coal resources and reserves (http://www.jorc.org/development.asp, "Early History of the JORC Code").
28. The speed at which the NSW coal industry mechanized its operations is even more impressive if one considers that the US coal mining industry relied on continuous miners to produce only 25% of its 1960 coal production.

29. The specific adjustments appear to be the hydraulic roof supports and in particular the hydraulic chocks that are designed to hold up the weight of the roof.
30. As of 1986, Queensland was just starting to apply the longwall mining system, and production data were not available for the years 1978 through 1985 for Queensland.
31. The loss of the town gas business started in the 1950s with the replacement of town gas with low-cost refinery gases (Australia Academy of Technological Sciences and Engineering (1978), p. 820). However, the ultimate demise of the town gas business resulted from the discovery, in 1967, of large oil and gas fields in the Gippsland Basin, which is located offshore from the state of Victoria. A gas pipeline system was developed, and natural gas supplies from the offshore Gippsland oil and gas fields and the onshore gas fields of the Cooper Basin were brought to most major cities (Australia). The development of the retail gas industry in NSW, Queensland, Victoria, and South Australia led to the shutdown of most town gas plants by 1975.
32. As a result of price shocks in 1973 (with the so-called Arab oil embargo) and 1978 (with the Iran oil crisis), Japan, and later Taiwan and Korea, came to see steam coal and uranium as the most cost-effective alternatives to petroleum products as a fuel source for power generation. Over the period from 1980 to 1995, these countries built many new coal-fired power plants, which expanded their demand for Australian steam coal.
33. Many Japanese customers held equity stakes in their sources of coal supply through a Japanese trading house. As buyers, the Japanese felt that their representation within the coal supply company would allow cost-plus pricing arrangements to be effectively monitored for fairness. The best examples from Queensland were the Thiess-Mitsui and the Utah-Mitsubishi partnerships from the 1960s, which led to the development of the massive coking coal deposits of the Bowen Basin. Today, the Utah-Mitsubishi partnership has morphed into the BHP-Billiton/Mitsubishi Alliance, or BMA.
34. ABARE, "Australian Commodity Statistics 2001," pp. 253 and 258, tables 253 and 257. ABARE commodity statistics for the period 1990–2000 indicate that the FOB price paid by Japan for Australian steam coal was higher than the average FOB prices paid by European and other Asian buyers of Australian steam coal. Moreover, ABARE commodity statistics for the period 1992–2000 indicate that Japan sourced 60%–70% of its steam coal and 45%–58% of its coking coal from Australia. Japan's heavy dependence on Australia for its black coal supplies probably acted to neutralize most of Japan's market power over Australia's black coal producers.
35. Over time, pricing provisions of term contracts were revised to allow annual negotiations of the price, quantity, and quality for any coal sales in each year. Certain contracts such as the Utah-Mitsubishi contracts with Japanese steel mills contained provisions that allowed significant price adjustments in the event that cost-plus prices diverged from future spot and term prices. Other

contracts had their price provisions revised to reflect market conditions either at the strong behest of the coal supplier or as a requirement of the Commonwealth government during the 1970s. Once the long-term contracts expired, the parties to these contracts had developed a sufficiently strong working relationship that they simply relied on annual price and quantity negotiations rather than long-term contracts to determine price and quantity of coal to be delivered in any year.

36. Queensland's expansion of its black coal production was largely the work of Utah International (Utah), an American mining company, which began producing high quality, low cost hard coking coals from open-cut mines in the Bowen Basin. Utah's development of the high quality coking coal deposits of Queensland occurred at the same time that Japan's steel industry was undergoing a rapid expansion. Between 1973 and 1982, Utah accounted for more than 70% of Queensland's black coal exports. Utah's share of Queensland's total coal exports dropped from a peak of 85% in 1978 to 39% by the end of the rapid growth phase. For a superbly researched book on the subject of Utah International's role in the development of Queensland's black coal industry, the reader is referred to Galligan (1989).

    NSW still remained the largest producer of saleable black coal until 1983 due to the large quantity of black coal used by NSW's steel-making and power-generating industries. In 1975, NSW consumed 19 Mt of black coal in domestic markets, whereas Queensland had a much smaller consumption of 5.3 Mt.

37. BPCT was conceived as a dedicated terminal for the export of Moura coal produced by the Thiess Peabody Mitsui (TPM) Coal Pty Ltd. The operation of BPCT commenced in the mid-1960s with first coal (1,600 tonnes) shipped from BPCT in 1967 (Gladstone Port).
38. RGTCT, which was originally named the Clinton Coal Facility, was approved for construction by the GPA in 1976 after BHP Co. Ltd. confirmed that it had secured long-term contracts for the sale of coking coal to Japanese steel mills. Stage one of the facility was completed in 1980 at a cost of A$45 million. During its first full year of operation, RGTCT exported 4.3 Mt of coal (Gladstone Ports Corporation Ltd., 2012).
39. The Hay Point Coal Terminal was originally constructed and owned by Utah International (Galligan, 1989, p. 65).
40. During the period 1850–1860, the state colonial governments were willing to allow the private sector to build, own, and operate rail systems. However, due to financial constraints, these private rail systems were within a short period of time taken over by government. For example, a private company, which attempted to build a railway in NSW in 1854, went bankrupt in 1855 and had to be taken over by the government before completion. South Australia's railways were government owned from the beginning, including a horse-drawn line that opened in 1854 and a steam-powered line that opened in 1856. In Victoria, private railways became insolvent soon after starting operations and were soon thereafter taken over by the colony government.

41. Figures for 2003 and 2006 for all of Australia were, respectively, 72.6 Mt, 90% of underground production, and 80.9 Mt, 91% of underground production.
42. The Queensland Mineral Resources Act 1989 covered all topics and regulatory issues related to the exploration for mineral resources, awarding of licenses, and development of mines. Health and safety matters were still handled under the Coal Mining Act 1925, which remained in force until 1999, when it was repealed and replaced by the Coal Mining Safety and Health Act 1999.
43. See Joint Coal Board/Queensland Coal Board, "1996 Australian Black Coal Statistics," p. 20, table 21. The JCB stopped publishing statistics on average earnings of mine workers in the NSW coal mining industry after 1982, and one can only speculate that productivity gains were offset by higher nominal wages.
44. Other supply-side factors, which contributed to the large price increases between 2007 and mid-2008, are the following:
    - The Indonesian government's removal in 2005 of price subsidies on diesel and fuel oil, which caused most Indonesian coal producers to experience significant increases in their costs of mining.
    - The decline in the US$ relative to the A$ and the Indonesian rupiah caused the US$ costs of Australian and Indonesian coal producers to increase. On average, over the period 2005 and 2008, the A$ and Indonesia rupiah appreciated by 60% and 15% respectively against the US$.
    - Most Indonesian coal producers had "sold out" their low-cost bituminous and sub-bituminous coal reserves and were starting to develop new resources with lower CVs, located further inland without transport infrastructure but with higher stripping ratios than existing coal reserves.
45. In late 2008, the US and Europe went into a deep economic recession due to the collapse of the housing and mortgage derivatives markets in the United States. For a short time, it led to a massive decline in the prices of coking and steam coals; the globalCOAL Newcastle Price Index (gcNEWC) dropped to $61.37 per tonne for March 2009 before starting a recovery that lasted through April 2012, with the highest price point being $129.99 on January 2011 – a far cry from the July 2008 peak price of $184 per tonne (Global Coal Limited, 2014).
46. After 2003, a very wide price differential developed in favor of coking coal. This increase in the coking to steam coal price differential caused supplies of PCI coals, which previously were being sold into the steam coal market, to be redirected back into the coking coal market due to the higher price they would earn if sold as PCI coals. These substitution effects added to the price volatility of steam coal over the period 2004–09.
47. In 2001, Lang Corporation changed its name to Patrick Corporation, which in 2006 was acquired by Toll Holdings in a hostile takeover. Toll then split itself into Toll and Asciano, an ASX listed company that holds the combined Toll and Patrick shares of PN.
48. See State Records Archive Investigator, "FreightCorp" (http://investigator.records.nsw.gov.au/Entity.aspx?Path=%5CAgency%5C1079), and History

of Patrick Corporation at http://www.patrick.com.au/index.php?id=49. Patrick Corporation was named Lang Corporation until October 2001.

49. PNR currently operates in all states and the Northern Territory, and prior to the November 2010, IPO of Queensland Rail was Australia's largest private rail freight company.
50. Shares sold at A$2.55 each, which was at the low end of the offer range (A$2.50–A$3.00) (Bloomburg Businessweek, "QR National Climbs in Sydney after A$4.05 Billion IPO," November 22, 2010 (www.businessweek.com/news/2010–11–22/qr-national-climbs)).
51. *The Australian*, "Queensland Sells Down Stake in Railways Carrier Aurizon," December 4, 2013. Available at http://www.theaustralian.com.au/business/companies/queensland-sells-down-stake-in-railways-carrier-aurizon/story-fn91v9q3–1226774573741#.
52. In 2010, Asciano, owner of PNR, signed a coal transport contract with Anglo American to haul 16.5 Mtpa of coal in Queensland with some deliveries beginning in 2010. It also announced plans to double its deliveries in Queensland by 2015.
53. Given the uncertainties concerning the water pollution risks from UCG, the Queensland government could be largely excused for proceeding with caution when dealing with the UCG industry. The CBM industry was also very highly developed by 2010 and had much greater political support in Queensland than the UCG industry.
54. This second, often heard "complaint" is probably of greatest concern from the perspective of sustainable coal supply from Australia because new mining projects will not go ahead without related port and rail projects being approved and under way.
55. NSW has recently been rocked by a corruption scandal involving the Minister of Labour (Ian MacDonald) from the former state Labor Government and a Labor MP (Eddie Obeid). In January 2014, the Independent Commission Against Corruption ruled that the Labor minister committed multiple corrupt acts in 2008 when he issued mining licenses to Eddie Obeid and other parties. The licenses have since been canceled by the current national government, but the message going out to the mining community is that NSW may have a pervasive culture of corruption along with an unfavorable regulatory regime (see http://www.smh.com.au/comment/culture-of-corruption-pervades-nsw-inc-20130801–2r23r.html" and http://www.smh.com.au/nsw/nsw-government-cancels-mining-licences-tainted-by-eddie-obeid-ian-macdonald-corruption-scandals-20140120–314iv.html).
56. In mid-2014, the Australian Parliament and Senate voted to rescind both laws, but prior to their repeal they were major investor concerns.
57. In 2000, Australia's production of saleable black coal was 257.5 Mt, of which 193.4 Mt was exported.
58. In particular, Indonesia's coals have low to medium CVs with high to medium moisture contents and low ash and sulfur contents, whereas Australian domestic coals have very low CVs, high ash and low moisture contents. Australian steam coals, which sell at an FOB premium to Indonesian coals, have a higher ash content (around 8%) than the

Indonesian coals (around 3%–5% by weight). As a result of this quality preference and the lower $/GJ price for Indonesian steam coals relative to Australian steam coals, Indonesia's exports of steam coal to India have increased smartly since 2005. Its exports to India are expected to increase at an even greater growth rate over the next decade.

59. More recent job estimates are not available.
60. The regulatory uncertainty created by Australia's CPM is only half the story. The impact of GHG policies of the governments of Japan, Korea, Taiwan, and the rest of Asia, which in 2012 accounted for 98% of Australia's steam coal exports and 72% of its coking coal exports, could significantly affect the fate of Australia's black coal industry.
61. The Australian government's official estimates of coal resources and reserves are published annually by Geoscience Australia (see Geoscience Australia, Canberra, "Australia's Identified Mineral Resources" for the latest report, which is available online at www.geoscienceaustralia.com.au). It provides estimates of demonstrated resources – both economic and sub-economic – as well as JORC reserves for all mineral resources, including black and brown coal and CBM. Resource estimates are first expressed as demonstrated resources and then separated into economic and sub-economic demonstrated resources. Economic demonstrated resources (EDR) are those resources that Geoscience Australia concludes have either near-term or long-term potential to be developed economically. EDR of black coal are approximately three times greater than JORC black coal reserves, which are reported annually by publicly traded mining companies.
62. Queensland government, Department of Mines and Energy, "Queensland's World-class Coals: Mine Production and Developments," November 2007, p. 1. Available at http://www.dme.qld.gov.au/zone_files/coal_files_pdf/wcc_nov_07.pdf.
63. This figure excludes production estimates for mines that will be producing only coking coal and also the large steam coal developments planned for the Galilee and Surat basins in Queensland and the Gunnedah Basin in NSW. A database showing the productive capacities of most steam coal mine expansions and new steam coal mines for NSW and Queensland was prepared in 2009 for PESD by Michael Friederich, an independent coal geologist from Queensland. It contains information on those steam coal projects that have advanced to the stage of construction. The database excludes all developments for the Surat, Galilee, and Gunnedah basins and mines in other basins that produce only hard coking coal. The production forecast assumes that each announced project is economically feasible and will go forward.
64. This forecast does not preclude the possibility of additional discoveries adding to the reserve figures, especially in the form of new underground mines in NSW.
65. Queensland government, Surat Basin Rail Joint Venture, "Rail Update Fact Sheet," November 2007. Available at www.suratbasinrail.com.au/files/SBR-fact-sheet-december07.pdf.

66. The Surat Basin coals have HGIs ranging between 32 and 35. (HGI, which is short for Hardgrove Grindability Index, is a measure of a coal's hardness, or more precisely the difficulty that a power plant's coal mills will have in grinding it.) Most power plants in Asia have mills designed to take coals having an HGI of 45 or higher. If lower HGI coals are used in those plants, the mills may run out of grinding capacity, and the plants' owners may need to derate the generating capacity of these plants.
67. Xstrata planned to develop the 30 Mtpa Wandoan mine over the next decade (2014–24) according to market conditions and development of new rail and port infrastructure. As part of the Wandoan project, Xstrata intended to develop a new 210 km rail line, known as the Surat Basin railway project, which would extend from the mine at Wandoan up to the existing Moura-Gladstone line. Xstrata and its joint venture partners received conditional approval of its EIS from the Commonwealth government in March 2011 and seemed on track to achieve a full mining license. However, in September 2012, in response to poor market outlook for thermal coal exports in Asia, Xstrata decided to put its Wandoan project on indefinite hold (Heber, 2013).

    In early 2009, Syntech Resources announced plans to develop its Cameby Downs steam coal project and estimated that the mine would cost A$250 million to develop with production commencing in 2010 at 1.4 Mt of saleable coal. Syntech then planned to quickly ramp up production to 20 Mtpa of saleable coal by 2013 with the Port of Gladstone (Wiggins Island Coal Terminal) as the eventual export port. This project was expected to share the costs of developing the Surat Basin railway project with Xstrata. Similar to Xstrata's Wandoan project, the Cameby Downs expansion project has been shelved due to poor market conditions. In addition, on August 5, 2011, Syntech sold out its interest in Cameby Downs to Chinese coal producer Yancoal, which has decided to reduce the project expansion plan to 8 Mtpa of saleable coal from the original plan of 20 Mtpa. Even mining at this new reduced rate will not occur in the foreseeable future due to lack of rail capacity (Queensland government, Department of Environment and Heritage Protection, "Cameby Downs Expansion Project," April 16, 2014. Available at http://www.ehp.qld.gov.au/management/impact-assessment/eis-processes/cameby_downs_expansion_project.html).
68. GVK Hancock Coal (GVK Hancock), the Adani Group, and Waratah Coal are leading the development of the three separate coal deposits in the Galilee Basin. They had publicly announced plans to bring anywhere from 60 Mtpa to 120 Mtpa of new steam coal deposits into production between 2013 and 2015. But as with the Surat Basin greenfield coal projects, the commercial operation dates for all Galilee Basin projects have been pushed back due to market conditions.

    GVK Hancock is developing two deposits, Alpha and Kevins Corner, which are located right on the far eastern edge of the Galilee Basin. Alpha will be developed as an open-cut mine, and Kevins Corner will support both open-cut and underground mines. Alpha and Kevins Corner have been extensively drilled and delineated according to JORC standards. As of

December 2013, GVK Hancock claims to have identified 7.9 Bt of JORC-compliant resources (measured, indicated, and inferred), of which at least 2.4 Bt are classified as measured and indicated resources. Development of the Alpha and Kevins Corner resources will require the construction of a new railway line between these two mine sites and a new coal terminal –Terminal 3 – with a coal handling capacity of 60 Mtpa within the existing Port of Abbot Point. GVK Hancock won conditional approval for T3 in October 2013 from the Commonwealth government under the Environment Protection and Biodiversity Conservation Act of 1999 (EPBC Act). GVK Hancock development plans will eventually lead to new coal mines capable of producing 60–80 Mtpa of ROM coal and exportable coal of 50 Mtpa. GVK Hancock estimates the total capital cost (inclusive of mines, rail, and port facilities) for its Alpha–Kevins Corner project to be A$10 billion. The GVK Hancock project is the most advanced of all of Australia's greenfield coal mine projects.

The Waratah project is being developed by Waratah Coal, which holds steam coal tenements that are contiguous to GVK Hancock's Alpha and Kevins Corner deposits. Waratah's tenements are considered an extension of Hancock's two deposits. Less information is available on the Waratah tenements and the company's development plans than is available for the Hancock projects. Waratah Coal plans to develop a mine with a ROM production capacity of 40 Mtpa that will generate saleable steam coal production after washing of 30 Mtpa. Waratah will either build its own 290 km rail link to Abbot Point or build it in cooperation with Hancock Coal. The "go-it-alone" cost of the railroad is estimated at A$2.1 billion (2008 prices), while the mine and port are estimated to cost A$3.18 billion and A$1.27 billion, respectively.

69. To put all these numbers in perspective, a well-managed port with adequate capacity should take no more than four days to turnaround a Panamax vessel and six days to turnaround a Capesize vessel. The four- to six-day allowance includes the time required to berth the vessel, load the coal into the holds of the vessels, and trim the vessel for sailing. At 2010's depressed vessel charter rates it cost only US$15,000 per day to charter a Panamax vessel with a 70,000 tonne cargo-carrying capacity under a three-year contract. Capesize vessels with a cargo carrying capacity of 160,000 tonnes cost no more than US$20,000 per day under a three-year contract. The potential demurrage charge for the Panamax vessel loading at DBCT would have amounted to US$315,000, or US$4.50/t, while the demurrage charge for the Cape at DBCT would have amounted to US$2.38/t. The Kooragang charges would have been much less; i.e., US$2.50/t for the Panamax vessel and US$1.33 per tonne for the Cape. It is likely that these demurrage charges were treated as noise by coal producers during those times when Newcastle coal prices were above US$90 per tonne. In today's depressed coal market it is likely that these demurrage losses are considered significant.

70. During 2007–08, the coal supply problems caused by the chronic transport constraints were exacerbated by a severe storm in June 2007 in the Hunter Valley, which reduced expected black coal output by more than 3 Mt.

71. In 2008, the Ministry of Industry issued a comprehensive report outlining its plans for expanding port and rail infrastructure to support the development of the Surat and Galilee basins and allow increased production from the Bowen Basin. This report, which is dated September 2008, was available through 2010 from www.transport.qld.gov.au but can no longer be found at that website. Note that this plan does not consider ongoing projects announced by GVK Hancock and Waratah Coal to build separate rail lines and coal terminals at Abbot Point and elsewhere, which could if realized lead to another 100 Mtpa of new nameplate port capacity being added by 2020.
72. At Newcastle, a new coal terminal, known as NCIG (stages 1–2) coal terminal have added 120 Mtpa of coal loading capacity as of December 2012 with an additional 66 Mtpa of capacity being added as stage 3 to NCIQ. There is also the possibility of a fourth terminal, with a capacity ranging from 65 to 120 Mtpa, being added by PWCS, which operates the Carrington and Koorangang Terminals. Newcastle is now able to load 211 Mtpa of coal, up from 125 Mt in early 2010. Gladstone recently expanded its RG Tanna and Barney Point terminals and can now handle up to 100 Mtpa, up from 75 Mtpa in 2010. It may eventually add the WICET coal terminal with a planned capacity of 90 Mtpa. With WICET, Gladstone will have a coal ship-loading capacity of 190 Mtpa, up from 75 Mtpa in 2010. (Newcastle Coal Infrastructure Group, 2013; Gladstone Ports Corporation, 2012.)
73. By 2015, NCIG is expected to reach a total coal handling capacity of 66 Mtpa, and the Port of Newcastle will have a total nameplate capacity of 192 Mtpa. The total cost of the NCIG terminal is estimated at A$1 billion including construction contingencies. NCIG will be owned by six coal producers: BHP Billiton, Centennial Coal, Donaldson Coal, Felix Resources, Peabody, and Whitehaven Coal.
74. The Surat Basin Rail Project, which was designed to transport 42 Mt of coal per year, is being implemented as a joint venture between Atec Rail Group, Aurizon, and Glencore/Xstrata. If it is ever built, it will connect the Surat Basin coalfields with the existing Banana-Gladstone line. As of November 2012, the project has been indefinitely delayed.
75. Friederich, Michael. Consulting Mine Geologists, private correspondence, June 2014.
76. These expansion plans do not take into account the potential development of the GVK Hancock and Waratah projects within the Galilee Basin, which will add an additional 60 Mtpa to 100 Mtpa of new steam coal export capacity over the next decade.
77. "UNFCCC" stands for the United Nations Framework Convention on Climate Change.
78. The cap-and-trade concept was first proposed during the late 1960s by the US National Air Pollution Control Administration (NAPCA), the predecessor to the United States Environmental Protection Agency's Office of Air and Radiation, as a practical emissions abatement method. At that time, NAPCA completed extensive computer simulation exercises, which showed that a "cap-and-trade" approach to pollution abatement was by far the least-cost solution for reducing emissions. The concept was not

applied until 1990, when it was used for reducing SO_2 emission as part of the Acid Rain Program of the 1990 Clean Air Act. The SO_2 cap-and-trade system reportedly led to a 50% reduction in the 1987 level of SO_2 emissions by 2007.

79. In Australia, a green paper is a national government report that contains a first proposal for a change in law without any commitment to action. It is a "first call for public inputs" and the first step taken before initiating a significant change in law. Green papers may result in the production of a white paper, which is an official statement of policy by the Australian government. It derived its name due to the historical fact that such papers were originally bound in white paper. The green paper supporting the CPRS can be found at www.climatechange.gov.au/~/media/publications/green-paper/greenpaper.ashx.
80. The white paper for the CPRS can be found at www.climatechange.gov.au/publications/cprs/white-paper/cprs-whitepaper.aspx.
81. It is most common for draft bills to be sent first to the House of Representatives, in part because that is where most of the ministers reside.
82. Bills are not deemed approved by either of the two houses until the bill has gone through three readings. The first reading simply introduces the bill and reads it into the record. The second reading involves the debate of policy and principles, and the third reading involves debating the details of the bill. Once it passes the third reading, the bill is typically passed from the originating house to the second house.
83. Australian government, Department of Climate Change, "Summary: Key Changes to the Carbon Pollution Reduction Scheme Legislation," May 2009.
84. Participants to the UN's Copenhagen Conference were expected to agree on a successor to the Kyoto Protocol and CO_2e emission reduction provisions that would commit all developed countries to reduce their CO_2e emissions in 2020 and 2050 below some baseline levels, and in the case of developing countries agree to a binding program for reducing the growth of their emissions. The conference was held between December 7 and December 19, 2009 in Copenhagen but failed to meet its lofty goals. A subsequent climate change conference organized under the auspices of the UN's COP15 organization was held in Cancun, Mexico, between November 29 and December 10, 2010. Only minor agreements were reached at that conference.
85. Deutsche Bank Australasia, "Australian Carbon Policy: Tax-then-Trade to Start July 2012," February 24, 2011.
86. The Gillard government argued that fixed carbon prices from July 2012 to July 2014 were not a tax because any payments from business to government would lead to the issuance of a permit, which under the carbon pricing mechanism would be assignable personal property with a tangible value to the holder of the permit.
87. Australian government, "Securing a Clean Energy Future – the Australian Government's Climate Change Plan," July 2011.

88. The Australian Coal Association forecast that eighteen gassy underground mines, primarily located in the Hunter Valley of NSW, could be forced to close by 2020 if the carbon pricing mechanism bill was passed into law. The Australian government has offered a A$1.3 billion assistance program for coal mines affected by the Carbon Pricing Mechanism of the Climate Change Plan (Platts International Coal Report, July 18, 2011).
89. With respect to power plant closures, two plants that are fired on brown coal were identified in the press as likely candidates for closure. They are Hazelwood (1,675 MW), which is located 185 km from Melbourne in the state of Victoria and which reportedly supplies 25% of Victoria's power requirements, and Playford (325 MW), which is located in South Australia. The newest unit at Hazelwood is forty years old, and the two oldest units are forty-seven years old. Its owners – International Power and SUEZ GDF – indicated in 2011 that they would be willing to close the plant and convert over to gas if a payment of reportedly A$2.5 billion was made to them for doing so. Playford was commissioned in 1963, making it at least forty-eight years old, but was refurbished in 2005, providing it with at least thirty years of additional life. The owners of Playford – Alinta Energy – requested A$250 million as a closure fee (Morton and Grattan, 2011).
90. EITEs would largely involve the free issuance of emission permits to eligible businesses, but in some cases could involve other direct financial support.
91. Australian Coal Association, "Letter to Greg Combet, Minister for Climate Change and Energy Efficiency titled Coal Industry – Interim Response and dated 11 May 2011." In addition to arguing that no other country had imposed a price or tax on methane emissions from underground and open-cut mines, the ACA noted in its letter to the Ministry of Climate Change that:
    - the methodology and procedures for measuring the amount of methane that escapes into the atmosphere from underground and open-cut mines have never been developed; and
    - technologies for combusting the ventilation air methane from underground mines are not commercially available.
92. Jack Saunders, "NSW Treasury Says Carbon Tax Is Job Killer," *Energy Publishing: Australian Coal Report*, August 10, 2011, p. 5.
93. As of November 2011, the Labor Party and its allies, the nine Greens senators and two independent Senators, commanded a majority of forty-two in the Senate, with a coalition of the Liberal, National, and Country Liberal parties controlling the remaining thirty-four senate seats. The Parliament (House of Representatives), which has a total of 150 members, was more evenly balanced, with both the Australian Labor Party and the Liberal – National Party coalition having seventy-two members each. In order to pass the CEA into law, Labor was forced to rely on the Green Party, which had a single MP (Adam Brandt), as well as the votes of three independents (Robert Oakeshott, Andrew Wilkie, and Tony Windsor). Other members of Parliament held

political views more closely aligned with those of the Liberal-National coalition.

94. Examples of some of the specific benefits that would accrue to low and middle income families under the Clean Energy Act were:
    - an increase in the statutory, tax-free threshold for individuals from A$6,000 per year to A$18,200 per year on July 1, 2012, and to A$19,400 per year by 2015;
    - a tax cut of A$300 for taxpayers with incomes above the tax-free threshold but lower than A$68,000;
    - an increase in annual pension support to pensioners by A$338 per year for singles and A$510 per couple; and
    - additional benefits for families with children.
95. Jack Saunders, "Fiercest Critics Prepare for Carbon Tax," *Energy Publishing: Australian Coal Report*, August 3, 2011, p. 3.
96. The original petroleum resource rent tax (PRRT) is a profit-based tax, which is imposed on all petroleum products recovered from Australian government waters (including crude oil, natural gas, LPG condensate, and ethane), except for petroleum products extracted from the North West Shelf project and the Timor Sea Joint Petroleum Development Area, and from processes that produce value added products such as LNG (Australian government, Department of Resources, Energy and Tourism, "Petroleum Resources Rent Tax." Available at http://www.ret.gov.au/resources/enhancing/taxation/prrt/Pages/PetroleumResourceRentTax(PRRT).aspx).
97. KPMG, "Minerals Resource Rent Tax: Exposure Draft," June 30, 2011. The draft legislation closely followed the recommendations made by a government-appointed advisory group, known as the Policy Transition Group (PTG). The PTG had as its chairman, Don Argus, the former chairman of the board for BHP-Billiton and the Minister of Energy.
98. Australian government, Treasury Department, "Explanatory Material for Exposure Draft Minerals Resource Rent Tax," June 10, 2011, pp. 12–18. The three allowances were:
    - *Royalty allowance*: Coal Miners are allowed to deduct from any mining profits all royalty payments made to state governments with an arrangement to carry-forward any royalty credits not taken in the year in which they are paid.
    - *Mining loss allowance*: If a mining operation incurs a loss in any MRRT year, it may use that loss credit to reduce its MRRT liability in future years.
    - Starting base loss allowance: For those mining operations that started prior to July 2012, the owners may deduct certain costs incurred prior to the commencement of the MRRT from any mining profits.

    Many of the allowances have uplift arrangements that preserve the present value of the portion of those allowances that are carried forward to future MRRT years.
99. KPMG, "Minerals Resource Rent Tax: Exposure Draft," June 30, 2011.

100. AMEC, "Response to the MRRT Legislation Exposure Draft and Explanatory Material Released on 10 June 2011," July 14, 2011, p. 4.
101. Based on financial modeling completed for AMEC by the University of Western Australia, AMEC argued that coal or iron ore projects that existed before May 2, 2010, would have an MRRT effective rate 4.3% lower than new projects which achieved commercial operations after July 1, 2012. Most of these established projects were owned by the "Big Four," making the MRRT in the eyes of AMEC "a biased and inequitable tax." The strongest opposition to the MRRT appeared to come from the small and medium-sized firms represented by AMEC, such as Fortescue Metals, an emerging iron ore miner, which took its opposition to the MRRT to a very personal level. Andrew Forrest, CEO of Fortescue, was quoted in early 2011 as stating that "BHP has literally written a tax for everyone else to pay" ("MRRT Remains Unfair: Forrest," *The Australian*, March 24, 2011).
102. The Coalition, which is composed of the National Party, the Liberal Party, the Queensland Liberal National Party, and the New Territory Country Liberal Party, took ninety seats in Parliament, while Labor took fifty-five, with one seat going to the Greens. The Senate had forty of seventy-six seats up for election, with the Coalition ending up with thirty-three seats, Labor with twenty-five seats, and crossbenchers, such as the Greens and Clive Palmer's United Party, controlling a record eighteen seats.
103. For example, the delicate political alliance of convenience between Labor and the Greens in the House and Senate could have fallen apart at any time if a single member of the Labor-Green coalition had passed away or resigned due to scandal before the clean energy legislation had been passed by both houses. Or the Greens could have over-reached in their efforts to shut down Australia's black coal industry and caused members of the Labor Party or the three independents to refuse to make any more concessions to the Greens on matters related to the coal industry. But the Alliance remained intact, and the two laws were passed.
104. Joint Coal Board, "Black Coal in Australia, 1986–87," p. 5, table 1; p. 97, table 100; and p. 99, table 101. Also, Annual Report of the Joint Coal Board, 1960–61, p. 134, table 10.
105. The decision by large mining companies to suspend new projects was also affected by the 2012 passage of the CPM and MRRT, but their inability to comprehend the coal price roller coaster of the volatile price phase was clearly an important factor in their decisions to delay new projects.
106. Economic demonstrated resources (EDR) estimates.
107. This estimate assumes that the rail and port infrastructure project outlined in Table 6.5 proceed according to schedule and that port capacity is used at 90% of its nameplate capacity.

References

ABARE [Australian Bureau of Agricultural and Resource Economics] (2001). *Australian Commodity Statistics, 2001*. December 19. Available at http://data.

daff.gov.au/data/warehouse/pe_abarebrs99000752/PC12239.pdf (accessed September 25, 2014).

ABARE (2009). *Australian Commodity Statistics, 2009*. December 3. Available at http://data.daff.gov.au/data/warehouse/pe_abarebrs99001676/acs_09.pdf (accessed September 25, 2014).

ABC News (2011a). "Gillard Vows to Fight After Poll Hammering." *ABC News*, March 8. Available at http://www.abc.net.au/news/2011–03–08/gillard-vows-to-fight-after-poll-hammering/2666198 (accessed September 26, 2014).

ABC News (2011b). "Windsor Savages Carbon Tax Strategy." *ABC News*, March 9. Available at http://www.abc.net.au/news/stories/2011/03/09/3158764.htm?site=news (accessed September 26, 2014).

AMEC (2011), "Response to the MRRT Legislation Exposure Draft and Explanatory Material." Released on June 10, 2011 (July 14, 2011).

The Australian (2013). "Queensland Sells Down Stake in Railways Carrier Aurizon," December 4. Available at http://www.theaustralian.com.au/business/companies/queensland-sells-down-stake-in-railways-carrier-aurizon/story-fn91v9q3–1226774573741#.

The Australian (2011). "MRRT Remains Unfair: Forrest," March 24.

Australian Academy of Technological Sciences and Engineering, and Australian Science and Technology Heritage Centre (2001). *Technology in Australia 1788–1988*. Melbourne: Australian Academy of Technological Sciences and Engineering, November 21, 1988. Available at http://www.austehc.unimelb.edu.au/tia/titlepage.html (accessed September 25, 2014).

Australian Coal Association (2011). "Letter to Greg Combet, Minister for Climate Change and Energy Efficiency titled Coal Industry – Interim Response," May 11, 2011.

Australians for Coal (2014). Available at http://www.australiansforcoal.com.au/ (accessed September 25, 2014).

Australian government (n.d.). "Climate Change." Available at www.climatechange.gov.au/publications/cprs/white-paper/cprs-whitepaper.aspx.

Australian government (n.d.). "Fact Sheet – Resource Super Profits Tax." Available at http://www.rooloose.com/file_download/7/10_Fact_sheet_Resource_Profit_Tax_Final.pdf (accessed September 25, 2014).

Australian government (n.d.). "Petroleum Resources Rent Tax." Department of Resources, Energy and Tourism. Available at http://www.ret.gov.au/resources/enhancing/taxation/prrt/Pages/PetroleumResourceRentTax(PRRT).aspx.

Australian government (2011). "Explanatory Material for Exposure Draft Minerals Resource Rent Tax," Treasury Department. June 10, 2011.

Bloomburg Businessweek (2010). "QR National Climbs in Sydney after A$4.05 Billion IPO," November 22, 2010. Available at www.businessweek.com/news/2010–11–22/qr-national-climbs.

Bowen, Bruce, and Peter Gooday (1993). *Coal: The Economics of Coal Export Controls*. Canberra: *ABARE*.

BREE [Bureau of Resources and Energy Economics] (2014). *Resources and Energy Quarterly, September 2014*. Available at http://www.bree.gov.au/sites/bree.gov.au/files/files//publications/res/quarterly_res_2014.pdf (accessed October 24, 2014).

BREE (2013). *Resources and Energy Statistics, 2013*. Report no. 3.1. April 12. Available at http://www.bree.gov.au/sites/bree.gov.au/files/files//publications/res/annual_res_2013.pdf (accessed September 26, 2014).
Bullimore, Kim, "1949 Coal Strike: Labour's 'Boots and All' Sell-out." Available at https://www.greenleft.org.au/node/19445 (accessed October 24, 2014).
Carswell, Andrew (2010). "Your Simple Guide to the Resource Super Profits Tax." *News.com.au*, June 17. Available at http://www.news.com.au/national/your-simple-guide-to-the-resource-super-profits-tax/story-e6frfkp9-1225880854023 (accessed September 25, 2014).
Coal Services Pty Ltd. *2013 Australian Black Coal Statistics*. Report.
Coal Services Pty Ltd. *2012 Australian Black Coal Statistics*. Report.
Coal Services Pty Ltd., and Queensland Department of Mines and Energy. *2006 Australian Black Coal Statistics*. Reports available for purchase from http://www.coalservices.com.au/blackcoalstatistics.aspx.
Commonwealth Bureau of Census and Statistics, "Official Yearbook of the Commonwealth of Australia."
Deery, Philip, ed. (1978). *Labour in Conflict: The 1949 Coal Strike*. Sydney: Hale and Iremonger.
Department of Foreign Affairs and Trade (n.d.). http://www.dfat.gov.au/facts/indigenous_land_rights.html.
Energy Publishing Inc. (2011). "Fiercest Critics Prepare for Carbon Tax." *Australian Coal Report*. August 3.
Energy Publishing Inc. (2011). "NSW Treasury Says Carbon Tax Is Job Killer," *Australian Coal Report*, August 10.
Energy Publishing Inc. (2010). *Australia Coal Report*, December 2010.
Energy Publishing Inc. (various years). *Australia Coal Report*. January 2009 to February 2010.
Energy Publishing Inc. (2010). *Coal Chain Australia*. Issue 0013, November.
Energy Publishing Inc. (2010). *Coal Chain Australia*. Issue 0014, December.
Energy Publishing Inc. (n.d.). *Coal Chain Australia Weekly*.
Environmental Protection Agency (2014). "About the EPA." *Epa.wa.gov.au*. Available at http://www.epa.wa.gov.au/AbouttheEPA/abouttheEPA/Pages/default.aspx?cat=About%20the%20EPA&url=AbouttheEPA/abouttheEPA (accessed September 25, 2014).
Fish, Brad (2010). "The Port of Abbot Point Multi Cargo Facility Project." Presentation to the Major Projects Conference, October 19. Available at http://www.dip.qld.gov.au/resources/major-projects/presentations/mpc10/brad-fish.pdf.
Friederich, Michael (2014). Consulting Mine Geologists, private correspondence, June 2014.
Galligan, Brian (1989). *Utah and Queensland Coal: A Study in the Micro Political Economy of Modern Capitalism and the State*. St. Lucia, Qld.: University of Queensland Press.
Garnaut, Ross (2011). *The Garnaut Review 2011 – Australia in the Global Response to Climate Change*. Report. July. Available at http://www.garnautreview.org.au/update-2011/garnaut-review-2011/garnaut-review-2011.pdf (accessed September 25, 2014).

Geoscience Australia (2009/2012/2013). "Australia's Identified Mineral Resources" Available at http://www.ga.gov.au/image_cache/GA16805.pdf (accessed September 25, 2014).

Gladstone Ports Corporation Ltd (2012). *50 Year Strategic Plan*. July. Available at http://www.gpcl.com.au/Portals/0/pdf/50%20yr%20plan/GPC_50yr_strategic_plan_2012_update.pdf (accessed September 26, 2014).

Gladstone Ports Corporation Ltd (2011). "History of the Gladstone Ports Corporation Limited." *Gpcl.com.au*. Available at http://www.gpcl.com.au/AboutGPC/HistoryofGPC.aspx (accessed September 25, 2014).

Global Coal Limited (2014). https://www.globalcoal.com/ (accessed September 25, 2014).

Heber, Alex (2013). "GlencoreXstrata Parks Wandoan Coal Project, Cuts Capex." *Miningaustralia.com.au*, September 11. Available at http://www.miningaustralia.com.au/news/glencorexstrata-parks-wandoan-coal-project-cuts-ca (accessed September 26, 2014).

House of Representatives (2011). "Exposure Draft – Clean Energy Bill 2011 No., 2011." *The Parliament of the Commonwealth of Australia*, July 28, 2011. (accessed September 26, 2014).

Illawarra Coal. "Technology – The History of Mechanisation." *Illawarracoal.com*. Available at http://www.illawarracoal.com/technology.htm (accessed September 25, 2014).

Joint Coal Board (1988–present). *Australian Black Coal Statistics* (from 1988 to 1997, a joint publication of JCB and the Queensland Coal Board).

Joint Coal Board (1980–1987). *Black Coal in Australia*.

Joint Coal Board (1947–1979). Annual Reports.

Joint Ore Reserves Committee (2009). "JORC – Its History, its Sponsors, and Current Committee," September 27, 2009, p. 1.

Joint Ore Reserves Committee of the Australasian Institute of Mining and Metallurgy, Australian Institute of Geoscientists, and Minerals Council of Australia (2004). *The JORC Code – 2004 Edition*. Publication. December. Available at http://www.jorc.org/docs/jorc2004web_v2.pdf (accessed September 25, 2014).

KPMG (2011). "Minerals Resource Rent Tax: Exposure Draft," June 30, 2011.

Leslie, Tim (2011). "Gillard Unveils Carbon Price Details." *ABC News*, February 24. Available at http://www.abc.net.au/news/2011-02-24/gillard-unveils-carbon-price-details/1955968 (accessed September 26, 2014).

Lucarelli, Bart (2010). *The History and Future of Indonesia's Coal Industry: Impact of Politics and Regulatory Framework on Industry Structure and Performance*. Working Paper no. 93. Stanford: Program for Energy and Sustainable Development, Stanford University.

Mimuroto, Yoshimitsu, Toru Kimura, and Koichi Koizumi (2004). *Restructuring of the Coal Industry in Australia*. Report. February. Available at http://eneken.ieej.or.jp/data/en/data/pdf/235.pdf (accessed September 25, 2014).

Moran, Kieran (2010). "Eight Sign On for WICET." *The Observer*, October 1. Available at www.gladstoneobserver.com.au%2Fstory%2F2010%2F10%2F01%2Feight-sign-on-for-wicet-wiggins-island-coal (accessed September 25, 2014).

Morton, Adam, and Tom Arup (2010). "PM Pledges 'People's Assembly' on Climate." *The Sydney Morning Herald*. July 23. Available at http://m.smh.com.au/federal-election/pm-pledges-peoples-assembly-on-climate-20100722-10myh.html (accessed September 26, 2014).

Morton, Adam, and Michelle Grattan (2011). "Death-knell for 'Dirty' Hazelwood." *The Age*. July 1. Available at http://www.theage.com.au/victoria/deathknell-for-dirty-hazelwood-20110630-1gt7t.html (accessed September 26, 2014).

Multi-Party Climate Change Committee (2011). "MPCCC Carbon Price Mechanism." *Scribd*. February 24. Available at http://www.scribd.com/doc/49435346/MPCCC-Carbon-Price-Mechanism (accessed September 26, 2014).

Newcastle Coal Infrastructure Group (2013). "Coal Export Terminal." Presentation. March. Available at https://www.ausimm.com.au/Content/docs/branch/2013/sydney_2013_03_presentation.pdf (accessed September 26, 2014).

Newcastle Port Corporation (2014). "Birth of the Port." Available at http://www.newportcorp.com.au/site/index.cfm?display=111710 (accessed September 25, 2014).

North Queensland Bulk Ports Corporation Limited (2012). "Our Organisation." *North Queensland Bulk Ports*. Available at http://www.nqbp.com.au/organisation/ (accessed September 25, 2014).

Official Yearbook of Australia (1947, 1951, 1956, 1990). *The Official Year Book of the Commonwealth of Australia*.

Pacific National (2014). "Pacific National." *Pacific National – Asciano*. Available at http://asciano.com.au/pacific-national (accessed September 25, 2014).

Patrick Corporation (n.d.). http://www.patrick.com.au/index.php?id=49.

Parliament of Australia (2000). "Bills Digest No. 42 2000–01: Coal Industry Repeal Bill 2000."

Patriot Coal (2012). "How We Mine." *Patriotcoal.com*. Available at http://www.patriotcoal.com/index.php?view=how-we-mine&p=3&s=52 (accessed September 25, 2014).

Platts (2011). *International Coal Report*, July 18, 2011.

Productivity Commission (1998). *The Australian Black Coal Industry*. Report no. 1. July 3. Available at http://www.pc.gov.au/__data/assets/pdf_file/0003/30576/coal1.pdf (accessed September 25, 2014).

QR National (n.d.). www.qrnational.com.au/CoalFreight/Pages/CoalFreight.aspxwww.freight.qr.com.au.

Queensland Coal (n.d.). www.qld.gov.au/assetssale/businesses/qr-national/docs/railing-queenslands-coal-section-2.pdf.

Queensland government (2014). "Cameby Downs Expansion Project," April 16, 2014, Department of Environment and Heritage Protection.

Queensland government (2014). "Department of Transport and Main Roads." Available at http://www.tmr.qld.gov.au/ (accessed September 25, 2014).

Queensland government (2007). "Queensland's World-class Coals: Mine Production and Developments," November 2007, Department of Mines and Energy, p. 1. Available at http://www.dme.qld.gov.au/zone_files/coal_files_pdf/wcc_nov_07.pdf.

Queensland government (2007). "Rail Update Fact Sheet," November, Surat Basin Rail Joint Venture. Available at www.suratbasinrail.com.au/files/SBR-fact-sheet-december07.pdf.

Queensland government (1925). *Coal Mining Act 1925*. Queensland Office of the Queensland Parliamentary Council. Available at https://www.legislation.qld.gov.au/LEGISLTN/REPEALED/C/CoalMiningA25_02_.pdf (accessed September 25, 2014).

Shaw, A. G. L., and Gordon R. Bruns (1947). *The Australian Coal Industry*. Melbourne: Melbourne University Press.

State Records Archive Investigator (n.d.). "FreightCorp." Available at http://investigator.records.nsw.gov.au/Entity.aspx?Path=%5CAgency%5C1079.

Sydney Morning Herald (2014). "NSW Government Cancels Mining Licences Tainted by Eddie Obeid-Ian Macdonald Corruption Scandals," January 20. Available at http://www.smh.com.au/comment/culture-of-corruption-pervades-nsw-inc-20130801–2r23r.html and http://www.smh.com.au/nsw/nsw-government-cancels-mining-licences-tainted-by-eddie-obeid-ian-macdonald-corruption-scandals-20140120–314iv.html.

Sydney Morning Herald (2013). "Culture of Corruption Pervades NSW," August 1.

Warner, E.M (1979). "*A History of American Continuous Miners*." Technical paper. International Conference on Mining Machines, Brisbane.

Wiggins Island Coal Export Terminal (2014). "Tripper Installed as WICET Stockyard Development Enters Next Phase." News release, April 11. Available at http://www.wicet.com.au/assets/docs/Tripper%20installed%20as%20WICET%20stockyard%20development%20enters%20next%20phase.pdf (accessed September 26, 2014).

Wiggins Island Coal Export Terminal (2011). *Information Sheet*. Available at Wicet.com.au.

7 Government as creator and destroyer

Indonesia's rapid rise and possible decline as steam coal supplier to Asia

Bart Lucarelli

7.1 Introduction

Over the past two decades, Indonesia's coal industry has transformed itself from being an unknown, minor player in Asia's coal markets into the world's largest exporter of steam coal. This chapter tells the story of how Indonesia created this world-scale industry despite challenges created by widespread government corruption, a weak legal system, the Asian Financial Crisis of 1997, and the fall of the Soeharto government in 1998. Indonesia's success has stemmed in large part from its ability, starting in the late 1960s, to create an attractive regulatory regime that laid the groundwork for rapid growth many years later. Geology and geography helped too. Large coal deposits were discovered near water, allowing transportation infrastructure to be built out at low cost as the industry grew. Indonesia's proximity to major Asian consumers gave Indonesian coal suppliers a significant cost advantage over their Australian and South African competitors, an advantage that continues to this day. But for all its strengths, Indonesia's coal industry today faces significant risks:

The author wishes to express his sincere thanks to Mark Thurber, associate director at PESD, and Richard Morse, affiliated researcher at PESD, for their expert reviews of this chapter and its earlier versions. Their many helpful suggestions on how to better organize the chapter, present data, and frame conclusions helped to make this chapter much more readable. He also wishes to thank Michael Friederich, coal exploration consultant with Coal Discovery, for taking so much of his valuable time to review earlier drafts and sharing his insights about the early development history of Indonesia's coal industry. Drawing on his years of coal exploration experience on Kalimantan and Sumatra, Mike provided details concerning the technical and economic limits of the coal resources of these two provinces, which were not available from publicly available information. The author wishes to express his special thanks to Stuart Murray, previously of London Commodity Brokers and now an independent coal consultant, for sharing his many insights on the coal markets of China, especially the reliability issues related to Chinese coal import data, and commercial and regulatory issues related to exporting Indonesian coal to China. Finally, the author extends his appreciation to Pat Markey, previously of globalCOAL, and Peter Ball, Marketing Director of BUMI Resources, who offered many helpful comments on drafts of this chapter when it was still a PESD Working Paper. To all of these reviewers, the author extends his heartfelt thanks.

notably, increasingly intrusive government intervention in the sector and a decrease in coal quality as higher rank deposits are depleted.

The remainder of this chapter is divided into three main sections. Section 7.2 describes the historical development of the Indonesian coal industry through 2009, focusing on the contributions of government regulatory framework, geology, and geography to observed results. The section starts with an examination of the supportive regulatory framework created for the mining sector during the Soeharto years. This key formative regulatory period (the Formative Period), which lasted from 1967 to 1988, catalyzed Indonesia's spectacular growth in coal production and exports between 1988 and 2009. This growth occurred as two distinct development periods:

- The Take-off Period (1988–1999), during which production grew at a torrid annual rate of 29%; and
- The Localization Period (2000–2009), when majority foreign owners were replaced by domestic investors and adverse political events and regulatory shifts first started to impact the coal industry.

The treatment of the Take-off and Localization Periods includes a review of Indonesia's coal supply fundamentals between 1988 and 2009 and the role that Indonesia's proximity to the major markets of Asia and attractive geology and favorable location of resources relative to coastal areas played in the rapid development of its coal resources.[1]

Section 7.3 discusses the current development period for Indonesia's coal industry, which is simply labeled the New Mining Law Period. The passage of the Mining Law of 2009 was the signal event that will influence the coal industry through the rest of this decade and perhaps the following decade. This section focuses on the government of Indonesia's (GOI's) efforts to reset the regulatory framework for its mining industry through this law and associated implementing regulations.

Finally, Section 7.4 assesses the critical challenges and uncertainties that Indonesia's coal industry is currently facing and will likely continue to struggle with over the next two decades. Specific challenges include the adverse development impacts of (a) the GOI's aggressive approach to implementing the new mining law and (b) the ongoing depletion of Indonesia's medium to high rank coal reserves near navigable rivers and coastal areas, and their replacement by low CV, high moisture coal resources located in less easily accessed areas. The section concludes by considering two possible futures for the Indonesian coal sector: one in which government policy makers build on past achievements and tread lightly to preserve the future vitality of the coal industry, and another in which they turn inward, try to hoard coal resources for domestic use, and end up undermining the success story of Indonesian coal.

7.2 The historical development of Indonesia's coal industry through 2009[2]

When compared with the coal industries of Australia, the United States, the UK, China, and India, Indonesia's coal industry is very young. Except for government-owned PT Tambang Batubara Bukit Asam (PTBA), Indonesia's major coal producers only started serious exploration work from the early 1980s, with commercial production at their mines commencing after 1988.[3]

Although Indonesia's coal industry has existed much longer than thirty-four years, the industry in its early years was small and fragile. During the Dutch colonial period (1849–1945), coal was produced in limited quantities from a few mines located on Kalimantan and Sumatra. The coal was sold primarily to the shipping industry as a fuel for marine propulsion, but by 1940, the shipping industry was about to shift from coal to petroleum as its preferred fuel (Friederich and Leeuwen, 2002, p. 3). Indonesia's annual coal production peaked at 2 million tonnes (Mt) in 1941, after which the coal industry entered into a prolonged and steep decline that lasted over three decades.

The immediate causes of the decline were (a) the adverse effects of World War II on coal export markets in Asia, (b) the impact of Indonesia's ongoing struggle for independence on coal mining activities within Indonesia, and (c) the advent of cheap oil as a replacement for steam coal in marine propulsion (Friederich and Leeuwen, 2002, p. 4). Indonesia's coal mining industry was later adversely affected by the ultra-nationalistic policies of President Sukarno, who ruled Indonesia from 1945 to 1965 and whose anti-Western rhetoric and incompetent management of the national economy discouraged foreign companies from making direct investments in Indonesia's mining sector from 1959 through 1965 (Schwarz, 2000, p. 18).

By 1972, Indonesia's coal industry produced only 200,000 tonnes of coal, or one-tenth of Indonesia's 1941 peak production (Friederich and van Leeuwen, 2002, p. 4). Asian export markets for steam coal were for all intents and purposes non-existent as of the early 1970s; the most significant coal exports in the region during this period were coking coals supplied out of Australia.

7.2.1 The formative period of development (1967–1988)

7.2.1.1 Building a favorable regulatory framework

It took a number of years for the various impediments to the development of Indonesia's coal industry to be removed. The first corrective action occurred on the political front in October 1965, when then Major General Soeharto[4] wrested political control from Sukarno after a failed coup attempt and created more favorable economic conditions for mining.[5] Soeharto was initially made chief of the army, or ABRI as it is known in Indonesia. He was made acting president by March 1967 and then officially appointed as president the following year. Sukarno was kept under house arrest until 1970 when he died of kidney failure.

One of the first steps taken by Soeharto after being named acting president was to stabilize the economy through the implementation of new laws that were meant to establish Indonesia as an attractive venue for foreign investment.[6] Changes included allowing foreign investment in the mining sector, reforming labor laws to standards acceptable to multinational corporations, providing foreign investors with legal guarantees of their rights to repatriate profits, and accepting funds from multilateral institutions such as the World Bank and the bilateral aid agencies of Western nations.

In 1967, the government of Indonesia (GOI) passed two laws specific to the coal industry that were aimed at encouraging foreign investment in this heretofore moribund sector. They were:

- The Mining Law of 1967, which allowed foreign companies to engage in mining activities, including coal mining, and to export their products;[7] and
- The Foreign Investment Act of 1967, which allowed foreign companies investing in Indonesia (known as Penanaman Modal Asing, or PMA, companies) to repatriate their earnings and be protected from expropriation without just compensation.[8]

Despite these investor friendly changes to the mining sector's regulatory framework, foreign mining company interest was initially muted due to low oil prices. Then, in 1973, the Arab Oil Embargo resulted in a tripling of oil prices, which, combined with the mining and foreign investment laws of 1967, led to renewed foreign interest in the development of Indonesia's steam coal resources. The interest at that time was largely export market driven – with Japanese, Korean, and Taiwanese power utilities requiring additional supplies of coal to fire new coal-fired power plants that would allow them to reduce their reliance on foreign oil.

The GOI initially tried to tie the development of the coal industry to the provision of fuel for new domestic coal power plants that would be built as replacements to existing oil-fired units. (This import substitution policy was abandoned in the latter part of the 1970s.) Due to the limited size of the domestic coal market in those days as well as the lack of a creditworthy buyer of coal in the Indonesian power sector, the GOI was only able to encourage two significant international companies – Shell Oil

Company affiliate Shell Mijnbouw and Rio Tinto Zinc (RTZ), now known as Rio Tinto – to invest in its nascent coal industry during the 1970s (Friederich and van Leeuwen, 2002, p. 4). These two companies, which in 1973 entered into Contracts of Work (COWs) with the GOI to "explore, extract and market coal," focused their exploration activities on the island of Sumatra.

Their exploration efforts were for naught. They discovered only low-rank coal that could not compete against high-quality Australian and South African bituminous coals. Due to the low quality of the coal that they discovered, both companies decided to relinquish their concession areas at the feasibility study stage. But these pioneer projects led to the extension to the coal mining industry of the contract of work concept, which had been successfully applied in Indonesia's other mining sectors, such as Freeport's Grasberg copper and gold mining project in Papua Province.[9] A decade later, this contract of work framework served as the basis for creating Indonesia's "world-scale" coal industry (Friederich and van Leeuwen, 2002, pp. 5–8).

7.2.1.2 Pioneering activities in Kalimantan under the Coal Contract of Work (CCOW)

The Coal Contract of Work (CCOW)[10] and a parallel framework open only to domestic investors, the Kuasa Pertambangan (KP), were created by Government Regulation 32/1969 (related to Law 11/1967) and later amended by GR 79/1992 and GR 75/2001. (Boxes 7.1 and 7.2 provide more complete explanations of the CCOW and KP, respectively.) The CCOW was arguably the most important regulatory instrument enabling the development of Indonesia's coal sector. Given the undeveloped and evolving nature of Indonesia's legal system during the 1960s and 1970s, the use of a bilateral CCOW, with the GOI and the foreign mining company as the contracting parties, was an absolute necessity for obtaining the investments and expertise of foreign mining companies.[11]

The modern coal industry of Indonesia really began to take shape in 1981, when PTBA, the government-owned coal company, and a number of foreign coal mining companies signed First Generation Coal Contracts of Work (1st Gen CCOWs) for concession areas on Kalimantan.[12] By 1990, when the 1st Gen CCOW program was discontinued, a total of ten contracts had been signed. Today, eight of those original ten companies are still Indonesia's most significant coal producers. Details concerning these eight companies and their 1st Gen CCOWs are shown in Table 7.1.

These eight pioneers used their first-mover status to dominate the coal industry of Kalimantan by tying up vast concession areas that later were

Box 7.1 The Coal Contract of Work explained

Only PMAs were allowed to apply for and operate under a Coal Contract of Work (CCOW). There are three generations of CCOWs in force in Indonesia:

- 1st Gen (issued from 1981 to 1990)
- 2nd Gen (issued only for 1994)
- 3rd Gen (issued from 1997 to 2000)

The differences between the different generation contracts are significant. But there are some important similarities.

First, CCOWs were direct agreements between the GOI and locally registered foreign companies, known in Bahasa as Penanaman Modal Asing and generally referred to in Indonesia as PMAs. By signing a contract directly with the GOI, the PMAs were awarded "lex specialis," or special law status under the Indonesian Constitution.[13] As a result of this special law status, holders of CCOWs were exempted from any changes in Indonesian laws, such as revisions to the general tax code, investment laws, and land use laws that occurred after the CCOW was signed and conflicted with the terms of the CCOW.[14] Each 1st Gen CCOW lays out in considerable detail the tax liabilities that a PMA will incur over the life of its CCOW and the land area that has been approved for its mining activities. The 2nd and 3rd Gen CCOWs have less detail and/or require the CCOW holder to accept changes in certain taxes and rights for repatriation of profits.

Second, CCOWs provide the PMA with "conjunctive title" protection, which allows PMAs operating under a CCOW to conduct all stages of mine development from exploration to production, transportation, and marketing without the need to go through a new tender or renegotiation of its CCOW. Conjunctive title meant that each CCOW holder was able to make long-term decisions to invest in infrastructure and equipment with the knowledge that its title to the concession area was secure.

Third, a CCOW of a specific generation has the same terms and conditions as all the other CCOWs from that same generation. Fourth, a fixed royalty of 13.5% for the term of the CCOW is charged based on the FOB price per tonne of coal sold. Fifth, and finally, the term for each CCOW is thirty years from the date of commencement of commercial production of coal from the area covered by the CCOW, and the term of the CCOW can be extended for two additional periods of ten years each.[15]

Box 7.2 The Kuasa Pertambangan explained

Kuasa Petambangan (KP) licenses, which were issued between 1990 and 2009, were available only to domestic investors and were for relatively small concession areas of 5,000 hectares. With the cancellation of the CCOW program in 2000 and passage of the Regional Autonomy Law in 1999, coal miners interested in developing new mining areas had to rely on KPs as their exclusive mining authorization. In 2009, when the Mining Law of 2009 passed into law, the GOI discontinued the issuance of KP licenses and required new investors to obtain coal mining licenses known as IUPs instead. The GOI also required holders of existing KP licenses to convert them into IUPs by May 2010. Most of those conversions have taken place and today KPs are considered null and void as mining authorizations.

A KP license granted exclusive mining rights to coal miners for a coal mining area for a specific stage of the coal mining program. It could be issued by any branch of government – regency (by the regency head), provincial (by the governor), or central government (by the Minister of Energy and Mineral Resources). The specific government level that issued the KP was determined by the location of the proposed mining area. If the mining area was contained within a single regency, the head of that regency was responsible for issuing the KP. If it overlapped two regencies but was located in one province, the provincial governor would have authority for issuing the KP. If it overlapped two provinces, the GOI's Minister of Energy and Mineral Resources issued the KP.

There were five different KPs that a company could hold depending on the stage of development of the coal mining operation. They were as follows:

(1) A General Survey KP, which had a term of one year with the possibility of a one-year extension. The maximum area allowed under this KP was 20,000 hectares.

(2) An Exploration KP, which was valid for a three-year period and could be extended twice by one year per extension. The maximum area allowed under an exploration KP was 10,000 hectares (Indonesian Coal Book, 2008/2009).

(3) An Exploitation KP, which has a term of thirty years with the possibility of two ten-year extensions. The maximum area allowed under this type of KP was reportedly 5,000 hectares.[16] The Exploitation KP could also contain terms related to processing, refining, transportation, and marketing. If these terms were not included as part of the Exploitation KP and the KP holder wished to engage in these activities later on, he was required to file for these additional KPs separately.

Box 7.2 (cont.)

(4) A Processing and Refining KP, which had a term of thirty years with two extensions of ten years each. This KP covered crushing, washing, upgrading, and drying of coal.
(5) A Transportation and Marketing KP, which had a ten-year term and was extendable for an indefinite number of five-year periods.

A review of the *Indonesian Coal Book 2008/2009* revealed that mining companies operating under KP licenses, in practice, tied up much larger areas at the general survey and exploration stages of KP mine development – in a number of cases 25,000 to 50,000 hectares (Indonesian Coal Book, 2008/2009). However, once these companies made their decisions to mine a particular concession area, they were required to relinquish most of that excess area such that there would be no more than 5,000 hectares per Exploitation KP license. To get around this restriction, a few KP holders attempted to divide their areas of interest into four to five separate General Survey and Exploration KPs. Once they completed the exploration phase of development, the KP holders would then consolidate their retained areas to allow economies of scale in their mining operations.

A KP was transferable to another party if the original holder of the KP could prove that the other party had the necessary financial strength and technical capability to carry the mining activities into commercial production. On this basis, many transfers of KP licenses occurred up until 2009 from the original holder of the KP to new companies with greater financial strength and mining capabilities. The main requirement for any transfer was that the proposed KP holder should be a domestic company, not a PMA.

Royalty rates paid by KP holders varied between open-cut and underground mines and the grade of coal being produced. For open-cut mines the royalty rate was 3% of the FOB price for each tonne of coal sold for coals with a Gross Calorific Value (GCV) of 5,100 kcal/kg and below and 7% of the FOB price for each tonne of coal sold for coals having a GCV of 6,100 kcal/kg and above.[17] Coal derived from underground mines had slightly lower royalty rates of 2% for coals with a GCV of 5,100 kcal/kg or lower and 6% for coals with a GCV of 6,100 kcal/kg and higher.

found to contain large deposits of low sulfur bituminous and subbituminous coals with low stripping ratios and locations close to navigable rivers or the coast.[18]

The pioneer mining companies operating under CCOWs chose Kalimantan over Sumatra for a number of reasons. First, the Kalimantan

Table 7.1 *Eight of the ten original 1st Gen CCOW companies are still producing coal as of January 2014**

Company	Original shareholder(s)	CCOW signing	Start date for commercial production	Expiration date	Initial area (ha)	Retained area (ha)	Resources/ reserves (Bt)
1. Arutmin	ARCO & Utah International	2 Nov 1981	Dec 1989	Nov 2019	1,260,000	70,153	2.5/0.5
2. Adaro	Enadimsa	16 Nov 1982	Nov 1992	Oct 2022	148,148	35,800	4.1/0.9
3. KPC	RTZ/BP	8 April 1982	1991	2021	790,900	90,938	7.5/1.4
4. Kideco	Korea- Indonesia Resources	14 Sept 1982	April 1993	March 2023	254,804	50,400	1.4/0.7
5. Berau	n/a	26 Apr 1983	1994	2024	487,217	118,400	1.3/0.5
6. Indominco Mandiri	Salim Group	8 Oct 1990	1997	2027	100,000	25,121	0.7/0.2
7. Tanito Harum	Kiki Barki Family	30 Jan 1987	1989	2019	125,412	35,757	n/a
8. Multi Harapan Utama	Swabara Australia (40%), PT Agrarizki Media (38%), Ibrahum Risjad 12%, PT Asminco (10%)	29 Nov 1986	1990	2020	189,954	46,063	0.3/0.1

* The other two 1st Gen CCOW companies – PT BHP Kendilo and PT Allied Indonesia – ceased production in 2006 and 2009, respectively.

Note: ha = hectares, Bt = billion tonnes

Source: All data taken from *Indonesia Coal Book 2012/2013*, published by Petromindo.com, except for Original Shareholders, which were obtained from on-line literature search.

resources were located within the provinces of South and East Kalimantan, which had good access to either navigable rivers or coastal areas, where ports could be built to load the coal onto sea vessels. Another location advantage of Kalimantan was its shorter sailing distance to the lucrative export markets of North Asia (Japan, Korea, and Taiwan) – a two-day shorter round trip by Panamax vessel compared with South Sumatra and a seven-day shorter round trip compared with Australian mines located in Queensland and New South Wales.[19] Second, Kalimantan coals for the most part had higher calorific values than Sumatran coals, along with the low sulfur and ash contents typical of most Indonesian coals. Within a few years, the legacy coal industry based on Sumatra and operated by PTBA was eclipsed by coal mining companies located on Kalimantan and operating under 1st Gen CCOWs.[20]

Due to the time required to complete exploration work and mine feasibility studies and then to actually bring the mines and related infrastructure into commercial production, Indonesia only started its relentless and rapid growth in production in 1988, at which point annual production stood at a mere 4.4 Mt. But once these 1st Gen CCOW producers started their commercial production in 1988, the Indonesian coal industry entered into the take-off stage of development.

7.2.2 The Take-off and Localization Periods (1988 – 2009)

Between 1988 and 2009, Indonesia's coal industry experienced significant growth in two distinct phases of development: a "Take-off Period," which lasted from 1988 through 1999, when the Regional Autonomy Law and the Forestry Law of 1999 were passed, and a "Localization Period," which lasted from 2000 until the passage of the Mining Law of 2009 and saw a transfer of mining sector ownership from foreign to domestic hands. The Take-off Period saw very rapid growth that built on the favorable conditions established in the formative period as well as Indonesia's advantages of geology and resource location. Robust growth continued in the Localization Period, but the government increasingly sought to assert more control over the sector – not only through the localization process itself but also through legally questionable attempts to change the investment terms of 1st and 2nd Gen CCOWs that had been so successful in making Indonesia a world-scale coal producer during the Take-off Period. In addition, Indonesia's coal industry was also threatened by a proliferation of local government interventions that focused on changing the fiscal terms of the CCOWs. In the remainder of this section, the author discusses, first, the production and export growth of the coal industry in the Take-Off and Localization Periods and how it was facilitated by geological and geographical factors, and, second, the various ways in which the regulatory environment changed during the Localization period.

7.2.2.1 Production and industry structure in the Take-Off and Localization Periods

Figure 7.1 illustrates production trends by type of mining authorization (1st, 2nd, and 3rd Gen CCOWs and KPs) for the period from 1982 through 2009.[21] Production increased at a 29% compound annual growth rate (CAGR)[22] during the Take-off Period. Absolute growth continued to be strong in the Localization period, but the CAGR fell to an average of 14%.

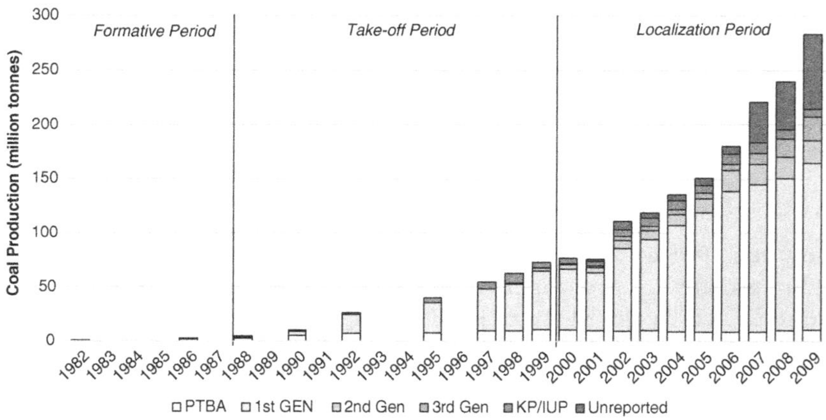

Figure 7.1 Indonesia's coal production by mining authorization type, 1982–2009. Sources: (1) Barlow Jonker, Coal Supply Series Indonesia 2005, 15th ed., 2006 for years 1982–1995; (2) Graeme Robertson, "Indonesian Coal Supply: Thriving in adversity," paper published in "Proceedings of the National Outlook Conference, Canberra, Feb 29–March 2, 2010, table 2, p. 217 for 1997 and 1998; Energy Publishing, "Indonesia Coal Report," Jan. to March issues for 2002 through 2010 for years 1999 through 2009. Unreported estimates for 2007 through 2009 are the differences between the total production estimates obtained from Energy Publishing data and the total production estimates provided by Bob Kamandanu (Chairman of the Indonesian Coal Mining Association) in a paper titled "Indonesian Coal Outlook," and presented at the Asia Pacific Mining Conference Indonesia 2012, October 10–11, 2012, Jakarta, Indonesia.

The primary drivers of coal production growth during the Take-off Period were the stable political climate until late 1998 and the "special law" protections provided to foreign mining companies under their CCOWs. In 1988, the first year of the Take-off Period, Indonesia produced 4.43 Mt, or 55% of total coal production, through state company PTBA, with KP holders accounting for 24% of production and 1st Gen CCOW holders for 21%. However, the 1st Gen CCOW holders rapidly grew to account for the majority of production.

Coal production by 2nd Gen CCOW companies commenced in 1997. By 1999, the combination of 1st Gen and 2nd Gen CCOW companies accounted for 76% of total coal production (72% by 1st Gen CCOWs and 4% by 2nd Gen CCOWs), PTBA's share had dropped to 14%, and KPs had increased their output to 9.1 Mt, or 10% of production. In total, private coal companies (CCOWs and KPs), mostly located on Kalimantan, accounted for 86% of Indonesia's coal output by 1999. Companies operating under 3rd Gen CCOWs only started to produce noticeable quantities of coal from their concession areas (> 5 million tonnes per annum [Mtpa]) around 2005.

Between 2000 and 2009, domestic investors gained majority ownership of Indonesia's largest coal producers.[23] Despite the localization of major producers, Indonesia's coal industry continued to grow over the Localization Period, at a slower yet still impressive CAGR of 14%.

In 2005, against all odds, Indonesia became the world's largest exporter of steam coal, with total steam coal exports of 111.6 Mt (*Indonesian Coal Book, 2008/2009*, p. 4) against "runner-up" Australia's 110.8 Mt of steam coal exports (Bureau of Resources and Energy Economics, 2013, p. 47, table 47).[24] By the end of 2009, Indonesia's exports of steam coal had increased to 198 Mt, while Australia's steam coal exports had only increased to 135 Mt due to chronic transport infrastructure constraints (*Indonesia Coal Book, 2010/2011*, p. 7; Bureau of Resources and Energy Economics, 2013, p. 47, table 47).

In each year of the Localization Period, companies operating under 1st and 2nd Gen CCOWs achieved a minimum 75% share of total coal production. It was also a time when 3rd Gen CCOW holders commenced production but at relatively low CAGRs. In 2009, companies operating under 1st, 2nd, and 3rd Gen CCOWs accounted for 66% of total coal production while PTBA and KPs had sunk to almost irrelevant shares of 5% and 4%, respectively. The remainder of the 2009 coal production is captured under the heading "Unreported," which is a catch-all coal production category that includes statistical discrepancies, KPs not reported to the GOI by local governments, and illegally mined coal ("Indonesia Coal and Power Report," May 2010).[25]

What made this achievement of companies operating under CCOWs even more impressive was that it occurred at a time when: (i) ownership of coal-producing companies operating under 1st Gen CCOWs was being transferred to local entrepreneurs (see Section 7.2.2.4.1), and (ii) Indonesia was experiencing considerable political instability as it struggled to transition from Soeharto's "New Order" government to a political process based on free elections, regional autonomy, and participatory democracy. These challenges and their effects are discussed further in Section 7.2.2.3. As seen in Figure 7.1, 2nd and 3rd Gen CCOW companies made modest contributions to Indonesia's coal production between 2000 and 2009, though nowhere near the level of 1st Gen CCOW holders. Eleven of the eighteen 2nd Gen CCOWs awarded in 1994 achieved commercial production over the next fifteen years. As of December 2009, eight were still in operation.[26]

Companies operating under 3rd Gen CCOWs did not achieve total production above 20 Mtpa until 2009, roughly ten years after their mining authorizations were issued. The 3rd Gen CCOWs were slow to achieve significant production levels for a number of reasons. First, most of the areas available under the 3rd Gen CCOW program were either low-quality deposits located within areas relinquished by 1st Gen and 2nd Gen CCOWs or were located very far inland and away from navigable rivers (Central Kalimantan). Second, most areas had little or no geological data to allow a value to be placed on them and allow for fast track development. Third, many of the 114 companies that signed up for 3rd Gen CCOWs had no background in mining. They were simply in the business of collecting success fees, obtaining free carried interests, and "flipping their assets" to others in an attempt to make some quick money (Barlow Jonker, 2006). (The differences in contract types are discussed in more depth in Section 7.2.2.4.2.)

By 2009, however, holders of 3rd Gen CCOWs achieved a total production of 21.7 Mt, and as of July 2013, thirty-five 3rd Gen CCOW companies were still in the exploration and feasibility stages of development.[27]

Finally, the KP mining companies remained minor players throughout the Take-off and Localization Periods, with relatively small coal production and coal export levels.[28] In general, the size of the mining area allowed under an exploitation KP (5,000 hectares) was too small to allow the most efficient mining methods to be applied. Companies holding these KPs also tended to be undercapitalized and lacked the technical capability to develop their resources in a timely and cost-effective manner.

Although companies operating under 2nd and 3rd Gen CCOWs increased their share of total production to 19% in 2009, they were still

Table 7.2 *Production per mine by type of authorization, 2009*

License type	2009 production	# of operating mines in 2009	2009 production/mine (Mt/mine)
1st Gen	154.23	8	17.14
2nd Gen	20.51	9	2.28
3rd Gen	21.52	17	1.27
KP	6.51	14	0.47

Note: The author identified fourteen KP mines that were producing coal in 2009 although it is highly likely that more than that number were operating in that year. Company-specific production data were obtained from Energy Publishing, "Indonesian Coal and Power Report."

mere shadows of the 1st Gen CCOW mining companies. This vast difference in scale of operations is reflected in Table 7.2, which shows the average mine size, in thousand tonnes of 2009 production per mine, for companies operating under different mining authorizations. The difference in scale between 1st Gen CCOW companies and all others was so vast that the 1st Gen CCOW companies appeared to come from an entirely different industry, if not country.

Indonesia's coal industry, in addition to being geographically concentrated on the island of Kalimantan, is also concentrated by producer. Indonesia's top six producers – all located on Kalimantan – produced between 78% and 86% of Indonesia's steam coal between 2000 and 2006 and between 60% and 68% between 2007 and 2009 (Table 7.3).

The Big 6 producers had annual growth in production that ranged from 10% to 13%. Barring some totally unexpected political event or a government regulation that places limits on their level of production, the outlook is for the Big 6 producers to continue to expand their levels of production by 10% per year over the remainder of this decade. For example, Bumi, Adaro, and Kideco announced separate mine expansion plans in late 2009 that, if successfully implemented, would increase their combined coal production from 122.8 Mt in 2009 to 272.8 Mt by 2015, which translates into a 12% annual increase from 2009 through 2015.[29]

7.2.2.2 *The balance of coal exports and domestic sales through 2009 and beyond*

Indonesia's modern coal industry has developed around exports to Asian markets. Figure 7.2 shows exports relative to domestic sales in the latter part of the Take-off Period continuing through the Localization Period.

Table 7.3 *The Big 6's share of Indonesian coal production declined from 86% in 2000 to 60% in 2009*

	Coal Production (Mt)									
Company	2000	2001	2002	2003	2004	2005	2006	2007	2008	2009
Bumi Resources (KPC + Arutmin)	20.8	25.1	28.2	29.8	36.3	44.6	51.6	54.9	52.0	57.5
Adaro	15.5	17.7	20.8	22.4	24.3	27.0	33.5	36.0	38.5	40.6
Kideco	8.0	10.4	11.5	14.1	16.9	18.1	18.9	20.6	21.9	24.7
Berau	4.9	6.8	7.1	7.4	9.1	9.2	10.8	11.8	12.9	14.3
Banpu (Indo Tambangraya Megah Tbk) (Indominco Mandiri, Jorong, Trubaindo, and Kitadin)	6.2	8.3	9.6	11.5	12.6	12.2	18.2	18.0	18.8	21.5
Bukit Asam (PTBA)	10.7	10.2	9.4	10.0	8.7	8.7	8.7	8.6	10.1	10.8
Subtotal	66.1	78.5	86.6	95.2	107.9	119.8	141.7	149.9	154.2	169.4
All others + unreported	12.1	24.2	24.5	23.7	33.6	31.1	38.4	71.2	85.8	113.8
Total production	76.9	94.4	111.1	118.9	135.6	150.9	180.1	221.1	240.0	283.2
Big 6 as % of total production	86%	83%	78%	80%	80%	79%	79%	68%	64%	60%

Source: Big 6 production figures from Energy Publishing, Indonesian Coal and Power Report February 2004 through February 2010 for 2002 through 2009 estimates. Total Production figures from Energy Publishing, "Indonesia Coal and Power Report" (2002–2006), and Bob Kamandanu, "Indonesia Coal Outlook," October 2012 (2007–2009).

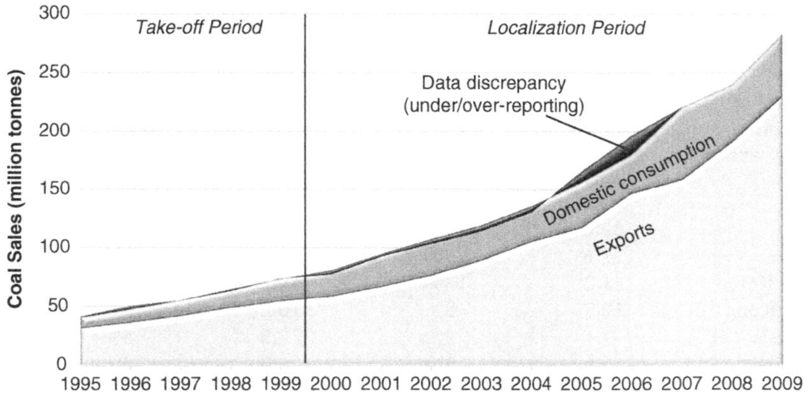

Figure 7.2 Export and domestic sales of Indonesian steam coal, 1995–2009. Source: *Indonesian Coal Book, 2004/2005* for years 1995–2003, *Indonesian Coal Book, 2006/2007* for years 2004–2006, and Bob Kamandanu, "Indonesian Coal Outlook," paper presented at Asia Pacific Mining Conference Indonesia (APMCI) 2012, October 10–11, 2012, Jakarta, Indonesia, for years 2007–2009.

In this section, the author focuses on trends in coal exports versus domestic sales through 2009 and the prospects that this balance will be altered by Indonesia's program to bring online a significant quantity of new coal-fired power plants.

Indonesia increased its steam coal exports by a minimum of 15% per year between 2000 and 2009, starting at 58.3 Mt in 2000 and reaching an estimated level of 230 Mt in 2009. Asia was the primary export market, with limited quantities being exported to Europe and the United States (Table 7.4). Japan was Indonesia's largest single customer until 2009, when China took over that distinction with imports of 34.3 Mt of Indonesian coal, replacing Japan by a wide margin as the largest importer of Indonesian steam coal. Japan dropped to third largest customer for Indonesian steam coal in 2009, with imports of 22.9 Mt, down from 30.1 Mt in 2007. India remained Indonesia's second largest customer for steam coal, and its share is expected to grow in importance over the next five to ten years ("Indonesia Coal Report," January 2009 and January 2010).

As with production, the Big 6 coal producers dominated the export trade, accounting for 79% of *reported* exports in 2009 (Table 7.5). Indonesia's largest coal exporter in 2009 remained Bumi Resources, the owner of KPC and Arutmin.[30] The Big 6 producers achieved annual growth rates in exports that ranged from 9% to 15%, with Bumi experiencing the lowest growth in exports and ITM the highest rate of growth

Table 7.4 *Indonesia's coal exports in 2007 vs. 2009 by country of destination (Mt)*

Country of destination	2007 coal exports	% of reported total	2009 coal exports	% of reported total
Japan	30.3	18%	21.0	14%
India	22.6	14%	25.9	17%
China	22.0	13%	32.2	21%
Taiwan	21.6	13%	19.1	12%
South Korea	21.2	13%	19.7	13%
Hong Kong	12.9	8%	10.2	7%
Malaysia	7	4%	7.5	5%
Thailand	6.6	4%	6.3	4%
Philippines	2.5	2%	2.1	1%
Europe	14.5	9%	10.7	7%
USA	3.9	2%	0.6	0%
Reported total	165.1	100%	155.3	100%
Unreported	−1.6	n/a	66.5	n/a
TOTAL	**163.5**	**100%**	**230**	**100%**

Note: Unreported amounts are those required to bring export totals in line with those shown on Figure 7.2. These entries show the level of uncertainty that exists when using GOI coal data reported by third-party sources.
Source: Energy Publishing, "Indonesia Coal and Power Report" (various issues for 2007, 2008, 2009, and 2010).

for exports. On average, the Big 6 producers increased their exports of coal by 11% per year between 2000 and 2009, compared to 15% annual growth in exports coming from all producers.

The percentage share of total production accounted for by domestic sales ranged from 17% to 30% over the Localization Period, with shares dropping to 17% to 20% from 2006 through 2009. The Big 6 producers dominated the domestic market during the first eight years of the ten-year Localization Period with shares over 80% in all of those years (Table 7.6).

Domestic sales were expected to represent a much larger share of total production as a result of PLN's 2006 program to construct thirty-three new coal-fired power projects with a total capacity of 9,483 MW on an expedited basis. PLN refers to this program as its First Phase, Fast Track Power Program ("Fast Track Program"). Most of the Fast Track power plants were designed to burn Indonesia's low CV, high moisture (low rank, or LR) coals. PLN planned for ten of those plants with a total capacity of 7,520 MW to be sited on Java. The remaining twenty-three plants with 1,963 MWs of capacity were to be sited on the outer islands. Table 7.7 lists the thirty-three coal-fired Fast Track power plants that

Table 7.5 *Indonesia's reported coal exports for six major producers, 2000–2009*

	Coal exports (Mt)									
Company	2000	2001	2002	2003	2004	2005	2006	2007	2008	2009
Bumi (KPC + Arutmin)	22.1	24.3	26.5	29.8	34.5	39.1	47.5	51.2	46.3	52.5
Adaro	9.7	10.7	12.7	14.2	16.5	17.3	24.7	27.3	30.2	31.6
Kideco	6.5	7.4	6.8	8.9	11.0	11.8	13.6	14.5	16.1	19.2
Berau	3.3	4.4	5.1	5.4	6.2	5.8	7.4	7.6	8.2	10.1
Banpu (Indominco Mandiri, Jorong & Trubaindo)	4.8	5.9	6.4	8.6	9.8	9.5	15.7	16.5	16.3	18.8
Bukit Asam (PTBA)	2.0	1.9	1.9	2.2	2.7	2.5	3.2	3.8	4.4	4.4
Subtotal (Big 6)	**48.4**	**54.6**	**58.4**	**69.1**	**80.7**	**86.0**	**112.1**	**120.9**	**121.5**	**136.6**
Others + unreported	9.9	11.9	17.9	20.0	24.7	31.2	34.9	42.6	69.5	93.4
Total exports	**58.3**	**66.5**	**76.3**	**89.1**	**105.4**	**117.2**	**147.0**	**163.5**	**191.0**	**230.0**
Big 6 exports as % of reported exports	**83%**	**82%**	**77%**	**78%**	**77%**	**73%**	**76%**	**74%**	**64%**	**59%**

Source: Energy Publishing, "Indonesian Coal and Power Report," February 2001 through February 2010 for 2000–2009 figures for Big Six Producers and Total Exports from 2000 through 2006. Bob Kamandanu, "Indonesia Coal Outlook," October 2012 for Total Exports 2007–2009.

Table 7.6 *Reported domestic coal sales for Indonesia's six major producers, 2000–2009*

	Domestic sales (mt)									
Company	2000	2001	2002	2003	2004	2005	2006	2007	2008	2009
Bumi (KPC + Arutmin)	0.5	0.9	1.1	0.8	1.4	6.8	2.1	4.8	4.5	5.5
Adaro	6.6	8.4	8.2	8.9	8.7	9.6	9.8	10.1	9.3	9.6
Kideco	1.7	2.9	4.5	5.3	5.7	6.2	5.5	6.0	5.6	5.6
Berau	1.4	1.8	1.8	2.3	3.0	3.8	3.2	4.4	4.9	4.1
Banpu (Indominco Mandiri, Jorong & Trubaindo)	0.4	1.2	1.3	1.0	1.0	0.8	1.6	2.3	1.8	3.1
Bukit Asam (PTBA)	9.0	8.3	7.3	7.7	7.1	7.3	6.7	6.9	7.9	7.6
Subtotal (Big 6)	**19.6**	**23.5**	**24.2**	**26.0**	**26.9**	**34.5**	**28.9**	**34.5**	**34.0**	**35.5**
Others + unreported	2.6	4.6	5.1	4.2	6.4	4.6	4.6	4.7	15.0	17.5
Total domestic sales	**22.2**	**28.1**	**29.3**	**30.2**	**33.3**	**39.1**	**33.5**	**39.2**	**49.0**	**53.0**
Big 6 domestic as % of reported domestic sales	**88%**	**84%**	**82%**	**86%**	**81%**	**88%**	**86%**	**88%**	**69%**	**67%**
Domestic sales as % of total production	**29%**	**30%**	**28%**	**26%**	**26%**	**24%**	**17%**	**18%**	**20%**	**19%**

Source: Energy Publishing, "Indonesian Coal and Power Report," February 2001 through February 2010 for 2000–2009 figures for Big Six Producers and Total Exports from 2000 through 2006. Bob Kamandanu, "Indonesia Coal Outlook," October 2012 for Total Exports 2007–2009.

Table 7.7 *PLN's 1st phase Fast Track coal-fired power plants*

On-Java plants			Off-Java plants		
No.	Project	Total MW	No.	Project	Total MW
1	Labuan, Banten	630	1	Meulaboh, NAD	220
2	Rembang, Jateng	630	2	Pangkalan Susu, Sumut	400
3	Indramayu, Jabar	990	3	Tarahan, Lampung	200
4	Suralaya, Bantem	625	4	Susel, Barru	100
5	Pacitan, Jatim	630	5	Kalteng 1, Pulang Pisau	120
6	Paiton, Jatim	660	6	Teluk Sirih, Sumbar	224
7	Teluk Naga, Bantern	945	7	Kalsel, Asam-Asam	130
8	Pelabuhan Ratu, Jabar	1,050	8	Katbar 1, Pant Baru	100
9	Tanjung Awar Awar, Jatim	700	9	Others (15 small plants < 100 MW)	469
10	Adipala, Jateng	660		**TOTAL**	**1,963**
	TOTAL	**7,520**			

Source: Energy Publishing, "Indonesian Coal Report, 2009."

either have been completed or are still being constructed both on and off Java.

PLN initially targeted 2009 as the year of completion for its Fast Track Program. It quickly recognized that this tight schedule was not possible and extended the date of completion for the 9,483 MW of new coal-fired capacity to 2012, which still proved to be an overly optimistic completion date. To speed things along, in 2008, the government of Indonesia required PLN to sign an MOU with a Chinese consortium consisting of China National Technology Import Export Corporation (CNTIEC), Harbin Power Company, Dongfang Electricity, and Shanghai Electricity Corporation to develop at least seventeen of the thirty-three coal-fired power plants with a combined capacity of 8,000 MW at an estimated value of US$5.6 billion. However, the Bank of China, China Development Bank, and CEXIM, which financed a large portion of the fast track power projects, required the GOI to guarantee PLN's credit risk. Finalizing a guarantee agreement with the GOI delayed the first phase of the Fast Track Program by at least one year.

PLN is reportedly still experiencing delays in achieving commercial operations for some of its 1st Phase Fast Track coal-fired plants for various reasons. For example, the Paiton Baru project had not achieved COD as of December 2013 due to soft soil conditions that were only discovered after construction had started.[31] Other plants have had generator failures that have delayed commercial operations. In another

Table 7.8 *Original projections of coal demand for domestic and export markets through 2015 (Mt)*

Coal market	2010	2012	2013	2015
Power	37	75	85	106
Other domestic	23	36	39	46
Total domestic demand	**60**	**111**	**124**	**152**
Exports	198	212	229	267
Total demand	**258**	**323**	**353**	**419**

Source: Roleva Energy for other domestic and exports; Nur Pamudji, "Securing Domestic Thermal Coal Needs of Indonesia's Power Industry," paper presented at IBC's 9th Annual Coal Markets Conference, Singapore, February 23, 2011 for power sector estimates. PLN's most recent power sector estimates for 2015 have been reduced to 90 Mt from 106 Mt but show domestic sales to the power sector increasing to 125 Mt by 2020 (Helmi Najamudin, "Coal Demand and Future Outlook for Indonesia Power Companies," paper presented at IBC Coal Markets Conference, Singapore, January 30, 2013, slide 33).

instance, PLN discovered, after it had secured low rank coal supplies for the Labuan and Indramayu plants, that the plants had in fact been designed for a higher-quality coal.[32] Until this coal procurement mistake is rectified, PLN will run these two plants at less than their rated capacities.

PLN will eventually succeed in commissioning all the plants proposed under its Fast Track Program, which according to PLN will lead to an increase in total domestic demand from 60 Mt in 2010 to 152 Mt by 2015, an increase of almost 100 Mt in five years or 17% per year (Table 7.8). There may be other large coal-fired plants built in the future, but any other new coal-fired capacity is likely to be much lower than the 1st Phase Fast Track Power Program quantity and will include IPP-built plants that are likely to be based on sub-bituminous coals. They will also likely apply supercritical technology, which will reduce the coal requirements per MW.

In summary, during the remainder of this decade, Indonesia's domestic sales of coal, as a percent of total coal production, should increase significantly as PLN eventually succeeds in commissioning all of its 1st Phase Fast Track power plants and IPPs bring new coal-fired capacity into commercial operation. However, even with the successful implementation of PLN's Fast Track Power Program, Indonesia's exports of coal are unlikely to fall below 65% of total coal production through 2020, unless the GOI ill-advisedly mandates that coal producers divert a greater share of its production to local markets than it has mandated thus far.

Alternatively, fearing a shortage of coal for the domestic economy after 2020, the GOI may mandate production quotas in order to conserve remaining reserves of coal.

The implications of PLN's Fast Track Power Program for domestic coal usage (and in particular Indonesia's consumption of low rank coals for power production) are summarized in Section 7.4.2.2.

7.2.2.3 The role of geology and resource location in facilitating production and exports

Indonesia's coal producers have achieved spectacular rates of growth in both production and exports over the Take-off and Localization Periods. While much credit goes to the entrepreneurial spirit of CCOW coal producers, one must also acknowledge the strong support role played by Mother Nature. In particular, the rapid growth in Indonesia's coal production and exports during the Take-off Period and its continuation during the Localization Period would not have been possible without easy and favorable mining conditions in the form of (i) soft overburden, (ii) low stripping ratios, and (iii) thick seams having shallow dips. Two other contributing factors were (a) the proximity of many of the early mining operations to the navigable Barito and Mahakam Rivers or the coast and (b) relatively placid weather conditions on Kalimantan and offshore.

These favorable natural conditions allowed Indonesia's coal miners to extract coal using simple truck and shovel surface mining methods, transport the coal by barge, and then load it into sea vessels using offshore transshippers. The soft and shallow overburden allowed the use of truck and shovel equipment with little or no blasting. Spare parts, operators, and maintenance staff for such equipment were (and still are) widely available in Indonesia on short lead times and at a low cost relative to other mining methods.

During the Take-off and Localization Periods, most Kalimantan coal deposits had very low stripping ratios and, until 2005, the GOI heavily subsidized the price of diesel, which made the truck and shovel approach the most cost-effective mining method in Indonesia by a large margin.[33] (Increases in the price of diesel in recent years have encouraged the development of alternatives for transporting coal within mines and to barge ports.[34])

Coal was transported from the mine stockpile to a barge port largely by truck and barged down either the Barito or the Mahakam rivers and on to either a fixed land port or an offshore, floating transshipment facility.

The main exceptions to this truck and barge dependent haul approach during the Localization Period were (1) KPC, which operated a 13 km electric overland conveyor[35] from its mine stockpile to the Tanjung Bara Coal Terminal (TBCT), which could accept barges and sea vessels up to 120 kilotonnes (kt), and (2) PTBA, which transported its coal by rail for distances of up to 416 km to the port of Tarahan in western Sumatra or via a shorter 167 km journey to Kertapati. Table 7.9 describes the inland transport arrangement adopted by major Kalimantan coal producers that were still in use as of December 2009.

The diesel-based truck and barge inland transport systems have allowed Indonesian coal producers to bring their coal reserves into production quickly without the associated transport infrastructure constraints experienced in Australia, South Africa, China, India, and Russia. This has enabled Indonesia's coal producers to beat their main competitors – Australia and South Africa – to new markets in China and India.

Barging coal down either the Barito or the Mahakam rivers is a much less costly means of transporting coal when compared with the cost of truck or rail transport, but there were historically some constraints. Namely, prior to January 2009, barging coal down the Barito River was limited to the high-tide period of the day, due to reduced draft at low tide on a 13 km channel from the river mouth out to the sea. A number of government agencies and the shareholders of Adaro Energy formed a company to fund the dredging of a new 13 km channel near the mouth of the Barito River. This dredging project, which was completed in January 2009, made it possible to operate barges up to 15,000 tonnes capacity, twenty-four hours a day, year-round.

Dredging the new channel also allowed Indonesia's coal producers, especially Adaro, which is totally reliant on the Barito River for barging its coals to domestic customers and transshipment terminals, to achieve a significant expansion of their coal exports at low first cost. As a result of the successful dredging of a new barge channel in 2009, Adaro stated that it intends to expand its coal production from 40.6 Mt in 2009 to 80 Mt by 2015. More than 30% of Indonesia's coal production is currently being transported by barge directly to customers on Java and as far away as the Philippines (Figure 7.3).

For a sizeable fraction of Indonesia's coal exports, floating transshipment facilities are a key link in the transportation chain. By the end of the Localization Period, Indonesia's coal exports were loaded onto seagoing vessels at six land-based ports, with a coal handling capacity of around 50 to 80 Mtpa, and, as of December 2008, at 34 floating transshipment facilities with a coal handling capacity of around 170 Mtpa. These transshipment facilities ranged in size from small floating cranes with rated capacities of 3 to 5 Mtpa to more advanced floating facilities that could handle up to 15 Mtpa (Figure 7.4).

Table 7.9 *Internal transport arrangements for Indonesia's six largest coal mines, 2009*

Company	2009 production (Mt)	2009 exports (Mt)	Mine site to barge port (km)	Barge port to T/S facility (km)	Remarks
KPC	38.2	35.3	13 (conveyor)	1/9 (conveyor)	Geared/gearless
Adaro	40.6	31.6	79 (truck)	250/450 (barge)	Taboneo anchorage/IBT
Kideco	24.7	19.2	39 (truck)	58 (barge)	From TMCT to FC via 8 kt–12 kt barges
Arutmin	19.3	17.2	7–18 (truck)	124/199 (barge)	Geared vessel/ Satui Port via 3.5 kt–7 kt barges to NPLCT
Berau	14.3	10.1	13 (truck)	74 (barge)	From Lati to Muara Pantai
Indominco	12.4	13.1	35 (truck)	0/9 (conveyor)	From Port to Bontang Coal Terminal

Note: The six largest mines should not be confused with the companies that form the Big 6. In this table, the author is referring to coal mines located within a single concession area and not aggregated coal production by coal mining companies. For example, the biggest coal mine operated by PTBA, which is one of the Big 6 coal producers, is Tanjung Enim, which produced only 7.6 Mt in 2009. In 2009, the six largest coal mines accounted for 53% of total production and 55% total exports.

Source: Production and export figures from Energy Publishing, Indonesian Coal Report, February 2011; industry sources for all other data.

Figure 7.3 A barge pulls Adaro Energy coal down the Barito River in South Kalimantan. Source: Reuters.

Figure 7.4 A barge with coal from Berau Coal approaches a floating transshipment facility in East Kalimantan that will transfer the coal to an oceangoing vessel for shipment to customers in Asia.[36] Source: Reuters.

Indonesia's coal suppliers rely heavily on floating transshipment facilities, instead of fixed land ports, for the following reasons:

(a) *Shorter gestation periods*: Transshippers and floating cranes have a lead time of around one to one and a half years from date of order to start of operation, while a land-based port can take seven to ten years to bring into operation due to the long lead times needed for acquiring land, obtaining necessary permits, and carrying out their construction.

(b) *Low first cost*: In 2009, transshipment facilities ranged in cost from around \$7 million for a 3 Mtpa (10 kt/d) floating crane facility to \$20 million for a 10 Mtpa floating loading facility (35 kt/d) with storage capacity. A fixed port facility with a capacity of 30 to 40 Mtpa would cost \$100 million plus. Although the fixed port will have larger total capacity, substantial coal storage capacity, and a longer operating life, the capital costs per tonne of capacity for transshippers and floating cranes are so much lower than those of fixed ports that they are the preferred transshipment technology in Indonesia.

(c) *Modular additions to ship loading capacity*: Transshippers and floating cranes offer the ability to add small increments of new ship loading capacity (as low as 3 Mtpa per facility) and to more closely match new transport infrastructure to growth in coal demand. Reliance on floating transshipment facilities also allows coal suppliers to locate ship-loading facilities closer to the sources of supply.

By 2009, coal producers located on Kalimantan had access to floating facilities located at 10 offshore anchorages, while Sumatran coal producers had access to two offshore anchorage areas.[37] The floating crane and transshipment facilities in those days had maximum throughput capacities ranging from 4 to 15 Mtpa. Most could load "Capesize" vessels. More transshippers and floating cranes have been added since 2009 and continue to enable Indonesian coal suppliers to meet their coal demands without the infrastructure-related shipping delays that have been experienced in Australia over the past decade.

In addition to its natural advantages in mining and transporting coal, Indonesia is very favorably located to the coal import markets of North, South, and Southeast Asia. Figures 7.5 and 7.6, respectively, show the distance advantage and, for like coals, the sea transport cost savings that have traditionally accrued to Indonesia's Asian customers for load ports on Kalimantan and Sumatra when compared against Australia's port of Abbot Point and RSA's Richards Bay.

For its traditional coals, Indonesia's coal customers achieved significant savings on sea transport costs as shown in Figure 7.6 under the heading "trad." for each source of coal. However, Indonesia is rapidly

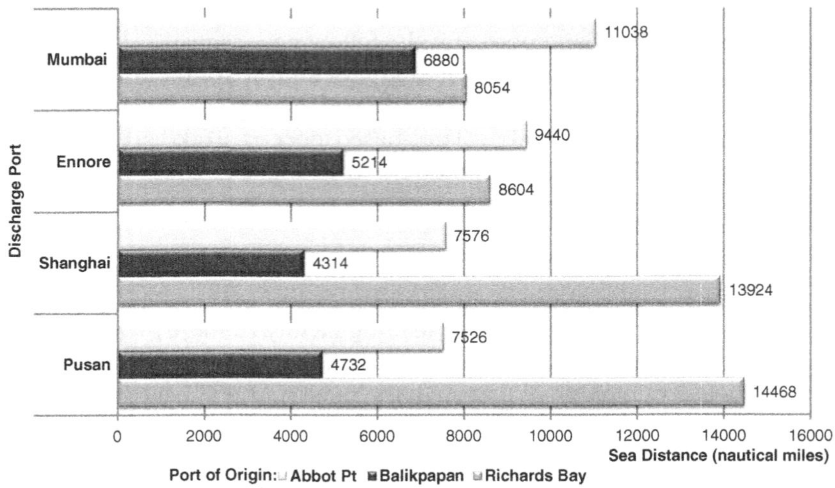

Figure 7.5 Sea distance to major Asian discharge ports from Abbot Point (Australia), Balikpapan (Kalimantan, Indonesia), and Richards Bay (South Africa). Consecutive round trip distances via Panamax vessel have been estimated in nautical miles using the NetPas Distance Calculator (https://www.netpas.net/).

depleting its middle and high range coals and its average coal quality is progressively degrading from a weighted average CV of 5,600 kcal/kg GAR as late as 2010 to a forecast level of 4,200 kcal/kg GAR by 2020 if not before. The end result is that Indonesia has effectively lost its sea transport cost advantage into most major markets in Asia.

7.2.2.4 The impact of political and regulatory changes

Even as production continued to grow strongly, the Localization Period was notable for the re-emergence of significant political and regulatory risk to investors in the coal sector. Following the resignation of Soeharto in 1998, Indonesia suffered through three weak and unstable governments from 1999 through 2004. These central governments were unable to control newly empowered regional governments and special interest groups, which led to the coal industry being subjected to numerous quasi-legal political events initiated by the central and local governments. These events also included GOI attempts to impose new taxes and fees on 1st and 2nd Gen CCOWs in violation of the special law protections afforded to these companies under their CCOWs. Many of these "political events" were clearly illegal and ended up being challenged by 1st and 2nd Gen

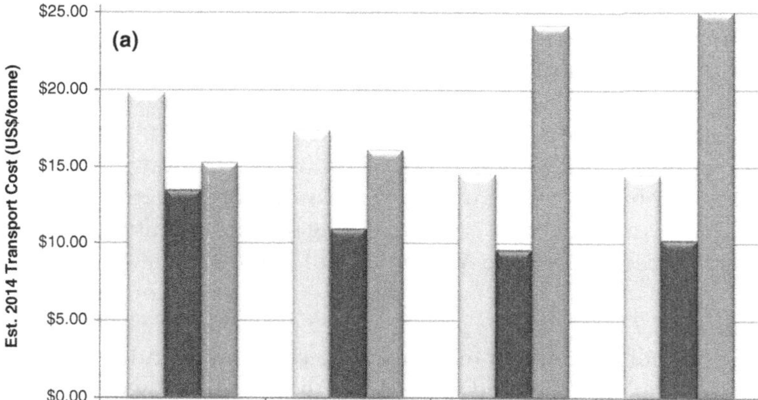

To:	Mumbai	Ennore	Shanghai	Pusan
From: Abbot Pt	$19.83	$17.40	$14.57	$14.49
From: Balikpapan	$13.51	$10.98	$9.61	$10.25
From: Richards Bay	$15.30	$16.13	$24.22	$25.05

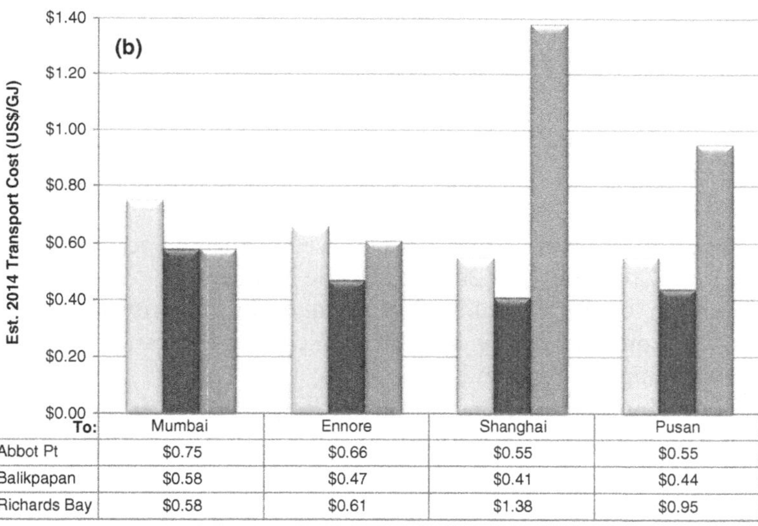

To:	Mumbai	Ennore	Shanghai	Pusan
From: Abbot Pt	$0.75	$0.66	$0.55	$0.55
From: Balikpapan	$0.58	$0.47	$0.41	$0.44
From: Richards Bay	$0.58	$0.61	$1.38	$0.95

Figure 7.6 Transport cost for traditional coals from Australia (Abbot Point), Indonesia (Balipapan, Kalimantan), and Richards Bay (South Africa) to major Asian discharge ports: **(a)** weight basis (USD/tonne); **(b)** energy basis (USD/GJ). Source: Roleva Energy calculations prepared for this chapter in July 2014 based on the following assumptions: (i) Panamax vessel leased over ten-year term at daily rate of $17K/day, (ii) IFO = $600 per tonne, (iii) MDO = $900 per tonne, (iv) cargo carrying capacity of vessel = 72.5 kt, load rate and discharge rate = 25kt/day. Coal CV = 6,300 kcal/kg, GAR for Abbot Point and Richards Bay and 5,600 kcal/kg, GAR for Balikpapan.

CCOW holders in Indonesian courts, where they were eventually overturned.

In the remainder of this section, the author assesses three broad categories of changes that significantly impacted coal investors during the Localization Period: (1) the localization process itself, (2) changes in contract terms, and (3) the proliferation of interventions by central and local governments. Following the discussion in this section, the author then devotes all of Section 7.3 to analyzing the most significant regulatory change in recent years, the Mining Law of 2009.

7.2.2.4.1 The localization process The Localization Period is so named due to the exit of a number of pioneer foreign investors in 1st Gen CCOW projects, who were required to sell down their majority ownerships by the tenth year of commercial operation. Rather than remain as minority shareholders, a number of foreign investors, such as Rio Tinto, British Petroleum, New Hope and BHP, decided to completely sell their equity stakes in 1st Gen coal projects.

Between 2002 and 2005, domestic investors either acquired majority ownership stakes or managerial control of Indonesia's four largest coal producers – KPC, Arutmin, Adaro, and Kideco – which in 2002 accounted for 57% of total coal production.

The most important sell-downs during this period were:

(a) BHP selling its ownership in Arutmin, Indonesia's fourth largest coal company, in 2002, to Bakrie Brothers;
(b) the partnership of BP and Rio Tinto selling their ownership in Kaltim Prima Coal (KPC) in 2003, to Bakrie Brothers;
(d) Samtan of Korea selling a portion of its majority share to the Sudwikotmono family in 2004, which put Samtan in a minority ownership position; and
(d) New Hope Mining of Australia selling its stake in Adaro to the Rachmat/Soerydjaya/Garibaldi families in 2005 and giving up managerial control of Adaro to its local owners.

In place of the pioneer foreign developers of Indonesia's coal industry, one now found domestic companies holding majority stakes in the largest Indonesian coal companies as well as most of the smaller companies. Foreign ownership still continued on a minority basis, with new foreign owners coming mostly from either regional power companies (e.g., Tata Power of India and Huadian of China) trying to secure access to Indonesian coal supply by taking partial equity stakes in Indonesian coal producers or from international investment banks and funds such as Goldman Sachs, Merrill Lynch, Saratoga Investments, and Farallon Capital Management.

Except in two cases, operational control of the mining operations also passed to the new Indonesian owners. The two exceptions were Kideco, which continued to be managed by Samtan of Korea, its previous majority owner; and Indominco Mandiri, which is a publicly traded Indonesian mining company but is effectively managed and controlled by Banpu of Thailand.

Expatriate managers, many of whom were long-term residents of Indonesia, continued to hold senior management positions throughout the localization period. In at least three instances – KPC, Arutmin, and Kideco – rights to market coal exports were held by foreign companies. For example, KPC contracted with Glencore to handle marketing of its coals into export markets under a ten-year marketing agreement that expired in 2013. BHP provided similar services for Arutmin under a ten-year marketing agreement that expired in 2011. Finally, Samtan had a contract allowing it to market all of Kideco's coal exports under a ten-year marketing agreement that expired in 2012. In return for their services, BHP, Glencore, and Samtan were paid a fixed commission for each tonne of coal sold.

One of the most salient aspects of the localization process has in fact been its relatively limited impact on the performance of the mining sector. Most mining companies now holding CCOWs, although localized, have preserved a strong international outlook and observe best industry practices with respect to environmental protection and community responsibility.

7.2.2.4.2 Changes in contract terms In 1990, the GOI stopped issuing new 1st Gen CCOWs. It was not until 1994 that the GOI offered investors a new bilateral contract, which is now commonly referred to as a 2nd Gen CCOW. The GOI only offered the 2nd Gen CCOW for one year (1994) before it too was withdrawn for unknown reasons. Despite being offered for only one year, the 2nd Gen CCOW has subsequently proved to be a moderate success. The GOI awarded 2nd Gen CCOWs to seventeen domestic companies (Table 7.2).[38]

In 1997, three years after discontinuing the issuance of new 2nd Gen CCOWs, the GOI decided to offer a 3rd Gen CCOW to both foreign and domestic investors. The 3rd Gen CCOW was offered to investors for three years, from 1997 to 2000 (Indonesian Coal Book, 2006/2007). The terms of the contract were structured to attract domestic investors more than foreign investors, but the program was open to both registered foreign companies (PMA) and 100% domestic companies. During that three-year period, 114 companies, mostly domestic firms, signed 3rd Gen CCOWs.

The 1st Gen CCOW is recognized as having more favorable terms than those of the 2nd Gen and the 3rd Gen CCOWs. The 3rd Gen CCOW is viewed by the coal industry as unattractive due to its lack of clarity on tax liabilities and the weak protections afforded to foreign investors for repatriation of profits. Table 7.10 provides a comparison of principal terms of the 1st Gen and 3rd Gen CCOWs. A cursory reading of this table might suggest that the 3rd Gen CCOW offered the better terms, but the differences of importance are the terms related to taxes and foreign exchange repatriation, which clearly favor holders of 1st Gen CCOWs.

The pioneer coal sector investors – those who signed 1st Gen CCOWs – had the significant advantage over later contract holders that they were able to choose the most attractive coal concession areas. They were also allowed to establish initial exploration areas that were larger than the concession areas permitted to later entrants. Their ability to tie up vast concession areas amenable to open-cut mining, and in many instances with locations close to the sea, allowed these pioneer investors to quickly achieve economies of scale in their production and inland transport systems. Their retained areas today are much smaller due to relinquishment requirements under their CCOWs, but they are still very large (25,000 to 118,000 hectares) when compared with the maximum retained area of 15,000 hectares that is allowed for IUP Exploitasi permits under the Mining Law of 2009 (see Sections 7.3.1 and 7.3.2).

Unfavorable foreign exchange and fiscal terms in 3rd Gen CCOWs – coupled with either poor quality reserves or unattractive (e.g., Central Kalimantan) locations for most concession areas and Indonesia's weak economic and political conditions from 1999 through 2004 – all contributed to the slow development of coal mines under this program. Yet, by 2009, twelve years after the signing date of most 3rd Gen CCOWs, companies operating under 3rd Gen CCOWs finally achieved significant levels of coal production (> 20 Mtpa), which in aggregate exceeded the total production level of 2nd Gen CCOWs for the first time. Moreover, if the infrastructure constraints of Central Kalimantan are ever resolved, there is the prospect that holders of some of these 3rd Gen CCOWs may transform Indonesia into an exporter of hard coking coal as well as contributing to Indonesia's position as the world's largest steam coal exporter.

7.2.2.4.3 Political and regulatory interventions by central and local governments through 2009 With the passing of the Soeharto regime in 1998, three decades of political stability and investment certainty came to an abrupt end. Over the next decade, the GOI, as well as Kalimantan's provincial governments, took a number of destabilizing actions that

Table 7.10 *Comparison of terms of 1st Gen and 3rd Gen CCOWs*

Terms	1st Gen CCOW	3rd Gen CCOW
Government counterparty	PTBA initially, after 1998, MEMR	MEMR
Mining authorization	Although separate PTBA/MEMR authorization is required for every stage of activity, CCOW holders are effectively given "Conjunctive Title" over their concession areas. PTBA authorization was largely a formality.	Conjunctive title or single "upfront" authorization given for all mining activities.
Exploration stage	Must be completed before exploitation.	Exploitation may commence during the exploration period.
Export approval	Government (MEMR) approval required for each export contract	Notification is only needed for long term export contracts of more than three years.
Domestic market obligation	Government (MEMR) may force CCOW holders to sell all or part of their coal in Indonesia if local demand is unmet.	Not certain
Levies and royalties	13.5% of production paid in kind or cash at the request of the GOI	13.5% of production paid in cash (negotiable for special cases such as low grade coal of UG Mining)
Advance payment	Yes, but not onerous.	None
Minimum expenditures	Nominal amounts must be spent during: General Survey Phase: $120/km^2$ Exploration Phase: $500/km^2$	No minimum expenditures required.
Regional taxes	Lump sum set at $100,000 per year.	None initially but subject to local government tax laws.
Taxation	Corporate tax rate fixed at 35% for first ten years and 45% for rest of CCOW term. CCOW holder either exempted from other taxes or tax rate stated in CCOW.	Corporate tax rate and other taxes, including local taxes, will vary from year-to-year based on changes to Indonesian tax regulations.
Tax exemptions and fiscal incentives	Exemption from duties, fees, and taxes on material, equipment, and supplies brought into or taken out of Indonesia	Same
	Investment allowance equal to 20% of total investment at rate of 5% per year.	Same
	FOREX repatriation guarantee for profits, loan repayment, depreciation, and share sale process	Requires approval of Parliament and President which is difficult to obtain

threatened to kill the coal industry goose that was laying golden eggs of employment, tax revenue, and royalty payments. Specific actions included attempts by the GOI and the regency and provincial governments of Kalimantan to rescind CCOWs and KPs and impose onerous new taxes on companies operating under 1st Gen and 2nd Gen CCOWs.

During the period 1999–2009, there were also numerous incidents of local government instigated work stoppages and illegal mining activities that many industry analysts claim were occurring due to either benign neglect on the part of local governments or the active connivance by these local governments in furtherance of other goals.

These and other government actions at times threatened the continuation of a number of coal mining operations on Kalimantan, frustrated industry representatives, and even prompted the US government to publicly call into question the GOI's commitment to a strong and growing coal export industry.[39]

Despite the many political challenges faced by Indonesia's coal producers during the localization period, they were able to effectively manage them. The level of political risk faced by each coal supplier in the past and its ability to mitigate those risks have been largely a function of the type of mining authorization (1st and 2nd Gen CCOWs vs. 3rd Gen CCOWs and KPs) held by the coal supplier.

Concessionaires such as KPC, Adaro, Kideco, Arutmin, and Berau, which operate under 1st Gen CCOWs, faced the lowest risk of supply disruption caused by government imposed changes in tax and land use policies. The highest risks were faced by companies operating under 3rd Gen CCOWs and KPs, which had to accept whatever changes were made to general tax laws and had fewer protections for expatriating revenues to pay offshore debt, operating expenses, and dividends to foreign investors.

The main complaints of the coal mining industry during the localization period focused on tax and fiscal policies. The problems caused by destabilizing GOI decisions on taxes and other matters have been compounded by the overlapping and sometimes competing political interests and legal jurisdictions of the MEMR, the Ministry of Forestry, the Ministry of Finance, and the Ministry of Trade and Industry. Although not currently a major problem, labor unrest has threatened to shut down a number of large coal mines in the past.

The background and status of a number of political risk events that have affected Kalimantan's coal producers during the localization period are summarized below. They will serve as useful historical background to the subsequent discussion in Section 7.3 of the impacts that the Mining Law of 2009 and its IR (implementing regulations) might have on the Indonesian coal industry over the next two decades.

Illegal mining Coal producers have had to cope with "unofficial" tolerance by local governments of illegal mining activities within their coal concession areas. Illegal mining operations were estimated to have "siphoned off" 4 to 8 Mtpa of coal from existing mines in 2004 and may have reached 20 Mtpa between 2005 and 2008, with most of the illegally mined coal coming from "operations adjacent to existing legal operations" (Baruya, 2009, p. 33).

The illegally mined coal was even identified by some analysts under the heading "KPs/others" by Barlow Jonker and its successor, Energy Publishing, when reporting production statistics for Indonesian coal (Barlow Jonker, 2006, p. 43). This practice of attempting to account for coal production from illegal miners was discontinued by Energy Publishing in 2009.

The primitive and short-term mining approaches followed by these illegal miners were reportedly causing permanent losses to remaining reserves, which the industry refers to as "the sanitizing of reserves." Illegal mining activities were reportedly most pronounced in the province of South Kalimantan, where the coal mining operations of Adaro and Arutmin are located, but were also a factor in East Kalimantan where KPC, Berau, and Kideco are located.

Illegal mining continued on both Kalimantan and Sumatra, with estimates of the "lost production" in 2009 and 2010 ranging from 20 to 30 Mtpa, with the caveat that such estimates remain highly speculative.[40] The illegal mining activities occurred within both unlicensed areas and mining concessions covered by a mining authorization.

Forestry Law 41/1999 In 1999, the GOI's Ministry of Forestry enacted Forestry Law 41/1999, which prohibited open-cut mining in areas designated as "forest conservation or protection areas." Officials within the Ministry of Forestry in cooperation with certain members of Indonesia's House of Representatives[41] tried to enforce this law upon all coal mining companies, even those operating under CCOWs that were signed long before the 1999 forestry law took effect, which, by law, should have exempted them from it.[42] Before this matter was resolved, Forestry Law 41/1999 forced the suspension of exploration activities within 150 mining concessions on the grounds that the areas being explored in these concessions were located in "protection forest areas." In response, more than 100 companies terminated their mining concessions. The fifty that stayed the course petitioned the Ministry of Energy and Mineral Resources to reconsider their cases.

Resolution of these cases was slow, with NGOs and industry groups lobbying both the Parliament and the Ministry of Energy and Mineral

Resources with their opposing views. The GOI tried to resolve the conflict in favor of a subset of affected mining companies through Presidential Decrees and Emergency Regulations (PERPU) such as:

- Presidential Decree 41/2004, which authorized thirteen companies affected by Forestry Law 41/1999 to continue their mining operations despite 41/1999 provisions.
- Emergency Regulation 1 (2004), which also authorized the thirteen affected companies to continue their mining operations in protected forest areas despite 41/1999.
- Forestry Regulation P.12/MENHUT-II/2004 (12/2004), which also stated that the thirteen affected companies could continue their open-cut mining operations, but all others must comply with 41/1999.
- Forestry Regulation P.43/MENHUT-II/2008 (Reg 43/2008), which clarified the actions that a mining company must take to allow it to conduct open-cut mining within protected forestry areas and the amount of compensation it must pay to the Ministry of Forestry for carrying out those mining activities. Special permits known as Rent-Use Permits were designed to standardize the permitting process.

Any forestry law issues related to CCOW holders appear to have been resolved through the application of the preceding regulations and decrees. However, new entrants to the industry who must comply with the Mining Law of 2009 and its implementing regulations (IRs) have been less fortunate (see Section 7.3.2).

Tax policy changes of Government Regulations 144/2000 and 65/2001
Regulation 144/2000 changed a long-standing government tax policy that allowed coal producers to recover their input VAT (value-added tax) from output VAT.[43] Regulation 65/2001 imposed VAT on equipment and materials that previously were exempted from VAT, including a 10% VAT on mining vehicles. These two Ministry of Finance regulations were said to have increased the cost of mining operations on Kalimantan by 6% to 10%. Although the Indonesian Supreme Court later nullified these two regulations, it did not set a date for their formal removal, so the GOI never considered this situation as being resolved.

Companies operating under 1st Gen and 2nd Gen CCOWs claimed that they were protected by very clear language in their agreements that indemnified them against any additional Indonesian taxes, duties, fees, and royalties levied by the government, except for those taxes and fees listed in their CCOWs. In this particular case, the eight largest coal producers – all operating under 1st Gen CCOWs – simply withheld the input VAT charges from the 13.5% royalty payments they owed the government. The Ministry of Finance argued for some time that the coal suppliers first needed to pay both their royalty bill and their input VAT bill to the responsible tax offices and then apply for a refund to the appropriate department within the Ministry of Finance. The 1st Gen CCOW companies argued that the whole exercise was moot based on the Supreme Court ruling on the matter in 2003 and continued to offset input VAT from their royalty payments.

The situation continued to fester until July 2008, when the GOI confiscated the passports of the directors of Indonesia's largest coal producers – Bumi Resources (the owner of KPC and Arutmin), Adaro, Berau, and Kideco. This extreme action by the GOI prevented these individuals from traveling overseas and disrupted their businesses. At the time, the GOI claimed that it would continue to hold their passports until the VAT offset matter had been settled to its satisfaction.

After making headlines for some months, the issue simply went quiet. The local press and industry newsletters speculated in late 2008 that the coal companies operating under 1st Gen and 2nd Gen CCOWs and GOI officials finally reached a settlement on the issue whereby these coal companies agreed to pay the VAT amount and the Ministry of Finance agreed to refund that amount. However, given that this event happened in mid-2008, after the Indonesian Supreme Court had ruled the regulation to be "illegal," it strongly suggests that political risk has started to reemerge in Indonesia. Unless rule of law prevails on such matters, investors may hesitate before making significant new investments in Indonesia's coal sector.

Imposition of a 5 percent export tax on the FOB price of exported coal The GOI, in October 2005, passed Regulation No. 95/PMK.02.05, which imposed a 5% tax on all coal exports. In late 2006, the GOI compounded the situation through the procedure it provided for estimating the amount of the export tax owed by each supplier. Instead of basing the amount of the export tax on the actual FOB price for each coal, the Ministry of Finance computed the tax against a hypothetical FOB coal price of US$32/tonne, which, at that time, was disadvantageous to exporters of lower rank coals. However, companies operating under 1st Gen and 2nd Gen CCOWs were exempted from this tax due to the tax indemnification provisions written into their agreements with the government. Nonetheless, this regulation acted to discourage parties to 3rd Gen CCOWs and KP permit holders from pursuing export opportunities since they would be at a 5% cost disadvantage relative to 1st Gen CCOW companies. In late 2006, the Indonesian Supreme Court struck down this export tax as illegal, and it was quietly dropped from further consideration by the GOI.

Provincial government interventions Some provincial governments have attempted to nullify concession permits issued to 3rd Gen CCOW companies. In other cases, provincial governments have attempted to impose new taxes and charges on the coal mine concessionaires. The coal mine concessionaires have thus far managed to defeat these provincial actions, largely because of strong support from the central government, which stated that existing contracts will remain in force and cannot be changed under international law. The Indonesian Supreme Court has also annulled a number of provincial tax laws as unconstitutional. At the moment, provincial government attacks on coal producers appear to be over. But a change in political control at the central government level could easily bring this issue back to center stage.

Industrial relations and wage rates Work stoppages are common in Indonesia, despite the fact that Indonesia has one of the highest unemployment rates in Asia. One complicating factor is that in 1999, the GOI transferred minimum wage level determination to the provincial governments, which imposed very large minimum wage increases such as a 67% increase in the minimum wage in East Kalimantan in 2002 followed by 8% per annum increases between 2003 and 2007. Labor laws remain in a state of flux, with "hire and fire" regulations changing frequently and open to liberal interpretations by most courts.

The government was supposed to have revised the Labor Law of 2003 to rectify the situation, but massive protests led to the government's delaying its submission to parliament of proposed revisions to the labor law. In this "worker-friendly" environment, labor unions and individual employees tend to test the limits of mining company resolve with occasional work stoppages and unreasonable wage and severance pay demands.

Thus far, union activity has not led to any disruptions in coal supply from Indonesia. Coal mining companies appear to have dealt with these labor issues without serious problems. Part of the explanation for their peaceful labor relations is that they contract out most of their labor-intensive mining and transport activities. Any labor dispute issues are handled by their contractors. In addition, most if not all coal mining companies are now majority-owned by Indonesians, who seem able to handle labor issues without causing disputes that lead to prolonged work stoppages.

Unauthorized police actions In July 2008, a rogue captain in the national police force shut down the entire operation of Kaltim Prima Coal (KPC), which, in 2007, was Indonesia's largest coal producer and owned by Abdul Rizal Bakrie, an influential businessman and also the minister of social welfare at the time of the police shutdown. The police occupied the mine site and established police cordons around mining equipment and coal haul trucks, effectively shutting down KPC operations for more than forty-five days, until the occupation was called off and the rogue captain transferred to an inactive posting elsewhere in Indonesia.

The expiration of an environmental permit by KPC was the reported reason for this action. KPC had filed for a renewal of this permit, but due to bureaucratic inertia, the new permit had not been issued prior to the old permit's expiration. Rather than declare force majeure and damage its reputation and that of the country, KPC continued to honor its contracts, arranging for alternative sources of coal and working with its customers to reschedule their orders.

This police action occurred at a time when coal prices were at their peak – gcNEWC having reached $184/t during the week of July 19, 2008. The action lasted for forty-five days and, according to reports from within KPC, it eventually cost the company tens of millions of dollars in lost sales revenues and demurrage charges on vessels that were forced to wait off-shore until the issue was resolved. Other companies were reported to have suffered shorter shutdowns of their operations causing significant financial loses, but none were as severe as in the case of KPC.

GOI-forced contract renegotiations In July 2008, the MEMR notified a number of coal companies that some of their contracts contained price terms that it deemed to be too low. The MEMR informed these companies that they had two choices: renegotiate the price terms of these contracts to conform to newly issued "guidelines" on pricing of coal under export contracts, or terminate the contracts. Adaro was one of the companies so notified. It agreed, under pressure from MEMR, to renegotiate twenty-three contracts, which at the time had low agreed-upon fixed prices relative to the spot prices for Indonesian coal. The reason offered by the MEMR for this action was "suspicions that a number of coal producers were self-trading with offshore affiliates," which allowed them to pay lower royalties and income tax on their net revenues.

Adaro dutifully attempted to renegotiate its fixed price Coal Sales Agreements (CSAs) with the unfortunate customers, which ranged from TNB in Malaysia to multiple customers in Spain and the United States. For those customers that refused to renegotiate, Adaro declared "an event of government force majeure" under their contracts and suspended shipments to those customers. Arutmin was also requested to renegotiate a fixed price contract that it had signed with TNB of Malaysia. It told TNB, however, that TNB had to renegotiate that contract directly with the MEMR. In the end, a number of the affected customers, such as

TNB, agreed to renegotiate their contracts. It appears that by the time the renegotiations were completed, 2008 coal prices were on a steep decline and the MEMR significantly softened its position on new price terms.

Letter of credit requirement The Ministry of Trade and Industry (MTI) imposed a new regulation in early 2009 that required foreign coal customers procuring coal shipments with a value above $1 million to arrange an irrevocable, at sight letter of credit (LC) through a domestic foreign exchange bank (Government of Indonesia, 2009). The LC requirement was imposed to ensure that exporters declared the full value of their exports for purposes of calculating company tax.

The list of affected customers included creditworthy power utilities from Japan, Korea, Taiwan, Hong Kong, and Malaysia, in addition to small traders and single plant IPPs. Past payment history had no bearing on the requirement to put an LC in place. This caused significant consternation on the part of the large power utilities. It also created concern among smaller coal buyers who had never issued such LCs before and did not know the mechanics of arranging an "at sight" LC in Indonesia and what their total financial exposure might be under the new regulation.

In response to industry complaints, the Ministry of Trade and Industry postponed the implementation of the LC regulation numerous times. It also issued an evergreen exemption to holders of CCOWs, as the imposition of the LC rule would conflict with the fiscal terms of their CCOWs. However, buyers of coal from KP holders were required to put irrevocable, at sight LCs in place in accordance with the MTI regulation. Similar to the experience with other ill-advised tax regulations, the LC regulation was officially revoked in early July 2010.

7.3 The new Mining Law of 2009 and recent regulatory issues

The latest and most significant regulatory change facing coal producers in Indonesia is the Mining Law of 2009. The GOI finally passed this law – officially known as Law No. 4 of 2009 on Mineral and Coal Mining (Law 4/2009) – in 2009 after years of aborted efforts. As of December 2013, fourteen implementing regulations (IRs) and one decree had been issued to support the implementation of this law. Of the fourteen IRs, nine have been issued by the Ministry of Energy and Mineral Resources and five by the GOI.[44] The nine MEMR IRs and one MEMR decree will most likely be supplemented by other regulations and decrees over time.

Given the often capricious regulatory actions taken by the GOI over the past decade, Indonesia's coal suppliers are understandably concerned about a number of aspects of the new mining law and its IRs, which have the potential to adversely impact production and exports. The main attributes of the new mining law and its supporting regulations are summarized below, along with a review of those aspects of the law and its regulations that are causing the greatest concern among coal producers. Following this review of the new regulatory framework, Section 7.3.2 discusses the possible short- and long-term impacts of the new mining law and its IRs on Indonesia's coal production, exports, and domestic sales.

7.3.1 Summary of the Mining Law of 2009 and its implementing regulations

Law 4/2009 made the following changes to the legal framework under which coal mining companies previously operated in Indonesia:

(a) Mining can only take place in zones designated by the national government as open for mining (Wilaya Pertambangan or WP).

(b) Mining zones can be of three types:
- Mining business areas (Wilayah Usaha Pertambangan or WUPs), which are areas open to private businesses on a competitive tender basis;
- Community mining areas (Wilayah Pertambangan Rakyat or WPRs), which are reserved for community mining activities; and
- State reserve areas (Wilayah Pencadangan Negara or WPNs), which are areas designated for the strategic national interest and reserved for government-owned corporations.

(c) The WPs shall be determined by the government after coordinating with relevant regional governments and the Indonesian House of Representatives. By establishing mining zones in advance following a transparent "zoning process," the government hoped to avoid the land use conflicts that resulted from the passage of Forestry Law 41/1999.

(d) New mining authorizations for private companies are now issued in the form of mining permits known as Izin Usaha Pertambangan (IUPs) with separate IUPs issued for exploration (IUP Eksplorasi) and for production (IUP Operasi Produksi).[45]

(e) Special permits, known as IUP Khusus (IUPK), are available on a preferred basis to government entities who wish to develop special mining areas, which are located in state reserve areas. IUPK permits are available as IUPK Eksplorasi and IUPK Operasi Produksi permits. If there are no government entities interested in developing these special mining areas, the MEMR is then free to offer them on a competitive tender basis to private companies.

(f) Small-scale mining permits, known as Izin Pertambangan Rakyat (IPR), allow communities to operate mines within a very small area of less than 10,000 hectares.
(g) Regional governments, mainly provincial governments, but in some cases regency governments, will be responsible for issuing most of the exploration permits with MEMR oversight or regulation to prevent abuses of process. The exception is where a proposed mining area overlaps two provinces. In these instances, the MEMR will be responsible for issuing the permit (Sullivan, 2010, p. 4).
(h) Exploration permits can only be issued through a transparent commercial tender. However, once a company is awarded such a permit, it is guaranteed the right to an IUP Produksi Operasi (production permit) without needing to go through a new tender as long as it has fulfilled the terms of its exploration permit.
(i) Companies that are in commercial production under terms of a CCOW will have their terms respected until the expiration of their initial contract period. However, the mining law and Government Regulation (GR) 23 also contain some confusing language to the effect that, "with the exception of state revenues," the terms of the CCOWs need to be "adjusted" within one year of the "promulgation" of Law 4/2009 (Article 169 of 4/2009).[46]
(j) Prior to the expiration of its CCOW, a company can negotiate a ten-year extension, but this will require the CCOW holder to accept an IUP-type production permit in its place. It is not clear how certain issues such as taxation and maximum size of the mining area will be dealt with when a CCOW is converted over to an IUP.
(k) KPs were required to convert to an appropriate IUP – either an IUP exploration permit or an IUP production permit – not later than May 2010.
(l) Within ten years of commercial production, foreign investment companies holding a majority ownership in IUP mining licenses are initially required to divest a minimum of 20% of the issued capital of the IUP or IUPK holder to Indonesian nationals. The divestiture requirement has subsequently been increased to 51% by the tenth year of commercial operations.[47]

7.3.2 Investor and miner concerns over the GOI's implementation of the new mining law

The coal mining industry and investors to the industry have raised a number of concerns about the specific rules and regulations that will be applied. These concerns are summarized in the following sections.

7.3.2.1 *Slowness in issuing regulations and lack of clarity for issued IRs*

The Indonesian regulatory process has never been a model of efficiency or predictability. But the pace with which the MEMR and the GOI in general have issued regulations and the weak and unclear wording of some of those regulations have exceeded the worst expectations of many industry analysts. The industry is also concerned by the issuance of regulations that limit the freedom of choice of miners when hiring mining service contractors, constrain their ability to tap long-term export markets, and change divestiture requirements for foreign investors in Indonesia's mining sector.

7.3.2.2 *Transitioning CCOWs to IUP permits*

Companies operating under 1st Gen CCOWs accounted for 64% of 2009 production. If one includes 2nd Gen and 3rd Gen CCOW holders in the total, the share of 2009 coal production jumps to 73%. These large mining companies are expected to hold their preeminent coal supply positions until the last of the 1st Gen CCOWs expires in 2025. Arutmin's CCOW will expire in October 2019, while KPC, Adaro, and Kideco will have their CCOWs expire in 2021, 2022, and 2024, respectively. With respect to 2nd Gen CCOWs, they will expire between 2025 and 2030, while 3rd Gen CCOWs will expire after 2030.

Under Law 4/2009, if PMAs operating under a CCOW wish to continue their mining operations for an additional ten years, they will be required to convert their CCOWs into production IUPs at least six months prior to the expiration dates of their CCOWs and to abide by new foreign investor divestiture requirements that were issued in 2013. It is unclear how the MEMR will deal with certain issues that could affect the sustainability of their future operations.

Specific areas of uncertainty include the following:

(a) *Limits to the maximum mining area*: CCOWs operate within mining areas that range between 25,000 and 140,000 hectares. Under Law 4/2009, companies operating under production IUPs will be limited to 15,000 hectares per IUP. This change could limit the ability of miners to achieve economies of scale.

(b) *Tax liabilities*: Revised tax imposts may increase the cost of coal production. This is an uncertainty more than a risk at this time. The 1st Gen CCOWs are now paying a 45% corporate tax rate, which is much higher than the general corporate tax rate of 25%. They also pay other taxes and fees as part of their CCOWs. So it is not clear that the change in tax regime will be an additional cost to CCOW holders.

(c) *Mining service companies*: Articles 124 to 127 under the new mining law require the permit holder to "only cooperate with local and/or Indonesian national mining services companies." As late as 2013, most mining services were being provided by large PMA service providers such as Leighton and Thiess, which do not qualify under the 2009 Mining Law as "Indonesian national mining services companies." A premature move to impose this provision could significantly increase the cost of mining and possibly cause coal supply shortages if labor shortages occur and/or if local mining contractors are moved prematurely into lead contractor roles without first demonstrating a strong capability in this area. The issues raised here about necessary transitions to a localized mining services industry will likely be resolved well before the first CCOW expires in 2019 (the Arutmin CCOW) but, if handled inappropriately, could damage the ability of Indonesian coal miners to expand their production as quickly and cost-effectively as they have in the past.

7.3.2.3 *Imposition of a domestic market obligation*

The MEMR issued regulation PerMen 34/2009 on December 31, 2009, which in accordance with the Mining Law of 2009, established procedures for the minister of MEMR to follow when setting a domestic market obligation (DMO) for each coal producer. The GOI later issued GR 23/2010, which provided the full legal basis for the issuance of this regulation and its contents.

PerMen 34/2009 established a DMO for each calendar year as a percentage of total production that each coal producer must make available to domestic customers. The process specified in PerMen 34/2009 requires the MEMR to survey domestic customers about their coal requirements for a particular calendar year by March of the preceding year. The minister is then required to issue a "DMO decree" by June of the preceding year, which provides the DMO for the next calendar year, along with the list of domestic customers and their requirements, which was used in setting the DMO. Mining companies are then required to submit production plans to the MEMR by November of the year preceding the year when the DMO will apply. Each producer must show in its production plan how it intends to meet its fair share of that year's DMO.

The MEMR indicated after issuing PerMen 34/2009 that it would not impose a DMO in any year that exceeds 35% of a producers' total production, which should not be a concern for most coal producers if domestic prices are competitive with export prices. Most coal suppliers appear to be working within the confines of PerMen 34/2009 already without any disruptions to their export business.

The problem with the DMO regulation is not the domestic requirement per se, as long as it is kept to a reasonable figure, but the lack of a contractual obligation on the part of buyers to take the DMO tonnage once offered. If a seller is unable to arrange the sale of its reserved DMO quantities within a certain time period, it will need to eventually sell those untaken amounts on the export market. However, the period of time a seller must wait before being released from its DMO has not been specified under PerMen 34/2009. This evergreen obligation to reserve a certain percentage of production for the domestic market without having firm contracts in place could lead to untaken DMO coal flooding the export markets and driving down the spot price for Indonesian coals, which would lead to lower royalties and corporate tax revenues for the GOI.

There are many other unresolved questions of detail surrounding the DMO such as:

(a) Should the DMO amount be based on tonnes of coal regardless of coal rank or should the DMO be broken into subcategories based on coal quality parameters such as CV, moisture, ash, and sulfur limits, which define the specifics of each customer's coal requirements?
(b) What happens if in any year a producer is unable to fulfill its DMO due to an event of force majeure? Is the producer relieved of its DMO or must the producer fulfill its DMO by reducing exports?
(c) In general, how will Indonesia retain its reputation as a reliable coal supplier in the region if its long-term customers feel that they can be cut off from their Indonesian supplies from year to year due to priority being given to domestic customers?

7.3.2.4 Regulation establishing reference prices for Indonesian coals

The Mining Law of 2009 and GR23/2010 state that the government, through the MEMR, will establish and publish each month reference coal prices that coal producers must charge as a minimum price to export and domestic customers. This requirement was included in the Mining Law because of GOI suspicions that coal exporters were basing their royalty payments on artificially low prices. According to the GOI, low announced prices were being agreed upon between an Indonesian coal producer and an offshore affiliate, which would later sell the Indonesian coal to export customers at a substantially higher price.

In September 2010, the MEMR issued PerMen 17/2010, which included an Annex providing a set of equations for determining benchmark FOB coal prices for individual coals, known in Bahasa Indonesia as Harga Patokan Batubara (HPBs). The original HPB price equations were revised at least five times by the MEMR, with many of the changes representing substantial revisions. Over the following three years, the MEMR spent additional time correcting numerous mistakes it had made in the online publication that presents the MEMR's price equations.

The price calculation methodology is needlessly complex and flawed in many ways.[48] As of December 2013, the MEMR has applied its complex price calculation methodology to estimate HPBs for seventy-one coal brands.[49] MEMR's website posting of the HPBs and their underlying methodologies is a welcomed effort by the GOI to be transparent about its reference coal prices. And, thus far, to its credit, the MEMR has not made unreasonable changes to its reference price methodology and, most importantly, has not required coal producers to use HPBs as the prices that will be applied under their commercial coal sales agreements. For the moment, the MEMR is only officially requiring coal producers to apply the HPBs for royalty calculation purposes.[50]

But if PerMen 17/2010 will only be used for calculating royalty payments, why did the MEMR put itself and the country's coal suppliers through all of this bother to develop the coal benchmarking methodology and data base? One possible explanation has to do with the fiscal provisions of the 1st and 2nd Gen CCOWs, which prescribe the taxes and fees that may be imposed on holders of those contracts. In particular, PerMen 17/2010 may be an attempt by MEMR to neutralize the negative impacts on royalty revenues caused by CCOWs self-trading in their coals by creating an objective and transparent mechanism for determining the deemed coal price for royalty payment calculations. A number of coal producers holding 1st Gen CCOWs believe that even this convoluted and complex method of determining their royalty payments violates fiscal terms of their CCOWs.

If the MEMR should decide to strictly interpret (and enforce) MEMR No. 17/2010 as requiring coal producers to price their coals at the relevant HPBs for their coals, this can easily lead to mandated coal prices that deviate substantially from true market prices. Due to the methodology used by the MEMR to calculate HPBs it is inevitable that the HPBs will deviate from time to time (if not always) from the true market prices that are being paid for the same grade or brand of coal. Moreover, calculating a regulated price for an individual coal based on quality parameters is, at best, an exercise in pseudo-mathematical certainty. Markets do not as a rule maintain a strict relationship between the quality of a coal and its price, especially relative to some contrived calculation formula. Coal producers also have justifiable business reasons for selling their coals at different prices from time to time based on short-term internal and external factors. To interfere with that process is to create added risk and uncertainty for both coal suppliers and buyers. In short, it is hard to see what the MEMR has gained from this regulation.

PerMen 17/2010 has other, possibly more damaging, flaws. In particular, it includes two "pork barrel" provisions, which, if not deleted from the regulation, may adversely impact both the security of coal supply and its delivered costs. These provisions require coal producers to (a) use only Indonesian-owned insurance companies to insure their sea cargos and then (b) ship those cargos on Indonesian-owned vessels. The regulation goes so far as to state that even if the buyer is arranging his own shipping, the coal seller must ensure that the buyer uses local shippers and insurers. Similar to the contract mining regulation, this pork barrel requirement is largely unworkable, as the local shipping and insurance industries do not have the capability to handle the large volumes and values of coal exports out of Indonesia.

As of December 2013, considerable uncertainty still exists about the MEMR's intention for PerMen 17/2010 and in particular whether HPBs should be used strictly for royalty calculation or for both royalty and CSA price determination. The MEMR stated in November 2010 that it would issue Implementation Guidelines (known as Juklak in Bahasa) on this matter. As of December 2013, the MEMR has failed to do so.

The result has been that some companies are just ignoring PerMen 17/2010 and selling their coals at prices they deem acceptable under current market conditions, the assumption being that it is easier to seek forgiveness than permission under such circumstances. Others have been selling their coals according to whatever price the market will bear and then informing the MEMR of what they have done and asking for MEMR's retroactive approval. And still others are informing coal buyers – at least initially – that they will only sell their coals at the appropriate HPB.

The possibility exists for this confused and unnecessary regulation to go the same way as the 5% Export Tax regulation, the MTI Letter of Credit Obligation decree and other ill-advised policies and regulations pushed by government officials and politicians with hidden agendas. But it is more likely that the MEMR will allow PerMen 17/2010 to continue in existence for some extended period of time, and then after struggling with its weak implementation, find a way to either render the regulation impotent or rescind it entirely.

7.3.2.5 New divestiture requirements for foreign investors

The MEMR issued two other ill-advised regulations, GR 24/2013 and MEMR 27/2013, which set onerous divestiture terms for foreign investors who hold majority positions in Indonesian coal assets. Once again, given the special law protections of CCOWs, it is unlikely that this regulation will have an impact on Indonesia's large mining companies, who for the most part are majority owned by Indonesians. However, IUPs that have majority ownership by foreigners are heavily disadvantaged under the terms of GR 24/2012 and MEMR 27/2013.

Detailed reviews of these two regulations have been provided by a number of international law firms with operations in Indonesia, such as Freehills and Baker and Mackenzie, and by Bill Sullivan of Christian Teo & Associates, who is a widely respected legal expert on topics related to Indonesia's Mining Law and its Implementing Regulations.[51] The consensus interpretation by these legal experts is that GR 24/2013 and MEMR 27/2013 have made changes to IUP divestiture requirements that will have a serious negative impact on Indonesia's reputation as a venue for mining sector investments.

The specific terms of the two regulations that have raised the concerns of usually staid and reserved international law firms are related to minimum share sell-down amounts and the price that a foreign owner might receive from its share capital sell-down. Under GR24/2012, foreign owners of IUPs are required to sell to domestic investors a minimum of 20% of their share capital in their IUP companies by the fifth year of commercial operations and by the end of the tenth year of commercial operations to have reduced their ownership to 49% or less of the company's share capital. This specific change to divestment requirements for IUPs came as quite a shock to some foreign investors, as a previous government regulation (GR23/2010) had more generously set the divestiture requirement at 20% of share capital by the end of the fifth year of commercial operations, without any further sell down required.

GR 24/2013 also set a priority list of possible Indonesian buyers of IUP shares being sold by foreign investors. The GOI and local governments were allowed first and second rights of refusal. If neither of those government entities have the financial resources and/or the interest in purchasing the share capital, then the foreign IUP is allowed to auction the shares off to the next priority stakeholder, which are state-owned enterprises such as PTBA and PLN. Finally, if that auction sale fails, the foreign investor may auction off its IUP shares to Indonesian private companies and individuals.

The MEMR followed up GR 24/20102 with MEMR 27/2013, which provided additional details on the sell-down process. One topic spelled out in considerable detail was the offering price for the shares that would be up for sale. In this regard, the MEMR has set share price conditions that only allow the foreign investor to sell its shares on a cost recovery basis, and then under unfavourable terms. MEMR 27/2013 sets a maximum Divestiture Price equal to the "Adjusted Replacement Costs," which MEMR defines as the "investment" during the period from the exploration stage up to the date on which the Divestiture Obligation "applies" (not necessarily the date the divestiture takes place) less accumulated depreciation, amortization, and any outstanding debts as of the end of the year when the Divestiture Obligation occurs.

If agreement between either the GOI or other priority buyers and the IUP is not reached on the amount of the IUP investment, the IUP will be required to contract with an independent appraiser to determine the investment amount. According to MEMR 27/2013, interested parties must indicate their willingness to procure the offered shares either in total or in part within sixty days of being informed by the selling IUP of its intention to divest and must pay for any shares they have indicated a willingness to buy within thirty days of its expression of interest, which does not sound practical.[52]

7.3.2.6 Forestry regulation for issuance of Rent-Use Permits

Indonesia's Ministry of Forestry (MoF) has issued highly problematic regulations on procedures for obtaining a permit to conduct surface mining on land that has been classified by MoF to be either productive or protected forest lands. These regulations have been particularly damaging to holders of the new IUPs established by the Mining Law of 2009. Since 2010, IUPs have been stymied in their efforts to develop new coal deposits by the Ministry of Forestry, which requires new entrants to apply for a Rent-Use Permit before exploring for and exploiting coal resources located underneath productive forestlands. The administrative and technical filing requirements are onerous and time consuming, and then the time before actual issuance of the Rent-Use Permit can best be described as indefinite.

To remove part of the delay in the issuance of the Rent-Use Permits, the Ministry of Forestry began in 2008 to issue Approval In-Principle Rent-Use Permits, which supposedly would allow a coal producer to conduct business as if a full Rent-Use Permit had been issued. However, the legal authority of the In-Principle Rent-Use Permit was never fully accepted by other government agencies, leading to a number of instances where coal mining companies were forced to cease further activity until the full Rent-Use Permit had been issued.

The issue was made even more complicated in May 2010 with the signing of the Oslo Accord by the GOI and government of Norway, under which the GOI agreed to impose a two-year moratorium on the further conversion of forested areas and peat lands into other uses. The delays related to the issuance of the Rent-Use Permits have slowed down exploration activities and the establishment of new mines in Indonesia by new entrants. Companies operating under CCOWs have not faced the same issue as they are protected by the special law provisions of the CCOW related to land allocation and usage.

The MoF Rent-Use Permit regulation became a source of particular controversy with the revocation in 2010 of the four mining licenses of Churchill Mining. The licenses were revoked by the Regent of East Kutai based on unsubstantiated allegations that Churchill was engaging in illegal mining within areas designated as productive forests (Sullivan, 2011, p. 34). The presumption was that Churchill may have been operating without a Rent-Use Permit or was operating under an In-Principle Rent-Use Permit that was later judged to provide insufficient authorization for the commencement of exploration or mining activities. Churchill applied to the Administrative Court in East Kalimantan for reinstatement of its licenses, but its application was rejected.

Rumors circulated in the coal industry that the revoking of the licenses had nothing to do with illegal mining and everything to do with a politically powerful Indonesian businessman wanting to acquire the Churchill mining areas without paying just recompense to Churchill's shareholders – and a pliant regency government and local administrative court that were willing to go along with this powerful politician.

If the Churchill case is not equitably resolved, this political event could have a strongly adverse effect on the sentiment of foreign investors toward new investments in Indonesia's mining sector.

7.3.2.7 "Possible" regulation to ban the export of low rank coals

Another regulatory matter that has recently received intense publicity (as of December 2013) was the imposition of a ban on the export of unprocessed metals and minerals. The policy to ban such exports was enshrined in the Mining Law of 2009 (Articles 102 and 170) and GR23/2010.[53]

The MEMR in 2013 issued a draft regulation (MEMR 644.K/30/DJB/2013), which, in response to GR23/2010, would have banned, by 2014, the export of any coals having gross calorific values less than 5,600 kcal/kg, GAD, which, assuming a total moisture content of 22% and an inherent moisture content of 14%, is equivalent to 5,000 kcal/kg, GAR. This draft regulation was withdrawn after "making the rounds" through the coal mining community and receiving a very strong negative response. If officially issued as a Ministry regulation, it would have likely been judged by Indonesia's courts as not applying to 1st Gen CCOW holders as it appears to violate section 12.5 of the 1st Gen CCOW, which states: "Contractor shall keep Batubara (the responsible MEMR agency) advised from time to time with respect to the technological and economical possibilities of Coal conversion processes in general. Upon Contractor's notification in writing to Batubara that in Contractor's opinion conversion of Coal from Agreement Area sources is a feasible proposition technically and economically the Parties will discuss the possibility of developing such conversion taking into account the technical, financial and economic aspects thereof."[54]

It is likely that 2nd and 3rd Gen CCOWs contain similar language that would exempt them from this now withdrawn regulation. Under such circumstances, the withdrawn regulation could only be enforced against the hapless IUPs, who are generally small, poorly capitalized coal miners with annual coal production of less than 5 Mtpa.

While it may seem far-fetched that the GOI and in particular the MEMR would implement such an inequitable regulation, which would have no meaningful impact on shipments of unprocessed minerals in any case, this is in fact exactly what it has done when it comes to exports of unprocessed metal ores. In December 2013, Commission VII of the Indonesian Parliament, which is the standing parliamentary body with oversight responsibility for mining, oil and gas, and natural resources, rejected a proposal from the MEMR to "allow individual mineral producers to continue processing unprocessed ores after 2014 so long as they could show a sufficiently strong commitment to eventually carrying out domestic processing and refining" of their unprocessed ores.[55] The export ban, which went into effect on January 12, 2014, was to affect exports of unprocessed copper, nickel, iron, bauxite, and gold ores, which are substantial export earners for Indonesia and strategic imports for a number of countries in the region, especially Japan and China (Jensen and Asmarini, 2014).

According to Reuters, the export ban was to be imposed in a discriminatory manner, with most holders of Contracts of Work (COWs) exempted from the export ban as a result of special law protection they have under the COWs. Two miners that were expected to benefit from such special law protections are (1) PT Freeport Indonesia, the operator of the Grasberg copper mine in Papua Province and (2) PT Newmont Mining, operator of the Batu Hijau copper mine. Together, these two copper mines accounted for 97% of Indonesia's 2012 copper production (Jensen and Asmarini, 2014). These two companies in 2013 processed about one-third of their raw ores, which is cited as the reason for their exemption, but the suspected reason was that the GOI would open itself up to international arbitration if it tried to impose this regulation on these two companies against the terms of their COW.[56]

IUPs, which do not have special law protections under their permits, are facing a dire situation. Many were expected to shutter their mines starting in January 2014 and be forced to lay off thousands of employees. According to the Indonesian Mineral Entrepreneurs Association, as many as 30,000 mine workers had the possibility of being laid off by January 2014 (BIORES, 2014). Ibris Nickel Pte Ltd is a specific case in point; in January 2014, the company was threatening to lay off 1,400 workers if the GOI persisted with its export ban regulation. One other nickel ore miner – Mobi Jaya Persada – claimed it would continue to export raw nickel ore and dared the GOI to come to its site and try to close its nickel mine down (Jensen and Asmarini, 2014).

7.4 Major conclusions and prospects for Indonesia's coal industry in the future

7.4.1 Lessons from the past

Over the past two decades, Indonesia's coal industry has transformed itself from being an unknown, minor player in Asia's coal markets into the world's largest exporter of steam coal. Indonesia was able to create this world-scale industry despite challenges created by widespread government corruption, a weak legal system, the Asian Financial Crisis of 1997, and the fall of the Soeharto government in 1998.

The history of Indonesia's post–World War II coal sector effectively began in 1967 with the replacement by military coup of the Sukarno regime with the government of Soeharto. This "Formative Period," which started with the passage of the Mining Law of 1967 and the Foreign Investment Act of 1967, lasted for twenty years, during which a functioning regulatory framework was put in place and the GOI signed "foreign investor-friendly" contracts with foreign mining companies.

Indonesia's coal production did not start to "take off" until 1988, a time when Indonesia's coal industry was still dominated by PTBA, the government-owned coal company operating from coal mines located in Sumatra. Between 1988 and 1999, Indonesia's coal production grew by an astonishing 29% per year from 4.4 Mt in 1988 to 73.8 Mt in 1999 (the Take-off Period). In the space of that one decade, Indonesia's coal industry shifted from being a Sumatran-based industry under the control of a government owned company to an industry located predominantly on the island of Kalimantan and comprising a diversified group of privately owned companies that were mostly run according to international standards by world-class mining and resource development companies.

The coal industry's growth during the Take-off Period was aided by a stable political and investment climate within Indonesia. It was also aided by strong growth in demand from Japan, Korea, and Taiwan for Indonesia's low sulfur, low ash steam coals.[57]

During the "Localization" Period (2000–2009), the expansion rates for the coal industry slowed considerably from the 29% annual rate of growth during the Take-off Period to 14%. But the slowdown was not due to the process of localization. It was mostly caused by (i) the "law of big numbers" coupled with the five-year economic slowdown caused by the Asian Financial Crisis; (ii) the political instability of Indonesia during this period and specific political events directed at the coal industry, which negatively impacted the expansion plans of some companies until 2004; (iii) the short-term, adverse economic impacts from SARs and bird flu; and (iv) until 2005, low prices for steam coal worldwide.

After 2004, the expansion of the Chinese economy started to have its salutary effects on other Asian economies, resulting in demand for new power plants and a concomitant increased demand for steam coal from Indonesia. However, toward the end of the Localization Period, the Global Financial Crisis struck, which also burst the Asian commodity bubble. In the final year of the Localization Period, the Indonesian parliament passed into law the Mining Law of 2009 and from that point, the next development phase (the New Mining Law Period) commenced. The story thus far for the New Mining Law Period is one of Indonesia's coal industry facing heightened regulatory and political risk amidst a resurgence of resource nationalism that smacks of the bygone Sukarno years.

Geological, locational, and economic factors continue to make Indonesia a theoretically attractive venue for foreign investments in the coal sector. Unfortunately, political risk and seemingly angry, ham-fisted government policies and regulations that smack of the economic nationalism of the bygone Sukarno era risk damaging its underlying fundamentals. Rogue actions by government agencies, including the national police, during the Localization Period and the first four years of the New Mining Law Period are especially troubling examples of political risk gone wild.

This experience with political risk events over the past decade does have a silver lining. Against the odds, coal mining companies managed to prosper during the Localization Period despite the numerous political risk events and economic challenges posed by the Asian Financial Crisis, SARS, and bird flu scares. It is against this backdrop that the next political saga – implementation of Indonesia's Mining Act of 2009 and its implementing regulations – is currently playing out. The degree of success in handling the challenges posed by this new regulatory system will largely determine whether Indonesia will play a leading coal supply role in Asia over the next two decades or will slowly drift into a secondary support role as has been the case in the oil and gas sectors.

The history of Indonesia's coal industry shows how a supportive regulatory environment is essential for cultivating a world-class coal industry. The successful establishment of a world-class coal industry was largely due to the regulatory framework created in 1967 by the Soeharto "New Order" government. It involved the establishment of direct contractual obligations between the GOI and foreign mining companies. The original terms of those agreements were in English and provided foreign investors with strong protections and incentives. Given the undeveloped legal system and the weak judiciary process in Indonesia at the time, it would not have been possible to jump-start the coal mining industry of Indonesia based on foreign investment and technology without such an arrangement.

Another crucial point that needs to be taken into account in policy making and regulation is that lead times for coal industry development are long. Even with "special law" arrangements and attractive contractual terms in 1967, it took more than twenty years before Indonesia's coal production started to take off. Arutmin, the first company to sign a 1st Gen CCOW, did so in 1981 but did not start producing coal in commercial quantities until early 1990, nine years after contract signing. Other 1st Gen CCOWs were signed with Adaro, KPC, and Kideco over the next two years. They too required about nine years from the date of CCOW signing to achieve commercial production. The long lead times resulted from the need to complete extensive drilling programs and detailed feasibility studies, to obtain financing for their mining projects, and then to develop the mines, construct the roads, and build the necessary ports and other supporting infrastructure. These staged, long lead-time implementation projects also had very high first costs and could have been canceled at any time if either market conditions or domestic political conditions had deteriorated during the period they were being implemented.

7.4.2 *Key issues affecting the future trajectory of Indonesia's coal industry*

7.4.2.1 *The effect of the Mining Law of 2009*

The immediate effects of the new mining law on Indonesia's coal production have been limited. From 2009–2012, production and exports have

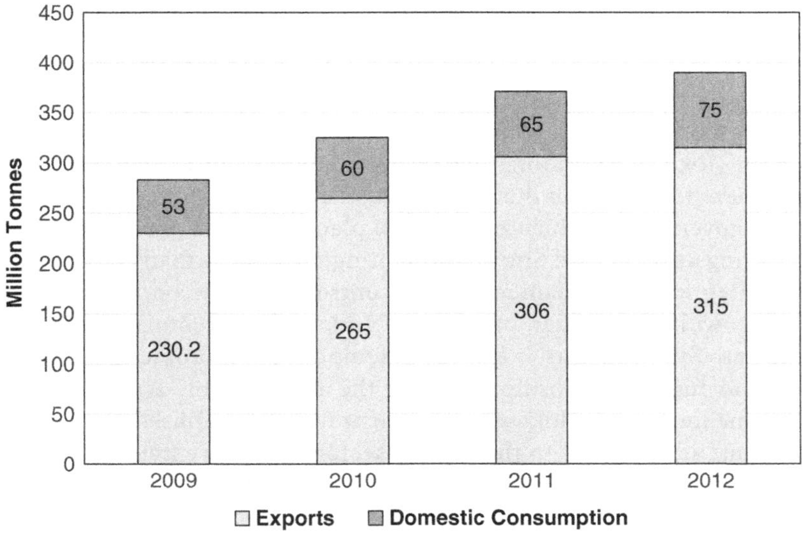

Figure 7.7 Indonesia's coal exports and domestic consumption, 2009–2012. Source: Bob Kamandanu, "Indonesia Coal Outlook," October 2012, slide 12 (2012 figures are best estimates).

continued to grow at a slower yet still impressive annual rate of 8.4% (Figure 7.7). Although this is significantly lower than the 14% annual growth rate for the Localization Period, it is still an impressive accomplishment given that in 2009, Indonesia (a) produced 390 Mt of coal, (b) had to weather the global financial crisis of 2009, and (c) was making a major shift in coal quality from medium rank sub-bituminous coals to very low CV (< 4,500 kcal/kg, GAR), high moisture (> 28% TM) coals.[58]

The bigger question, however, is what effect Indonesia's mishandling of its new mining law and associated IRs will have over the longer term. Optimists are quick to note that Indonesia has always been a difficult place to do business, but, in the end, the GOI has always found a way to neuter those laws and IRs that threaten the Indonesian economy. They argued in late 2013 that many of the ill-advised regulations that smacked of resource nationalism and overreach by government regulators were largely a product of politicians playing to the electorate in advance of Indonesia's presidential election, which took place in July 2014. Now that the election is over, so the story goes, a more balanced and less vitriolic approach will be followed by the GOI and MEMR with respect to mining sector regulations. It will also be a time when many of the regulatory extremes over the past year will be corrected.

These optimists may yet prove to be correct over the long term. However, the GOI's ill-advised regulatory initiatives, starting with the passage of the Mining Law of 2009, have effectively shattered Indonesia's reputation as a reliable venue for making mining sector investments. This author does not believe that the election of Joko Widodo (Jokowi) as Indonesia's new president in July 2014 will lead to any near term fixes to the mining sector regulatory system. The new Jokowi government, which has been in place since October 2014, was still feeling its way at the time of this writing. Among its many challenges are dealing with a Parliament that is controlled by the opposition and working with the leader of its own Indonesian Democratic Party, Megawati Sukarnoputri, who has attempted to promote unpopular figures to high-level positions within the government. It will clearly take some time for the Jokowi government to fully establish itself before significant adjustments to the mining sector regulatory framework can be made. There is also no guarantee that Jokowi and his cabinet will not eventually succumb – for reasons of political expediency – to the siren call of resource nationalism and follow the path of the previous SBY government.

Instead of the new, popularly elected government being a quick fix to Indonesia's regulatory morass, it seems more likely that ill-advised mining regulations will continue to be issued by the MEMR while old, ill-advised regulations passed during the SBY administration remain in force due to the regulatory inertia that prevails in the face of splits between the presidency and Parliament. New foreign investment in Indonesia's coal industry is also being adversely affected by the depressed coal prices that have prevailed since February 2012, when the globalCOAL Newcastle price index (gcNEWC) stood at $117.52 per tonne. In January 2015, the gcNEWC price was $62.80 per tonne, a decrease of 47%. The outlook is for depressed coal prices to continue through 2016 and perhaps through 2018.

It is worth noting that Indonesia has always been considered a very difficult, if not downright undesirable, place to make new mining investments. But between 2009 and 2013, things moved from bad to worse. The Fraser Institute, a respected Canadian public policy research and educational organization, ranked Indonesia as the "least desirable place for making exploration investments in the mining sector" in its 2012/2013 Annual Survey of Mining Companies (Fraser Institute, 2013, p. 5). This last-place ranking out of a sample of ninety-six jurisdictions (countries, states, and provinces with large mining sectors) put Indonesia below foreign investment pariahs such as Venezuela, the Democratic Republic of the Congo, Zimbabwe, Kyrgyzstan, and Bolivia. The authors of the survey report noted that Indonesia "dropped the most in the rankings for [countries within] Oceania" (Fraser Institute, 2013, p. 6). In the Fraser Institute's 2013 Survey of Mining Companies (Fraser Institute, 2014, p. 10), which was released in March 2014, Indonesia was in the bottom ten countries out of a sample of 108 – still below countries such as Vietnam, the Democratic Republic of Congo, Greece, and Bolivia, which are not noted for their investor-friendly governments.

The next Fraser Institute Survey on Mining Companies was due to be issued in April 2015. It will be interesting to see whether the perception of Indonesia's climate for mineral exploration investments registers any improvement.

7.4.2.2 Ability to meet demand from domestic and export markets: perceptions and reality

Indonesia's politicians and bureaucrats, when they initially saw PLN's forecast of 2015 coal requirements for its 1st Phase Fast Track Power program (~90 Mt), concluded that Indonesia and the region were heading for a coal supply shortage. But this has not proven to be the case. Indonesia's coal reserves and resources as of December 2009 appear to be more than adequate to meet the needs of both domestic and export customers over the next two decades (see appendix at the end of this chapter). Total production growth has thus far outstripped both domestic and export sector demands, and PLN's Fast Track Power Program has actually encouraged increased investments in low rank (LR) coal projects on Kalimantan and Sumatra.

Short-term supply shortages are, of course, possible in one or two years over the next decade. But given Indonesia's large untapped LR coal resource base and its proven ability to rapidly add infrastructure and increase output by up to 29% in any one year, Indonesia's coal industry should not have any difficulty meeting new coal requirements of both domestic and export customers over the next decade.[59]

But this demonstrated reality is still not being accepted by many Indonesian government bureaucrats and PLN executives. Rather than seeing the additional demand for LR coal that is being generated by PLN's fast track program as competition for export coal markets, PLN and others should view this increase in domestic demand as a market-based mechanism for expanding the pool of large-sized (> 10 mtpa) LR coal producers from just a handful to perhaps fifteen to twenty located on the islands of Kalimantan and Sumatra. Indeed, it is providing the impetus for doing just that by encouraging coal and power producers to develop the vast LR coal resources of Sumatra, which have not been developed thus far due to transportation infrastructure constraints and the lack of a market.

What is currently being viewed by PLN executives and GOI bureaucrats as an "either-or" situation with Indonesia's coals going to either the domestic market or the export market, should instead be viewed as an enabling program that will enhance and strengthen the development of the Kalimantan and Sumatran coal supply chains, resulting in increased investments in inland transport facilities (roads, barge ports, and overland conveyors) and offshore coal loading facilities.

Assuming that the GOI does not impose production quotas on its coal producers, the only reason for a possible slowdown in Indonesia's coal exports is a major reduction in coal imports to either the Chinese or Indian markets and/or a prolonged continuation of today's depressed coal prices beyond 2016. In this regard, questions have been raised about the permanence of the 2009 Chinese export market, which burst onto the Indonesian coal supply scene almost overnight in 2007 (Morse, 2010; Parker, 2013). As explained in Chapter 9 in this volume, the growth in Indonesia's coal exports to China in 2009 was not in fact the result of a structural production shortfall in China, but was due to Indonesian steam coals in that year having a significant price advantage into southern Chinese ports when compared with the landed costs of domestic coals into the same southern Chinese ports.

There is a distinct possibility that the Chinese coal suppliers, with Chinese government prodding or insistence, may from 2015 onward agree to delivered coal prices with buyers that are lower than the delivered prices for Indonesian coals (Morse, 2010). Moreover, China's coal industry is undergoing a major consolidation that will result in the shut-down of many small mines that are unsafe, undercapitalized and with very high cash costs. Once this ongoing consolidation of the Chinese coal industry is completed over the next three years, China's ability to produce steam coal at a more competitive price might once again allow it to become a net exporter of steam coal into North and Southeast Asia. At the very least, it will allow China to reduce its imports of Indonesian coals. If either of these two market developments should occur, they would result in a significant reduction in Indonesia's coal exports to China and perhaps to the rest of Asia (Parker, 2013).

India's imports of Indonesian coal, on the other hand, can be viewed as more structurally sound and of a more permanent nature due to chronic and seemingly unending shortages of steam coal to support the expansion of India's power supply system. But India's ability to increase steam coal imports by very large amounts will be restrained by controlled domestic power prices that make it impossible for many State Electricity Boards and even IPPs to afford unlimited amounts of imported coal, even low cost Indonesian LR coals.

7.4.2.3 Shift to low CV coals and impacts on inland transportation costs

Prior to 2007, most of Kalimantan's coal suppliers offered significant quantities of bituminous and high rank sub-bituminous coals (5,400 kcal/kg, GAR and above) under competitively priced long-term contracts. Starting in 2007, Indonesian coal suppliers were no longer willing to offer their traditional price discounts of 15% to 20% against the Japanese Benchmark Price (now called the JPU Reference Price).

Instead, Indonesian sub-bituminous coals, with CVs above 5400 kcal/kg, GAR, are now being priced on an energy-adjusted par with globalCOAL's Newcastle Index for Newcastle coal with a typical CV of 6,000 kcal/kg, NAR (6,300 kcal/kg, GAR). For the past four years, the market has been witnessing the emergence of low CV, high-moisture Indonesian sub-bituminous and lignitic coals with CVs < 4,500 kcal/kg, GAR and TM > 28% as received, as the long-term replacements for Indonesia's traditional medium rank, sub-bituminous and high rank, bituminous coals (Wood Mackenzie, 2013).

This ongoing shift in Indonesia's coal supply is related to the success with which Indonesia's coal suppliers have marketed their supplies of higher rank coals. It is also a function of the coal industry's belief that PLN will eventually succeed in bringing into commercial operation all of the coal-fired power plants built for the 1st Phase of its Fast Track Power Program. Finally, most large coal producers on Kalimantan, operating under 1st Gen CCOWs, have either sold all of their remaining reserves of higher rank coals under multi-year contracts that will last up to the expiration dates of their CCOWs or will shortly reach this point.

As a result of these factors, low CV, high moisture (LR) coals are expected to account for at least 40% of Indonesia's coal production (in raw tonnes) by 2017. Wood Mackenzie (2013), a reputable coal analysis group, goes one step further with its forecast that exports of Indonesian LR sub-bituminous coals, which it chooses to call lignites, "will jump seven times to 520 Mt by 2035, accounting for 80% of Indonesia's thermal [coal] exports." Except for the timing of the transition to low rank coals, this news should not be a surprise. Forty-six percent of Indonesia's remaining 2009 coal reserves were classified as "low CV" coals with CVs < 4500 kcal/kg, GAR (see appendix at the end of this chapter). Moreover, the GOI through PLN has created a large domestic market for these low rank coals by setting LR coal as the design coal for

PLN's 1st Phase Fast Track Power Program. The real issue is whether the export markets can accept this forecasted level of LR coal.

A number of Indonesia's coal producers have attempted, at the demonstration project level, to "upgrade" their LR coals before transporting them to barge ports. Upgrading high-moisture, low CV coals requires coal producers to employ coal drying technologies that are either still at the pre-commercial stage of development or are not cost-effective when compared against consuming the LR coals in their raw form in power plants or factories (Lucarelli, 2013). The demonstration results for promising pre-commercial technologies have, thus far, not been encouraging. Either the costs of upgrading have been too high or the upgraded product has not been stable enough for overland and sea shipping. In most cases, the upgraded product either crumbles and forms excessive fines or reabsorbs moisture during handling and storage (Lucarelli, 2013). Barring an unexpected technological breakthrough in the area of coal upgrading, one can expect that most of Indonesia's LR coal will be sold in raw form through the next decade.

With the shift to low CV, high-moisture coals, one can also reasonably expect significant growth in the total tonnage of coal that will be shipped over each coal supplier's inland transportation route. This growth in tonnage will be required to meet the growing requirements of coal consumers for energy over time as opposed to raw tonnes of coal.

This increase in raw tonnage will also result in increased use of diesel for mining and transport unless comprehensive mine site electrification programs are implemented. And most importantly it will lead to a significant loss of Indonesia's sea transport advantage over Australian coal suppliers. Figure 7.8 shows sea transport cost data (on an energy basis) for both traditional and "new" (LR) coals. As one can see the shift to LR coal will result in Indonesian coal producers losing most of their sea transport cost advantage into the south and north Asian markets.

7.4.2.4 Alternative Indonesian coal futures

The Indonesian coal industry's ability, over the past fourteen years, to navigate around the shoals of Indonesian political and regulatory risk continues to provide many industry observers with confidence that the coal industry of Indonesia will weather the regulatory storm it currently faces and continue to prosper over the next two decades. Geological, locational, and economic factors have until recently made Indonesia an attractive venue for foreign investments in the coal sector. Even in the midst of considerable economic and political uncertainty between 1999

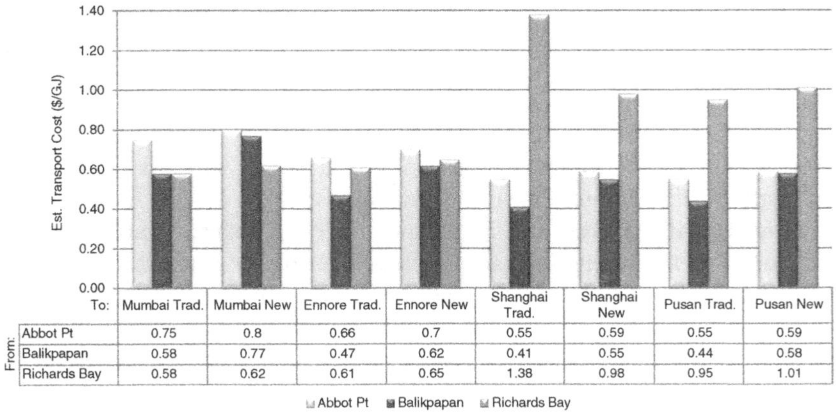

To:	Mumbai Trad.	Mumbai New	Ennore Trad.	Ennore New	Shanghai Trad.	Shanghai New	Pusan Trad.	Pusan New
Abbot Pt	0.75	0.8	0.66	0.7	0.55	0.59	0.55	0.59
Balikpapan	0.58	0.77	0.47	0.62	0.41	0.55	0.44	0.58
Richards Bay	0.58	0.62	0.61	0.65	1.38	0.98	0.95	1.01

Figure 7.8 Comparison of sea transport costs for traditional and "new" (low-rank) Indonesian coals with traditional and new coals from Australia and South Africa (USD per GJ). *Assumed ports of origin*: Abbot Point (Australian coal), Balikpapan (Indonesian coal), Richards Bay (South African coal). *Traditional coals*: Indonesia 5,600 kcal/kg, GAR, Australia and South Africa 6,300 kcal/kg, GAR. *New coals*: Indonesia 4,200 kcal/kg, GAR, Australia and South Africa 5,900 kcal/kg, GAR.

and 2005, Indonesia's coal industry grew at a respectable rate of 14% per year between 2000 and 2009.

However, political events and regulatory overreach since 2000, and especially since passage of the Mining Law of 2009, have created legitimate concerns that the GOI and its partners in local government may take ill-advised actions that will eventually kill the coal mining goose that is laying golden eggs of royalty and tax payments and creating considerable levels of employment for the otherwise depressed outer islands of Kalimantan and Sumatra.

Success in handling the challenges posed by this new regulatory system will largely determine whether Indonesia will continue to play a leading coal supply role in Asia over the next two decades. For the new mining law and its IRs to work in an efficient and equitable manner, the GOI and its partners in local governments will need to refrain from precipitating the types of political events that have occurred over the past decade.

If the GOI changes direction and starts to implement the new mining law in an impartial manner, one that encourages broad foreign and domestic investor participation in the mining sector and allows flexibility in the size of mining areas and the use of foreign mining contractors, one can have a high degree of confidence in the ability of existing coal mining companies and new mining industry entrants to expand Indonesia's coal production to much higher levels of sustainable production. Coal resources and reserves on Kalimantan and Sumatra appear to be more than adequate to meet domestic and export demand for Indonesian coal into the foreseeable future. But the GOI will need to take a vow of non-interference and at times provide enhanced policy support for its coal industry for those potentials to be realized.

However, if the GOI and its partners in local government choose to behave in the manner they have behaved over many of the past fourteen years, the Mining Law of 2009 may end up destabilizing the Indonesian steam coal industry. In its place will be an industry that is a fragmented and disoriented shell of its former self, similar to the hapless state today of Indonesia's oil and gas industries.

These are two very discrete, mutually exclusive paths that the GOI can choose to follow. One path is for the GOI to adopt paranoid, inward-looking coal policies that encourage, if not force, the husbanding of its remaining coal resources for domestic use with the "leftovers" sent to export customers. This inward looking path would be based on the mistaken perception that Indonesia's exploitation of remaining coal resources represents a zero-sum game in which sufficient resources are not available to meet the needs of both domestic and export customers. The siren's call of economic and resource nationalism has been heard by enough observers of the Indonesian coal mining scene that the alarm has been sounded in the international investment community.

One can only hope that the GOI will choose a second, more enlightened path to coal sector development, which is inclusionary and allows foreign and domestic investors to co-exist, guided by fair and consistent rules and regulations. This path would also encourage the increased integration of Indonesia into the broader Asian market and allow Indonesia to play a stronger role as supplier of energy resources to Asia. It is the path that the author of this chapter hopes will be adopted, as it is the path that will best serve the interests of the Indonesian people and the economies of Asia.

Appendix Resources and reserves during the Take-off and Localization Periods

The GOI provided its first publicly available resource and reserve estimates in 1999.[60] The estimates were largely, if not entirely, compiled from data supplied to the GOI by coal miners operating under 1st and 2nd Gen CCOWs. The GOI's approach of relying exclusively on data from coal producers had the weakness of underestimating the amount of

Table 7.11 *Indonesia's coal resources and mineable reserves by type of mining company, 1999 (Mt)*

	Resources			
Company type	Measured	Indicated	Total	Mineable (proven) reserves
PTBA	1,902	4,657	6,559	2,804
Contractors/CCOWs	8,998	22,185	31,183	2,054
KPs and Coops	584	442	1,026	504
Total	**11,484**	**27,284**	**38,768**	**5,362**

Source: Embassy of the United States of America, Jakarta, "Coal Report: Indonesia 2000" October 2000, p. 2. (with Embassy citation of GOI's Directorate of Coal as its primary data source).

coal resources and reserves located in areas outside the concession areas of CCOW holders and also within their concession areas due to lack of exploration activity.

The 1999 resource and reserve estimates, which are presented in Table 7.11, are disaggregated by type of mining authorization – PTBA, CCOWs, and others (KPs and cooperatives). At that time, estimates were limited to (a) measured and indicated resources (38.8 Mt) and (b) mineable reserves (5.4 Mt). No estimates were provided of inferred and hypothetical resources and probable reserves. The main take-away is the overwhelming resource advantage held by CCOWs, which held 80% of measured and indicated resources by 1999. Although PTBA had larger mineable reserves that advantage was soon to be lost as companies operating under a CCOW conducted more exploration within their concession areas.

In 2003, the GOI through its newly established Directorate of Mineral Resources Inventory provided updated resource and reserve estimates (Indonesian Coal Book, 2004/2005, pp. 8–9). The updated estimates did not provide a breakdown by type of coal miner, PTBA, CCOW, KP, or Cooperative, but provided more interesting resource and reserve breakdowns by (i) quality (CV range) and (ii) provincial location. The updated resource estimates were expanded to include inferred and hypothetical resources along with the customary measured and indicated resources. Reserves however, were still limited to Mineable, which it relabeled as Proven Reserves in accordance with standard JORC reserve classification (Tables 7.12 and 7.13). For unknown reasons, the Directorate of Mineral Resources Inventory did not provide an estimate of Probable Reserves. It is believed that the 2003 estimates were also derived from data submitted

Table 7.12 *Government of Indonesia's coal resources and reserves estimates by CV range, 2003 (Mt)*

	Quality/CV range[61]		Resources					
Class	(GAD)	(GAR)	Hypothetical	Inferred	Indicated	Measured	Total	Proven reserves
Low	< 5,100	< 4,500	0	1,512	9581	4,021	15,114	3,452
Medium	5,100– 6,100	4,500–5,800	476	17,463	10,255	4,997	33,191	1,809
High	6,100–7,100	5,800–6,500	57	4,888	696	3,329	8,970	1,662
Very high	> 7,100	> 6,500	0	452	2	119	573	59
TOTAL			533	24,315	20,534	12,466	57,848	6,982

Notes: Original CV basis was GAD, which the author converted for this chapter into a more widely recognized GAR basis. Subsequent Resource and Reserve tables will only show the GAR CV range.
Source: Indonesian Coal Book 2004/2005, p. 8.

Table 7.13 *Government of Indonesia's estimates of coal resources and reserves by province, 2003 (Mt)*

	Resources						
Province	Hypothetical	Inferred	Indicated	Measured	Total	% of total	Proven reserves
Sumatra	34	4,555	20,047	2,755	27,391	47%	2,735
Kalimantan	499	19,494	487	9,690	30,170	52%	4,246
Other	0	266	0	21	287	1%	1
Total	533	24,315	20,534	12,466	57,848	100.0%	6,982

Source: Indonesian Coal Book, 2004/2005, pp. 8–9.

by the mining companies only and did not include independent GOI estimates of coal resources and reserves located in areas not already under contract or license (Indonesian Coal Book, 2004/2005, pp. 8–9).

With respect to coal quality, Indonesia, unlike Australia, which has vast reserves of high-quality bituminous thermal and coking coals with higher heating values that range between 6000 and 7,500 kcal/kg (GAR), has coal resources and reserves that are ranked as mostly medium rank sub-bituminous (4,500 – 5,800 kcal/kg, GAR) and low CV, high moisture coals (< 4,500 kcal/kg, GAR) (see Tables 7.12 and 7.13).[62] Around 84% of resources and 75% of proven reserves fall in the low to medium CV classification.

The GOI's 2003 reserve estimate was 30% higher than its 1999 reserve estimate while Indicated and Measured Resources decreased to 33 billion tonnes (Bt) suggesting that over time, coal producers had reclassified a significant amount of its measured resources into proven reserves. The 50% increase in total resources was entirely due to the GOI decision to start reporting inferred and hypothetical resource categories along with measured and indicated resources.

In 2003, the Japanese Government funded a joint study between the New Energy Development Organization of Japan (NEDO) and the GOI's Geological Agency within the Ministry of Energy and Mineral Resources (MEMR) (the Joint Study). The purpose of the Joint Study was to develop a GIS database system for storing and updating information on Indonesia's coal resources and reserves, compiling existing coal resource and reserve data provided by coal producers to GOI and other information acquired by GOI, training staff within the MEMR's Geological Agency to conduct future resource and reserve assessments, and produce an updated set of coal resource and reserve estimates.[63]

The GOI first announced the resource and reserve estimates resulting from the Joint Study in 2007 but then, in 2009, issued a significant update to the initial resource and reserve estimates (*Indonesian Coal Book, 2008/2009*, pp. 3–4; *Indonesian Coal Book, 2009/2010*, pp. 4–5). The updated GOI estimates, which are presented on Tables 7.14 and 7.15, indicate that, in 2009, Indonesia had coal resources of 105 Bt and reserves of 19 Bt, which represent an 82% increase in resources and a huge 171% increase in reserves when compared to the GOI's 2003 estimates (*Indonesian Coal Book 2004/2005*).[64]

At the time, the increases were not unjustified given the large increases in regional coal prices experienced between 2003 and 2008. However, given the significant decline in Asian coal prices from February 2012, the size of Indonesia's coal reserves and resources may need to be adjusted downward until prices recover.

Table 7.14 *Classification of Indonesia's coal resources and reserves by CV range, 2009 (Mt)*

		Resources					Reserves		
Coal quality	Calorific value (kcal/kg) GAR	Hypothetical	Inferred	Indicated	Measured	Total	Probable	Proven	Total
Low	< 4,500	5,058	6,586	3,721	5,816	21,181	6,705	1,359	8,064
Medium	4,500–5,800	27,772	18,961	11,008	11,995	69,736	5,871	2,718	8,589
High	5,800–6,500	1,695	6,173	1,063	4,040	12,971	829	1,327	2,156
Very high	> 6,500	103	498	12	440	1,052	73	126	199
TOTAL	n/a	34,628	32,218	15,804	22,291	104,940	13,479	5,530	19,008

Source: Indonesian Coal Book, 2010/2011.

Table 7.15 *Government of Indonesia's estimates of coal resources and reserves by province, 2009 (Mt)*

	Resources					Reserves		
Province	Hypothetical	Inferred	Indicated	Measured	Total	Probable	Proven	Total
Sumatra	20,154	13,949	10,634	7,699	52,535	10,645	905	11,550
Kalimantan	14,378	18,051	5,137	14,536	52,102	2,832	4,625	7,458
Other	97	218	33	55	403	0	0	0
TOTAL	34,628	32,218	15,804	22,290	104,940	13,479	5,530	19,008

Source: Indonesian Coal Book, 2010/2011, p. 4.

The increases were a reflection of increased exploration activities that helped to discover additional coal deposits. They are also a reflection of the large increase in Hypothetical and Inferred Resource estimates and the inclusion of Probable Reserves as a second reserve category. Without the increases in Hypothetical and Inferred Resources and the inclusion of Probable Reserves, the Joint Study would have resulted in very small change to the estimate of total coal resources and a decline in total coal reserves.

Another important finding of the GOI's 2009 reassessment of coal resource and reserves is that low and medium CV coals (< 5,800 kcal/kg, GAR) comprise 87% of resources and reserves, which in the case of resources was up slightly from the 84% share of low and medium CV coal resources in 2003 but up significantly from 75% share of 2003 reserves classified as low and medium CVs (Table 7.14). With only 13% of Indonesia's remaining coal resources and reserves classified as having high and very high CVs, it is clear that any growth in Indonesia's coal production and exports will primarily come from the large and growing resources and reserves of low and to a lesser extent, medium CV grade coals. A number of Indonesian coal analysts would argue even more strongly that the growth in Indonesian coal production and exports will come exclusively from its low CV coal resources and reserves. This shift in coal quality also means that the amount of tonnage being mined, trucked, and barged will increase over time if Indonesian coal suppliers wish to maintain existing energy equivalent levels of coal exports over the next decade.[65]

Sumatra and Kalimantan account for almost 100% of Indonesia's 2009 coal resources, with higher CV resources and reserves largely found on Kalimantan (Table 7.14). Only about 0.4% of the total resource base is located on islands other than Kalimantan and Sumatra and this meager position of the other islands is unlikely to change over time.

If one looks only at proven reserves, Kalimantan has a clear advantage, accounting for 83% of total proven reserves with Sumatra accounting for the remaining 17%. However, if probable reserves are included as an indicator of future production potential, Sumatra has the advantage at 61% of total reserves, with Kalimantan accounting for the remaining 39% of reserves (Table 7.8) (*Indonesian Coal Book, 2010/2011*, p. 4).

In summary, although the GOI's estimates of Indonesia's remaining coal resources and reserves are plagued by geological uncertainties, a lack of government transparency about underlying assumptions used in the Joint Study, and the looming threat that the depression in world coal prices could force a downward revision in reserves, they still indicate that Indonesia should be able to maintain its position as a dominant coal exporter for the foreseeable future.[66]

Notes

1. Because of the more extensive sources of data and the better data quality from 2000 onward, the author has focused his discussions about coal industry fundamentals such as levels of production, exports, and domestic sales, on the period starting in 2000 and ending in 2012.
2. Indonesia's coal resources are largely steam coal resources (sometimes referred to in the literature as "thermal coal"), and when the author refers throughout this chapter to the Indonesian coal industry, he is referring to an industry based on the production of steam coal. The country produces only small amounts of coking or metallurgical coals, and those coking type coals are known as semi-soft and PCI coals, which are lower end metallurgical coals.
3. PTBA was previously known as PNTABA (Perusahaan Negeri [State Company] Tambang Arang Bukit Asam), which was established in 1950. In 1981, PNTABA was converted into a limited liability corporation and its name changed to PT Tambang Batubara Bukit Asam, or PTBA for short. (Source: www.ptba.co.id)
4. English-language articles and books about Soeharto's life and presidency tend to use the spelling "Suharto," even though the Indonesian government and media continue to follow the original spelling "Soeharto." He also preferred "Soeharto" when signing his name. Soeharto's full name was Mohammed Soeharto, later expanded to el-Haji Mohammed Soeharto, which designated him as one who had performed the Haj, but like most Javanese, he only used his last name, Soeharto. See Romano (2003), p. ix.
5. In October 1965, army troops led by Major General Soeharto put down a coup attempt by rebellious troops who referred to themselves as the 30 September Movement. The coup attempt resulted in the kidnapping and murder of six generals in the Indonesian army, loyal to then-President Sukarno, an ultra-nationalist who was fervently opposed to foreign investment in the mining and resource sectors of the economy. The blame for the coup attempt and the murder of the six generals was placed on the shoulders of the Communist Party of Indonesia (PKI). The army and Islamic sympathizers went on a rampage and, according to whose figures you choose to believe, killed between 100,000 and 1 million members of the PKI and its supporters over a one-year period. The extermination of the PKI also led to the imprisonment of up to 1.5 million others, many of whom were not released from prison until 1979. The details of the coup and its immediate aftermath were captured in the novel *The Year of Living Dangerously* by Christopher Koch, Peter Weir, and David Williamson, which was later made into a motion picture. Many books over the years have speculated about the perpetrators of the coup, their motives, and the number of Indonesians killed in 1965–66. For a thoroughly researched and balanced account of that period, read Schwarz (2000).
6. During Sukarno's last two years in power (1963–65), the Indonesian economy was on the brink of collapse with inflation running at 1,000% and foreign investors having all but abandoned the country.
7. The Indonesian Constitution of 1945 stated that "Indonesia's natural resources are to be controlled by the State and must be used for the maximum benefit of the Indonesian people." The Mining Law of 1967 (Law

11/1967) interpreted the word "controlled" to mean "owned" and therefore allowed companies to gain rights to mine and sell coal and other minerals but never to own those mineral deposits outright.

8. Specifically, the Foreign Investment Law of 1967 (Law 1/1967) and its subsequent amendment (Law 11/1970) guaranteed that the government of Indonesia:
   (i) will not nationalize a foreign investment or revoke rights granted to a PMA unless appropriate compensation is paid to the owners of that investment;
   (ii) will allow foreign investors to repatriate funds as profits and for payment of expenses including debt and foreign denominated operating expenses; and
   (iii) will allow PMA companies to appoint their management team.
9. Papua Province was up until 2002 known as the province of Irian Jaya.
10. In Bahasa a CCOW is known as a PKP2B (short for Perjanjian Karya Pengusahaan Pertambangan Batubara) and is a special purpose contract for foreign investment companies, which are referred to in Indonesia as Penanaman Modal Asing (PMA).
11. Foreign mining companies initially signed CCOWs with PT Tambang Batubara Bukit Asam (PTBA). In 1997, PTBA and the companies operating under CCOWs agreed to transfer the rights of PTBA to the Ministry of Energy and Mineral Resources.
12. For unknown reasons, the GOI initially referred to these Coal Contracts of Work (CCOWs) as Coal Cooperation Agreements (CCAs). Today they are referred to as CCOWs by some producers and as CCAs by others. For ease of presentation, the author refers to all such bilateral contracts between coal producers and the GOI as CCOWs.
13. Phone conversation with William Sullivan, licensed foreign advocate with Christian Teo & Associates, Jakarta, August 12, 2009.
14. Phone conversation with William Sullivan, licensed foreign advocate with Christian Teo & Associates, Jakarta, August 12, 2009.
15. With the passage of the Mining Law of 2009, any extension offered to a company operating under a CCOW will be based on its CCOW being converted into a new mining license known as an IUP Produksi Operasi on terms and conditions that are consistent with the provisions of the implementing regulations (IR) for the Mining Law of 1999 that are in force at the time that the CCOW expires.
16. A review of the KPs listed in the *Indonesia Coal Book 2008/2009* indicated that some holders of Exploitation KPs have land areas well in excess of 5,000 hectares.
17. It is not clear from the KP regulations if the GCV levels were on an as-received or an air-dried basis.
18. The eight companies listed in Table 7.1 accounted for 70% of production and 73% of exports in 2008. The two companies excluded from Table 7.1 are BHP Kendilo and Allied Indonesia Coal, which discontinued their mining operations in 2007 and 2008, respectively.
19. The round trip distance by sea from Balikpapan, East Kalimantan, to Japan is 760 nautical miles shorter than a sea journey originating at Palembang in East

Sumatra. This shorter travel distance translates into a two-day savings on the round trip sailing time, which would have resulted in a transport cost savings of $0.60 to $0.90/t depending on the charter rates and bunker fuel price adopted for the cost comparison.

20. During the Localization Period (2000–2009), approximately 90% of Indonesia's coal production and exports were sourced from Kalimantan. The predominance of Kalimantan as a source of coal supply has been maintained through 2013. (Source: *Indonesian Coal Books, 2004/2005, 2006/2007,* and *2008/2009*)
21. Reliable data on Indonesia's coal exports and domestic coal sales became available in 1995. Data used to produce Figures 7.1 and 7.2 can be found in the appendix at the end of this chapter. From 2007 onward, the GOI data appear to seriously understate actual production and exports. The unreported tonnes are acknowledged by the GOI, which claims it is the result of both unreported, illegally mined coal, and legally mined coal by KPs (now IUPs), which were not reported as production data to the GOI by local governments responsible for administering the KPs. The unreported production, labeled as "Unreported" on Figures 7.1 and 7.2, reached 69 Mt in 2009. This category of coal production has continued to grow exponentially between 2010 and 2012. Data on actual production and exports were obtained from Kamandanu (2012).
22. All growth rates cited in this chapter, unless otherwise noted, are "compound annual growth rates," or CAGRs, defined as the rate of growth which, if applied annually would convert a known present value into a known future value. It is sometimes described as the geometric mean growth rate on an annualized basis. The equation for calculating the CAGR plus an example is:

    $$CAGR = (FV/PV)^{(1/n)} - 1 = [(200/100)^{(1/5)} - 1] = 14.87\%$$

    Where:
    PV = present value = 100
    FV = future value = 200
    n = no. of time periods = 5
23. Under the terms of their CCOWs, 1st Gen and 2nd Gen CCOWs were required to offer and sell to domestic entities a minimum of 51% ownership in their companies by the tenth year after they started commercial coal production. Domestic entities were defined to include the central government, local governments, Indonesian citizens, and domestic companies. For the top four coal producers (KPC, Adaro, Arutmin, and Kideco), which operated under 1st Gen CCOWs, those sell-by dates fell between 1999 and 2005. Due to the Asian Financial Crisis, the GOI agreed to delay the first sell-down date until 2002 for Arutmin but insisted that KPC, Adaro, and Kideco complete their sell-downs by 2005.
24. Australia coal export data were obtained from Coal Services Pty Ltd (www.coalservices.com). Interesting side note: If one were to compare the coal export levels of both countries on an energy adjusted basis, it is likely that Indonesia would not have surpassed Australia as the world's leading steam coal exporter until 2007. It should also be noted that Australia had total

"black coal" exports of 240 Mt in 2008, consisting of 115 Mt of steam coal and 125 Mt of coking coals. Australia is in the process of completing a major program to expand the capacities of its rail networks and coal handling ports in Queensland and NSW, which may eventually allow Australia to support total coal exports of 500 Mt and steam coal exports of at least 250 Mt between 2015 and 2017. Port and rail infrastructure capacity is forecast to increase to 600 Mt by 2020, which will support a minimum level of steam coal exports of around 300 Mtpa (see Chapter 6).

25. Company specific data on production were obtained from Energy Publishing, "Indonesia Coal and Power Report," May 2010, p. 9. The author then identified each company according to its mining authorization and then summed the production amounts by mining authorization to obtain totals for each mining authorization. Total production for 2009, which includes "unreported" amounts, was taken from Kamandanu (2012). Some of the unreported amounts may also be caused by omissions by Energy Publishing when reporting production of CCOWs.
26. Four of these eight 2nd Gen CCOW companies had established coal mining operations that were, through 2009, considered significant contributors to Indonesia's coal production and exports. They were PT Gunung Bayan Pratama, PT Jorong Barutama Greston, PT Trubaindo, and PT Mandiri Inti Perkasa. However, two of the four top 2nd Gen producers (Gunung Bayan and Jorong) are very likely to cease production at their mines sometime between 2015 and 2020 due to insufficient mineable reserves.
27. Eleven of the thirty-five companies still at the exploration-feasibility stages of development are clustered in the Muarateweh area of Central Kalimantan, which is known to contain high-quality coking coals. Seven of those eleven companies (Juloi Coal, Kalteng Coal, Lahai Coal, Maruwai Coal, PT Pari Coal, PT Ratah Coal, and Sumber Barito Coal) are owned jointly by BHP Billiton and Adaro Coal Indonesia. These 3rd Gen CCOW companies are currently attempting to develop the high-quality coking coal deposits found within their concession areas. However, infrastructure constraints, such as draft limits on the northern branch of the Barito River and the unwillingness of the East Kalimantan provincial government to support a project to construct a railway line from Muarateweh to Balikpapan, have constrained the development of their high-quality coking coal resources. However, if successfully developed, the BHP-Adaro consortium and the other 3rd Gen companies located in Central Kalimantan may change Indonesia from being a producer of almost 100% steam coal to a balanced supplier of both steam and high-quality coking coals.
28. The MEMR in late 2009 speculated that the KP contribution to Indonesia's coal production was much greater than official production statistics indicate – perhaps as much as 20 Mtpa higher. Local governments, which are responsible for collecting and reporting KP production data to the MEMR, may not have collected and reported the monthly production data from each KP holder, leading to the very low KP production figures (Source: Interview with Agus Yulianto of MEMR). One will never know if this was the case as all KPs were converted into IUPs as of July 2010.

29. Author's survey of these four producers in 2009.
30. KPC experienced a significant but temporary decrease in production and exports between 2007 and 2008 partly due to a political event that forced the closure of its mine for forty-five days (discussed in Section 7.4) and partly due to weather-related supply disruptions.
31. This information was obtained from various sources within PLN and private companies active in Indonesia's power sector who wish to remain anonymous.
32. This information was obtained from various sources within PLN and private companies active in Indonesia's power sector who wish to remain anonymous.
33. With the decontrol of diesel prices and the depletion of reserves with low stripping ratios, some industry experts felt that by early 2008, Indonesia's costs of production were dangerously close to those of Australian coal producers. However, due to other favorable mining conditions, the reduction in oil prices from their 2008 peak and the savings in sea transport costs when shipping coals from Kalimantan to markets in Asia, Indonesian coal producers continue to maintain their cost advantage over Australian coal producers.
34. After the extraordinary jumps in the price of diesel in 2008 and its continued high level through July 2011, most large Indonesian coal suppliers are now considering the construction of on-site coal-fired power plants with low rank coal as the design coal. The electricity produced from these mine mouth power plants will be used to power "in-pit" crushers and conveyors and electric overland conveyors for transporting coal from the mine site to the barge ports, instead of using trucks and shovels running on diesel.
35. The KPC conveyor was powered by electricity from a 2 × 5 MW coal-fired power plant located at the Tanjung Bara Coal Terminal.
36. Transshippers load coal from barges into hoppers located on the transshipper vessel. The coal is then transferred to the sea vessel via a single conveyor. Transshippers have much faster loading rates than floating cranes, cost less per tonne of handling capacity, and are less expensive to operate.
37. The only drawbacks to these mobile loaders are their inability to operate continuously between December and March when storms or rough seas occur and the slower loading rates for floating cranes, which in 2009 had loading rates of around 8,000 tonnes per day (tpd). Fixed land ports are able to load vessels at rates ranging from 35,000 to 45,000 tpd. However, floating loading facilities are now being used, which have loading rates of up to 30,000 tpd. Over time, as FLFs replace the smaller floating cranes, the loading speed advantage of fixed ports will disappear.
38. It is not clear whether the 2nd Gen CCOW program, which lasted for one year only (1994), was closed to foreign investors (PMA companies). However, not a single 2nd Gen CCOW was awarded to a foreign company. Either foreign investors were explicitly excluded or made to feel unwelcome by the 2nd Gen CCOW terms. A number of foreign companies, such as Banpu of Thailand, eventually acquired the shares of companies holding 2nd Gen CCOWs.

39. During the Soeharto era, the central government exercised complete regulatory authority over the mineral extraction industry and collected all taxes and royalties. All this changed in 1999, during the short interim presidency of Jusuf Habibie, who signed into law two acts: the Regional Autonomy Act of 1999 and the Forestry Act of 1999. The Regional Autonomy Act gave provincial governments an equal share of both the decision-making authority and the taxes and royalties collected from coal mining companies. The Forestry Act of 1999 gave the Ministry of Forestry the right to close down mining operations if they were "encroaching" on protected forest areas that were only defined after the Forestry Act of 1999 was passed. As a result of these new central government laws and ad hoc actions by provincial and regency governments between 1999 and 2004, the US Embassy issued a report in 2005 that it posted to the embassy website lambasting the GOI policies toward the coal-mining sector.
40. Discussion with MEMR official who wishes to remain anonymous, Jakarta, April 2010.
41. The official name of the Indonesian House of Representatives is "Dewan Perwakatan Rakyat" (DPR), which means People's Representative Council.
42. The holders of 1st Gen CCOWs have "special law" protections against the Forestry Law 41/1999. It is believed that 2nd Gen and 3rd Gen CCOWs also have special law protections against Forestry Law 41/1999 but a legal review of the 2nd Gen and 3rd Gen CCOWs is needed before a definitive statement can be made.
43. According to Ministry of Finance interpretations of Regulation 144/2000, sales of unprocessed coal were, as of the year 2000, a zero-rated VAT commodity, and all coal suppliers, including companies operating under CCOWs, could no longer offset their input VAT for imported and domestically procured materials and services against the output VAT they had previously charged to customers on their coals.
44. The list of significant Implementing Regulations for the New Mining Law up to December 2013 are:
    (a) Government Regulations ("GR")
        1. GR No. 22 of 2010 dated February 1, 2010, on Mining Area Determination;
        2. GR No. 23 of 2010 dated February 1, 2010, on Mineral and Coal Mining Enterprise Activities;
        3. GR No. 55 of 2010 dated July 5, 2010, on Mining Direction & Supervision; and
        4. GR No. 78 of 2010 dated December 20, 2010, on Reclamation & Post Mining Activities.
        5. GR No. 24 of 2012 dated February 21, 2012.
    (b) Minister of Energy and Mineral Resources ("MEMR") Regulations
        1. MEMR Regulation No. 18 of 2009 dated August 19, 2009, on Investment Changes Pursuant To Contracts of Work and Coal Contracts of Work;
        2. MEMR Regulation No. 28 of 2009 dated September 30, 2009, on Mining Services for Minerals and Coal;

3. MEMR Regulation No. 34 of 2009 dated December 31, 2009, on Prioritization of Coal and Mineral Supply for Domestic Interest (Domestic Market Obligation); and
4. MEMR Regulation No. 17 of 2010 dated September 23, 2010, on Procedures for Coal and Mineral Benchmark Price Determination.
5. MEMR Regulation No. 12 of 2011 dated August 11, 2011, on Guidelines on Determination of Mining Business Area and Coal and Mineral Mining Territory Information System.
6. MEMR Regulation No. 7 of 2012.
7. MEMR Regulation No. 11 of 2012 dated May 16, 2012
8. MEMR Regulation No. 20 of 2013 dated August 6, 2012
9. MEMR Regulation No. 27 of 2013 dated September 13, 2013, on Procedures for the Divestment and Share Pricing and Changes to Capital Investment in Mineral and Coal Processing Businesses

(c) MEMR Decree No. 2360 K/30/MEM/2010 dated August 31, 2010, on Determination of Minimum Domestic Coal Sales Requirement for Year 2011.

45. Traders, although they do not control the coal areas from which their sales coal is obtained, are still required to hold an IUP Produksi Operasi in order to export their sales coal. Until late February 2011, the MEMR had not issued any of these "trader-only" IUP Produksi Operasi permits, leading to the Ministry of Trade and Industry (MTI) blocking the export of coal from Indonesia. The MTI action reportedly caused a long queue of vessels offshore of Kalimantan and major financial losses to many coal traders in January 2011. According to MEMR, the issue was resolved during the 1st quarter of 2011 and traders are exporting coal again.
46. The first and second drafts of the first set of IRs to the Mining Law of 2009 made no mention of any changes in CCOW terms that may be requested by the MEMR. Then, on June 16, 2009, the MEMR issued a list of ten changes it wanted to negotiate with companies operating under CCOWs. The list included immediate changes and changes that would be made once the CCOW expires. The companies operating under a CCOW have agreed to deal with these issues as a united front through the Indonesian Coal Mining Association. All CCOW companies have reportedly told the MEMR that they have no intention of changing the terms of their CCOWs prior to their expiration.
47. Initially the New Mining Law and GR 23 stated that only 20% needed to be divested by the end of the sixth year of commercial operation. Later under GR 24, the GOI changed the divestiture requirements of foreign investors in the mining sector to read: 20% by the end of the sixth year of commercial operations, 30% by the end of the seventh year, 37% by the end of the eighth year, 44% by the end of the ninth year, and 51% by the end of the tenth year of commercial operations. In addition, GR 24 specifies less favorable price terms for foreign investors when they engage in a sell-down. See Sullivan (2012).
48. The HPB for each coal brand is calculated each month in two steps. First, the MEMR calculates the simple average of four independent index prices:

(a) the Barlow Jonker Index (now called the NEX index), which reports spot prices for Newcastle grade coal;
(b) the globalCOAL Newcastle Index (gcNEWC), which is also a Newcastle linked spot price index;
(c) the Platts Indonesia Coal Index for spot sales of coal with a CV of 5,900 kcal/kg; and
(d) the ICI 6500, which is produced by Argus Media in partnership with an Indonesian organization called Coalindo, for spot sales of coal with a CV of 6,500 kcal/kg.

After establishing this "benchmark coal price," which MEMR refers to as the Harga Batubara Acuan (HBA), the MEMR calculates the HPB for each coal brand by applying one of three price equations, which calculate the HPB as a function of the HBA and the typical CV, moisture, sulfur, and ash of each coal brand. The MEMR methodology links specific coal brands to one of the three price estimation equations according to the CV and TM of each coal brand. See Lucarelli (2012) for a detailed review of the deficiencies of the MEMR coal price methodology.

49. The equations and resulting coal price calculations are contained on the MEMR/ESDM website (www.esdm.go.id). One must select "Mineral Batubara and Panas Bumi" from the top of the first page and then "Informasi Harga Batubara (download PDF)" from the next page and then either know a fair bit of Bahasa Indonesia to figure out what has been written or use Google translate to convert the Bahasa into English.
50. However, some large coal producers, such as Adaro, Kideco, and Bumi, have interpreted the Reference Price regulation as requiring them to price their coals at the relevant HPB. They claim that they are requiring all customers to accept the HPBs for their coals as the commercial selling price in their Coal Sales Agreements. Moreover, the MEMR has apparently pursued "behind the scenes" discussions with a number of large coal producers encouraging them to use the HPBs for their specific coal brands as the selling prices in their commercial coal sales contracts, even export sales contracts.
51. See Sullivan (2013), pp. 100–107; Freehills (2012); Baker McKenzie International (2012).
52. Details, such as what constitutes an investment cost and what isn't an investment cost, have not been spelled out in MEMR 27/2013. Other details worth noting: (a) a listing of a PMA company on the Indonesian Stock Exchange will no longer satisfy the IUPs divestiture requirement; (b) once the shares are transferred to an Indonesian government agency or to an Indonesian company they cannot be diluted by future share offerings or encumbered by debt or through assignment. See Sullivan (2013), pp. 100–107, for a thorough legal review of MEMR 27/2013.
53. See Sullivan (2013/2014). According to Bill Sullivan, Articles 102 and 170 of the Mining Law, "although very poorly worded, are generally understood as meaning there is an obligation to commence domestic processing and refining, including smelting, within 5 years or by January 2014." He also notes that GR23.2010, because it has been issued as a government regulation, has the same status as a law.

54. Author's review of a 1st Gen CCOW whose holder wishes to remain anonymous.
55. See Sullivan (2013/2014). The Indonesian Parliament is oftentimes referred to by its abbreviated name, DPR, which is short for Dewan Perwakilan Rakyat or People's Representative Council.
56. Recent reports in the public media indicate that both Newmont and Freeport have agreed to amend their COW but the actual terms of those agreements are not publicly known as of July 2014. See Sullivan (2014) and Russel (2014).
57. Because Indonesia's coal mining companies were still unproven in export markets as reliable and efficient suppliers, during the take-off period they offered their customers significant FOB price discounts, equal to 15% to 20% of the Japanese Benchmark Price, the prevailing price benchmark for all export coals in Asia at that time. The large price discounts also aided the exceptional rates of growth achieved by Indonesia's coal industry during the take-off period.
58. When assessing the trends in coal production, exports, and domestic sales during the New Mining Law period, the author, by necessity, needed to rely on highly aggregated coal statistics, which appear to be accurate at the level of total production and exports but not adequate for purposes of presenting the production and export levels of companies operating under different mining authorizations. Once again, the database weakness appears to be caused by illegal mining, MEMR omissions, and incomplete reporting of production and sales by regency and provincial governments.
59. In 2010, PLN and the GOI expected domestic sales of coal to reach 35% of total production by 2015, but that forecast was on the assumption of much lower export totals than Indonesia's coal industry achieved in 2012. It was also based on the assumption that PLN would have all of its 1st Phase Fast Track coal-fired power plants in operation. But delays in commissioning and continued high production growth rates have resulted in the domestic share of total production reducing to 18% in 2012. It is unlikely that domestic sales of coal will exceed 25% of total production by 2020 unless the GOI forces the coal industry to cut back on its level of production.
60. See a report published by the Embassy of the United States of America, Jakarta, entitled "Coal Report Indonesia 2000" (October 2000, p. 2).
61. The *Indonesian Coal Book 2004/2005* identifies the reporting basis for its CV ranges as "adb," which more accurately should be stated as gross, air-dried (GAD). The author has assumed that the CV ranges for resources and reserves as provided in the *Indonesian Coal Book 2004/2005 and 2010/2011* have been reported on a gross, air-dried (GAD) basis. To convert these assumed GAD units to gross as received (GAR) units, the author first identified representative coals listed in the *Indonesian Coal Book* with GAD CVs close to the values shown in the calorific value column of Tables 5 and 7 and compiled data on total moisture (TM) and inherent moisture (IM) for those coals. He then converted the GAD values shown under the calorific value column to GAR values using the following equation:

$$CV(GAR) = CV(GAD) \times [(100 - TM)/(100 - IM)].$$

The quantities reported remain air-dried measures but the CV ranges now represent units easily recognized by industry analysts and energy researchers.

62. The resource and reserve data reported in this Indonesian Coal Book 2010/2011 are segmented by calorific value (CV) range. However, the publisher did not clarify as a footnote to its tables whether the reporting basis for these CV ranges was "gross as received" (GAR) or "gross air-dried" (GAD). Based on a footnote to similar data tables presented in the Indonesian Coal Book for 2004/2005, the author has assumed for this chapter that the CV ranges for resources and reserves are reported on a GAD basis. The author converted these GAD ranges into GAR range values by first identifying representative coals in the *Indonesian Coal Book* that have GAD calorific values that fall within the CV ranges, compiling data on total moisture (TM) and inherent moisture (IM) for those coals and then converting the GAD CVs to GAR CVs using the following industry-accepted ASTM basis conversion equation:

$$CV(gar) = CV(gad) \times [(100 - TM)/(100 - IM)].$$

The author has not formed an opinion on whether tonnages were also reported on an air-dried basis or whether the data have been compiled according to the 2004 JORC or other standard.

63. The author has made numerous requests to MEMR for a copy of the Joint Study but for unknown reasons was refused access to one. The only public document that the author could obtain when completing research for this chapter was a NEDO presentation issued in 2007 and titled: "Coal Resource and Reserve Evaluation System: The Joint Study on Evaluation of Coal Resources and Reserves in Indonesia (Project Outline), February 7, 2007." The Joint Study is titled: "Joint NEDO-MEMR Coal Resources and Reserves Study, 2007/08."

64. More recent GOI estimates of Indonesia's coal resource and reserves indicate even higher totals. For example, in 2010, the MEMR published a map showing total coal resources of 120 Bt and total reserves of 20.4 Bt but these revised estimates were not disaggregated by CV range or province. Dr. R. Sukhyar, the head of the Geological Agency of ESDM, in a recent presentation at Coaltrans Asia estimated Indonesia's coal resources to be 119.4 Bt and reserves at 31.8 Bt (see Sukhyar, 2013). The level of confusion created by the competing estimates of Indonesia's remaining coal resources and reserves, with some of these competing estimates coming from the same agency, is astonishing if not breathtaking. Therefore, one should not place too much importance on the accuracy of Indonesia's resource and reserve estimates but one can have some confidence that the estimates of measured and indicated resources and proven reserves are reasonable ballpark figures based on their consistency with earlier estimates of resources and reserves.

65. Given that Indonesia's coal mining and transport operations are almost 100% reliant on diesel (trucks, barges, generator sets and pumps), the amount of diesel consumption in the coal mining sector is expected to climb to record levels unless electricity and alternative fuels, such as CNG, are used as substitutes for diesel.

66. As already mentioned in note 59, the Joint NEDO-MEMR Study is not publicly available and therefore many of the underlying assumptions for the public resource and reserve estimates such as reporting basis are not known. It is an uncertainty worth noting as it can lead to mistaken conclusions about the ability of Indonesian coal suppliers to meet future demand forecasts and can also lead to inappropriate government policies and regulations for the coal industry.

References

(Personal communications and other sources are described in notes.)

Baker McKenzie International (2012, March). "Client Alert: Game-changing amendments to Mining Regulations issued!" Available at www.bakermckenzie.com.

Barlow Jonker (2006). "Coal Supply Series Indonesia, 2005" (15th edition).

Baruya, Paul (2009, June). "Prospects for coal and clean technologies in Indonesia." IEA Clean Coal Centre, Report # CCC/148.

BIORES (2014, 14 January). "Indonesia Ushers in Mineral Export Ban." Available at ictsd.org.

Bureau of Resources and Energy Economics (2013). "Resources and Energy Statistics, 2013."

Coal Services Pty Ltd (2006). "2006 Australian Black Coal Statistics" (16th edition). Available at www.coalservices.com.

Energy Publishing (multiple years/months). "Australian Coal Report." Monthly publication available from www.coalportal.com.

Energy Publishing (multiple years/months). "Indonesia Coal and Power Report." Monthly publication available from www.coalportal.com.

Fraser Institute (2014). "Annual Survey of Mining Companies 2013." Available at http://www.fraserinstitute.org/uploadedFiles/fraser-ca/Content/research-news/research/publications/mining-survey-2013.pdf (accessed on February 18, 2015).

Fraser Institute (2013). "Annual Survey of Mining Companies 2012/2013." Available at http://www.fraserinstitute.org/research-news/display.aspx?id=19401 (accessed October 30, 2014).

Freehills (2012, 7 March). "New Indonesian mining regulations impact foreign investors." Available at www.freehills.com.

Friederich, Michael, and T. M. van Leeuwen (2002). "Coal Exploration in Indonesia."

Government of Indonesia (2009). "Ministry of Trade and Industry Regulation 01/M-DAG/PER/1/2009 as revised by the Ministerial Regulation 10/M-DAG/PER/3/2009 ("LC Regulation")."

Government of Indonesia, Ministry of Energy and Mineral Resources (2007/2008). "Joint NEDO-MEMR Coal Resources and Reserves Study, 2007/08."

Indonesian Coal Mining Association (various years). *Indonesian Coal Book.* Available at www.petromindo.com.

Jensen, Fergus and Wilda Asmarini (2014). "Indonesia mineral export ban uncertainty starts to bite." *Reuters*, January 7, 2014.

Kamandanu, Bob (2012). "Indonesian Coal Outlook." Paper presented at the Asia Pacific Mining Conference Indonesia 2012, October 10–11, 2012, Jakarta, Indonesia.

Lucarelli, Bart (2013). "Coal Upgrading Plants Mine vs. Power Plant Siting?" Paper presented at the IBC Coal Upgrading and Conversion Conference, April 23–26, 2013, Singapore.

Lucarelli, Bart (2012). "Review of Indonesia's Coal Pricing Regulation (Permen 17/2010)." Paper presented at IBC Coal Markets Conference, February 21–24, 2012, Singapore.

McLeod, Ross (2000). "Government-Business Relations in Soeharto's Indonesia." In Peter Drysdale, ed., *Reform and Recovery in East Asia*. London: Routledge Press.

Morse, Richard K. (2010). "The Chinese Factor: Short- and Long-Term Drivers of China's Coal Imports." Paper presented at Coaltrans Asia, May 30–June 2, 2010, Bali, Indonesia.

Pamudji, Nur (2010). "The Role of Coal in Indonesia's Future Domestic Energy Mix." Paper presented at Coaltrans Asia, May 30–June 2, 2010, Bali, Indonesia.

Parker, Michael (2013). "Chinese Coal Miners ... No Diamonds, Plenty of Rough." Paper presented at 2nd Coaltrans Asia Coal Trading Forum, February 20, 2013.

PricewaterhouseCoopers (2010). "Energy, Utilities & Mining: New Mining Regulations." NewsFlash* Special Edition, March 2010, #34.

PricewaterhouseCoopers Indonesia (2010). "Mining in Indonesia: Investment and Taxation Guide." May 2010.

Romano, Angela Rose (2003). *Politics and the Press in Indonesia*. New York: RoutledgeCurzon.

Russel, Clyde (2014). "Indonesia Plays Winning Hand with Global Miners." *Reuters*, September 1, 2014.

Schwarz, Adam (2000). *A Nation in Waiting: Indonesia's Search for Stability*. Boulder: Westview Press.

Sukhyar, R. (2013). "Status of Coal Resources and Reserves in Indonesia." Paper presented at Coaltrans Asia, June 2–5, 2013, Bali, Indonesia.

Sullivan, William (2014). "COW/CCOW Renegotiations and International Arbitration – A Tale of 2 (Very Different) Cities." *Coal Asia*, August–September 2014.

Sullivan, William (2013/2014). "Indonesia's Mining Regulatory Regime – Is the Export Ban Really Going to be Enforced in January 2014 After All." *Coal Asia*, December 2013–January 2014.

Sullivan, William (2013). "Indonesia's Mining Regulatory Regime – Foreign Investors Experience a King Canute Moment." *Coal Asia*, November 23–December 15, 2013, pp. 100–107.

Sullivan, William (2012). "Major Changes to Indonesia's Mining Regulatory Regime Become a Reality – Mostly Bad News for Foreign Investors and for Indonesia's International Reputation." *Coal Asia*, March–April 2012.

Sullivan, William (2011). "The Churchill Mining Debacle: A Wake Up Call for Mine Owners and Potential Investors." *Coal Asia*, June 1, 2011.

Sullivan, William (2010). "Summary and Analysis of Key Articles of the Implementing Regulations for Indonesia's 2008–2009 Mining Law." Paper available from Christian Teo & Associates Law Offices, Jakarta, or online at http://news.mitraismining.com/Mining-Regulations-Documents/10WAS028-01-Office-Implementing-Regulations-re-Mining-Area-Determination.pdf.
UBS Investment Research (2009). "Indonesia Market Strategy." July 21, 2009.
Wood Mackenzie (2013). "Thermal Trade Coal Market Report." June 2013.

Part III

Understanding the international coal trade

8 US coal to Asia

Examining the role of transportation constraints in energy markets

Mark C. Thurber

8.1 Introduction

In the face of declining coal consumption in the United States, producers are seeking to export coal from the Powder River Basin (PRB) in Montana and Wyoming to Asia. Economic modeling in Chapter 10 of this volume suggests that PRB export volumes could reach several hundred million tonnes per year if new coal port capacity is built in the Pacific Northwest, but at present a lack of port capacity limits exports. Modest amounts of PRB coal already reach Asia via capacity-constrained ports in British Columbia, but proposals to develop new coal terminals in Oregon and Washington to handle larger export volumes have run into fierce opposition on environmental grounds despite the local jobs and tax revenue they could bring.[1] At the time of writing, low global coal prices were an additional factor dimming the near-term prospects for coal port investment in Oregon and Washington.[2] Of the six ports in these two states that were originally proposed, three had been shelved at the time of writing (Learn 2013) and a fourth had recently been denied an export permit by the state of Oregon, although this did not definitively foreclose the possibility that it would go forward (Davis 2014).[3]

This chapter uses the particular case of PRB coal exports from the Pacific Northwest – a situation that is evolving rapidly and may look different in a few years – as a vehicle to examine larger issues around how transportation constraints shape energy markets and their environmental impacts. Much of the distinct character of global coal, natural gas,

An earlier version of this piece was developed as a conference paper for the 2014 Pacific Energy Forum in Seattle, Washington, hosted by the National Bureau of Asian Research (NBR). The author thanks Meredith Miller and Clara Richardson-Barlow at NBR for providing the opportunity to do this work and for their feedback on earlier drafts. Bart Lucarelli provided extremely valuable insights into how the Newcastle coal price affects the economic prospects for shipping PRB coal to Asia. The author also appreciates helpful suggestions from Frank Wolak, Michael Miller, and Richard Morse. Any errors are the author's alone.

and oil markets is related to the nature of the transport infrastructure needed for these respective commodities, and how and to what extent it is constrained. Constraints may arise (and persist) when the needs of the market evolve more rapidly than political and stakeholder processes allow new transport infrastructure to be sited and built.

Certain questions characteristically come up in policy debates around new energy export infrastructure – whether it be the Keystone XL oil pipeline, LNG export terminals, or Pacific Northwest coal ports. First, will energy resources find another way to market if a given transportation infrastructure project does not go forward? Second, will removing a constraint to energy exports harm consumers in the exporting country, and if so how should consumer interests be weighed against producer interests? Third, does the removal of a transportation constraint have a net positive impact over the longer term? Given the firmly established economic benefits of free trade, it might seem that the answer to the latter question should always be yes. However, if markets do not price in negative externalities such as climate change, there are scenarios in which allowing more of an energy commodity to reach faraway markets could be a net negative due to its environmental impact.

The global environmental implications of eliminating a major barrier to the transportation of an energy commodity can be surprisingly difficult to assess. Coal emits more carbon dioxide (CO_2) per unit energy produced than any other major fuel, so at first glance it would seem obvious that hindering transport of PRB coal to countries with robust demand like China would be a win for climate change mitigation. However, on closer examination it becomes apparent that the net impact of Pacific Northwest coal exports on greenhouse gas emissions depends on a number of factors. These include the nature of the demand for coal and alternatives (both in the United States and in major importing countries like China, India, Japan, Korea, and Germany), whether and how decisions to build new power plants and associated infrastructure might be influenced by the availability of PRB coal resources, and whether it is even feasible to block exports of PRB coal.

This chapter provides a general review of how transportation constraints affect energy markets while at the same time offering a framework for thinking more specifically about the greenhouse gas implications of building new coal ports on the West Coast. It is structured as follows. Section 8.2 explains how transportation constraints have shaped the respective markets for coal, natural gas, and crude oil. Section 8.3 explores how transportation constraints arise and why they sometimes persist. Section 8.4 considers the conditions under which an economically valuable resource can be blocked from reaching a source of demand by policy and regulatory barriers. Section 8.5 identifies the factors determining the effect of PRB exports on global greenhouse gas emissions in the near term, and Section 8.6 considers harder-to-quantify effects that PRB exports might have on global greenhouse gas emissions over the longer term. Finally, Section 8.7 concludes with some broader thoughts about the factors that will determine if new coal ports are built in the US Pacific Northwest and the conditions under which exporting coal would be compatible with climate change goals.

8.2 The role of transportation constraints in shaping markets

An integrated energy market is one in which there are no significant transportation constraints, so that demand can be served with the lowest-cost supply sources available anywhere (taking into account the cost of transportation). By contrast, wherever transportation infrastructure is insufficient – or too costly – to bring supply from one region to demand in another, a larger market breaks into smaller ones with prices that may not track each other.[4] Key energy transportation links include electricity transmission lines, oil and natural gas pipelines, oil tankers, natural gas liquefaction and regasification plants, liquefied natural gas (LNG) tankers, railroads (critical for coal but also sometimes used for carrying oil), and barges and ocean bulk carriers for coal.

The markets for oil, natural gas, and coal all have distinct characteristics that are shaped by their transportation linkages. By the 1970s, transportation infrastructure for moving crude oil around the world and refining it was well developed (Jacoby 1974). As a result, oil in recent decades has behaved very much like an integrated global market, with price indices at different locations that differ only slightly as a function of crude grade and transportation cost (see Figure 8.1). From the beginning of 2011 through the middle of 2013, however, the surge of production from the Bakken shale in the United States and the oil sands in Canada created a disequilibrium in which more light crude was available in Cushing, Oklahoma – where the West Texas Intermediate (WTI) benchmark is priced – than could be shipped by pipeline to the US Gulf Coast.[5] This shows up in Figure 8.1 as a delta between the WTI price and the price of Light Louisiana Sweet (LLS) crude at the US Gulf Coast. The WTI-LLS differential had narrowed by mid-2013 due to additional pipeline capacity that had come online to carry crude from Cushing to the Gulf Coast. In the second half of 2013, however, a gap opened up between the LLS price and the Brent benchmark in the North Sea, in part because of the ban on US crude oil exports – another important transportation constraint.

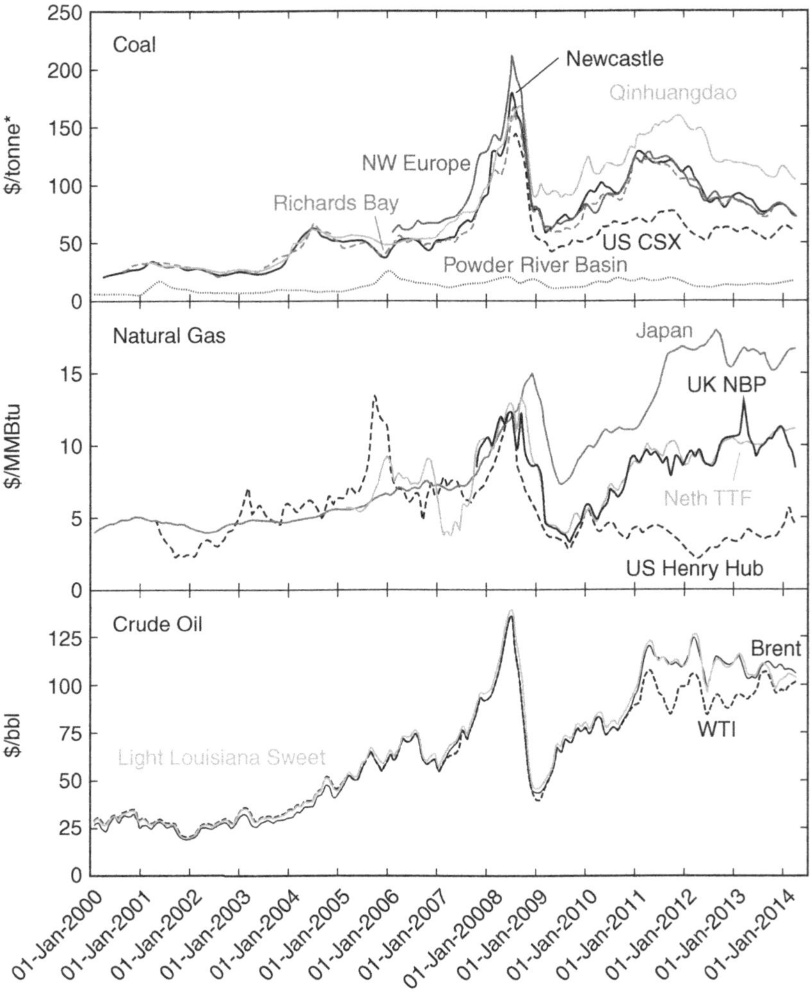

Figure 8.1 Monthly averaged benchmark prices for coal, natural gas, and crude oil. To allow approximate comparison on an energy basis, US$/tonne figures for Qinhuangdao, US CSX, and Powder River Basin coal are linearly adjusted by energy content to match the 6,000 kcal/kg standard used for Richards Bay, Newcastle, and Northwest Europe (Amsterdam/Rotterdam/Antwerp, or ARA) price indices. All prices are spot except where otherwise noted. *Coal*: Qinhuangdao (FOB, 5,800 kcal/kg net as received, adjusted here to 6,000 kcal/kg, from IHS McCloskey); Richards Bay (FOB, 6,000 kcal/kg net as received, from IHS McCloskey); Newcastle (FOB, 6,000 kcal/kg net as received, from IHS McCloskey); Northwest Europe (delivery into ARA hub,

The importance of transportation constraints is illustrated vividly by natural gas markets. Unlike for oil, an integrated global market for natural gas does not yet exist. Infrastructure for natural gas transportation by pipeline or in the form of LNG is extremely capital-intensive and, as a result, has rarely been built in the absence of long-term contracts with customers that guarantee cost recovery over many years (Victor et al. 2006). As late as 2008, it appeared that increased global movement of gas in the form of LNG was beginning to create a global gas market, with prices tracking each other ever more closely (see Figure 8.1). However, the explosion of US shale gas production in the years that followed caused Asian, European, and North American prices to split apart in dramatic fashion. Prices at Henry Hub in Louisiana fell to rock-bottom levels in the face of an oversupply of gas, while Asian prices trended higher and Europe, with its wider array of cost-effective supply options, settled in the middle. (The use of oil-indexed contracts in Europe and Asia also caused prices in these regions to move up with the oil price, unlike Henry Hub prices, which respond more directly to natural gas supply and demand.) Time will tell if planned US LNG export facilities will be able to narrow the price gap with Asia, although a significant transportation differential will persist due to the high cost of LNG, at least until Asia develops its own shale gas.

Transportation infrastructure is just as pivotal to coal markets, but the patterns are distinct from those in either oil or gas. Price indices at major international coal ports (Newcastle in Australia, Richards Bay in South Africa, and the Amsterdam/Rotterdam/Antwerp hub in northwest Europe) track each other, but coal at locations that are not international, liquid trading hubs can see significantly different prices due to rail and port costs and constraints (see Figure 8.1). In China, rail capacity constraints among other factors have helped keep the benchmark price of coal at the port of Qinhuangdao high in comparison with prices at major international

---

Caption for Figure 8.1 (cont.)

6,000 kcal/kg net as received, from IHS McCloskey); Powder River Basin (at mine mouth, 4,900 kcal/kg, adjusted here to 6,000 kcal/kg); US CSX (one month ahead, FOB onto barges on the Ohio or Big Sandy rivers, 6,700 kcal/kg, adjusted here to 6,000 kcal/kg). *Natural gas*: Japan (average imports including freight); UK National Balancing Point; Netherlands TTF (one month ahead, ICE-ENDEX); US Henry Hub. *Crude oil*: Brent; West Texas Intermediate (Cushing, Oklahoma); Light Louisiana Sweet. Data source: Bloomberg.

trading hubs. In the United States, the PRB price at mine mouth is largely isolated from, and far below, international market prices. BNSF and Union Pacific are the only real options for moving coal out of the PRB, and rail rates for shipping across the country are high, as reflected in the difference between the PRB mine-mouth price and the US CSX price for thermal coal delivered onto river barges at the junction of Ohio, Kentucky, and West Virginia.[6] The less rail-intensive option of shipping PRB coal from the US Pacific Northwest to Asia is ruled out at present by the lack of adequate port infrastructure.

8.3 How transportation constraints arise (and persist)

Price differentials across locations create an economic incentive for developing new energy transportation links, but actually mobilizing investment, obtaining regulatory approval, and building capital-intensive infrastructure can be time-consuming. Transportation constraints arise when the location and quantity of supply and demand shift more quickly than new transportation links can be built. The WTI-LLS spread in 2011 and 2012 arose because new infrastructure projects between Cushing and the Gulf Coast, including the reversal of the Seaway pipeline and the completion of the southern segment of the Keystone XL pipeline, were playing catch-up with the rapid growth in crude oil supply from US shale and Canadian oil sands (Borenstein and Kellogg 2012).

Regulatory processes for siting and approving needed infrastructure are characteristically contentious and can on occasion be intractable. The Keystone XL pipeline from Alberta to the US Gulf Coast, whose cross-border segment awaited approval from the US State Department and President Obama at the time of this writing, has become a political flashpoint over the question of whether it will significantly increase greenhouse gas emissions. Enbridge's Northern Gateway pipeline, which would take oil from Alberta to the coast of British Columbia for export, is facing stakeholder challenges from First Nations groups. Siting interstate (and even within-state) electricity transmission lines in the United States is notoriously difficult, which has led to persistent transmission limitations that constrain additions of renewable energy.

Approval processes can be particularly challenging when multiple jurisdictions have authority, many stakeholders are able to pursue challenges, and powerful stakeholders have divergent interests. For example, siting new interstate electricity transmission lines in the United States typically requires the involvement of the public utilities commissions (and sometimes other agencies) in each state as well as federal bodies like the Federal Energy Regulatory Commission and the

Department of Energy. Transmission lines face many stakeholder challenges, especially by environmental groups and local communities that might be affected. Moreover, there are inherent difficulties in aligning incentives because ratepayers in states with more expensive electricity will tend to benefit from the transmission line, whereas those where electricity is cheaper are likely to see rates go up.

The fact that new energy transportation infrastructure creates not only winners but also significant losers is a common political economy problem that can hinder approval. For example, US natural gas producers are pushing for permits to export LNG and access higher international gas prices, while industries that consume natural gas are fighting exports to keep their input prices low (Krauss and Schwartz 2013). Such cases can produce odd political bedfellows, such as the alignment of anti–fossil fuel environmental groups with large chemical producers lobbying against LNG exports (Harder 2014).

Energy transportation infrastructure may also fail to be built simply because it presents too great a commercial risk. This is especially true before new kinds of infrastructure have been shown to be financially viable. Investors only pursued early LNG plants in Sumatra after guarantees from the Japanese government dramatically lowered the financial risk involved (Von der Mehden and Lewis 2006). But even after a method of transporting energy is proved financially viable, investors need to be certain that market conditions will allow for cost recovery on the infrastructure over a long period of time. This requires a stable political and regulatory climate in the location where the energy is produced as well as a reliable source of demand at sufficiently high prices. In an effort to ensure the latter, LNG and gas pipeline developers have traditionally locked in long-term, "take or pay" contracts before proceeding with construction.

Investments in US coal and natural gas export infrastructure could become unprofitable if the price arbitrage opportunity that currently exists between US and foreign markets does not persist over the long term and volumes dry up as a result. LNG regasification plants built in the early 2000s, when the United States foresaw a large supply gap, are a cautionary tale; their business models evaporated due to the shale gas boom (Wolak 2012). At the time of this writing, Newcastle thermal coal prices were below US$80 per tonne, significantly off of January 2011 highs of over US$140 per tonne. While energy consultancy Wood Mackenzie forecasts long-term prices that would support exports of PRB coal, environmental groups have tried to play up the financial risks of port investments (Bernton 2013). In the case of natural gas, the price spread between US and foreign markets could narrow over time if a large number of export projects go forward and if other countries develop their own unconventional gas resources.

8.4 Is it likely that valuable resources would end up stranded?

Notwithstanding current transportation constraints, a commonly expressed view is that it is unrealistic in the long term to expect that abundant US coal resources will not find their way to growing Asian markets in desperate need of energy. A similar line of reasoning was used in the US State Department's environmental impact statement for the Keystone XL pipeline, which argued that the pipeline would have little ultimate effect on oil sands production, and thus greenhouse gas emissions, because the oil would find other routes to consumers even if Keystone XL approval were denied (US Department of State 2014).

Modest quantities of PRB coal are already exported to Asia from coal terminals in British Columbia, and the desire to export more has helped animate planned expansions of these ports (Buchsbaum 2012). There is some resistance in British Columbia to new coal port development (Jang 2014), although as yet it does not seem to have risen to the level of that in Washington and Oregon. At the time of writing, approvals had been granted for an expansion of Vancouver's Westshore terminal as well as a new facility on the Fraser River, both of which would handle PRB coal (Storrow 2014). This suggests that BC ports may shape up as at least a partial substitute for port capacity in Washington or Oregon, but they are unlikely to be as economic and therefore to handle comparable volumes.

There are no significant examples to date of governments intentionally closing off export routes for their own fossil fuels on climate change grounds. The climate-oriented policies that have been put in place thus far in fossil fuel exporting countries have regulated emissions inside the country. Such policies may lead to reductions in greenhouse gas emissions associated with domestic fossil fuel production, but they do not address the usually larger greenhouse gas emissions associated with burning the fuel that is exported. For example, the short-lived carbon tax in Australia (see Chapter 6) would have tended to increase the costs of Australian coal production by increasing the price of the coal-fired electricity used by the industry. Norway, another major fossil fuel exporter, has had a stiff carbon tax on offshore oil and gas operations since 1991. This tax has helped to lower domestic emissions – for example, by encouraging national oil company Statoil to inject CO_2 separated from natural gas from the North Sea's Sleipner field into a saline formation deep under the seafloor rather than release it into the atmosphere. However, the tax does not act directly on oil and gas exports. The increased cost of fossil fuel production associated with carbon pricing in a producing country is likely to have at most a slight effect on the international price of a fossil fuel by making that country's supply more expensive and perhaps discouraging new investment in marginally economic supply projects.

While explicitly climate-related prohibitions on fossil fuel exports appear unprecedented (though theoretically possible), it is common for policy and institutional roadblocks to cause plentiful resources to stay in the ground. The policy and institutional environment does not have to explicitly bar exports; it just has to make them expensive and risky enough that the large investments needed to develop transportation infrastructure do not take place. For example, international sanctions and a generally risky investment climate have dramatically slowed development of Iran's South Pars gas field even as Qatar has developed major LNG exports from its side of the same geological formation (known as North Dome). In many countries, policies favoring domestic use of gas (at reduced prices) over export applications have led to much slower overall resource extraction (Thurber and Chang 2011).

In almost all global energy markets, institutional factors within producing countries play a significant role in shaping the aggregate supply curve and influencing international prices. (The policy and regulatory environment also has an important effect on the demand curve – for example, through energy consumption subsidies.) In the United States, too, it may turn out that non-climate-oriented policies and stakeholder processes will have the determining say as to whether and when PRB coal is exported. Local impacts like increased rail traffic or release of coal dust could prove more salient in building grassroots opposition to new ports than the more diffuse impacts of climate change. The structure of the rail market may help make exports of PRB coal from more distant ports on the East and Gulf Coasts uneconomic, while nearer port alternatives in British Columbia may prove too capacity-constrained to ship massive amounts of PRB coal.

8.5 The short-term impact of removing constraints to exports

Assuming that regulatory barriers are overcome and transportation constraints are alleviated in a particular energy market, the following effects might typically be expected in the near term.[7] Prices for the exported commodity go up in the country where it is produced as some supply is diverted to export. To the extent that demand is downward-sloping, less of the exported quantity is consumed in the exporting country and there may be a shift to substitutes. This can have a positive or negative environmental effect in the exporting country depending on the environmental attributes of the exported commodity relative to its substitutes.

Prices of the energy commodity in the international market tend to go down overall as additional supply becomes available. This decrease in international prices causes countries with more elastic demand to increase consumption of the commodity in question relative to alternatives. Consumption is not significantly affected in countries with inelastic demand. The net environmental impact of removing the transmission constraint is the sum of the impacts in the exporting country and those in all importing countries.

The net environmental impact of removing the transportation constraint thus depends in a complex way on the nature of demand in exporting and importing countries. Even for a product with negative environmental externalities like coal, it should not automatically be assumed that facilitating exports will increase global greenhouse gas emissions. In fact, preliminary modeling work by Frank Wolak and Michael Miller suggests that, in the short term, removal of the coal port constraint in the US Pacific Northwest may not appreciably increase global greenhouse gas emissions and could even reduce them under certain assumptions (Wolak and Miller 2014). Their econometric analysis found downward-sloping short-term demand for coal in the United States and Europe because these regions are able to switch between coal and natural gas in the power sector; it assumed less elastic short-term demand for coal in China and other major Asian consumers because of the present lack of affordable substitutes (and of infrastructure for transporting and consuming them at scale) in these countries.

With the assumptions they used, Wolak and Miller found that PRB coal exports increase the price of coal in the United States, decrease it slightly in the Pacific Basin, and either decrease or increase the price in Europe depending on exact model assumptions.[8] The increased price of coal in the United States leads to reduced coal usage, cutting US greenhouse gas emissions. Europe experiences a reduction in coal consumption if its coal price increases and an increase in consumption if its coal price decreases. The slightly decreased price of coal in the Pacific Basin does not significantly affect coal consumption in Asian markets because of the relatively inelastic demand from these countries. In net, the model predicts that the reduction in US coal consumption due to PRB exports predominates over any increases elsewhere, resulting in a modest global reduction in coal usage and therefore in greenhouse gas emissions.

While this is a plausible short-term outcome, the predicted results depend significantly on model assumptions. If coal demand becomes more elastic in coal-importing countries – for example, due to the further development of their domestic gas resources or gas import capacity – this tends to reduce any favorable effect of US coal exports on global greenhouse gas emissions. Additionally, if US railways reduce their rates to keep shipped volumes high in the event of increased exports, there may be less fuel switching to natural gas in the United States than predicted with the model's current assumption of fixed transport costs.

8.6 The long-term impact of removing constraints to exports

Even if building Pacific Northwest coal ports is neutral or favorable for greenhouse gas emissions in the short term, it is possible that the increased access of PRB coal to world markets could have negative environmental impacts over the longer term. For example, the higher prices fetched by PRB coal as a result of new port capacity might stimulate investment in PRB production that would moderate coal price increases and dampen fuel switching to natural gas in the United States.

It is also possible that access to the enormous reserves of PRB coal would encourage consuming countries in Asia to build more coal power plants by providing confidence that they will have an affordable source of coal over the long term. This was one of the concerns cited by Washington governor Jay Inslee and Oregon governor John Kitzhaber in a letter in which they urged the Obama administration to review the climate impacts of PRB coal exports from the West Coast (Bernton 2013).

How realistic is the concern that planners in Asia might factor PRB coal availability into their decisions to invest in coal plants and coal infrastructure and perhaps forgo investments in cleaner alternatives like natural gas or nuclear power? On the one hand, coal planners in major importing countries like Japan and Korea (which already receive some PRB coal through the Westshore terminal in British Columbia) would certainly welcome more availability of PRB coal as a way of diversifying supply (Kyungro 2014; Davis 2013). On the other hand, major investments in new coal power plants are already being made, and it is unclear if the enhanced availability of PRB coal would significantly increase such investment in the near term. China, with its significant domestic reserves, probably views coal as a secure long-term fuel irrespective of whether PRB supply is available – and in any case seems unlikely to trust its energy security to US supplies. Japanese planners look to coal as an attractive option for meeting supply needs in the face of idled nuclear capacity and expensive natural gas; they already appear to perceive coal as being available at good prices from reliable suppliers (Iwata 2014). Korea also values coal for its cost advantage over gas, although a planned tax on coal use in power generation could somewhat blunt this advantage (Han and Humber 2012).

On the other hand, it is possible that global coal supply will become tighter over the longer term if demand from India continues to grow (Chapter 4) and supply and transportation constraints worsen in key exporting countries (Chapters 5 to 7). Under such a scenario, the availability of PRB coal could conceivably become a salient factor in power plant planning around the world.

Beyond the direct effect of PRB coal availability (or lack thereof) on coal-plant investments, might there be an indirect, psychological effect on world markets if the United States were to deliberately strand a massive energy resource on climate grounds? There is a new strain of climate activism that seeks to build moral opprobrium against the use of fossil fuels. A centerpiece of this movement has been a campaign exhorting universities to divest their endowments from fossil fuel companies (Revkin 2013). Arguably there would be few stronger statements than the United States making a conscious choice not to export coal. However, such a move to deliberately "shut in" resources because of the climate impacts of consuming them abroad would seem to be unprecedented. Even if the United States were to pursue such a policy, its effectiveness as a statement could well be undermined by international skepticism about the true motivations behind it.[9]

8.7 Coal exports and the climate

It is this author's view that the federal government is unlikely to block coal export terminals in the Pacific Northwest on climate change grounds. Rather, their fate is likely to hinge on how much cost, delay, and uncertainty are added by stakeholder processes at various levels of government, and how these interact with the global coal market to influence decisions on whether to go ahead with investment. Environmental groups have increasingly learned that they can be just as effective in blocking a given project – in any energy market – by adding cost, delay, and uncertainty as by having the project expressly prohibited. Indeed, if there is a larger lesson about new transportation infrastructure projects in energy markets, it is that, because of the huge investments and market risks involved in building out a global supply chain, they are exquisitely sensitive to the regulatory and investment environment.

Determining the net environmental impacts of Pacific Northwest coal ports requires a more comprehensive and sophisticated analysis than has been performed to date. Such an analysis will need to incorporate models of demand for coal and its alternatives both in the United States and abroad, such as are being developed by Wolak and Miller (2014). (As mentioned previously, one counterintuitive finding of initial modeling work is that there could be scenarios in which coal exports could actually reduce global greenhouse gas emissions, at least in the short term.) Critically, the analysis will also need to be informed by detailed insight into the planning process for new power plants and transport infrastructure in key consuming countries – and especially into how planners develop and incorporate long-term energy supply and demand forecasts when deciding on their choices of fuel and generation technology.

In theory, technologies that improve the efficiency of coal power plants, reduce local pollution, and capture and store CO_2 have the potential to make coal use more compatible with environmental objectives, thereby removing a potential objection to US exports. The biggest challenge is cost. The most attractive feature of coal is its low cost, and "clean coal" technologies that substantially increase upfront and operating costs risk erasing the economic edge of coal relative to alternatives like natural gas and nuclear. High-efficiency coal plants using "ultra-supercritical" technology are starting to be deployed in some locations, with the attraction that the efficiency improvement not only reduces emissions but also may help pay for itself over the long term through fuel savings. Technologies to reduce local pollution, such as flue gas desulfurization for sulfur dioxide (SO_2) and selective catalytic reduction for nitrogen oxides (NO_x), are well established and can be implemented relatively cost-effectively. (In fact, wider use of the PRB's low-sulfur coal around the world could itself reduce SO_2 emissions problems.) But carbon capture and storage (CCS) remains extremely expensive in power plant applications and yields no significant benefits apart from reducing greenhouse gas emissions.[10] Due to the absence of sufficiently compelling carbon prices or other incentives, there have as yet been no demonstrations at scale of CCS on coal power plants. Until demonstrations and development programs prove out the technology at scale and reduce its costs, and carbon pricing provides a genuine spur to implementation, CCS will not be able to assuage concerns about the climate change impacts of US coal exports.

Notes

1. These proposed ports have aroused concern on various grounds: the effects of increased rail traffic on local communities, possible local environmental hazards from coal dust released in transit or from stockpiles, and the implications for global greenhouse gas emissions of burning the coal in Asia. On the other hand, building coal ports in the Pacific Northwest would generate significant tax revenue and both temporary and permanent jobs.
2. Premium Newcastle coal prices, as estimated by globalCOAL in its industry benchmark price index gcNEWC, have stagnated below US$70 per tonne since July 2014, making it very difficult for shipments of PRB or other US coals to Asian destinations to compete on cost. Between January 2011 and June 2013, the period when there was the greatest US coal industry interest in building new coal terminals on the West Coast, gcNEWC was trading in the range of US$80 per tonne to US$130 per tonne. If today's coal price environment persists for an extended period of time, it seems unlikely that new West Coast coal terminals will be built even if environmental issues are resolved in favor of companies interested in building these new terminals. (Lucarelli [2012] provides insights into the kinds of competitive dynamics that could be observed between PRB coal terminals, if they are built, and coal suppliers in Indonesia, Australia, and South Africa.)
3. The two projects that seemingly remained under active consideration at the time of writing were both in Washington state: the Gateway Pacific Terminal in Cherry Point, in northwest Washington, and the Millennium Bulk Terminals, on the Columbia River. The terminal that was denied an export permit was to be in Boardman, Oregon, also along the Columbia River.
4. In an intermediate case, capacity constraints on the most cost-effective form of transportation may force a shift to more costly modes (e.g., shipment of oil by rail rather than pipeline), increasing the price differential between regions. Price co-integration analysis is one means of statistically characterizing the degree of integration observed in a market. For a coal-market example, see Zaklan et al. (2012).
5. Crude prices in Alberta, which were even more transport-constrained, showed an even deeper discount to international prices.
6. US railways typically price to compete with the next best competitor fuel source at the destination point. See, for example, Busse and Keohane (2007).
7. As will be discussed later, there are a number of complexities that can cause actual effects to differ.
8. The reason that prices in Europe can increase under certain model scenarios is that some non-PRB coal that would have been exported to Europe from the East Coast stays at home as a result of the higher US coal price.
9. While it is hard to see how keeping coal resources in the country would provide a significant boost to US industries, there could certainly be questions, for example, about whether any effort to block LNG exports on environmental grounds was in fact an attempt to help gas-consuming industries in the United States gain an advantage over international competitors.
10. The main exception to this is niche cases where enhanced oil recovery (EOR) applications create demand for CO_2.

References

(All web links verified to be functional as of September 25, 2014.)

Allchin, Joseph (2014, March 4). "Bangladesh's coal delusion." *New York Times*. Available at http://nyti.ms/1n63Ylg.

Bernton, Hal (2013, March 25). "2 governors wade into coal-export controversy." *Seattle Times*. Available at http://seattletimes.com/html/localnews/2020640596_coalexportsinsleexml.html.

Bernton, Hal and Brian M. Rosenthal (2013, September 3). "Demand cools as fight rages over coal-export terminals." *Seattle Times*. Available at http://seattletimes.com/html/localnews/2021747781_coalexportmarketsxml.html.

Borenstein, Severin and Ryan Kellogg (2012). "The incidence of an oil glut: Who benefits from cheap crude oil in the Midwest?" Energy Institute at Haas Working Paper 231. Available at http://ei.haas.berkeley.edu/pdf/working_papers/WP231.pdf.

Buchsbaum, Lee (2012, March 20). "While Canadian terminals expand export capacities, many U.S. producers are still going to be short of space." *Coal Age*. Available at http://www.coalage.com/features/1808-while-canadian-terminals-expand-export-capacities-many-us-producers-are-still-going-to-be-short-of-space.html#.VQNiuI7F-Ag (accessed March 13, 2015).

Busse, Meghan R. and Nathaniel O. Keohane (2007). "Market effects of environmental regulation: Coal, railroads, and the 1990 Clean Air Act." *RAND Journal of Economics* 38(4): 1159–1179.

Davis, Rob (2014, August 18). "Oregon Department of State Lands rejects Ambre Energy coal export permit, dealing major blow." *The Oregonian*. Available at http://www.oregonlive.com/environment/index.ssf/2014/08/oregon_department_of_state_lan_1.html.

Davis, Rob (2013, December 13). "Viability of Oregon, Washington coal export terminal threatened by falling Asian prices." *The Oregonian*. Available at http://www.oregonlive.com/environment/index.ssf/2013/12/viability_of_oregon_washington.html (accessed March 13, 2015).

Dickinson, Tim (2014, February 3). "How the U.S. exports global warming." *Rolling Stone*. Available at http://www.rollingstone.com/politics/news/how-the-u-s-exports-global-warming-20140203.

Han, Sangim and Yuriy Humber (2012, April 13). "Nuclear halt in South Korea seen boosting coal: Energy markets." *Bloomberg*. Available at http://www.bloomberg.com/news/articles/2012-04-13/nuclear-halt-in-south-korea-seen-boosting-coal-energy-markets (accessed March 13, 2015).

Harder, Amy (2014, March 18). "Are natural-gas exports the next Keystone?" *Wall Street Journal*. Available at http://blogs.wsj.com/washwire/2014/03/18/are-natural-gas-exports-next-keystone/.

IEA (2014). "IEA coal information statistics." Available at http://www.oecd-ilibrary.org/energy/data/iea-coal-information-statistics_coal-data-en.

Iwata, Mari (2014, March 27). "Japan's answer to Fukushima: Coal power." *Wall Street Journal*. Available at http://on.wsj.com/1dzFpwD.

Jacoby, Neil H. (1974). *Multinational oil: A study in industry dynamics*. New York: Macmillan.

Jang, Brent (2014, February 9). "B.C. coal export plan faces resistance." *Globe and Mail*. Available at http://www.theglobeandmail.com/report-on-business/industry-news/energy-and-resources/bc-coal-export-plan-faces-resistance/article16770342/.

Krauss, Clifford and Nelson D. Schwartz (2013, August 15). "Foreseeing trouble in exporting natural gas." *New York Times*. Available at http://www.nytimes.com/2013/08/16/business/energy-environment/foreseeing-trouble-in-exporting-natural-gas.html.

Kyungro, Lee (2014, January 13). "Outside view: Asia's growing coal markets." *UPI*. Available at http://www.upi.com/Top_News/Analysis/Outside-View/2014/01/13/Outside-View-Asias-growing-coal-markets/28461389589380/.

Learn, Scott (2013, May 8). "Another Northwest coal export project falls by the wayside; Kinder Morgan drops Oregon terminal plan." *The Oregonian*. Available at http://www.oregonlive.com/environment/index.ssf/2013/05/another_northwest_coal_export.html.

Lucarelli, Bart (2012). "Competitive Threats to Indonesia's Coal Industry." Presentation at IBC Kalimantan Coal Conference, Balikpapan, September 4–5, 2012.

Rai, Varun, David G. Victor, and Mark C. Thurber (2009). "Carbon capture and storage at scale: Lessons from the growth of analogous energy technologies." *Energy Policy* 38: 4089–4098.

Revkin, Andrew C. (2013, October 4). "A closer look at Harvard's choice on fossil fuels." *New York Times*. Available at http://dotearth.blogs.nytimes.com/2013/10/04/a-closer-look-at-harvards-choice-on-fossil-fuel-divestment/.

Spiegel Online (2014, January 7). "Green revolution? German brown coal power output hits new high." Available at http://www.spiegel.de/international/germany/researchers-alarmed-at-rise-in-german-brown-coal-power-output-a-942216-druck.html.

Storrow, Benjamin (2014, September 2). "With Oregon and Vancouver decided, coal export fight moves to Washington." *Casper Star Tribune*. Available at http://trib.com/business/energy/with-oregon-and-vancouver-decided-coal-export-fight-moves-to/article_eed6489e-c617–5dbc-8e77-d45bc65569ca.html.

Thurber, Mark C. and Joseph Chang (2011). "The policy tightrope in gas-producing countries: Stimulating domestic demand without discouraging supply." Advance Summit paper from the 2011 Pacific Energy Summit. Jakarta, Indonesia, February 21–23, 2011.

US Department of State (2014, January). "Final Supplemental Environmental Impact Statement for the Keystone XL Project," Chapter 1. Available at http://keystonepipeline-xl.state.gov/finalseis/.

Victor, David G., Amy M. Jaffe, and Mark H. Hayes, eds. (2006). *Natural gas and geopolitics: From 1970 to 2040*. Cambridge: Cambridge University Press.

Von der Mehden, Fred and Steven W. Lewis (2006). "Liquefied natural gas from Indonesia: the Arun project." In D. Victor, A. Jaffe, and M. Hayes, eds., *Natural gas and geopolitics: From 1970 to 2040*. Cambridge: Cambridge University Press.

Wirth, Harry (2014, March 19). "Recent facts about photovoltaics in Germany." Fraunhofer ISE paper. Available at http://www.ise.fraunhofer.de/en/publications/veroeffentlichungen-pdf-dateien-en/studien-und-konzeptpapiere/recent-facts-about-photovoltaics-in-germany.pdf.

Wolak, Frank A. (2012, November). "Careful what you wish for: The shale gas revolution and natural gas exports." SIEPR Policy Brief. Available at http://siepr.stanford.edu/?q=/system/files/shared/pubs/papers/briefs/PolicyBrief11_2012.pdf.

Wolak, Frank A. and Michael Miller (2014, February 10). "Modeling the world coal market: The impact of China's behavior on world coal consumption." Draft paper.

Yeo, Sophie (2014, March 24). "Netherlands to stop funding overseas coal power plants." RTCC. Available at http://www.rtcc.org/2014/03/24/netherlands-to-stop-funding-overseas-coal-power-plants/.

Zaklan, Alexandar, Astrid Cullmann, Anne Neumann, and Christian Von Hirschhausen (2012). "The Globalization of Steam Coal Markets and the Role of Logistics: An Empirical Analysis." *Energy Economics* 34: 104–116.

9 The world's greatest coal arbitrage

China's coal import behavior

Richard K. Morse and Gang He

9.1 Introduction

In 2009, the global coal market witnessed one of the most dramatic realignments it has ever seen – China, long a net exporter of coal, suddenly imported a record-smashing 126 million metric tons (Mt) (see Figure 9.1).[1] This inversion of China's role in global coal markets meant that Chinese imports accounted for nearly 15% of all globally traded coal, and China became the focal point of global demand as traditional import markets like Europe and Japan stagnated in the wake of the financial crisis. By the first quarter of 2010, even Colombia was defying established trade patterns by sending cargoes to China despite its massive geographic disadvantage when it comes to exporting coal into Asian markets. The middle kingdom's appetite for imported coal seemed insatiable, and the "China Factor" appeared to have ushered in a new paradigm for the global coal market.

But China didn't "need" the coal. The world's largest coal producer cranked out 2.96 billion metric tons (Bt) of production in 2009, backed up by 114.5 Bt of reserves.[2] Although the world's other fastest growing

A related study (He and Morse, 2014) that uses the same modeling approach as this chapter but Chinese data sources appeared in *Globalization, Development and Security in Asia Volume 3: The Political Economy of Energy* (Singapore: World Scientific Publishing, 2014, pp. 69–85). The authors would like to thank a number of people who made this analysis possible. Our colleagues at PESD, including Frank Wolak and Mark Thurber, as usual provided invaluable insights. We also greatly benefited from discussion with many people in the coal and financial industries. The Macquarie Securities team, including Adam Worthington, Yeeman Chin, Jon Windham, and Albert Saputro, shared their own insightful analyses of these issues for which we are quite grateful. IHS McCloskey generously shared pricing data for the purposes of this research. Bart Lucarelli also helped sharpen many of the ideas here. Stu Murray of London Commodity Brokers, who is unmatched in his insight into Chinese buying behavior, shared his experiences with us. Thanks also to many Chinese scholars and analysts, including Yang Xianfeng, Li Hongjun, Hu Pingfan, Cai Guotian, Wu Lixin, Zhang Guoqiang, Wu Wenhua, Li Hong, and Gao Shixian. Finally, thanks to Kathy Lung and Sunny Wang for their support. Any errors are ours alone.

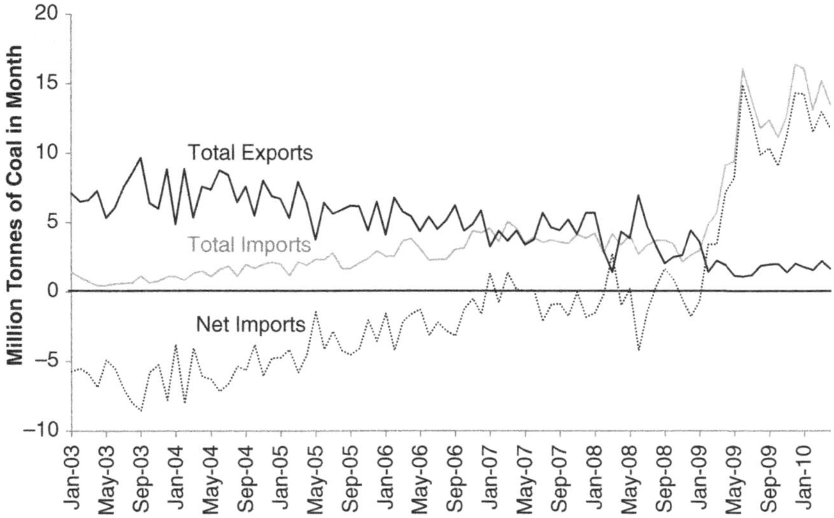

Figure 9.1 China's coal import surge. Source: IHS McCloskey.

importer, India, is plagued by a growing gap between coal supply and power demand that it is unable to fill domestically (see Chapter 4), this is not the case in China. The spike in Chinese demand for imported coal is therefore a more complex (and less easily predictable) phenomenon that requires careful examination if the world is to understand what impact China might have on global energy markets in the coming decade.

In this chapter we outline a model that explains Chinese coal import patterns and that can allow coal market participants to understand, and to some degree predict, China's coal import behavior. We argue that the unique structure of the Chinese coal market creates a series of key arbitrage relationships between Chinese domestic coal markets and international coal markets that determine Chinese import patterns. Based on this theory of China's import behavior, we construct an arbitrage-based import model and show how it explains the dramatic shift in Chinese net imports that occurred between mid-2007 and mid-2010. (Arbitrage relationships have fluctuated somewhat since then but have mostly conformed to the early 2010 pattern favoring imports; as shown by He and Morse (2014), actual imports have continued to track those predicted by the model.)

The implications of this model are significant for the development of the global coal trade in the coming decade. First, we find that China's import behavior does not represent a "structural shift" in global markets. China, as a participant in the global coal market, is a cost minimizer that will be both a buyer and seller in the global market as key price relationships fluctuate. Second, and perhaps most importantly, the arbitrage relationships that we describe directly link the domestic price of coal in China to the global price of coal. This linkage has significant implications for the use of coal and the cost of generating power globally in the coming decades. Developments in China's domestic coal market will be a dominant factor determining global coal prices and trade flows (and by implication power prices in many regions). This makes understanding the domestic Chinese coal market, which operates according to a unique economic and political logic (see Chapters 2 and 3 and in the Appendix), crucial for any participant in the global markets.

9.2 Geographic fundamentals of the Chinese coal market

China's coal reserves and production are concentrated in the north and west of China. Three provinces in these regions – Shanxi, Shaanxi, and Inner Mongolia – had 69% of the country's proven reserves and were home to half of national production in 2009, as China was first becoming a net importer of coal. That same year Inner Mongolia surpassed Shanxi to become the largest producer at 637 Mt; Shanxi produced 615 Mt, and Shaanxi produced 296 Mt.[3] Seventy percent of the production in these three provinces is transported outside of the producing province to help satisfy China's coal demand, which is most heavily concentrated along the eastern and southern coasts. Figure A.1 in the Appendix illustrates the basic geography of Chinese coal regions.

Northern coastal Chinese coal demand is served by a network of truck routes and railways that move coal east and south from western and northern production centers. (Figure A.16 in the Appendix illustrates China's major coal transport infrastructure segments.) But rail and truck capacity to supply coal to southeast China is both insufficient and prohibitively expensive (see Appendix). Therefore coal supply for southeast China is first transported east on rail lines like the Da-Qin and Shuo-Huang to eastern ports like Qinhuangdao, Huanghua, and Rizao; next loaded onto boats; and finally shipped south via sea routes. This rail-to-sea link is much cheaper than moving coal overland from north to south. However, the costs associated with this rail-to-sea transport route are still high: the transportation cost of moving coal from Shanxi to Guangzhou can be as much as 50%–60% of the price of coal delivered to Guangzhou.[4] The high cost of moving coal to the heavily industrialized coastal area that includes the Pearl River Delta and the Yangtze River Delta opens windows for import coals to compete with domestic coals.

Southeast China is also the closest region in China to two major global coal exporting nations, Indonesia and Australia. Coal buyers in southeast China therefore are often confronted with two options: buy domestic coal delivered by sea from northern Chinese ports, or buy international coal. This arbitrage opportunity allows Chinese coal buyers to take advantage of price differentials between domestic Chinese coal and international coal prices. Until 2009, those differentials had not favored imports.

9.3 The arbitrage model of Chinese coal imports

We argue that modeling arbitrage spreads between domestic and international coals for coal buyers in southern China explains China's 2009 import spike and can also be used to reliably analyze China's national import behavior under future market conditions. In this section we describe the parameters of our model, called the China Coal Import Arbitrage Model (ChinaCoalArb for short), and demonstrate how its results can be used to interpret China's import trends.

Chinese coal buyers in southeastern China can buy coal from multiple markets, and price discrepancies between different markets create profit opportunities. For a portion of spot market demand, buyers will compare the CIF[5] cost of coal landed in Guangzhou from multiple destinations and, all other things equal, will take the cheapest coal.[6] The differentials between CIF Guangzhou coal prices from multiple origins therefore create arbitrage opportunities for Chinese buyers who can shift their purchasing patterns to capture the differential between domestic and international markets under different conditions. The model calculates these arbitrage relationships of domestic to key international coals. We then compare these arbitrage relationships to historical imports, demonstrating that import levels have broadly tracked these arbitrage trends, increasing where price spreads favored international coals over domestic coals.

9.3.1 Supply points and FOB prices

Chinese domestic prices in ChinaCoalArb are represented by FOB prices at the Qinhuangdao port. Qinhuangdao port is mainly supplied with coal from Shanxi, Shaanxi and Inner Mongolia and is China's largest coal port. Qinhuangdao throughput in 2009 was 206.33 Mt,[7] and the total throughput of the seven major coal ports serving northern China was 433Mt.[8] Coal loaded in Qinhuangdao and delivered to Guangzhou is transported south down the Chinese coast via maritime shipping.

Table 9.1 *China's coal imports by source, 2009 (in Mt)*

Australia	Indonesia	Vietnam	Russia	Mongolia	Canada
44.602	30.461	23.932	11.787	6.002	4.093
North Korea	USA	South Africa	New Zealand	Other	TOTAL
3.599	0.805	0.732	0.303	0.145	126.491

Source: IHS McCloskey.

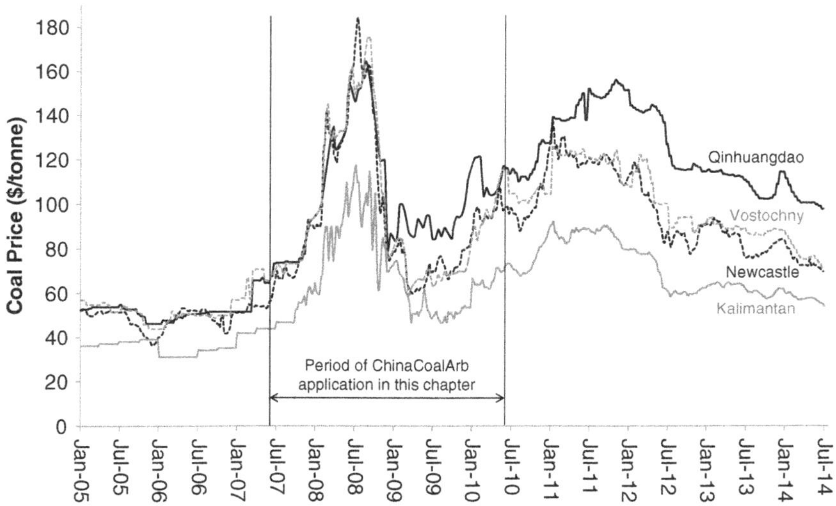

Figure 9.2 Major coal price indices in Asia. All price indices are FOB markers from IHS McCloskey: Qinhuangdao, China (5,800 kcal/kg NAR); Vostochny, Russia (6,700 kcal/kg GAD); Newcastle, Australia (6,000 kcal/kg NAR); Kalimantan, Indonesia (Indonesian sub-bituminous 4,900 kcal/kg NAR). Source: IHS McCloskey.

We select three key international coal supply countries as the model's suppliers: namely, Australia, Indonesia, and Russia, which were the largest exporters of coal to China in 2009.[9] All three countries are major exporters to the international market and benefit from reasonable proximity to Chinese import markets. Table 9.1 shows 2009 total Chinese imports from these origins. FOB coal prices in each of these markets are derived from bids and offers at those locations.[10] Figure 9.2 shows the historical development of prices in Qinhuangdao as well as at the major ports in Russia, Indonesia, and Australia from which coal is exported to China.

Table 9.2 *China's coal imports by province, 2009 (in Mt)*

Guangxi	Guangdong	Fujian	Zhejiang	Shanghai	All other	TOTAL
14.14	33.72	12.00	10.26	3.13	52.75	126.00

Source: Calculated from China Customs Statistics.

9.3.2 Import demand centers

The port of Guangzhou in Guangdong province functions as the demand center of our arbitrage model. Guangdong is a heavily industrialized zone that has historically been the largest coal importing province. Coal imports by province for 2009 are shown in Table 9.2. Guangzhou port's coal handling capacity reached 56.5 million metric tons per annum (Mtpa) in 2008. Nearby import centers witness similar price relationships between domestic and international markets and exhibit similar import patterns. Arbitrage relationships for Guangzhou can therefore be used as a proxy for arbitrage relationships for all of southeastern China.[11] Other major coal ports of southeastern China include Shanghai, Ningbo, Fuzhou, Xiamen, Quanzhou, Shantou, and Beihai.

9.3.3 Freight prices

The delivered cost of coal in Guangzhou (CIF) is calculated by adding freight costs between loading and discharge ports to the FOB coal cost.[12] Dry bulk freight rates from Indonesia, Russia, or Australia into China largely track international dry bulk freight markets, which are volatile, internationally traded commodity markets (see historical freight rates in Figure 9.3).[13] The model uses specific freight prices quoted by AXS Marine from each FOB loading port delivered into Guangzhou port.

Freight prices for the Chinese maritime coastal shipping market are not fully connected to international freight prices, however. Smaller boats are typically dedicated to the domestic market, but larger boats can switch into the international shipping markets when prices are attractive.[14] The Shanghai Shipping Index measures the cost of sending coal from Qinhuangdao to Guangzhou (see Figure 9.3).[15] Model freight price assumptions assume capesize vessels for all international shipping routes, and 40,000–50,000 dead weight ton (DWT) vessels for Chinese coastal shipping prices. Representative port handling charges for loading and discharge are added for all ports.[16]

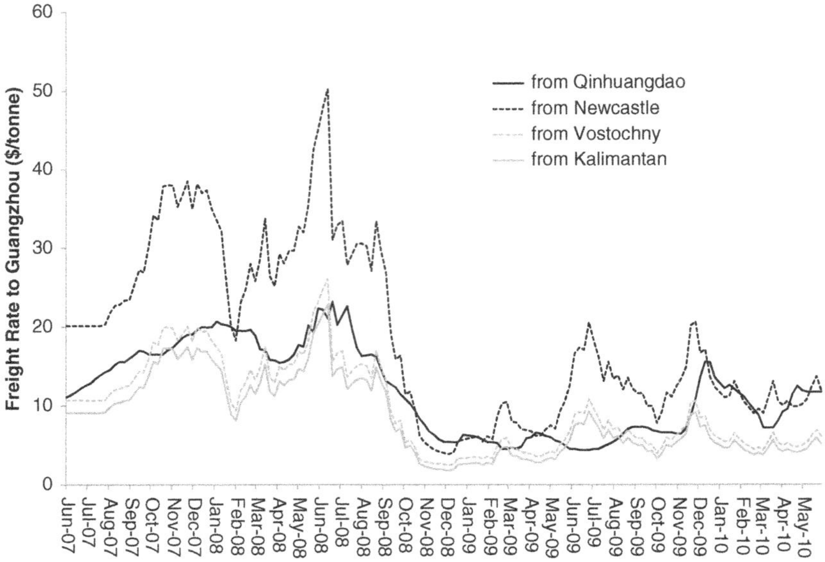

Figure 9.3 Dry bulk freight rates from FOB ports to Guangzhou port in ChinaCoalArb model. Data source: AXS Marine, Reuters, and authors' analysis. Rates are based on historical quotes for specific shipping routes. All international quotes are based on capesize vessels, China domestic quotes are based on 40,000–50,000 DWT vessels.

9.3.4 *Adjustments: energy content, taxes, exchange rates, transaction costs*

We make the following adjustments in order to more accurately reflect real market conditions:

- Coal buyers are buying energy. We therefore adjust all FOB prices so they are energy equivalent to Newcastle coal (see the caption of Figure 9.2 for the energy contents and price basis – gross air dried (GAD) or net as received (NAR) – of the coal price markers we use here).[17]
- Relevant taxes are added to all coal prices. Chinese VAT of 17% is added to the CIF price of all coals. We do not adjust for China's import tax because it had been phased out by 2007, before China's import surge.[18]
- All coal and freight prices are adjusted for historical exchange rates between RMB and USD.
- Chinese buyers face increased transaction costs when purchasing coal from the international market as compared to the domestic market.[19]

Transaction costs for Chinese buyers associated with import vs. domestic coal include raising letters of credit, dealing with foreign sellers, and more onerous contracts. Though it is impossible to know precisely the increased transaction costs for individual buyers, we add a $3/tonne advantage to Qinhuangdao coal over all international coals.[20]

9.3.5 Implications of key ChinaCoalArb assumptions

Several necessary assumptions in ChinaCoalArb may impact the performance of the model and should be noted when comparing model results to real market outcomes.

First, the model does not explicitly separate thermal from coking coal.[21] Though some coals can switch between these markets, making the distinction blurry at times, buying behavior in these markets will differ under certain conditions. Most worth noting is that some demand for high-quality material is likely to be less responsive to price movements than demand for thermal coal because high-quality coking coal supplies are much tighter in China and internationally. We suspect this accounts for a large share of Australian imports into China pre-2009 when CIF prices for most Australian coals compared to Qinhuangdao coals were significantly higher (see Figure 9.4).

Second, by using the Qinhuangdao spot price to represent the Chinese domestic price, the model does not take into account discrepancies between spot prices and "term prices" for power generators in the Chinese market. Although in theory the two should be tightly correlated since the deregulation of all coal prices in 2006 (spot was deregulated in 2002, but the NDRC still directly capped term prices for power generators until 2006), this is not always the case in practice. As evidenced by the June 2010 NDRC price cap on term prices,[22] China can in reality have a "two-tiered" coal market under certain conditions. If demand for import coal in the power sector can replace either spot domestic coal or term domestic coal, any price discrepancies between these two domestic markets could affect arbitrage relationships. Though we believe that the current model capably represents Chinese buying behavior in the aggregate, under certain circumstances modeling of power generators' buying behavior may be improved by using the NDRC capped term price instead of the spot price as the domestic price.[23]

Third, China announced that it would relax foreign exchange controls in June 2010, which has led to minor RMB appreciation. Possible RMB appreciation going forward could increase China's buying power for foreign coal and thus make imports more attractive.[24]

Fourth, while the indicative energy-equivalent coal tonne that we have created reflects differences in energy content, the relationship of coal price to energy content may not always be linear (as we have already discussed). Thus lower CV coals that are priced at a deeper discount may present a greater arbitrage opportunity than our model indicates.

Fifth, although energy content is arguably the most significant variable impacting coal pricing, the model does not reflect several other coal properties that can impact price, such as ash, moisture, volatiles, and sulfur content. Our energy-equivalent coal tonne should be broadly indicative of buying behavior but cannot capture all variables that impact coal purchasing and pricing.

Sixth, while we argue that Guangzhou is the best proxy to represent China's national import behavior, other importing locations will witness slightly different arbitrage relationships that might impact purchasing decisions. For instance, Shanghai is slightly farther from Indonesia and Australia than Guangzhou and slightly closer to Qinhuangdao and Russia. This will increase freight costs from the former two sources and favor coal shipped from the latter two.

9.3.6 Relevant external factors not captured by ChinaCoalArb

Two major macro-level factors that could impact coal imports should be noted.

First, there are technical limitations in coal-consuming applications that could prevent a full switch to imports, at least in the short term. Power plant boilers are designed to burn particular specifications of coal, which is almost always domestic coal for energy security reasons. Power generators can blend imports and domestic coal supplies, but for each boiler there will be a technical limitation on the amount of blending that can occur.[25] Other industrial applications, like cement or steel making, could also be designed to burn domestic specifications of coal and may not be able to fully switch to imports.

Second, import port capacity could be a theoretical limit on Chinese coal imports, albeit one that has not yet been tested.

9.3.7 Model results

Figure 9.4 displays the results of the arbitrage model. We compare arbitrage relationships to total monthly imports in order to demonstrate the relationship. The left axis indicates the delivered price advantage on a \$/tonne basis of import coals landed in Guangzhou compared to domestic

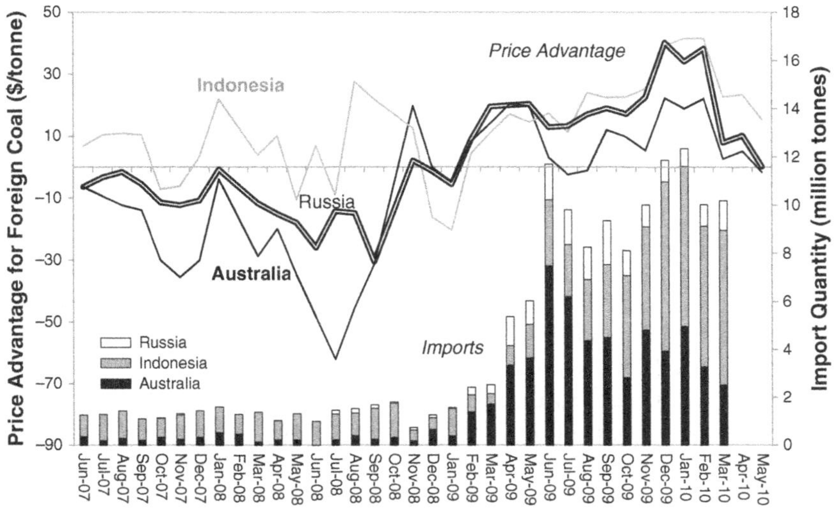

Figure 9.4 ChinaCoalArb model results. The figure shows how arbitrage relationships (the delivered price advantage of foreign coals with respect to Qinhuangdao coal – shown by the lines, with values on the left axis) drive Chinese coal imports from different countries (the columns, with values on the right axis). All price advantage values are the CIF Guangzhou from Qinhuangdao minus CIF value of imported coals. Source: McCloskey, Reuters, AXS Marine, Shanghai Shipping Index, and authors' analysis.

coals from Qinhuangdao landed in Guangzhou. Negative values indicate a delivered price advantage for domestic coals and positive values indicate a delivered price advantage for imports. The right axis indicates million tonnes of national coal imports by month.

The results explain the dramatic shift from China importing 40.4 Mt in 2008 to importing 126 Mt in 2009. Prior to the fourth quarter of 2008, international coals had a price disadvantage compared to domestic coal. In summer 2008 Australian and Russian imports were out of the money against Qinhuangdao by as much as $65/t and $30/t respectively. While minor amounts of Australian material were still imported – likely specific qualities of coking coal not readily available in the Chinese domestic market – China's imports were negligible because importing was not profitable in general. The partial exception was the case of Indonesian coal, which came in and out of the money against Qinhuangdao in southern China. This can be attributed to two primary factors, both of which give Indonesia a competitive advantage exporting into China.

First, Indonesia's geographical proximity to Chinese markets means that it pays a smaller freight penalty than Australia and Russia (assuming Russian material is delivered into southern China, which in reality it may not always be). Second, Indonesian FOB prices were historically slightly lower than Australian and Russian prices on an energy-adjusted basis. Thus as freight rates and domestic and international coal prices fluctuated, import windows opened for Indonesian coal in southern China.

At the end of 2008, this historical relationship of domestic to international coal changed dramatically. In the wake of the global recession, the historical relationship of domestic to international coals in southern China inverted and a massive arbitrage opportunity arose. By late 2009, Indonesian coal was as much as $40/tonne more affordable than domestic coal, Australian coal was as much as $29/tonne more affordable, and even Russian coal – which suffers from a huge rail transport penalty due to the cost of moving coal from central Russia to eastern ports – was pricing into southern China against domestic coal at $40/tonne better than Qinhuangdao. Imports skyrocketed, cresting in winter 2009–2010.

This model provides a credible explanation of China's coal import behavior in general and in particular of China's record imports in 2009. We conclude that China's coal buying behavior indeed follows the logic of a "cost minimizer" and that China's coal imports will continue to fluctuate according to the arbitrage differentials between domestic and international coal prices.

As shown in Figure 9.4, the "arbitrage window" seemed to briefly close in the summer of 2010. However, Qinhuangdao prices started in 2011 to move higher against foreign coal prices (see Figure 9.2), which, consistent with model predictions, has helped ensure a strong role for imports (especially from Indonesia) in recent years (He and Morse 2014; IEA 2014).

9.3.8 Drivers of arbitrage inversion in 2009

Analyzing causes of the dramatic inversion of the historical relationship between China's domestic coal market and the international coal market provides a more detailed understanding of the market conditions that caused this shift, and thus enables coal market observers to examine how likely these conditions are to carry forward. We argue that in the wake of the global financial crisis of 2008, four principal drivers caused the inversion of historical price relationships between international and domestic Chinese coal prices.

First, the global financial crisis had a comparatively smaller impact on China than on many other coal consuming nations. China's GDP growth

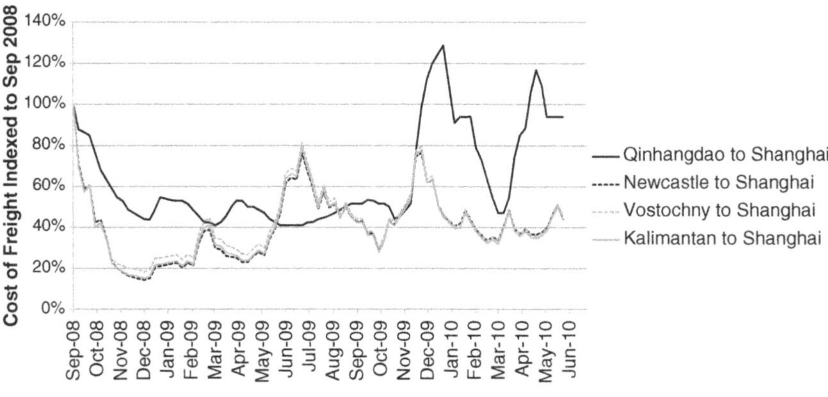

Figure 9.5 Cost of freight indexed to pre-financial-crisis levels (September 2008). Sources: Shanghai Shipping Index, AXS Marine.

rate was 9.6% in 2008 and 9.1% in 2009, declining by only 0.5% (NBS, 2011). By contrast, real global GDP growth was estimated to be a negative 2.1% in 2009 (IMF, 2009). This meant that Chinese macroeconomic activity sustained a comparatively high level of energy demand relative to other coal importing economies. And in China energy means coal.

Second, after the global financial crisis, Chinese domestic freight prices remained higher than international freight prices, giving imports an advantage over domestic coal. (Figure 9.5 shows Chinese domestic freight compared to international freight rates as a percentage of their pre-financial crisis levels.[26]) This meant that the freight component of the delivered price of imports decreased relative to its domestic counterpart.

Third, international FOB prices declined more than Chinese domestic FOB prices when measured as a percentage of their pre-financial crisis levels (see Figure 9.6). This shift, combined with the relative freight advantage, put imports at a significant advantage to domestic coal. The relative strength of China's domestic coal price was largely due to a series of regulatory events in key production regions that curtailed supply and supported prices. Consistent with national coal mining policies, Shanxi province embarked on a major campaign of mine consolidation. The government began a program of closing small mines or consolidating them into larger mines and implementing more rigorous safety standards. The targets of that program are shown in Table 9.3. The result was a shut-in of traditional supply that helped keep prices high. (There is evidence that the program was not completely successful and that government officials may re-implement similar measures.[27]) Temporary factors in

Table 9.3 *Mining consolidation targets in Shanxi*

	2008 plan	2009 plan	2010 plan
Mine number	2600	1053	1000
Average mine size	300,000 t/yr	900,000 t/yr	900,000 t/yr
Number of firms	2200	130 4 at 100 Mt/yr scale 3 at 50 Mt/yr scale	100

Source: Government of Shanxi.

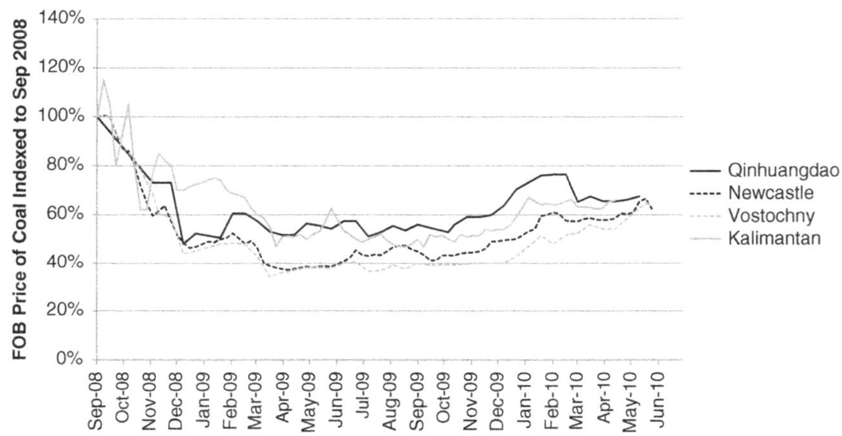

Figure 9.6 FOB coal prices indexed to pre-financial-crisis levels (September 2008). Sources: IHS McCloskey, Reuters.

China like weather interruptions of transport and weak hydro generation due to droughts also contributed to higher domestic coal prices.

Fourth, the traditional negotiations for term coal contracts between coal producers and power generators, which in 2008 allocated 1.1 Bt of coal (40% of China's total coal consumption that year), broke down in 2009 when an agreement on price could not be reached (see Chapter 3). This led some coal buyers that would otherwise have purchased domestic coal to look overseas.

Fifth, Chinese national policy on resource use facilitated increased imports. The so-called Two Markets, Two Resources[28] policy encourages coal buyers to import coal when the economics justify it.

There are a number of long-term factors that will determine the price relationship of Chinese coal to international coal. Chapters 2 and 3 and the Appendix of this book discuss some key issues and reforms that will impact China's domestic market and thereby affect this relationship. These include reforms to vertical integrate China's coal and power industries in the form of major coal-power bases, the consolidation of coal production into larger mines, and the adoption of more efficient power generation technologies.

9.4 Conclusions and implications for the global market

Once a largely isolated coal market, China now plays a central role in determining global trade flows and prices. Understanding Chinese import behavior under current and future market conditions is therefore imperative for any analysis of the global coal trade. We have put forward a theory of Chinese import behavior based on arbitrage relationships between China and the global market and validated that theory through the ChinaCoalArb model.

Our findings indicate that China is a "cost minimizer" in the international market that will import heavily when the price is right – as it was in 2009 due to a confluence of circumstances we have described here – and largely rely on domestic coal when imports are economically unattractive.

The nature of Chinese demand for international coal is therefore fundamentally different from that of India, the other source of dramatic demand growth in international coal markets. India is structurally short on coal because its demand growth (mainly for power) has outstripped its ability to tap domestic coal resources (see Chapter 4). China, on the other hand, is now the world's largest coal arbitrage trader. This means that the relationship between China's domestic coal price and the international coal price will be one of the key factors determining global trade flows in the coming decade. Depending on the nature of this relationship, China could just as easily buy 15%–20% of internationally traded coal as it could buy very little.

China's role as the world's largest coal arbitrageur has a hugely significant implication for the global coal market: it links the international price of coal to China's domestic price. China's buying and selling activity on the margins of its massive domestic coal market bring domestic and global prices closer to parity, though at present not to complete parity. In other words, what happens in the mines of Shanxi will impact the price of power in Munich. The unique politics and economics of the Chinese coal market are now therefore by necessity the politics and economics of the global market, and whether or not China imports coal in a given year, the "China factor" will increasingly define how the world sells, buys, and uses coal.

Notes

1. According to the National Energy Administration of China, 103 Mt net imports.
2. This reserve number is widely used by IEA, BP Energy Statistics Review, etc. China's Ministry of Land and Resources shows 183 Bt in its updated "Mineral Resources Reserves Classification." See more details in Wang, Qingyi (2013), China Energy Data 2013.
3. National Bureau of Statistics of China, available at http://www.stats.gov.cn/english/Statisticaldata/AnnualData/.
4. This is an estimate. Costs fluctuate according to the price of coal and the price of shipping coal.
5. "CIF" is a coal market term indicating that a price is for the delivered location, and thus includes all freight costs.
6. For purposes of modeling, we have slightly simplified the dynamics of Chinese coal buying behavior to focus on the domestic vs. import tradeoff. For instance, we have not included term contracts or prices at port stockpile in our model.
7. Qinhuangdao Port, China Coal Transportation and Distribution Association.
8. These ports include Qinhuangdao, Tangshan, Huanghua, Tianjin, Rizhao, Lianyungang, and Qingdao.
9. China has historically imported coal from Vietnam, but due to increased Vietnamese domestic consumption and price increases, imports from Vietnam are expected to decline. Russian imports are not necessarily delivered into southern China as the port of Vostochny is north of the Chinese border. But even though Russian ports are north rather than south, the general arbitrage principle applies and thus we have included Russia in our model even if Russian material is not always imported into Guangzhou. Russian exports to China are still comparatively small, but increased from 0.76 Mt in 2008 to 11.8 Mt in 2009.
10. There are multiple price indices used in the coal market. We have chosen to use indices provided by IHS McCloskey, which we believe are of high quality.
11. The basis difference between Guangzhou and other cities in southeast China will be roughly the freight cost differential between those two locations. Thus, while Guangzhou is the center of coal imports in China and is indicative of market conditions for imports in general, some minor basis differentials will exist for other locations.
12. There are a few other adjustments, which we describe later.
13. The Baltic Dry Index, which is widely used to indicate the cost of chartering dry bulk freight vessels, reached highs of 11,459 in 2008 before crashing to lows near 670 in late that year. The Index is comprised of charter rates for four types of ships: capesize, panamax, supramax, and handysize. For further details, see the Baltic Exchange (http://www.balticexchange.com). A price history is available from Bloomberg at http://www.bloomberg.com/quote/BDIY:IND/chart (accessed September 20, 2014).
14. We benefited from discussion with Jon Windham of Macquarie Securities who offered his views of the Chinese domestic shipping market and generously shared his research and data.

15. More information on the Shanghai Shipping Index is available at http://en.chineseshipping.com.cn/html/index.asp.
16. We assign port fees based on rates provided by AXS Marine.
17. We assume a linear relationship between price and energy content, though in some cases pricing may not exactly follow this method, especially pricing for lower CV coal. For instance, historically, lower CV Indonesian coals have traded at a deeper discount to higher CV material (see Chapter 7).
18. China's coal import tax for all coals with the exception of coking coal was 6% prior to April 1, 2005, 3% until Nov. 1, 2006, and 1% until May 31, 2007. The import tax was abolished after June 1, 2007. Coking coal import taxes were 3% until January 1, 2005, at which point they were abolished.
19. Stuart Murray of London Commodity Brokers shared his insights with us on what transaction costs Chinese buyers of international coal face.
20. Results are not highly sensitive to this assumption because shifts in arbitrage relationships that we describe here are much larger than $3/tonne. The overall impact is to make domestic coal slightly more favorable in all circumstances. For instance, the highest import disadvantage prior to 2009, which was for Newcastle against Qinhuangdao, would only drop from highs of $70/tonne to $67/tonne. Conversely the highest import advantage after 2008, which was for both Indonesian and Russian material, would only increase from highs of about $43/tonne to about $46/tonne. Domestic coals at the margin that are less than $3/tonne more profitable than import coal may become unprofitable against imports if this assumption is eliminated.
21. In 2009, total coking coal imports were 34.5 Mt out of 126 Mt total imports.
22. See NDRC: People's Bank Announced Further Reform of Renminbi Exchange Rate: http://xwzx.ndrc.gov.cn/ywdt/201006/t20100624_356342.html (accessed June 15, 2011).
23. This is complicated because even coal buyers that can acquire coal at below-market, government-capped prices may still have to buy a fair amount of their supplies from the spot market. We estimate that 40%–50% of coal supply for key state-owned enterprises is under term contract (and thus impacted by the recent cap), and the remainder is purchased on the spot market.
24. Albert Saputro and Adam Worthington of Macquarie Securities helpfully highlighted the importance of this issue to us, which they have analyzed in their own research.
25. From interviews with coal and power experts at Yudean Group in Guangdong. Blending is a complex engineering issue; although we will not address it in detail here as there are multiple variables that impact blending ratios at coal plants, we do want to note that some blending limitations may exist.
26. We choose September 2008 as the pre-financial-crisis point, as equities markets began their precipitous drop and Lehman Bros. filed for bankruptcy in that month. For a history of the performance of the S&P 500, see Bloomberg, http://www.bloomberg.com/quote/SPX:IND/chart (accessed June 10, 2011).
27. Research by Macquarie Securities in June 2010.
28. "Two Markets, Two Resources" is a Chinese term reflecting a strategy to encourage Chinese companies and Chinese industry "walking out" of the country to explore both domestic markets and international markets.

References

BP (2010). *BP Statistical Review of World Energy, June 2010*. Available at www.bp.com/statisticalreview (accessed March 4, 2011).

He, Gang and Richard Morse (2014). "China's Coal Import Behavior and Its Impacts to the Global Energy Market." In *Globalization, Development and Security in Asia Volume 3: The Political Economy of Energy*. Singapore: World Scientific Publishing.

IEA (2014). *Coal Information 2014 (with 2013 Data)*. Paris: International Energy Agency.

IMF (2009). World Economic Outlook (WEO): Sustaining the Recovery. Available at https://www.imf.org/external/pubs/ft/weo/2009/02/index.htm (accessed March 5, 2011).

NBS (2011). 2010 National Economy and Social Development Communique. Available at http://www.stats.gov.cn/tjsj/tjgb/ndtjgb/qgndtjgb/201102/t20110228_30025.html (accessed September 20, 2014).

Wang, Qingyi (2013). China Energy Data 2013. Available at http://www.efchina.org/Attachments/Report/reports-20131212-zh/reports-20131212-zh (accessed September 20, 2014).

10 The COALMOD-World model

Coal markets until 2030

Franziska Holz, Clemens Haftendorn, Roman Mendelevitch, and Christian von Hirschhausen

10.1 Introduction

Developing a structural model of international steam coal markets is a true challenge in an energy world at a crossroads, where the last years have been characterized by some fundamental structural changes and the future perspectives are unclear. There is a strong trend toward renewables in the OECD (Organization for Economic Cooperation and Development) countries, in particular in Europe, where climate change awareness has induced a number of climate policies. At the same time we observe an increasing energy demand in Asia and other emerging economies due to continuous economic growth. This and other factors have led to a considerable volatility of energy prices in the last five years or so. All this has coincided with the "shale gas revolution" in North America which has made more American coal available for export markets. The strong demand increase driven by Asia and the subsequent capacity expansions have led to a consistent cost escalation affecting the fundamentals of the coal markets, too. Hence, it has been a challenge to develop a structural model and database for the year 2010.

While many features of the 2006 market presented in Haftendorn and Holz (2010) and Haftendorn et al. (2012) are still valid, new developments have taken place: a shift from Europe to Asia as coal consumption center, which leads the previous swing supplier South Africa to increasingly focus its exports toward Asia; a shift in relative costs which, among

We would like to thank Richard K. Morse, Mark C. Thurber, and the entire PESD team at Stanford University, as well as Steven A. Gabriel, Ruud Egging, Lars Schernikau, Philipp M. Richter, and Daniel Huppmann for their helpful comments. Moreover, we thank Andreas Tissen, Christoph Boden, Michaja Pehl, and Michael Rose for research assistance. We would also like to thank the participants of the PESD 2009 seminar on global coal markets for the fruitful discussions. We gratefully acknowledge funding by PESD at the initiative of David G. Victor for the development of the model as well as by the German Ministry of Education and Research in the funding program "Economics of Climate Change" for the continuous update and use of the model. All remaining errors are ours.

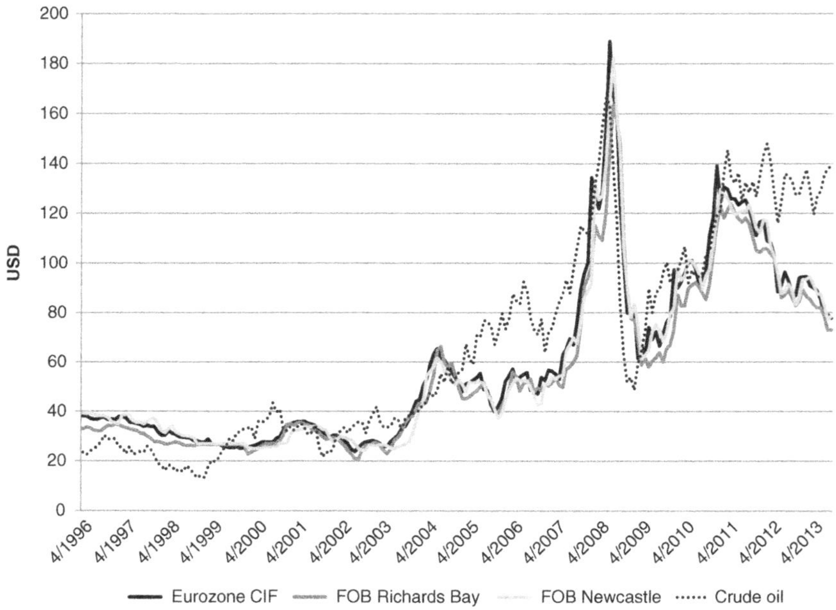

Figure 10.1 Monthly prices for steam coal in USD per tonne (CIF Eurozone, FOB Richards Bay, and FOB Newcastle) and crude oil in USD per barrel (crude oil index) between April 1996 and April 2013. Source: HWWI commodity prices in the Thompson Reuters Datastream database.

other things, has made exports from the traditional large exporter Australia more and more expensive; and the emergence of the US Powder River Basin as a prospective exporter via the US West Coast, to name just a few.

When coal market modeling came back to the surface of the academic literature, after its first "boom" in the 1980s, coal markets had seen some stable years with continuous low prices, more or less competitive trading relations, and a stable increase in seaborne trade (Haftendorn and Holz, 2010). However, coal markets have been caught in the turmoil of a general energy price and cost increase since 2007. They experienced a similar price peak in 2008 as oil markets (see Figure 10.1), and they compete with other mining sectors for qualified labor, mining services, and machinery, which has led to a substantial cost increase in many producing countries.

In this chapter, we report on the modeling effort undertaken within the framework of the "Global Coal Market" project led by the Program on

Energy and Sustainable Development (PESD) at Stanford University. PESD invited the modeling team at DIW Berlin to expand their model of the seaborne export market (Haftendorn and Holz, 2010) into a comprehensive model of the global steam coal market that additionally includes national and regional trade flows. Henceforth, the COALMOD-World model was developed to analyze the future evolution of the international steam coal market until 2030. It covers the two large pillars of the international steam coal market: international trade flows and domestically produced and consumed coal in large countries. This chapter presents an update of the model presented in more detail in Haftendorn et al. (2012).

International coal trade has an important role to play in current and future global coal markets. Although the share of internationally traded coal in total world coal consumption is relatively small – approximately 15% (IEA, 2013) – it has a major impact on the evolution of domestic markets and therefore plays a central role in the analysis. As is discussed in the previous chapters of this book, the international steam coal trade volume has increased in the last years and this trend is expected to continue. A rising global trade means that the centers of production and consumption continue to diverge and more countries rely on imports. Identifying how the trade flows will develop and where steam coal will come from in the future can help us better assess possible energy security issues. Since the developments on domestic markets will also affect the future global trade, in particular in countries that are at the same time coal producers and consumers (e.g., China, the US, South Africa), they are incorporated in the model, too.

COALMOD-World is a multi-period model that simulates yearly market outcomes, trade flows, and prices for the years 2010, 2015, 2020, 2025, and 2030 as well as investments in the coal sector value chain. We assume profit-maximizing players who optimize their expected and discounted profit over the total model horizon. The model results are a cost-efficient outcome that abstracts from the short-term real world frictions and cycles but gives a valuable indication of future trends.

The value chain of the steam coal sector will be reflected in the model setup. Various types of players are involved at each stage. *Producers* can be large national and sometimes state-owned companies. There are a few large multinational coal companies but also many small companies, usually operating in one country only. *Transport infrastructure* in the production countries can be built by the mining company or by another entity. Often, it consists of rail infrastructure but in some countries trucks or river barges are used. *Export ports* can be dedicated to one company or utilized by multiple companies. *Traders* as intermediaries also play a role in this market that is characterized by bilateral relationships. They can be vertically integrated or contractually connected to every stage of the industry.

In Haftendorn and Holz (2010), we provide an analysis of the market structure of the global steam coal trade and simplify the value chain for the modeling purpose. We give some evidence that, contrary to the oil market, the international steam coal market tends to be competitive.[1] This result allows us to make some simplifying assumptions for the COALMOD-World model: since in a competitive market prices equal marginal costs, we can simplify the role of the players in the value added chain to obtain two types of model players: the producers and the exporters.

Recent research on international coal markets has pointed out that the traditional separation of the Pacific and the Atlantic market has faded (e.g., Ellerman, 1995; Warell, 2006; Li et al., 2010). In our model, we therefore consider the global market as one integrated market, albeit not neglecting spatial aspects of a market where transport costs play an important role in determining the trade relations. Moreover, we take into account the reserve constraints of the individual producing regions given that coal is a finite, non-renewable resource.

Simultaneously to the trend of global market integration, we have also observed a trend toward commoditization of coal in increasingly liquid market places. This means that coal is traded more and more as a homogeneous good, which is reflected by the creation of price indices with standardized coal qualities and an increasing volume of paper trade (Zaklan et al., 2012). However, the energy content of steam coal varies depending on its origin. The differences in coal qualities are not as large on the international market as with the coal types that are produced and sold domestically but there are still some significant variations.[2] This suggests the incorporation of different coal qualities with their respective energy contents and the use of both dimensions, the mass and the energy content, in the model.

The remainder of this chapter provides a description of the COALMOD-World model structure and shows the perspectives of the world coal markets until 2030. Coal will no doubt play an important role globally and in particular in Asia in the next decades. However, climate policy efforts may lead to a slower growth in coal consumption, not only in the OECD but also in non-OECD countries, as well as to a shift of consumption regions. We will therefore explore the supply and trade patterns in two contrasting scenarios, based on "current policies" or "new [energy and climate] policies" (IEA, 2012d). As the fossil fuel with the highest carbon content, coal may be more than proportionally affected by climate policies targeted at reducing CO_2 emissions. At the same time, climate policies are implemented with different magnitudes in the different world regions, and we will therefore see effects on the coal trade patterns.

The next section details the structure and assumptions of the COALMOD-World model in light of other coal market modeling efforts. We also describe our effort of data collection, partially carried out in cooperation with the country teams of this broader project. We then present the results of the two scenarios in Section 10.4, specifying the supply and trade patterns as well as investment needs until 2030. We observe a significant reorganization of steam coal trade, which is rapidly diverted from the Atlantic to the Pacific basin, following the trends of the last years, and involves substantial exports from North America in addition to the traditional world market suppliers. (Chapter 8 considers in more detail the prospects that current coal port constraints on the west coast of North America will be eased.) The Moderate Growth scenario shows the same trends as the Base case with an even lower coal consumption in OECD countries. Section 10.5 concludes that more climate policy efforts are needed if CO_2 emissions from coal consumption is to be reduced significantly.

10.2 The COALMOD-World model

10.2.1 Coal market models: a recent renaissance

Little research on the specifics of the coal sector has been carried out. Only lately have economists looked into the market mechanisms of the steam coal sector, often motivated by the fact that, despite climate awareness and support to renewables, coal will remain a major primary energy carrier, in particular in the developing world but also in OECD countries. Understanding the supply mechanisms in coal markets will help to predict the short-term and long-term market reactions to climate and energy policies.[3]

An extensive review of the – sparse – coal-market specific modeling literature until 2010 or so is provided in Haftendorn and Holz (2010). There were some early modeling efforts applied to the US and international coal markets in the 1970s and 1980s (see Shapiro and White, 1982; Kolstad and Abbey, 1984). Often, coal modules are part of larger energy system models, as in most of the models applied in the Energy Modeling Forum number 2, *"Coal in transition 1980 – 2000"* (EMF, 1978). However, both the situation in the international steam coal market as well as modeling techniques have evolved since the 1980s. For other energy and resource markets, such as natural gas, multi-period models with endogenous investments were developed during the 2000s (e.g., Hartley and Medlock, 2006; Egging et al., 2010, for world natural gas markets). We argued that there was a need for a similar model of the global steam coal market.

In Haftendorn and Holz (2010), we developed a simple model of the international coal trade, still neglecting the domestic trade flows as well as investments in production and transport capacities. With this work we followed the stream of literature of detailed equilibrium (complementarity) models of other energy resource markets. The development of the COALMOD-World model is rooted in the previous static, one-period model "COALMOD-Trade" (Haftendorn and Holz, 2010), as well as in the multi-period modeling experience with other markets (Egging et al., 2008, 2010; Huppmann and Holz, 2012).

Complementarity models are numerical models that provide solutions to optimization problems under constraints (Cottle et al., 1992). The complementarity format can be used to model games, in particular non-cooperative market games such as a Cournot game. It represents the simultaneous behavior of several players and calculates their mutual best response functions. The complementarity model gives the Nash equilibrium solution, which is why they are also called equilibrium models. They are formulated by using the optimality conditions (called Karush-Kuhn-Tucker conditions, or KKT) of the optimization problems under constraints.[4] Often, the optimization problems are profit maximization problems of representative players with some economic and technical constraints. For a tractable model, some assumptions such as perfect and complete information of all players and over all model periods are generally made.

Two modeling teams have been the major contributors to the recent renaissance of coal market modeling applying the equilibrium technique, coincidentally both from Germany where steam coal imports have an important place in the generation mix: we have developed first the COALMOD-Trade model (Haftendorn and Holz, 2010) and then the COALMOD-World model (Haftendorn et al., 2012). Trüby and Paulus (2012) developed a one-period trade model similar to COALMOD-Trade and tested the market structure until 2008. They also used it to investigate the interaction between the Chinese and the world coal market (Paulus and Trüby, 2011). At the same time, optimization models continue to be in use, often as part of energy system models such as the National Energy Modeling System (NEMS) in the US Department of Energy or the PRIMES model used by the European Commission. These optimization models often have a different focus than supply and trade patterns, for example, optimization of coal qualities for certain power plant requirements, which is an aspect that we neglect here.

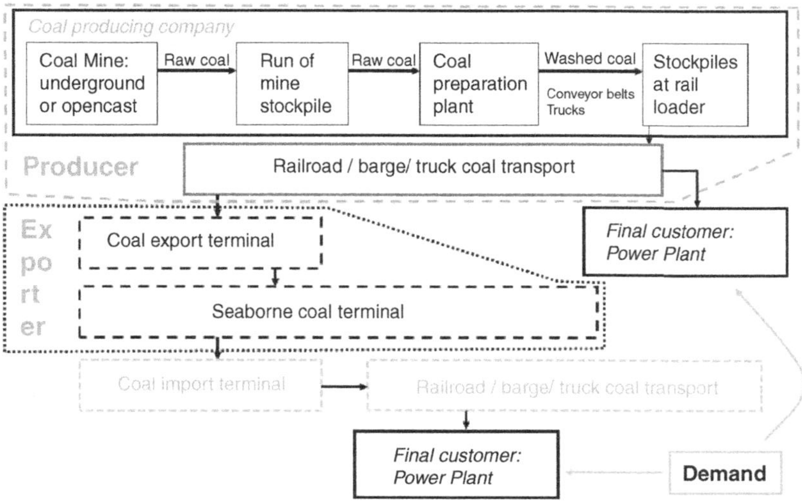

Figure 10.2 Model players in the steam coal value added chain.

10.2.2 Model structure

COALMOD-World is a multi-period equilibrium model of the global steam coal market with two types of players: producers and exporters facing consumers represented by a demand function. The stages of the real-world value chain that are included in the model are represented in Figure 10.2 by small rectangles inside the larger producer and exporter boxes. We exclude coal import terminals and the subsequent land transport link to the final consumer because their capacities are assumed to be sufficient. By assumption, we situate demand for seaborne import coal close to the import port. The second type of demand node can be reached by a land link directly from the producer. The producer player includes the coal mining company and also the land transport links. The exporter operates the export terminal and also pays for the seaborne transport. All players are aggregated on a national or regional (subnational) level.

The model producers and exporters represent stylized players defined for aggregated production, export and consumption nodes primarily determined using geographical parameters. A production node represents a geographically restricted area (mining basin) and aggregates the mining companies present in that area into one player called "producer." In the model, production node and producer are equivalent terms. Production nodes are defined based on the geography of reserves, type of coal, and production cost characteristics.

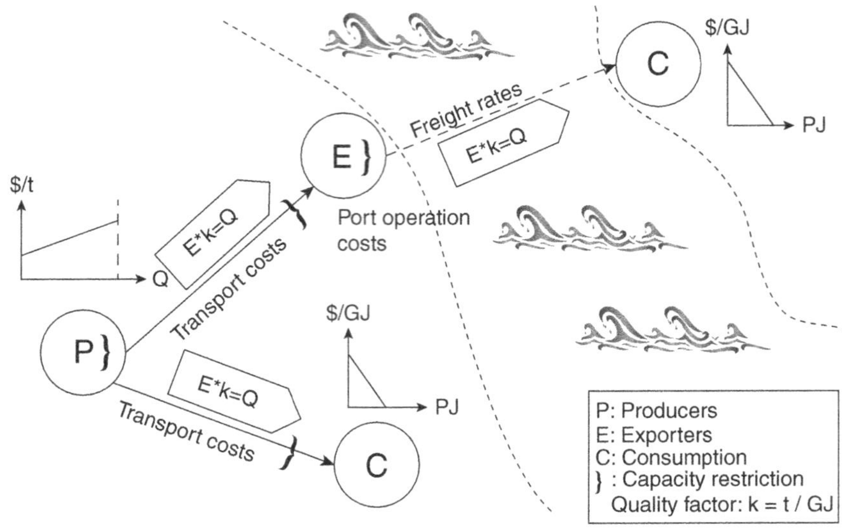

Figure 10.3 COALMOD-World model structure.

An export node represents the coal export terminal of one region and aggregates the real world coal export harbors present in that region into one model player called "exporter." Here again, export node and exporter are used synonymously. The export nodes are primarily defined based on geographic factors. A demand node represents a geographic area where the coal is consumed. It aggregates the consumption by the coal-fired power plants in a region. It can have access to seaborne coal through a port or only be supplied domestically. The demand nodes are primarily defined based on geographic factors, but other factors may come into play such as the connection to a port or the presence of mine-mouth power plants.

Figure 10.3 shows the model structure and the relationships between producers, exporters and demand. The producers extract and treat (i.e., produce) the coal under some production capacity constraint. They can sell it either to local demand nodes or to linked exporters. They bear the production and the inland transport costs. Further, they can invest in additional production capacities and in transport capacities to local demand or to the exporter. These investments are, in turn, also subject to constraints. Moreover, a constraint on reserves is applied over the total model horizon. Shadow prices (dual variables) are obtained for all constraints and provide an incentive to expanding capacities.

In the same logic as the producers, the exporters maximize their profit. Each exporter is linked to a maximum of one producer. The profit for each year is defined by the revenue from sales net of the costs of purchasing the coal at the FOB price from the producer, the costs of operating the export terminal, the costs of transport (shipping) to the final market, and finally the potential costs of investing in additional export capacity. An exporter can only sell to a demand node with a port. Since each model exporter has a dedicated model producer, the energy content factor of the exporter is equal to the energy content of the producer that supplies it for any given year.

Final demand is located at a consuming node and represented by a linear inverse demand function, that is, a willingness to pay function. We construct a different demand function for each demand node using their reference prices and reference demand values for the model starting year 2010 and use demand growth projections for future years. Moreover, we make assumptions about the demand elasticities. The producers can be in indirect contact with the final demand through their exporter or sell directly to their domestic demand. Prices are expressed in USD per GJ because we concentrate on the demand for energy embodied in the coal.

The model runs until 2040 and calculates yearly equilibria for the energy quantities sold in the years 2010, 2015, 2020, 2025, 2030, 2035, and 2040, which can be called "model years." Also, the players can decide on investments in each model year that will be available in the next model year.[5] Thus, the model does not only calculate an equilibrium within each model year but also over the total model horizon regarding optimal investments. For the years between the model years (e.g., for 2011, 2012, 2013, 2014, between 2010 and 2015), we interpolate the produced quantities since they are necessary to model the reserve depletion. We assume that production and other capacities will be made gradually available in the years between the model years to reach their new value in the following model year.

Both producers' and exporters' problems are profit maximization problems over the entire model horizon. The players have perfect foresight, meaning that they choose the optimal quantities to be supplied in each model period and the investments between model periods under the assumption of perfect information about current and future demand. Thus, the model simulates how demand should be served optimally given that the players behave rationally using all the information that is available to them. The full model formulation can be found in Haftendorn et al. (2012).

It is important to note that the traded quantities are the quantities of energy contained in the coal and expressed in petajoules (PJ). Whenever

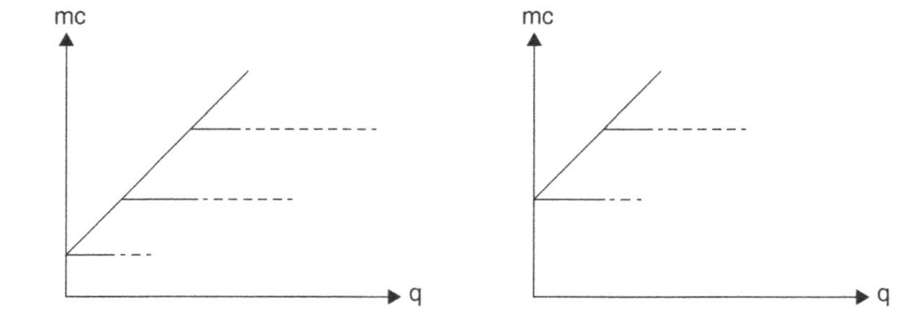

a. Aggregation from the individual mines' marginal costs to the model producers' (mining basin) marginal cost curve.

b. Effect of cumulated production on the intercept of the model producers' short-run marginal cost curve

Figure 10.4 Production cost mechanism for a model node.

the model needs to deal with mass quantities in million tonnes of coal (for the costs, capacities and investments) these energy quantities are converted to mass using a conversion factor in tonnes per gigajoule (GJ) that varies by producer (see Figure 10.3).

We assume short-run production cost functions for a year for each producer that can vary over time. Until recently, resource market models have often used the same short-run costs for every model period (e.g., Egging et al., 2010). This is not a realistic solution for a model of the coal market since there are many potential factors that influence future costs and change the short run costs. Other models only use the long run marginal costs (e.g., Lise et al., 2008). This is also problematic for a model of the coal market since the short-term marginal costs determine the prices in each period and, as we have seen in our previous static modeling work (Haftendorn and Holz, 2010), enable us to represent the trade flows accurately.

Figure 10.4a shows the logic of aggregation of individual mines in a mining basin to form the model producers' marginal cost curves. We assume that a specific mine with a certain geological setting operates at constant marginal costs. The horizontal line together with the dashed line represent the reserves of a mine. The horizontal line represents the production capacity at a given point in time. Thus, in order to obtain the aggregated cost curve in one period we add the production capacities on the q-axis and connect them with their respective marginal costs on the mc-axis.

After this static consideration, let us consider how this cost function might evolve over time. This effect of cumulated production is illustrated in Figure 10.4b. We assume that, even if all the mines along the cost curve may produce coal in one period, the cheapest mines are depleted first. Thus, we follow the rules stated by Hotelling (1931) that for exhaustible resources the cheapest deposits are extracted first. This is due to the fact that the cheapest mines are usually the deposits that are the easiest to access and are, hence, the oldest ones in operation. The effect of cumulated production from one model period to another is then that the cheapest mines are mined out and that the cheapest producer in Figure 10.4a disappears from the cost curve. We call this effect "mine mortality."

Mine mortality causes the intercept of the cost function to increase as shown in Figure 10.4b. The mine mortality factor gives the position on the cost curve of the previous period to determine the new intercept. Graphically, this is the passage from Figure 10.4a to Figure 10.4b. If the mine mortality factor is equal to one (i.e., the mine mortality rate is equal to 100%) it means that the cumulated production leads to a complete depletion of the oldest, cheapest mines. This may happen for mature and old mining basins. On the contrary, a factor close to zero means that the mines situated on the low cost segment of the basin's cost curve still have significant reserves and will only be depleted in the mid- to long term.[6] Since the slope of the cost curve indicates the relation of cheap and more expensive mines in the production region, the slope is the second factor in determining the new intercept's position.

10.3 Model specification and data

In Section 10.2.2 we introduced the concepts of nodes and model players. The model simulates the market on an aggregated basis in that we do not include individual mines or coal-fired power plants separately. However, the spatial characteristics of the market and the associated transport costs make it necessary to define aggregated nodes in the different producing and consuming countries. Section 10.3.1 describes our choice of model countries and nodes, and then we provide a detailed overview of our data in subsequent sections. The data collection required a major effort since there is no central source available. We generally collected data from publicly available sources; however, there is scarcity of data in the public domain and improvements could be achieved by using more detailed data. In order to remove inconsistencies in the data and to properly represent the base year 2010 we carried out a calibration of the data. Thereby, the COALMOD-World database is able to provide realistic runs and give insights into the future developments of the global steam coal market as is shown in Section 10.4.

10.3.1 Countries and nodes definition

We include all countries that were either consuming at least 5 million tonnes per annum (Mtpa) or producing and exporting at least 5 Mtpa in 2006, at the time of the development of the model. Some additional countries that

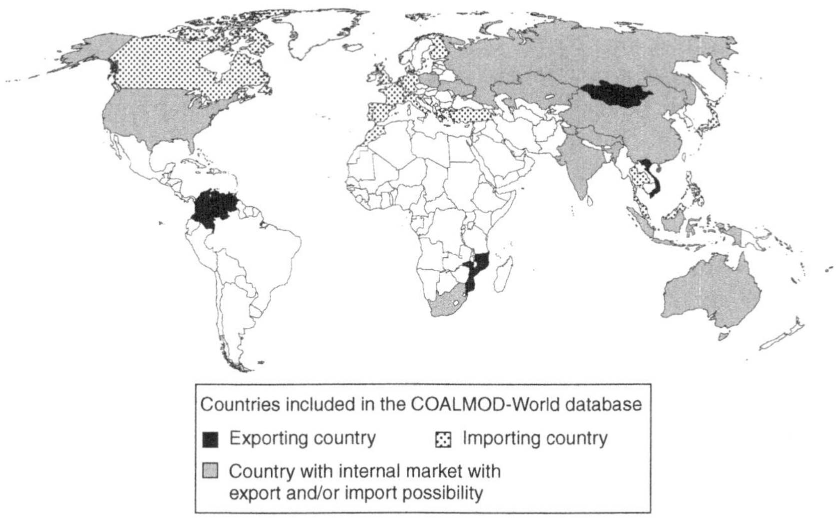

Figure 10.5 Countries included in the COALMOD-World database.

are becoming relevant players on the global market are included too (e.g., Mongolia and Mozambique). The world map in Figure 10.5 shows the represented countries, indicating their role on the world steam coal market (importer, exporter, or both).

In our data set, we distinguish production and consumption nodes. Hence, a country that only produces for export is represented in the data set with a production node from which it also exports (e.g., Colombia). A country that only imports and consumes coal is included with a consumption node (e.g., Italy and Turkey). For a country in which production takes place and that also consumes coal, we include at least one production node and one consumption node. For larger countries, there can be more than just one production and demand node; this is the case for the US, China, India, and Australia. The complete list of countries and nodes in the model can be found in Section 10A.1 in the appendix at the end of this chapter.

Producing nodes are generally defined by mining basins that are restricted by geological realities. The location of power plants is more dispersed as it relates to human settlement patterns. This makes it more difficult to locate our consuming nodes. For the consumers that can only be reached via an import harbor we define the demand as being located close to the port. For consumers that can be reached by land we aggregate regional data on capacities to form the demand node and define an average for the transport costs.

10.3.2 Production, costs, and reserves

The cost data is based on Baruya (2007) as well as several recent IEA sources and publications by the German Association of Coal Importers (VDKI). For each export country, Baruya (2007) provides estimates for the low and high average costs; the other sources usually follow the same methodology. This information is used to construct the producers' cost functions for the base year. We assume that the average low and high costs also represent unit costs for the cheapest and the most expensive mine. We construct a marginal cost function using the low estimate to determine the curve intercept. We place the second point at the intersection of the high cost estimate and the maximum production capacity in order to obtain a linear marginal cost curve.

In Figure 10.6 we depict the marginal cost curves for the base year 2010. One can see that the different production regions in a large country (e.g., in China) can well have different production costs. As would be

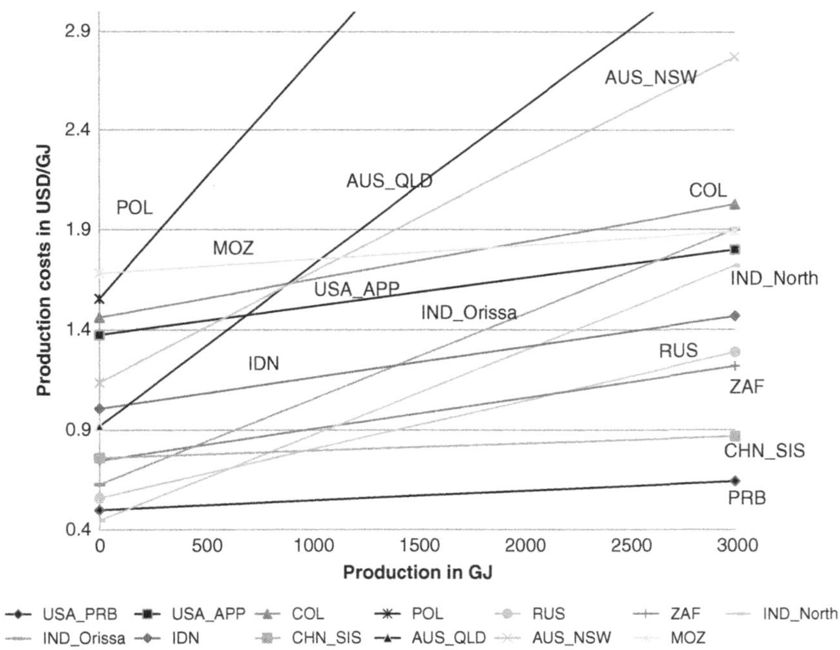

Figure 10.6 Marginal cost curves (2010) for selected production nodes. Source: Authors' work based on EIA (2012); NSWDPI (2009); QLDDME (2009); Rademacher (2008); Baruya (2007); NBSC (2007); Ritschel and Schiffer (2007).

expected, Poland is the most expensive producer on the world market, while some Chinese regions (Shanxi, Shaanxi, Inner Mongolia – abbreviated SIS), the Powder River Basin (US), and South Africa are on the lower end.

Clearly, there has been a substantial cost escalation in the last years that is evident when comparing to our earlier 2006 base year (Haftendorn et al., 2012). However, this cost increase did not hit all producers in the same way and proportionally across regions. While some producers such as Australia and also Indonesia have experienced substantial cost increases, others like the US Powder River Basin have almost maintained their costs of four years earlier. This has led to a shift in the relative cost structure: for example, with increasing export capacities (cf Figure 10.6) Powder River Basin coal is competitive with all other producers in all consuming regions, while the more expensive Australian coal is only competitive to ship within its Pacific home region (see, e.g., IEA, 2011; VDKI, 2013).

For the producers from the CIS (Russia, Kazakhstan, and Ukraine), Colombia, Venezuela, South Africa, and Indonesia, the cost data and the parameters of the marginal cost function as well as the capacities are updated based on IEA (2011). Countries with more than one production node require more detailed data on production capacities.[7] This data is used to build a merit order curve of costs using production data for each model producer and so estimate the producers' cost functions. In order to determine the cost functions in the long run, some assumptions had to be made about mine mortality in each five-year period and the associated rate of growth in the intercept. Table 10.11 shows these assumptions and the values of the intercept and slope for each producer.

The investment costs are a major input to the multi-period model since they determine the investment decisions. For the value-added chain from production to the export terminal, the IEA estimates investments costs of 50 USD (2007) per tonne of annual capacity addition (USD/tpa) and for some new projects this number goes up as high as 80 USD/tpa (IEA, 2008). Rademacher (2008) finds average investment costs of 62 USD/tpa with a wide range from 15 USD/tpa for some Australian opencast mines to 130 USD/tpa for new underground mines in Ukraine and Mozambique. But investment costs in Australia can also exceed 100 USD/tpa if the project includes new transport and washing facilities.

We therefore assign values from 40 to 80 USD/tpa to the different producers' investment costs for production capacity based on information about the country and mine mortality rates. The assignment is based on factors such as the prevalent type of mining, geology, and the state of technology. Unit investment costs and the production capacity for the base year and every production node are shown in Figure 10.7.

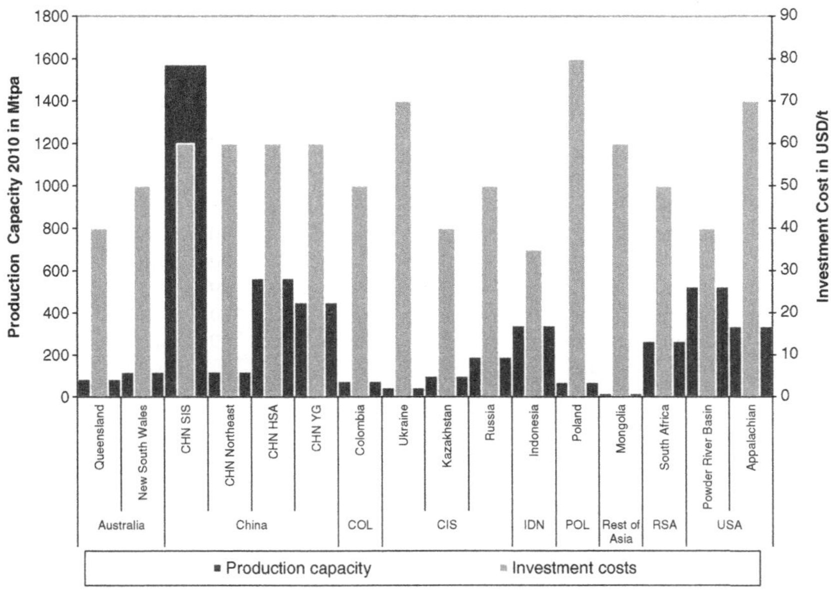

Figure 10.7 Capacity and investment costs for selected production nodes in the base year. Source: Authors' work based on IEA (2008) and Rademacher (2008).

Investments are restricted per period; this restriction reflects the players' limited ability to add more production capacity (and also export capacity, see Section 10.3.4). The data in Table 10.1 is based on historical data and country reports, such as the chapters on Indonesia and South Africa in this volume. For all countries with several nodes, the value gives the aggregate of all producing nodes.

Another important parameter for a multi-period model is the discount rate applied to the profit functions of the producers and exporters. We use the costs of capital to determine the discount rate. The database of A. Damodaran at the New York University's Stern Business School provides estimates of the costs of capital. For the coal industry, using data from eighteen US coal companies including major ones like Peabody, Massey Energy, or Arch Coal, he reports an average cost of capital of 10.3%.[8] For the model, a discount rate of 10% is used for both producers and exporters.

Producers may, in theory, be limited by their available reserves, and we want to be able to capture possible incentives for producers to modify their behavior in reaction to constrained reserves. We follow the standard definition of reserves from the World Energy Council: "proved

Table 10.1 *Assumed production capacity expansion limitations per five-year period*

Country	Production expansion limitation (Mtpa)	Country	Production expansion limitation (Mtpa)
Russia	51	Poland	5
Ukraine	10	Kazakhstan	15
Venezuela	10	Colombia	30
China	450	Mongolia	20
Vietnam	10	Indonesia	90
Australia	40	India	110
Mozambique	15	South Africa	40
US	224		

recoverable reserves are the tonnage within the proved amount in place that can be recovered (extracted from the earth in raw form) under present and expected local economic conditions with existing available technology" (EIA, 2008).

In Figure 10.8 we depict the distribution of global steam coal reserves by major global producing regions. In order to use consistent reserves data, we base ourselves primarily on one source (EIA, 2008, table 8.2) but complement it with other sources for some individual nodes (e.g., DERA, 2012). This data is aggregated on a national level; thus to get the distribution on a sub-national level other sources had to be used. For the US, EIA data (2008, table 15) was used; the reserve distribution for the Indian production nodes is based on GSI (2010) and that for the Chinese producers on NBSC (2007).

Using a static reserves number is, of course, a limitation of our analysis, since real world observations show that the reserve assessments in the coal sector are sparse and often limited in geographical scope. Indeed, as a result of a large available reserve base and of costly exploration, the new delineation of resources that may potentially be mineable reserves is generally sluggish (Rogner et al., 2012). No structural analysis of resource to reserve conversion of the coal sector is available in the literature to our knowledge. However, since we do not want to neglect this fundamental aspect of production of a non-renewable resource, we opt for including regional reserve numbers.

The coal quality data is shown in Table 10.2.[9] The Energy Watch Group (2007) provides evidence that coal quality is generally decreasing over time as reserves are mined. According to this study the decline in coal quality is not only due to a shift toward lower rank coals, like

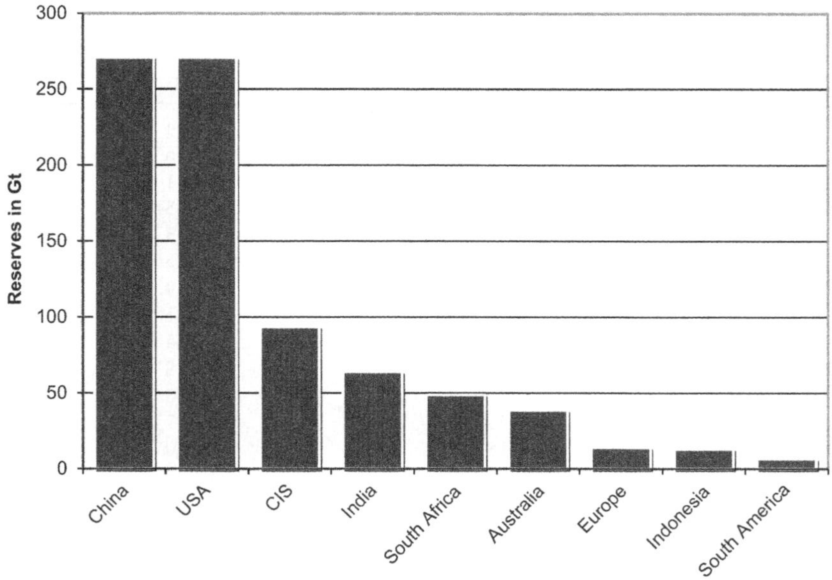

Figure 10.8 Reserves of major countries in COALMOD-World. Source: Authors' work based on DERA (2012), EIA (2008), GSI (2010), and NBSC (2007).

sub-bituminous coals, but also to a quality decline within each class. The model captures some of this effect through the different coal qualities of the producers of the larger countries. For example, if the recent developments in the US continue with more (lower grade) coal from the Powder River Basin being produced, the overall quality of US coal will decrease.

10.3.3 Land transport

Land transport costs and capacities are associated with the transport from a producer to either a local demand node or to an exporter. In case of transportation to a demand node, the center of the demand region is used to calculate the distance; in the case of an exporter, the location of the main harbor is used. Land transport represents mainly transport by train but can also include road transport on trucks and in certain cases river transport by barges or overland conveyor belts. The transport costs are assumed to be constant over time and the capacities can be expanded by investments.

The transport costs for Colombia, Venezuela, South Africa, Indonesia, China, and Australia are based on Baruya (2007). For these countries,

Table 10.2 *Energy content of coal by production node*

Node	Calorific value (in kcal/kg)	Energy content (in GJ/t)
US PRB	4,781	20.004
US Rockies	6,338	26.516
US Illinois	6,226	26.051
US Appalachia	6,949	29.075
Colombia	6,375	26.673
Venezuela	6,375	26.673
Poland	6,300	26.359
Ukraine	6,200	25.941
Kazakhstan	6,000	25.104
Russia	6,400	26.778
South Africa	6,400	26.778
India North	4,717	19.737
India Orissa	4,187	17.520
India West	5,209	21.793
India South	4,866	20.360
Vietnam	7,000	29.288
Indonesia	5,450	22.803
China SIS	6,597	27.600
China Northeast	5,154	21.565
China HSA	6,118	25.598
China YG	6,074	25.413
Australia Queensland	6,500	27.196
Australia New South Wales	6,300	26.359
Mongolia	6,100	25.522
Mozambique	6,400	26.778

Source: Authors' work based on Platts (2009), Indian Bureau of Mines (2012), Ritschel and Schiffer (2007), and Tewalt et al. (2010).

transport capacity data is based on relevant production, consumption, and export data. For the US, data for the transport costs is based on EIA (2011) for the transport costs. The transport capacities inside the US are determined using actual flow data given in EIA (2011, Spreadsheet "Domestic Distribution of U.S. Coal by Origin State, Consumer, Destination and Method of Transportation"). The land transport cost data for the CIS is from Crocker and Kovalchuk (2008) as well as IEA (2011) and the capacities are determined using relevant production, consumption, and export data. This method is also used to estimate the transport capacities in Vietnam and India. The Vietnamese costs were based on relevant price data. The Indian transport cost data is based on Indiastat (2012, Spreadsheet "Railway Freight on Coal in India"). Investments in additional overland transport capacity are set in a range between 10 and 55 USD/tpa depending on distance, landscape and topography, and if the project is mostly greenfield or not.

10.3.4 Export ports

The data for the export ports includes the export capacity in the starting year and the port handling costs as well as the investment costs and investment limits per five-year period. For mathematical reasons, in the complementarity modeling format, we need to include a separate exporter for each producer in each exporting region. The coal of the dedicated exporter has the same calorific value as the producer and, thus, avoids the so-called "pooling problem" of mixing coals of different qualities. This way, we have, for example, an exporter "PRB" (Powder River Basin) next to an exporter "Rockies" at the West Coast of the US. We needed to decide on the allocation of the total available capacities to these two export players. Similarly, where steam coal and coking coal exports use the same harbor facilities we had to decide on the capacity available to steam coal exports (e.g., in Australia). These decisions were taken case-by-case keeping in mind that capacities included in 2010 can be used at relatively low costs throughout the model's time horizon without bearing investment costs for capacity expansions.

Cost data for most regions is based on Baruya (2007). For Colombia, the CIS, South Africa, and Mozambique, the capacity data in the starting year is taken from IEA (2011, 2012a). For Venezuela the costs are assumed to be similar to Colombia, and the capacities are determined using relevant export data. For the United States, the capacity data comes from VDKI (2012) as well as company information. VDKI (2011) provided information on Australia. Chinese port capacities are provided by the NBSC (2007). The costs for Poland are taken from Ritschel and Schiffer (2007) and the capacity is based on export data. Investment costs for additional export capacity are set between 10 and 30 USD/tpa depending on the country and the preexisting infrastructure. Data on allowed maximum expansion per five-year period is mostly from VDKI (2008, 2011) and IEA (2011). Figure 10.9 shows the unit costs of expanding export capacity together with the exporting harbor capacity in the base year.

Constraints in export capacity expansion are particularly notable in global trade patterns. As above for the production investment limitations, the data in Table 10.3 is based on historical data and country reports, such as the chapters on Indonesia and South Africa in this volume. For all countries with several nodes, the value gives the aggregate of all export ports. This means in particular that in the US (a country with four producing regions and three export ports, each possibly having several

Table 10.3 *Assumed export capacity expansion limitations per five-year period*

Country	Export expansion limitation (Mtpa)	Country	Export expansion limitation (Mtpa)
Russia	65	Poland	5
Ukraine	10	US	115
Venezuela	10	Colombia	40
China	50	Australia	108
Indonesia	50	Mozambique	15
South Africa	10		

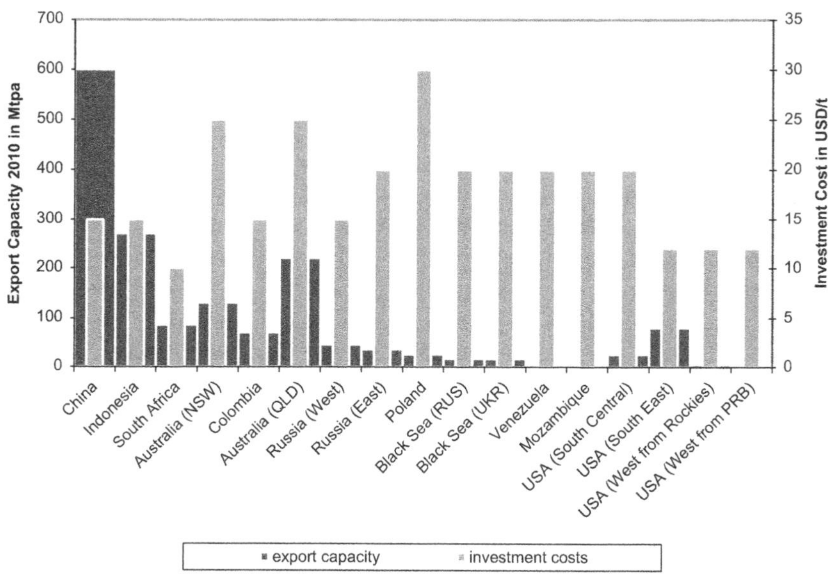

Figure 10.9 Capacity and investment costs for all export nodes in the base year. Source: Authors' work based on IEA (2008) and Rademacher (2008).

dedicated producers) the individual export limitation is lower. For example, the US West exporter has a total expansion possibility of 35 Mtpa per five-year period, but the PRB exporter via the West Coast has only 30 Mtpa investment capacity, the other 5 Mtpa of allowed investments on the West coast being allocated to the US Rockies producer.

China's politically determined export restriction is assigned to the Chinese exporter. For 2010 we use a value close to the actual exports of

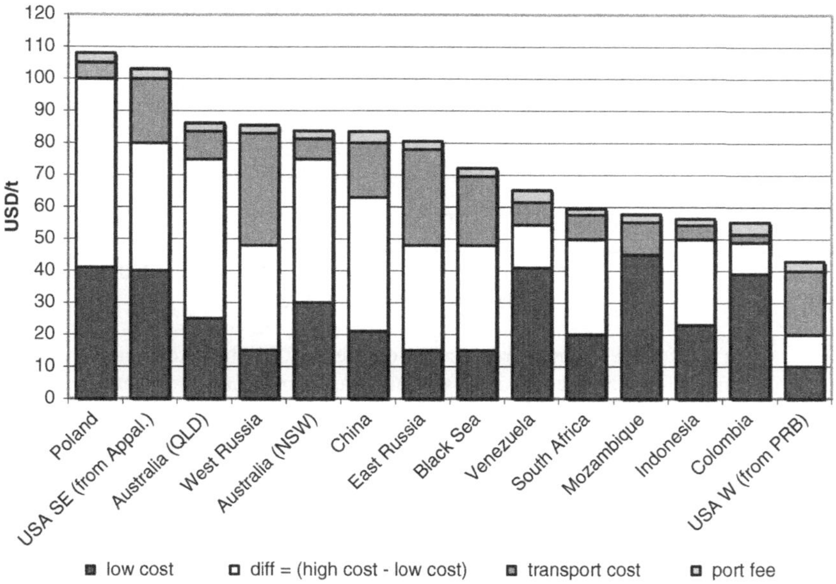

Figure 10.10 FOB costs (2010) for the export countries in COALMOD-World. Source: Authors' work based on IEA (2012a); IEA (2011); Baruya (2007); NBSC (2007); and Ritschel and Schiffer (2007).

20 Mt. Forecasting the level of future export licenses is difficult, and there are no such projections available. For the base case we assume the following values: 2015: 80 Mt; 2020: 90 Mt; 2025: 100 Mt; 2030, 110 Mt; 2035: 120 Mt; and 2040: 130 Mt.

In sum, the cost of a ton of exported coal is the sum of production costs, land transport costs, and the export fee. This is shown in Figure 10.10 for each exporter. In this figure, we also include the range of production costs in the respective production region. This cost range is represented by a white bar in the figure; it is calculated by subtracting the lowest average costs (black bar) from the highest average costs of the production region. (Positive shadow values of binding constraints on production, transportation, or export constraints may also increase effective costs in the model but they depend on the market outcome and are therefore not depicted here.)

10.3.5 Freight rates

Overseas shipment is a cost to the coal importers that we approximate by the freight rates paid to the shipping companies. Freight rates result from

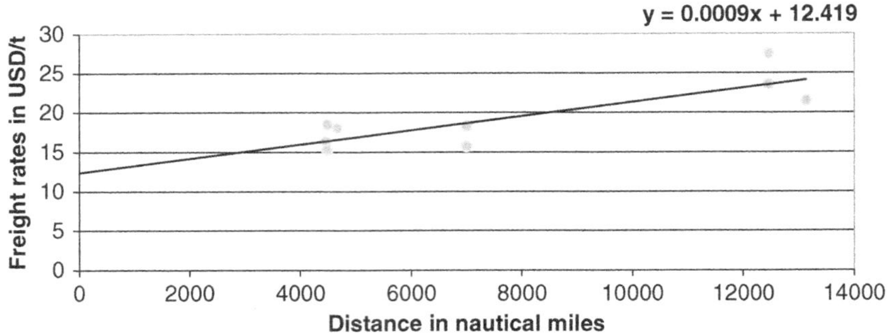

Figure 10.11 Linear regression of average freight rates between 2002 and 2009. Source: Authors' work based on *Platts* newsletters 2002–2009.

the supply-demand equilibrium in the dry bulk carrier market and their quotations have been very volatile in the past.[10] In general, the freight market behaves cyclically. This makes it difficult to predict future freight rates, which are needed as a transportation cost input for the model. Moreover, for the same route there is a difference between Capesize and more expensive Panamax freight rates; the capacity of Capesize ships is higher but Panamax vessels are used more often on shorter routes. In the model, we assume the freight rate (transport cost) to be dependent on distance to reflect the spatial character of the international coal market.

Given historical information on weekly freight rates on all available routes, we specify a linear regression using distance as the explanatory variable. This is necessary because freight rates are only reported for the major shipping routes and not for all exporter-importer pairs that we include in the model. We specify a different regression for 2010, which is based on 2010 observed data (obtained from *Clarksons* for all available coal shipping routes, including Panamax and Capesize), than for the subsequent model periods 2015–2040. For all future periods, we use data from 2002 to 2009 as input to the regression, to obtain long-run average costs.[11]

The regression equation for freight rate as a function of distance determined using observed freight rate averages between 2002 and 2009 is $y = 0.009x + 12.419$. The computed values y are used as shipping costs and are set constant from 2015 until 2040. The shipping costs between every export node and every import node with import possibility are calculated using this equation by plugging in the corresponding distance x.[12] Table 10.4 gives calculated freight rates for some main routes – for example, from South Africa (Richards Bay) to Northern Europe (Rotterdam). Freight rates on each route, in addition to export port

Table 10.4 *Freight rates for selected routes (in USD/t)*

	2010	2015–2030
From	**Australia – Queensland**	
To		
Rotterdam	20.47	23.72
Japan	14.97	15.69
From	**Australia – New South Wales**	
To		
Rotterdam	20.33	23.52
Korea	15.38	16.42
From	**Colombia – Puerto Bolivar**	
To		
Rotterdam	15.44	16.5
From	**South Africa – Richards Bay**	
To		
Rotterdam	16.94	18.65
Chennai	15.09	16.01
From	**US West – Portland, OR**	
To		
Rotterdam	18.08	20.29
Shanghai	15.72	16.91
From	**Indonesia – Banjarmasin**	
To		
Rotterdam	19.84	22.82
Guangzhou	13.76	14.09

Source: Authors' work based on *Platts* newsletters, 2002–2009.

fees, are added to the FOB costs depicted in Figure 10.10, which gives the CIF costs as shown in Figure 10.12. This figure does not include possible additional shadow values of restricted capacities (of production or export capacities) that may be computed in the model runs.

10.3.6 Demand

For the specification of the demand function of each consumption node, we need the "reference price" and "reference quantity" point for each model year. To obtain a consistent demand database for all countries in the model we use data from the International Energy Agency (IEA,

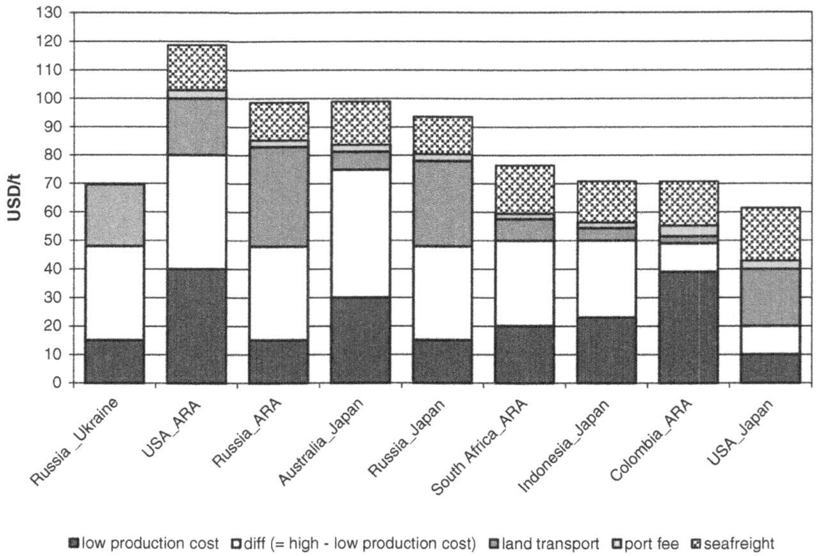

Figure 10.12 CIF costs in 2010 for selected routes (in USD/t).

2012b, c, d). IEA (2012d) reports on two scenarios which we use and contrast in the model: the Current Policies scenario in which coal demand continues to grow strongly; and the New Policies scenario in which coal demand grows more slowly due to some additional climate policy efforts. Our "Base Case" is based on the Current Policies scenario, while the New Policies scenario defines our "Moderate Growth" scenario.

The IEA (2012b, c) reports demand data for 2010 of each consumer in Mtpa. This was converted to Petajoules using the calorific values for the main supplier from Table 10.2. IEA (2012d) gives coal demand projections until 2035 in the two scenarios. This is regionally aggregated data and steam coal is not distinguished from coking coal or lignite. Hence, we apply the regional growth rates between 2010 and each reported later period to the 2010 demand data obtained from IEA (2012b, c) instead of using the reported consumption values directly. We generally use the growth rate of the reported coal consumption for power generation, except for China and India where we use the growth rate of the TPED (total primary energy demand) for coal.[13] This is due to the fact that China and India currently still use a lot of steam coal in other consumption sectors (households, industry) and that, while we have to include this consumption, we need to take into account its falling tendency as reflected in the projected TPED growth rates.

Table 10.5 *Reference consumption in 2010, 2020, and 2030 by IEA region for Base Case and Moderate Growth scenarios (in Mtpa)*

		2020		2030	
Regions	2010	Base Case	Moderate Growth	Base Case	Moderate Growth
World	4,947	6,491	5,871	7,784	6,011
OECD	1,346	1,348	1,259	1,342	1,099
OECD Americas	871	862	818	883	747
United States	862	852	810	873	739
OECD Europe	207	205	183	193	136
EU	182	179	159	166	115
OECD Asia Oceania	267	281	257	265	215
Japan	120	126	115	119	96
Non-OECD	3,600	5,010	4,597	6,000	4,893
East Europe/CIS	99	105	98	115	94
Russia	97	113	103	137	110
Non-OECD Asia	162	255	226	334	253
China	2,493	3,343	3,064	3,780	3,075
India	555	938	867	1,376	1,123
Africa	191	254	236	255	236

Price data was taken from the IEA (2012b, CIF prices and steam coal prices for electricity generation). The prices in 2010 to 2040 in each scenario were calibrated such as to obtain the reference consumption in all demand nodes (with a margin of 1% or less) based on a cost-minimization approach for supply. The calibration resulted in an average price increase between 2010 and 2030 of 1.89% p.a. in the Base Case scenario and 1.06% p.a. in the Moderate Growth scenario. Clearly, this is a much stronger growth of prices than foreseen by studies in the last ten years or so (e.g., EC, 2008). However, we believe that the ongoing demand growth will lead to increasing prices over time on average even though there may be significant short-term fluctuations.

Own-price elasticities of coal demand are part of our demand curve definition. However, empirical research on elasticities, especially for coal, is scarce and the results are often not very satisfying. Dahl (1993) estimates short run elasticities of coal between -0.55 and -0.3. Aune et al. (2004) use a value of -0.19 for the short run elasticity of coal demand in their model. Liu (2004) yields a rather peculiar result of a zero elasticity that is of rather limited use for defining demand functions for the model. We conclude that the price elasticity of coal demand is rather inelastic and assign elasticity values of -0.1, -0.2, or -0.3 to the model consumers, based on the percentage of coal use in the total power generation mix: the more dependent a country is on steam coal use in its electricity sector, the less elastic its demand is assumed to be. We take into account a higher long-run elasticity of demand, which allows a structural change in the demand for coal depending on the expected future prices, by gradually increasing the price elasticities of each period over time: the values reached –0.4 in all countries by 2020, –0.5 by 2025, and –0.6 by 2030 and subsequently.

All consuming countries that are modeled as importers without domestic production are only included with their import demand by subtracting the respective local production from demand in the 2010 data. Where 2010 demand has to be allocated to several consumption nodes in one country, country-specific information is used.[14]

10.4 Modeling results until 2030

10.4.1 Scenario assumptions: high coal demand or climate policies with moderate demand growth

For each model year, the COALMOD-World model delivers results for the inland and seaborne trade flows, the prices, the level of investments and the value of the dual variables of the constraints that indicate if and how strongly a specific constraint is binding. The results for the last two model years 2035 and 2040 are not presented as there is a risk of distortion because there is less incentive to invest without any possible revenue after 2040. For convenience, we only present the results for the years 2010, 2020, and 2030 here.

Our results are based on the assumption of competitive and liberalized markets.[15] We also assume that the markets are fully integrated, that is, when a demand node can be reached by different producers or can import coal from overseas, it can fully substitute between the different sources. We assume that no fundamental structural change in the coal market will happen before 2030. The model results can be called "ideal" results, as they tell us how future demand should be served optimally and in which countries investments should take place. We further assume that there are no policy-based restrictions on trade apart from the case of Chinese exports. The demand side of the Base Case is based on the IEA's *World Energy Outlook*'s Current Policies scenario (IEA, 2012d), implying that there is no strong climate policy affecting steam coal demand. We discuss how the global coal market would develop if more climate policies were implemented as in the IEA (2012d) New Policies scenario with its more moderate growth in coal demand.

The Current Policies and the New Policies scenarios differ in the energy mix induced by climate policy efforts and, hence, in CO_2 emissions. Globally and from all sectors, annual energy-related CO_2 emissions increase by 36% in the Current Policies scenario between 2010 and 2030 and by 20% in the New Policies scenario. (This translates into a doubling of emissions relative to 1990 levels in the Current Policies scenario and a rise of 76% in the New Policies scenario.) The relative contribution of coal to total CO_2 emissions by 2030 is forecast to stagnate in the New Policies scenario (about 42%) and even increase slightly in the Current Policies scenario (46%).[16] In other words, at the global scale no structural transformation involving a drastic shift away from coal is expected in either of the two scenarios. Development between 2010 and 2030 will rather stem from changes in regional energy mix structures.

Regional consumption patterns are affected in the two scenarios by the instruments and relative stringency of climate policy across regions. The IEA scenarios assume a carbon price in the EU, Australia (perhaps unwisely), Korea, California (from 2013 on), China (2020), and Japan (loosely, starting in 2015). Additional national or regional CO_2 intensity targets are assumed in the New Policies scenario (e.g., a 20% reduction in 2020 compared to 2005 in India and a 40% reduction in China) along with renewable targets. All of these assumed carbon policies influence the predicted coal consumption patterns in the demand countries.

We incorporate these national differences in the scenario assumptions in COALMOD-World by using the coal consumption projections from the respective IEA scenarios. Specifically, we use the demand growth projections for coal for power generation in the respective scenarios, with the exception of China and India where the growth rates of Total Primary Energy Demand (TPED) from coal are used (for more details, see Section 10.3.6 and, in the appendix at the end of this chapter, Section 10A.2). This gives the global coal demand for each scenario as shown in Figure 10.13.[17]

We assume that there are restrictions on export capacity expansion for technical and economic reasons. These restrictions are based on historical experience as well as on planned and forecasted expansions. They range from 5 to 30 Mtpa of additional capacity that can be added over a five-year period depending on the country. We do not impose restrictions on expansions of inland transport capacity due to a lack of detailed data on this for all modeled countries.

The model results presented here serve as a starting and reference point for the implementation of scenarios; future applications of the model could refine these scenarios to look at other questions of interest. Restrictions on investments and sales could be imposed according to

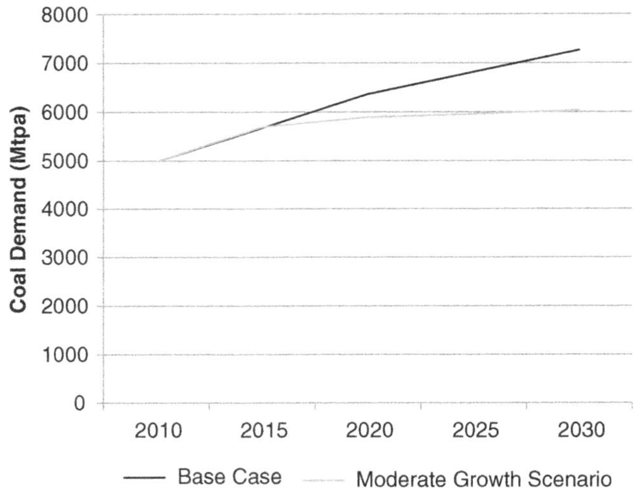

Figure 10.13 COALMOD-World results: development of yearly global coal demand in both scenarios until 2030 (in Mtpa).

the technical and political frameworks of specific countries. On the demand side, stronger climate policies could be implemented. Given the number of parameters in the model there are many possible scenarios that could be explored. In Haftendorn et al. (2013) we present some climate policy scenarios and show how both the demand and the supply side can be important.

10.4.2 Overview of results: the Asian "hunger" for coal

Figure 10.14 and Figure 10.15 give an overview of the evolution of consumption in the world steam coal markets in the Base Case and the Moderate Growth scenario and provide more detail on the supply structure. The total surface of this graph represents total consumption, and the different areas decompose the consumption by its origin, seaborne trade or local domestic supply, with a special focus on India and China. In the Base Case, global coal consumption will rise to more than 7,000 Mtpa in 2030, which is a growth rate of almost 2% p.a. on average between 2010 and 2030. The fastest growth in consumption will take place until 2020, when global coal demand will already reach 6,400 Mtpa, with a demand growth of 2.5% p.a. from 2010.

In contrast, the Moderate Growth scenario will see approximately 6,000 Mtpa coal consumption by 2030, that is 17% less than in the

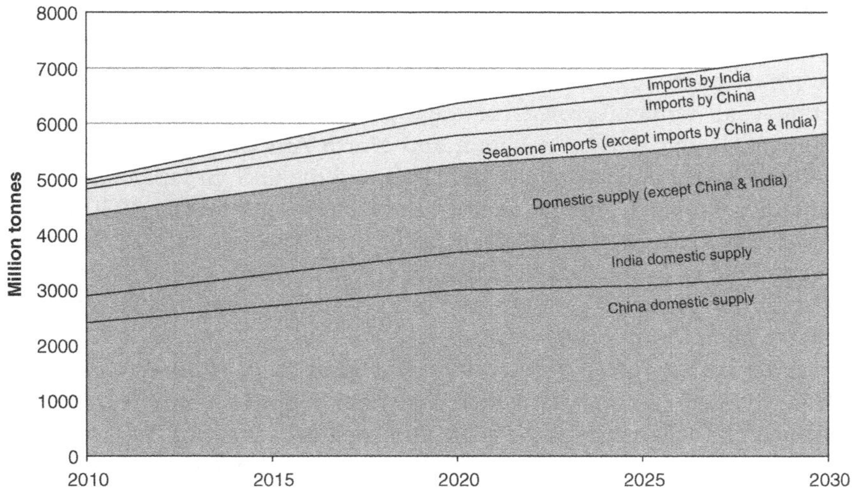

Figure 10.14 Global COALMOD-World results: aggregated consumption and imports in the Base Case (in Mtpa).

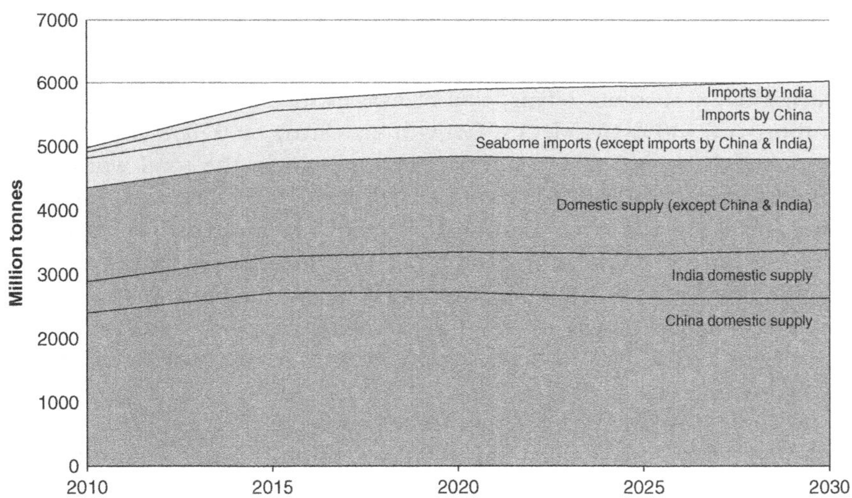

Figure 10.15 Global results: aggregated consumption and imports in the Moderate Growth scenario (in Mtpa).

Base Case, but still a growth of 1% p.a. on average. After a similar development in both scenarios until 2015, consumption growth slows down between 2015 and 2020 and stagnates thereafter. This more moderate coal demand growth in the world results from a reduction of demand in the European Union and North America, and from a considerable reduction of growth in Russia and the CIS, in Japan and, most notably, China. China alone consumes 650 Mtpa less in 2030 in the Moderate Growth scenario than in the Base Case due to the assumed carbon policies after 2020 – a major impact on the world consumption, but hardly an impact on the world market as the import level is approximately the same in both scenarios.

In both scenarios, a quarter of total steam coal consumed is imported from the seaborne market in 2030, up from about 15% in 2010. In both scenarios, the largest share of consumption takes place in Asia, in particular in China and India. Asia[18] accounts for about three-quarters of global consumption across all scenarios and years, with China alone consuming about 50% of the global coal supplies.

While China continues to rely largely on domestic production to satisfy its demand (between 84% and 95% of its consumption, depending on the model years and scenario), India depends more heavily on imports. In the model results it further increases its dependence on imports from 11% in 2010 to 30% or more in 2030 depending on the scenario. In absolute terms, this translates into an increase in Indian imports of 500% (Moderate Growth scenario) or 670% (Base Case) between 2010 and 2030, compared to growth in the total world seaborne trade of 190% (Moderate Growth) or 220% (Base Case). Indian imports represent more than a quarter of the international trade by 2030. Imports to China and India together account for more than half of the international trade already by 2020 and for more than 60% by 2030 (in both scenarios). China has stable imports across the two scenarios and adjusts its domestic production in response to a demand slowdown.

Section 10A.3 in the appendix at the end of this chapter shows that, in addition to the big players China and India, other Asian countries also increase their coal demand to a varying extent in the two scenarios. Indonesia, to date mostly known as one of the largest exporters, is expected to see a strong growth of domestic coal consumption – a development which may also influence its exporting behavior if new mining areas are too costly to be added to the reserve base. Moreover, the traditional importers – Japan, Korea, and Taiwan – will keep their consumption levels stable, and the smaller consumers – Thailand, the Philippines, and Malaysia – will increase their demand.

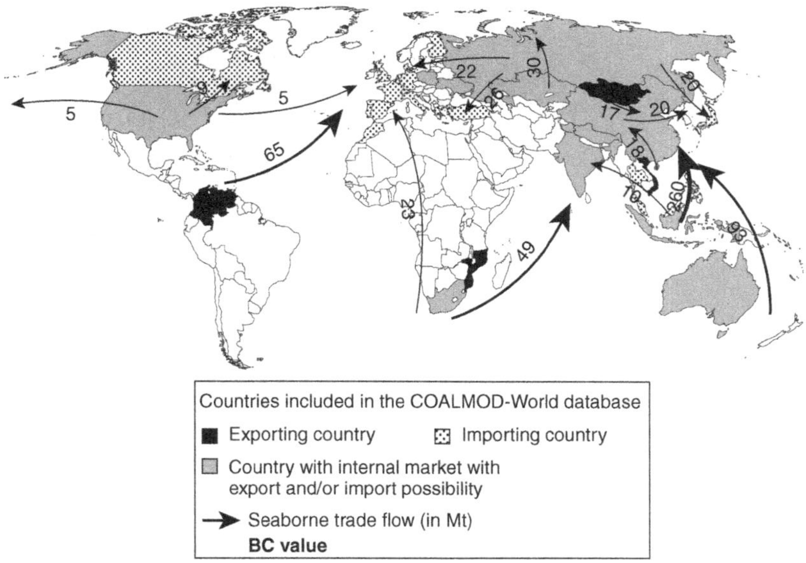

Figure 10.16 Global results 2010: seaborne trade flows (in Mtpa).

10.4.3 Global trade results

10.4.3.1 A shift to Asia

The major focus of the modeling effort is on investigating international trade flows. We obtain long run equilibrium results which abstract away from the volatility of prices or costs, which may cause coal trade flows to diverge from real-world observations in the short run. Our modeling approach implicitly assumes that the production and investment quantities as well as the trade flows of the equilibrium path will be reflective of long-run trends.

The results for 2010 show a notable similarity with the actual observed trade pattern. This is an important achievement, given that we simulate not only the trade flows shown on the maps in Figure 10.16 to Figure 10.18 but also internal markets. We have a global integrated market with flows from South Africa and North America, traditional suppliers of the Atlantic basin, to Asia in the Pacific basin. The direction and relative amounts of the trade flows are a reasonable match to actual trade flows.

Figure 10.17 and Figure 10.18 show the trade flows in 2020 and 2030 in a stylized and simplified representation, distinguishing Base Case and Moderate Growth scenario results. Most obviously, the demand

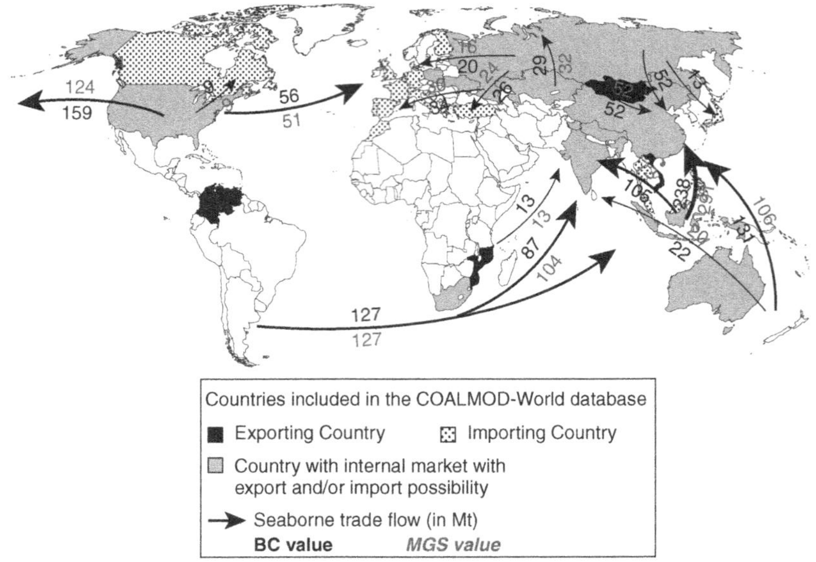

Figure 10.17 Global results 2020: seaborne trade flows (in Mtpa).

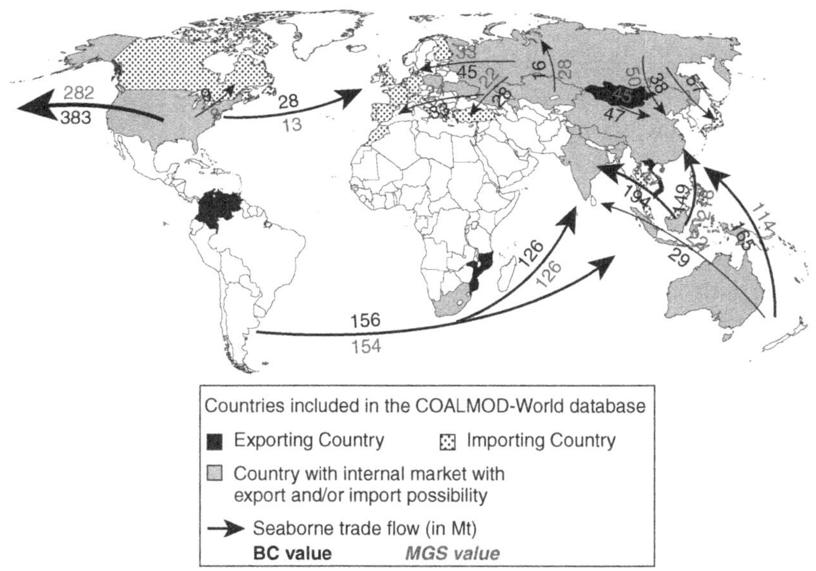

Figure 10.18 Global results 2030: seaborne trade flows (in Mtpa).

reduction in Europe and North America and the consumption increase in Asia lead to a noticeable shift of international trade flows to Asia. This is the case not only in the high-growth Base Case but also in the Moderate Growth scenario. This is due to the concentration of climate and energy efficiency policies in Europe and North America, while Asia (including OECD Asia) pursues a carbon-intensive development path. The model predicts that the overall picture for the global market will have significantly changed by 2030: the US, Russia, and Poland are the only suppliers to Europe; South Africa, Europe's traditional supplier, will have become a major supplier to India and the Pacific market; and Colombia also mainly delivers to the Pacific market.

Australia and Indonesia have traditionally been the key players in the Pacific market. Their exports further increase by 50% and 25%, respectively, and reach the high levels of up to 200 Mtpa for Australia in 2030 and more than 340 Mtpa for Indonesia. Indonesia has been the most dynamic player in the past years, with a higher export level today than Australia. Our model replicates this trend for 2010 and forecasts that Indonesia will consolidate its role as the leading steam coal exporter, ahead of Australia and the US. Relatively low production costs and flexible, low cost investments are the main reasons for this development (see Chapter 7).

However, in the Base Case, Australia will be the more dynamic player, with more than a doubling of its exports between 2010 and 2030. In the case of Moderate Growth on the world markets, Australia would fail to achieve much of this export growth, while Indonesia would be able to increase its exports to even higher levels than in the Base Case (365 Mtpa in 2030). Australia is the coal exporter that would be most affected by the implementation of climate policies (i.e., when comparing Base Case and Moderate Growth scenario) due to its high production costs: its exports are highly vulnerable to future demand in Asia, especially in China.

The US will join this group of large exporters as soon as sufficient export capacities for Powder River Basin (PRB) coal from the US West Coast are available.[19] Some PRB coal may even be exported via the Gulf or East Coast ports, if sufficient domestic railroad capacities in the US are available. Due to its substantial cost advantage, PRB is the most dynamic coal production region in terms of both domestic supply and exports. The role of PRB has increased considerably in recent years, partly due to the low sulfur content that has made it attractive in the US power sector in times of tightening environmental regulation and partly because of its low cost. The model results suggest a strong economic logic for coal ports in the US Pacific Northwest if environmental opposition can be overcome (see Chapter 8).

Moreover, the strong global demand, in particular in Asia, also leaves space for the higher-cost production regions of the US. For example, Appalachian coal is exported to Europe in all years and scenarios because the traditional suppliers South Africa and Colombia will divert their exports to Asia. This continues the recent renaissance of US exports to Europe of coal that has been crowded out domestically by cheap US natural gas.

The third most important exporter in 2010 was South Africa. Despite a further increase of its exports after an expansion of its critical railway capacities to the ports, South Africa is predicted to slip several positions in its export ranking, sitting behind the US, Indonesia, Australia, and possibly Russia and Colombia in later periods.[20] After years as swing supplies between Europe and Asia, we expect South African exports quickly to be redirected to Asia alone, and especially to India. We can see the emergence of a third market (in addition to the traditional Atlantic and Pacific markets) that could be called the "Indian market" and where South Africa will become the key player. However, South African exports are slowed down by the strong demand increase in its domestic market, with an expected consumption of 240 Mtpa after 2020.

10.4.3.2 Other trends

Colombia, and with a similar pattern the smaller Latin American exporter Venezuela, has traditionally been a supplier to the Atlantic basin. In 2010, Europe was its main export destination. However, Colombia also has export capacity on its Pacific coast and therefore relatively easy access to the Asian market. By 2020, the model scenarios predict that it will completely abandon the low-price markets in Europe and turn to Asia where it can obtain higher prices. Colombia has repeatedly faced infrastructure bottlenecks in the past (e.g., in its railroad transportation). Our model results report that if these bottlenecks are eased then Colombia can considerably increase its market share and become the third largest exporter in the world (in the Moderate Growth scenario) thanks to its relatively low production costs.

Russia strengthens its position as an important supplier in the European market, taking over market share from South Africa and Colombia, which turn more to the Asian market. By 2030, approximately 80% of European steam coal demand is served by Russia in both scenarios. Against the background of concerns about security of supply from Russia this might not be a favorable development for European consumers. While the model forecasts a price differential of about 5% between Europe and prices paid in India and China, this gap might be smaller if the

Europeans opt for a self-imposed diversification of suppliers. In the high demand Base Case, Russia also exports some small volumes to Asia, in particular to Korea and China.

In the model scenarios, China will cease its exports to other countries due to its strong domestic demand. Hence, the export restriction, which was still binding in the last decade, will play no role any more. China increases its import levels in the next decades on different paths in the two scenarios: in the Base Case, imports gradually increase from just under 100 Mtpa in 2010 to 450 Mtpa in 2030; in the Moderate Growth scenario, the 2010 and 2030 import levels are the same but there is a steep increase before 2015/2020 and relatively flat import growth thereafter. The congruence of the 2030 import levels between the two scenarios shows that China adjusts its domestic production levels in response to the demand path.

Interestingly, China will rely on coal supplies from Mongolia in addition to imports from the seaborne market that can land in the demand regions along the coast. Mongolia has recently started to scale up its coal production and will continue to do so according to the model by relying on the stable market for its coal in China. It can export the same amounts to China in both scenarios (see Section 10A.3 in the appendix at the end of this chapter). Table 10.6 shows that Mongolia can quickly take market share and become the seventh largest exporter in the world by 2020.

The highest willingness to pay in the global market throughout the model years is found in Japan and Korea, which continue to rely on coal for a large share of their energy mix in both scenarios (cf IEA, 2012d). The shift in their suppliers highlights the shift that takes place at a global scale. In 2010, Japan and Korea purchase coal from Australia, Russia, and China. In the next period, the US invests in export capacities on the West Coast to ship Powder River Basin coal to Asia, and Colombia turns to the Asian market too. By 2020, the US exports large amounts of coal to Japan and Korea: PRB is the lowest cost supplier, followed by Colombia and then Russia. By 2030, the US exports almost 380 Mtpa to Asia, still followed by Colombia and Russia. Australia and Indonesia focus their exports on India, China, and the smaller Asian importers such as Malaysia and Thailand.

These supply relations in Asia also explain the expansions of export capacities of the major suppliers and their sensitivity to demand changes. Table 10.7 reports the difference of export capacity expansions in the Moderate Growth scenario compared to the Base Case for some major exporters and the total of all exporters (including the smaller ones). As could be expected from the assumption of policies that curtail coal demand after 2020 in all countries except India, only Indonesia expands

Table 10.6 *Share and rank in international trade flows of major exporters in both scenarios and over time*

			Base Case				Moderate Growth			
	2010		2020		2030		2020		2030	
Region	Share of int'l trade	Rank	Share of int'l trade	Rank	Share of int'l trade	Rank	Share of int'l trade	Rank	Share of int'l trade	Rank
US	3%	8	19%	2	27%	1	16%	2	23%	2
Russia	10%	3	11%	4	12%	4	10%	4	9%	6
South Africa	10%	3	7%	6	8%	6	9%	6	10%	5
Indonesia	39%	1	29%	1	22%	2	32%	1	28%	1
Australia	13%	2	13%	3	13%	3	11%	3	10%	3
Mongolia	2%	10	4%	7	3%	8	5%	7	3%	7
Colombia	9%	5	9%	5	9%	5	10%	5	10%	3

Table 10.7 *Expansion of export capacity: results of Moderate Growth scenario compared to Base Case (in Mtpa)*

	2015	2020	2025	2030
US	−14	−43	−83	−104
Russia and Ukraine	0	−16	−41	−70
Indonesia	23	22	22	22
TOTAL	**15**	**−34**	**−101**	**−151**

Table 10.8 *Expansion of production capacity: results of Moderate Growth scenario compared to Base Case (in Mtpa)*

	2015	2020	2025	2030
US	−23	−50	−89	−180
Russia and Ukraine	0	−39	−74	−116
South Africa	0	0	−18	−18
India	−46	−83	−113	−112
China	−87	−258	−387	−654
Indonesia	23	12	2	−7
Australia	0	−34	−52	−69
TOTAL	**−128**	**−451**	**−734**	**−1,159**

its capacities even further in this scenario. Indeed, Indonesia is a major supplier to India, and Indonesian exports can be higher in the Moderate Growth scenario because the domestic demand is smaller than in the Base Case. Over all model years, however, Indonesia is limited by its binding reserve constraint and cannot respond indefinitely to high prices in its export markets.[21]

The smaller demand in the Moderate Growth scenario is, of course, reflected in slower production growth. Less new production capacity has to be built to accommodate the lower consumption levels, as reported in Table 10.8, which shows the difference between capacity expansions in the Moderate Growth scenario compared to the Base Case.[22] Only Indonesia expands its production capacities in excess of Base Case expansions in the first years while demand still grows at a significant pace. This is due to the lower production level in the earlier periods of the Moderate Growth scenario, which leave more coal reserves to be produced in later periods. In fact, Indonesia as the only producer with a binding coal reserve constraint until 2040 produces the same cumulated amount between 2010 and 2040 in both scenarios.

10.4.4 Domestic markets

Modeling the interaction between imports and domestic supply is a difficult task. For those demand nodes that can rely on both sourcing possibilities we assume total substitutability. This means that domestic producers and exporters compete directly in a market if the technical capacities allow for it. However, substitutability may in reality not always be the case; for example, some power plants may be specifically designed for domestic coal or, conversely, some coastal power plants may not have the infrastructure to receive domestic coal.

Table 10.9 suggests that producers generally serve their domestic markets first before turning to the export market. Large shares of the total production in a region with domestic coal markets may go toward local consumption. However, the very high shares of domestic production used at home in 2010 go down for most exporting producers over time, reflecting the increasing share of trade in total consumption. For example, in 2030, only about 70% of US production is consumed domestically (up from 98% in 2010), with the remaining 30% being exported. Similarly, Russia and Poland reduce the share of their domestic consumption in order to serve the European market. Likewise, South Africa and Australia decrease their share of production directed to their domestic markets to supply the high-priced Asian demand. In Russia and South Africa, the smaller share of domestic production goes hand in hand with an increase in demand on their domestic markets, showing an expansion of production that is overweighted to increasing exports.

We observe that India is much more dependent on imports than any other major producing country: in all scenarios, India imports about a third of its total consumption in 2030. For reasons discussed in Chapter 4 that have little to do with coal reserves per se, India's domestic production, all of which is consumed in India, only covered about 90% of the country's demand in 2010, and the model forecasts this figure to drop to 67% by 2030. Hence, India will be more affected by possible price volatility in the world market.

In Indonesia, domestic demand will rise strongly as the country builds more coal-fired power plants (see Chapter 7). Indonesia's coal demand doubles between 2010 and 2030 in the Base Case and increases by 50% in the Moderate Growth scenario. These trends have led to some calls for prioritizing coal for domestic use, possibly squeezing exports, but since the extent of any possible export restrictions is unclear, we have decided against including any of these, and our results represent a best-case policy environment for Indonesian production and exports. At the same time,

Table 10.9 *Domestic consumption and share of production consumed domestically by region and scenario (selected countries)*

			Base Case		Moderate Growth	
	2010		2030		2030	
Production region	Domestic consumption (Mtpa)	Share of domestic production	Domestic consumption (Mtpa)	Share of domestic production	Domestic consumption (Mtpa)	Share of domestic production
US	886	98%	921	69%	798	72%
Poland	71	100%	65	97%	45	90%
Ukraine	19	56%	14	22%	6	14%
Kazakhstan	61	67%	70	81%	58	67%
Russia	69	49%	120	40%	84	41%
South Africa	187	72%	249	66%	232	65%
India	488	100%	867	100%	758	100%
Indonesia	58	18%	119	26%	91	20%
China	2,385	99%	3,238	100%	2,584	100%
Australia	63	40%	63	25%	52	28%
TOTAL	4,287	86%	5,726	79%	4,708	78%

Indonesia faces a stringent reserve restriction and a rapid decrease of its coal quality, which is reflected in the increase in its production costs (see Figure 10.20 and Figure 10.21).

In the United States, coal from the Powder River Basin is pushing toward the East Coast and partially substitutes for supplies that were formerly exclusively served by Appalachian coal. Based on current cost data the model results anticipate a further reduction in Appalachian coal supply and an increase of Powder River Basin's share in domestic supplies. Appalachia will continue to serve the European market throughout 2030, however.

10.4.5 Price analysis

10.4.5.1 Higher prices to be expected

COALMOD-World calculates prices in each demand region. The price is the fundamental signal that attracts coal into a market and reflects that market's willingness to pay. In our perfect competition setup, the price equals the production and transportation costs of the highest cost supplier (the so-called marginal supplier, i.e., the supplier of the last, marginal unit) plus any possible shadow prices of the various constraints. Hence, prices may vary significantly between regions depending on the willingness to pay (demand function) and the availability of high-cost and low-cost suppliers. Regional (nodal) prices are calculated in the model calibration mechanism so as to obtain the consumption levels set by the reference data. In other words, the price results show what price level is necessary to reach a given consumption level. Our model cannot predict short-term price volatility but gives long-term price trends based on the fundamentals of the market.[23]

Figure 10.19 shows the price evolution for some major regions in the Base Case. Globally, prices show an upward trend over the time period 2010 to 2030. We observe continuous price increases of 1.89% p.a. on average in the Base Case and 1.06% p.a. in the Moderate Growth scenario until 2030 – that is, an increase of almost 50% or 25%, respectively, over the time horizon. The lowest prices (which also show the lowest price increases over time) are the domestic prices in South Africa, the US, and Russia. These demand nodes are close to large and cheap sources of supply and are not connected by imports to the global market; thus, they are less affected by higher prices in other regions. However, the increase in 2030 of domestic prices in Russia-Central and the US is higher than in the years before due to the increased exports of their producers to Europe and Asia in that year.

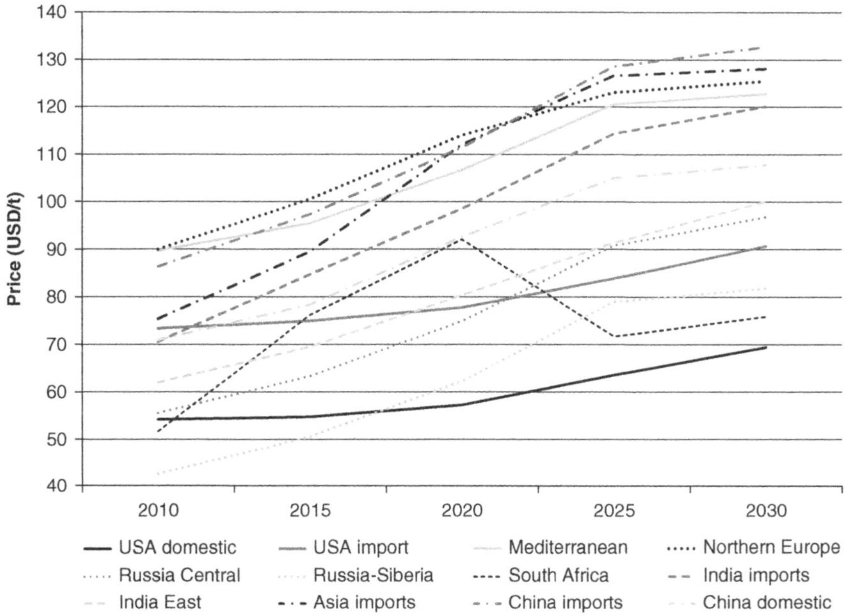

Figure 10.19 Average prices of selected regions for all model years (in USD/t) in the Base Case.

The highest prices are the import prices in Europe, the Mediterranean countries, and in particular in Asia due to the long transport distances (with high transport costs) and the high willingness to pay because of the lack of domestic resources. The highest increase over time can be seen in the Indian and Chinese prices, both domestic and import. This is due to an increased integration into the world market through increased imports.

The kink in the South African price curve directly reflects the cost evolution of the South African producer, which in the absence of import options is the only (and, hence, marginal) supplier in this country. In the first periods, the South African producer carries out significant investments in new production capacity – to replace old mines and, more importantly, to serve the strong demand in Asia – as well as in land transportation capacity. These investment costs affect the final price to consumers in these periods.

It is noteworthy to look at the development of production costs in Figure 10.20 and Figure 10.21 since the prices reflect the costs along the value chain of the marginal supplier in the perfectly competitive

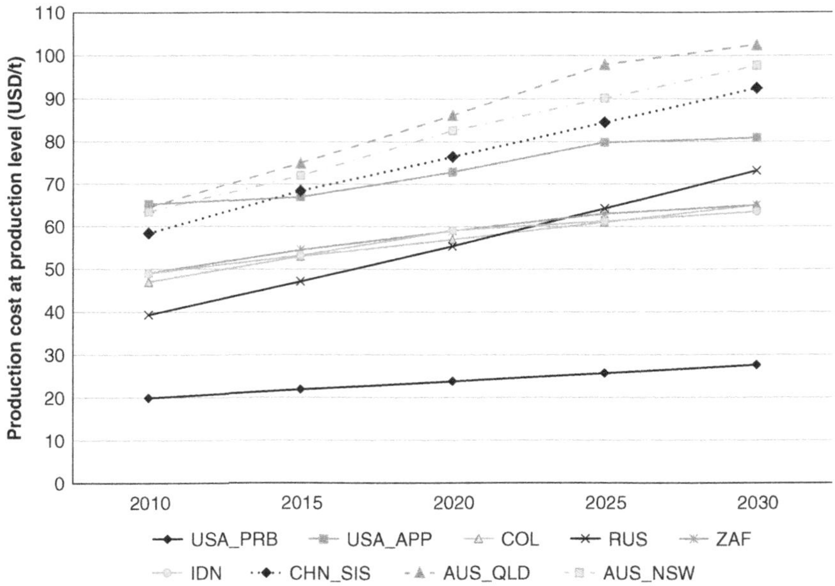

Figure 10.20 Production costs at production level for selected producers over time in the Base Case.

market (in addition to shadow values of binding capacity constraints). The production cost function changes over time (i.e., shifts upward) due to the mine mortality mechanism in the model, which leads to an upward movement of the cost curve's intercept. The upward trend is more or less fast for each producer, for example, it is considerably slower for the Powder River Basin than for Colombia or Australia. The decisive point of the cost curve is the cost level of the last produced unit; this is what is depicted in its development over time in Figure 10.20 and Figure 10.21 (compare to the 2010 cost curves in Figure 10.6 in Section 10.3.2).

Both figures confirm the picture of the base year (2010) cost order for the following years: Australia is the most expensive producer, followed by China SIS, and US Appalachia. The US Powder River Basin is the cheapest supplier on the world market. The different mine mortality parameters and production capacity levels affect the increase of the production cost of the last unit differently for each producer: some producers see their production costs rise quite steeply due to a high mine mortality and expansion investments (e.g., Australia's Queensland and New South Wales producers), others only register a slow increase of the costs of the last unit because of little mine mortality (e.g., Powder

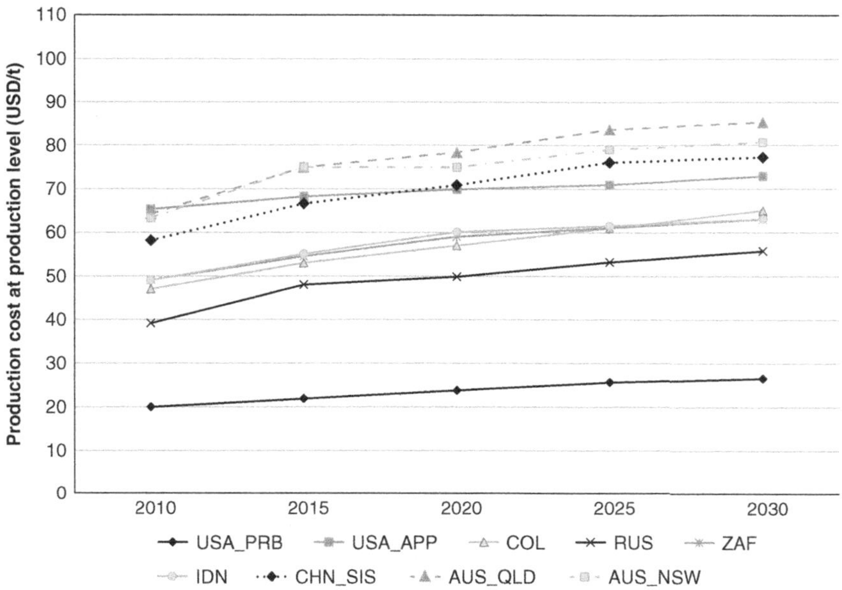

Figure 10.21 Production costs at production level for selected producers over time in the Moderate Growth scenario.

River Basin) or due to the fact that the production level hardly increases (e.g., Appalachia).

The lower consumption levels of steam coal in the Moderate Growth scenario lead to a lower cost (and price) level than in the Base Case. This is because the (increasing) cost function in each year intercepts at a lower point. While in the Base Case in 2030 some production costs may exceed 90 and even 100 USD/t, the cost level in the Moderate Growth scenario stays considerably below this threshold.

10.4.5.2 A resilient, low-risk market

COALMOD-World reports other variables calculated in each run that are not reported here but may be worth further investigation, for example, the shadow values of the production and transportation constraints. While the two scenarios here are presented as a broad indication of the future development of global coal markets, many other scenarios can be thought of to explore the sensitivity of trade flows and investments to different variables. The results and data presented here show that plenty of supplies are potentially available in the global coal market under the condition of competitive pricing and little (if any) restrictions to investments. We conclude that the global coal market would be rather resilient to local supply curtailment shocks as a large number of alternative suppliers can fill in a gap in the seaborne market. This is even more the case because of our assumption that low-cost Powder River Basin coal is free to reach the global market via the Pacific Northwest. However, the supply situation may be more vulnerable for some local markets that are hardly or not at all linked to imports from the international market (e.g., the consumers in exporting regions such as South Africa which are directly exposed to all cost variations).

10.5 Conclusions: plenty of coal supplies require new approaches to climate policy

In this chapter, we have presented a tool of analysis for analyzing the future global steam coal market, the COALMOD-World model. We have shown how we can model this market and its future development using a large-scale equilibrium model that relies on microeconomics and game theory. The combination of model theory and detailed market analysis provides the ground for the development and the implementation of the model.

The model results are based on the assumption that international steam coal consumption will continue to grow. In the Base Case, global coal consumption will rise by almost 50% to more than 7,000 Mtpa in 2030, which is a growth rate of about 2% per annum on average between 2010 and 2030. The growth of demand comes with a rise of international trade due to the increasing mismatch between supply and demand regions. The projected growth of 130% (i.e., more than a doubling) between 2010 and 2030 in the Base Case is mainly driven by fast growing economies like India and China that will multiply the volume of their imports by five. Imports from other countries are projected to increase by only 20% during the same time.

The increasing trend of coal consumption and trade would also persist in a more climate-friendly development path, as shown by our Moderate Growth scenario. Here again, China and India are the main drivers of global coal demand, together with smaller Asian importers in OECD and non-OECD countries. The major trend of Asian demand has substantial effects on the world market since it also affects the Atlantic basin. With falling demand in Europe in the future, many traditional Atlantic coal suppliers turn to Asia. This trend starts very soon, with South Africa supplying India, and continues in the model with exports from Colombia and most notably the US. In addition to the Powder River Basin, which can cheaply supply large volumes of coal as soon as adequate export facilities on the North American west coast are available, even East Coast suppliers would have an incentive to supply the Asian market in the long run.

In a global perspective, the most significant result of our modeling exercise is the shift of trade flows toward the Asia/Pacific markets coupled with the entrance of US PRB coal into the Pacific export market. The picture from the last decade of a globally integrated market with trade flows from the Pacific market to the Atlantic market will change markedly in the next years. In our model results, we observe a rapid shift eastwards until 2020, with flows from South Africa being directed toward Asia and especially India. The US and Russia replace South Africa as the key supplier to Europe; in addition, Colombia turns to the Asian market and supplies Japan and Korea, resource-poor countries with high demand.

This, of course, assumes a strong willingness to pay on the part of Asian consumers to cover the costs along the value chain for higher-cost producers. In order to reach the high coal consumption values of the IEA's *World Energy Outlook* scenarios, a sustained price increase seems necessary, as cheap coal mines will be mined out over time. This also implies that the relatively low prices of the last decades might not be seen again. While we are not arguing that there is any reason to believe the high prices seen in 2008 will reappear anytime soon, we model continuous price increases of 1.89% p.a. in the Base Case and 1.06% p.a. in the Moderate Growth scenario until 2030 – which is an increase of almost 50% or 25%, respectively, over the model's time horizon.

In some producing countries there are currently discussions about restricting exports, for example, to protect the local demand for coal. However, our model results suggest that it would be beneficial for the producers not to restrict their exports so as to benefit from the high prices in Asian import markets. This is especially true for Indonesia and South Africa, which are major suppliers to the international steam coal market. This type of restrictive coal market policy could be explored with the model in future work. Also, restrictions on investment in production capacity in major markets could have a significant effect on global trade and increase the import dependency of India and China. In contrast to our earlier studies, we are now faced with the potential for a lively revival in US exports. These large volumes, which would mostly come from the Powder River Basin if new coal port capacity is built, would somewhat alleviate the pressure on other large suppliers in the next decades and make the global steam coal trade resilient to localized supply shocks.

As the most carbon-intensive fossil fuel, coal is a massive contributor to global CO_2 emissions and, hence, anthropogenic climate change. Thus, climate policies based on a carbon price achieve their results in significant part by reducing coal consumption. We have tried to shed some light on the influence of some (moderate) climate policies by comparing our Base Case to a scenario based on the New Policies scenario of the IEA *World Energy Outlook 2012* (IEA, 2012d). We observe that moderate climate policies will only reduce the pace of the coal demand growth that goes along with economic growth. 2030 coal consumption would still be about 20% above the 2010 levels even under the New Policies scenario, but the regional differences are large. In particular, large developing countries such as India and Indonesia will continue to increase their coal use (+100% and +50%, respectively, with respect to 2010 levels), while the climate policy efforts that reduce coal demand will mainly be undertaken by OECD countries (resulting in a projected reduction of coal use of 20% with respect to 2010 levels).

While one would expect a decrease in coal consumption following stringent climate policies, the successful implementation of carbon capture and storage for coal-fired power plants might in theory change this reality after 2020. CCS could make coal attractive again for the developed countries with ambitious emissions reduction goals that could also afford such technologies. This would boost global trade but also increase competition between importers. Conversely, developed country climate policies in the absence of CCS could lead to a "carbon leakage" effect by causing low global coal prices that provide an incentive for developing countries to consume more coal since they are not bound by climate policy commitments (see Haftendorn et al., 2013). The extent of carbon leakage will depend on the price effect of the climate policy and the substitution possibilities on the global market. The richness of the COALMOD-World model allows the investigation of both effects.

Clearly, supply constraints, though they may surge temporarily in individual producing regions, will not restrict a continuous growth of coal consumption. Hence, to achieve a positive climate outcome, coal consumption needs to be reduced by effective policies on the supply and/or demand side. In Haftendorn et al. (2013), we analyzed the implementation of a withholding of resources by Indonesia compensated by a climate-fund as one possible "unconventional" supply-side policy.[24] More of such policies need to be considered and researched. For example, export taxes on coal could be levied by governments willing to earn additional revenue that compensate foregone profits by climate-constrained exporters. The COALMOD-World model can be used for the implementation of a wide range of scenario model runs to investigate the interaction of the national and international steam coal markets under various energy and climate policy scenarios.

Appendix Model details: node structure, data, and results

10A.1 Nodes of COALMOD-World

Country	Producers		Exporters		Consumers		Port
Canada	*P_CAN*				*C_CAN*	Ontario	No
USA	*P_USA_PRB*	Powder River Basin	*E_USA_SC*	Houston, TX	*C_USA_W*	AK, AZ, CA, CO, ID, MT, NV, NM, OR, UT, WA, WY	No
	P_USA_Rocky	Rocky Mountains	*E_USA_SE*	Norfolk, VA	*C_USA_NC*	IL, IN, IA, KS, MI, MN, NE, MO, ND, OH, SD, WI	No
	P_USA_ILL	Illinois Basin	*E_USA_W*	Portland, OR	*C_USA_SC*	AR, LA, OK, TX	No
	P_USA_APP	Appalachian			*C_USA_SE*	AL, DE, DC, FL, GA, KY, MD, MS, NC, SC, TN, VA, WV	Mobile
					C_USA_NE	CT, ME, MA, NH, NJ, NY, PA, RI, VT	No
Colombia	*P_COL*		*E_COL*	Cartagena			
Venezuela	*P_VEN*		*E_VEN*	Puerto la Cruz			
Morocco					*C_MAR*		Mohammedia
Portugal					*C_PRT*		Sines
Spain					*C_ESP*		Gijon
UK					*C_GBR*		Immingham
Netherlands/ France/ Belgium					*C_NFB*	Netherlands, France, Belgium	Rotterdam

(cont.)

Country	Producers		Exporters		Consumers		Port
Germany					*C_DEU*		Rotterdam + land transport
Denmark					*C_DNK*		Aalborg
Finland					*C_FIN*		Kotka
Italy					*C_ITA*		Taranto
Poland	*P_POL*		*E_POL*	Gdansk	*C_POL*		No
Turkey					*C_TUR*		Mersin + Black sea port
Israel					*C_ISR*		Ashdod
Ukraine	*P_UKR*	Ukrainian/Russian Donetsk	*E_Black_Sea_UKR*	Odessa	*C_UKR*		No
Kazakhstan	*P_KAZ*	Ekibastuz	*E_RUS_West*	Ust-Luga	*C_KAZ*		No
Russia	*P_RUS*	Kemerovo/Kuznets	*E_RUS_East*	Vladivostok	*C_RUS_Central*		No
			E_Black_Sea_RUS	Odessa	*C_RUS_Siberia*		No
South Africa	*P_ZAF*		*E_ZAF*	Richards Bay	*C_ZAF*		No
India	*P_IND_North*	Jharkhand, Madhya Pradesh, Chhattisgarh, West Bengal			*C_IND_East*	Bihar, Chhatisgarh, Jharkhand, Orissa, West Bengal	No
	P_IND_Orissa	Orissa			*C_IND_North*	Delhi, Punjab, Rajasthan, Uttar Pradesh	No
	P_IND_West	Maharashtra			*C_IND_West*	Gujarat, Madhya Pradesh, Maharashtra	Kandla
	P_IND_South	Andhra Pradesh			*C_IND_South*	Andhra Pradesh, Karnataka, Tamil Nadu	Chennai

Thailand					*C_THA*		Bangkok
Malaysia					*C_MYS*		Port Klang
Vietnam	*P_VNM*						
Indonesia	*P_IDN*		*E_IDN*	Surabaya	*C_IDN*		No
China	*P_CHN_SIS*	Gansu, Inner Mongolia, Hebei, Ningxia, Shaanxi, Shaaxi, Shanxi	*E_CHN*	Qinhuangdao	*C_CHN_Northeast*	Heilongjiang, Jilin, Liaoning	No
	P_CHN_Northeast	Heilongjiang, Jilin, Liaoning, Tianjin			*C_CHN_Main*	Beijing, Tianjin, Hebei, Henan, Shandong, Tianjin	No
	P_CHN_HSA	Anhui, Bejing, Fujian, Henan, Jiangsu, Jiangxi, Shandong, Zhejiang			*C_CHN_Eastern*	Anhui, Jiangsu, Hubei, Shanghai, Zhejiang	Shanghai/ Ningbo
	P_CHN_YG	Chongqing, Guangxi, Guizhou, Guizhou, Hubei, Hunan, Sichuan, Yunnan			*C_CHN_South*	Chongqing, Fujian, Guangdong, Guangxi, Guizhou, Hong Kong, Hunan, Jiangxi, Sichuan	Guanzhou
					C_CHN_SIS	Shanxi, Shaaxi, Inner Mongolia	No
Korea					*C_KOR*		Ulsan
Japan					*C_JPN*		Yokohama
Taiwan					*C_TWN*		Kaohsiung
Philippines					*C_PHL*		Manila
Australia	*P_AUS_QLD*	Queensland	*E_AUS_QLD*	Gladstone	*C_AUS*		No
	P_AUS_NSW	New South Wales	*E_AUS_NSW*	Botany Bay			
Mozambique	*P_MOZ*		*E_MOZ*	Maputo			
Mongolia	*P_MNG*						

10A.2 Data of COALMOD-World: reference demand data

Table 10.10 World Energy Outlook *demand projections for coal for power generation in the scenarios (converted to PJ)*

		2020		2030	
Regions	2010	CPS	NPS	CPS	NPS
World	94,161	123,552	111,746	148,171	114,425
OECD	36,635	36,718	34,122	36,341	28,931
OECD Americas	20,725	20,641	19,594	21,269	17,752
United States	19,385	19,176	18,213	19,636	16,622
OECD Europe	9,253	9,085	8,081	8,457	5,862
EU	8,876	8,290	7,494	7,243	4,982
OECD Asia Oceania	6,657	6,992	6,406	6,615	5,359
Japan	2,596	2,554	2,219	2,512	2,052
Non-OECD	57,485	86,834	77,665	111,829	85,494
East Europe / CIS	5,736	6,071	5,652	6,657	5,443
Russia	2,973	3,433	3,140	4,187	3,349
Non-OECD Asia	48,860	76,660	68,119	100,483	76,158
China*	67,076	89,937	82,442	101,702	82,735
India*	11,849	20,014	18,507	29,351	23,950
Africa	2,638	3,517	3,266	4,019	3,433

* For China and India the table shows the respective values for Total Primary Energy Demand (TPED) from coal. Besides the power generation sector, in these two countries steam coal is also heavily used in industrial processes. Reference consumption values for 2010 are based on total steam coal consumption. Taking this fact into account we use TPED as a basis for forecasting future steam coal demand in India and China.

Source: Authors' work based on IEA, 2012d.

Table 10.11 *Various input parameters for COALMOD-World: production nodes*

	Prod. capacity 2010 (Mtpa)	Reserves (Mt)	Investment costs ($/tpa)	Max. capacity expansion per 5 year (Mtpa)	Prod. cost function intercept 2010 ($/t)	Prod. cost function slope ($/t×t)	Mine mortality rate (%)
US PRB	525	112,555	40	100	10	0.019	8
US Rocky	79	20,704	60	26	30	0.0633	1
US ILL	115	82,887	50	34	40	0.1128	3
US APP	336	54,572	70	64	40	0.119	10
Colombia	75	6,229	50	30	39	0.1333	15
Venezuela	10	479	50	10	41	1.346	6
Poland	71	13,997	80	5	41	0.831	2
Ukraine	45	16,271	70	10	31	0.6889	7
Kazakhstan	100	28,145	40	15	15	0.147	20
Russia	190	49,078	50	51	15	0.1737	4
South Africa	267	48,740	50	40	20	0.1124	2
IND North	281	35,663	40	60	9	0.1709	1
IND Orissa	123	14,416	40	30	11	0.1302	2
IND West	53	7,134	40	10	22	0.0759	5
IND South	58	6,755	40	10	23	0.344	5
Vietnam	62	150	60	10	12	0.327	60
Indonesia	340	13,000	35	90	23	0.0794	3
CHN SIS	1573	213,400	60	300	21	0.0267	0.8
CHN Northeast	121	15,900	60	50	26	0.2	1
CHN HSA	564	4,700	60	70	30	0.0514	1
CHN YG	450	36,800	60	30	25	0.1111	14
AUS QLD	85	24,764	40	20	25	0.5882	5
AUS NSW	119	13,829	50	20	30	0.3782	8
Mongolia	17	1,170	60	20	30	0.2	10
Mozambique	5	212	80	15	45	0.05	10

10A.3 Results of COALMOD-World

Table 10.12 *Results of COALMOD-World: consumption, domestic supply, and imports by consuming country and scenario in 2010, 2020, and 2030*

	Base Case						Moderate Growth			
	2010		2020		2030		2020		2030	
Consumption region	Imports (Mt)	Dom. supply (Mt)	Imports (Mt)	Dom. supply (Mt)	Imports (Mt)	Dom. supply (Mt)	Imports (Mt)	Dom. supply (Mt)	Imports (Mt)	Dom. supply (Mt)
US	0	886	0	897	0	921	0	861	0	798
Canada	9	0	9	0	9	0	9	0	8	0
Poland	0	71	0	70	0	65	0	62	0	45
Ukraine	19	19	26	14	30	14	30	7	30	6
Kazakhstan	0	61	0	64	0	70	0	60	0	58
Russia	30	69	29	84	16	120	32	72	28	84
South Africa	0	187	0	249	0	249	0	232	0	232
India	63	488	222	684	423	867	207	629	320	758
China	116	2,385	405	2,943	495	3,238	413	2,671	497	2,584
Indonesia	0	58	0	91	0	119	0	81	0	91
Australia	0	63	0	67	0	63	0	61	0	52
Korea	83	0	82	0	77	0	75	0	63	0
Japan	123	0	140	0	148	0	130	0	127	0
United Kingdom	20	0	18	0	19	0	16	0	12	0
Netherlands, France, Belgium	22	0	20	0	20	0	18	0	14	0
Germany	31	0	29	0	29	0	26	0	20	0
Taiwan	62	0	80	0	109	0	75	0	81	0

Denmark	4	0	4	0	4	0	3	0	3	0
Finland	5	0	4	0	4	0	4	0	3	0
Spain	10	0	9	0	9	0	8	0	6	0
Morocco	4	0	6	0	7	0	5	0	6	0
Portugal	4	0	4	0	4	0	4	0	3	0
Italy	14	0	15	0	13	0	13	0	9	0
Turkey	16	0	17	0	19	0	16	0	16	0
Israel	9	0	9	0	9	0	8	0	6	0
Thailand	17	0	22	0	29	0	20	0	22	0
Malaysia	21	0	33	0	43	0	29	0	33	0
Philippines	11	0	15	0	19	0	13	0	15	0
TOTAL	**693**	**4,287**	**1,198**	**5,163**	**1,535**	**5,726**	**1,154**	**4,736**	**1,322**	**4,708**

Table 10.13 *Results of COALMOD-World: domestic supply and exports by producing country and scenario in 2010, 2020, and 2030*

	Base Case						Moderate Growth			
	2010		2020		2030		2020		2030	
Production region	Export (Mt)	Dom. supply (Mt)	Export (Mt)	Dom. supply (Mt)	Export (Mt)	Dom. supply (Mt)	Export (Mt)	Dom. supply (Mt)	Export (Mt)	Dom. supply (Mt)
US	19	886	224	897	420	921	184	861	303	798
Poland	0	71	0	70	2	65	0	62	5	45
Ukraine	15	19	36	14	49	14	35	7	37	6
Kazakhstan	30	61	29	64	16	70	32	60	28	58
Russia	72	69	135	84	182	120	116	72	122	84
South Africa	72	187	87	249	126	249	104	232	126	232
India	0	488	0	684	0	867	0	629	0	758
Vietnam	18	0	0	0	0	0	0	0	0	0
Indonesia	270	58	343	91	343	119	365	81	365	91
China	20	2,385	0	2,943	0	3,238	0	2,671	0	2,584
Australia	93	63	153	67	194	63	125	61	136	52
Mongolia	17	0	52	0	47	0	52	0	45	0
Colombia	60	0	112	0	136	0	112	0	136	0
Venezuela	5	0	15	0	19	0	16	0	18	0
Mozambique	4	0	13	0	0	0	13	0	0	0
TOTAL	**695**	**4,287**	**1,199**	**5,163**	**1,534**	**5,726**	**1,154**	**4,736**	**1,321**	**4,708**

Table 10.14 *Trade flows in COALMOD-World (in Mtpa)*

			2015		2020		2030	
From	To	2010	BC	MGS	BC	MGS	BC	MGS
US Appalachia	Canada	9	9	9	9	9	9	8
US Appalachia	Europe	5	53	60	56	51	28	13
South America	Europe	65	23	2	0	0	0	0
Kazakhstan	Russia (Siberia)	0	0	0	0	0	0	0
Kazakhstan	Russia (Central)	30	30	35	29	32	16	28
Russia & Ukraine	Ukraine	19	23	22	26	30	30	30
Russia	Europe	22	10	23	20	16	45	33
Russia	Turkey	26	26	26	26	24	28	22
Russia (via Far East)	OECD Asia	20	50	48	13	0	57	0
South Africa	Europe & West Mediterranean	23	4	0	0	0	0	0
South Africa	East Mediterranean	0	0	0	0	0	0	0
South Africa	India	49	80	94	87	104	126	126
Vietnam	China	18	21	21	0	0	0	0
Indonesia	India	10	55	29	105	73	194	147
Indonesia	Asia (except India)	260	241	291	238	291	149	218
China	S. Korea	20	0	0	0	0	0	0
Australia	India & Thailand	0	18	18	22	20	29	22
Australia	Asia (except India & Thailand)	93	103	110	131	106	165	114
Mongolia	China	17	35	35	52	52	47	45
Mozambique	India	4	12	15	13	13	0	0

Table 10.14 *(cont.)*

			2015		2020		2030	
From	To	2010	BC	MGS	BC	MGS	BC	MGS
South America	Asia	0	83	107	127	127	156	154
US West	Asia	5	40	55	159	124	383	282
Poland	Europe	0	0	0	0	0	2	5
RUS	China	0	0	2	52	50	38	50
Black Sea	Europe & West Mediterranean	0	21	21	34	30	33	24

Notes

1. Note that this does not hold true for the coking coal market in which a small number of suppliers can exert monopolistic market power (Trüby, 2013), and several consumers have for a long time been organized in oligopsonies (Graham et al., 1999).
2. Typically, low quality coal such as lignite or low-calorific sub-bituminous coal is sold domestically. The relatively low energy content per tonne compared to steam coal makes long-distance international transport uneconomic.
3. See Haftendorn et al. (2013) for an example of an analysis of climate policy effects with the COALMOD-World model.
4. Put simply, these are the first order conditions of the optimization problems.
5. We interpret the results only until 2030 because of a risk of distortion of the investment results given the short payback period after 2030.
6. For the coal sector, in which reserve assessments are sparse, another direction of the change of the cost function intercept is possible: the intercept could decrease due to the discovery and opening of lower-cost mines. In Haftendorn et al. (2012), we discuss this possibility and propose an extended endogenous cost mechanism. Moreover, technological progress in coal extraction methods could reduce costs over time even in existing mines (see Chapter 6, for example, for a discussion of how mining technology improvements at different junctures boosted productivity and kept costs in check in Australia). However, we lack empirical data to include these mechanisms for cost decreases in the model structure.
7. For the United States, EIA (2012) gives production information; for China, data comes from the NBSC (2007); for India, from Indiastat (2012, Spreadsheet "*Average Cost of Production of Coal*") and Indian Bureau of Mines (2012, in the chapter title "Prices"); and for Australia, updated data from Rademacher (2008, p. 78), Bayer (2012), NSWDPI (2009), and QLDDME (2009) are used. For Vietnam, the production capacity is taken from Rademacher (2008); since there was no cost data available for Vietnam, these were determined using relevant price data. For Poland, the costs are based on Ritschel and Schiffer (2007).
8. See http://pages.stern.nyu.edu/~adamodar/New_Home_Page/datafile/wacc.htm (accessed on January 27, 2010).
9. Coal quality data is based on Platts (2009) for the US, Colombia, Venezuela, Poland, Russia, South Africa, Indonesia, and Australia. For China, it is based on Tewalt et al. (2010), and for India on Indian Bureau of Mines (2012, chapter titled "Coal and Lignite"). For Vietnam the quality data is taken from Ritschel and Schiffer (2007).
10. Dry bulks include commodities such as iron ore, coal, or grain.
11. Sources for weekly freight rates (end-of-week quotations) are *Platts* newsletters from 07/2002 to 10/2009 (where extreme values have been removed from the data).
12. Distance is calculated using the *PortWorld* online distance calculator; see http://www.portworld.com/map/.

13. In this, our approach is consistent with IEA (2012d, chapter "Coal Market Outlook").
14. For the US, data from the EIA (2010, Spreadsheet *Domestic Distribution of U.S. Coal by Origin State, Consumer, Destination and Method of Transportation*) is used. For Russia, the allocation of demand is based on the regional location of coal-fired power plants given by EFA (2008). For China, the coal flows reported in Mou and Li (2012) are used to obtain the regional breakdown of demand. For India, Indiastat (2012, Spreadsheet *Region/State-wise Linkage, Receipt, Import and Consumption of Coal in Various Thermal Power Stations*) gives regional information.
15. We are aware that not all countries currently have fully liberalized domestic markets (e.g., India). However we assume that the markets' structure or outcomes will move toward competitiveness in the future.
16. In the IEA scenarios we consider, CCS will play at best a marginal role in reducing CO_2 emissions from coal consumption. IEA (2012d) is not explicit about CCS in its scenario descriptions, but the slight reduction of CO_2 emissions per unit of energy of coal consumed implies that a very small use of CCS is assumed in regions where such projects were planned in 2012. Only the stringent climate policy scenario (450 ppm) includes a large deployment of CCS (IEA, 2012d). See Hirschhausen et al. (2012) as well as Chapters 11 and 12 in this volume for perspectives on the prospects for CCS technology.
17. COALMOD-World is an energy-based model that calculates trade flows in Petajoules. For better representation, the results shown are aggregated and expressed in Mtpa. These values are calculated using the relevant quality factors of each producer. Detailed results are reported in Section 10A.3 in the appendix at the end of this chapter.
18. Without Israel, Central Asia and Russia, and Oceania.
19. In the real world, this capacity could also come from ports in British Columbia (see Chapter 8).
20. As of 2012, South Africa had already fallen behind Russia and Colombia in total steam coal exports (see Chapter 1).
21. In our model results, the Indonesian reserves constraint is binding until 2040. In other words, all steam coal currently known to be mineable in Indonesia will be exploited until 2040, and even more will be produced if the reserve base increases. One can reasonably extrapolate from past observations that more exploration activity and an increase of the reserve base will take place in Indonesia once the remaining reserves have passed below a certain threshold.
22. Note that this table only reports on the expansion of production capacities and does not include those investments in new mines that replace old mining capacities in line with the mining mortality mechanism.
23. Short-term volatility can be caused by extreme weather events, strikes, infrastructure problems, etc. on the supply side. On the demand side, we can consider shocks that come from the energy system such as short-term problems with nuclear reactors or low water level of hydropower plants.
24. This is similar to the proposed (and failed) climate fund for the oil in the Ecuadorian Yasuní reservation.

References

Aune, Finn Roar, Rolf Golombek, Sverre A. C. Kittelsen, Knut Einar Rosendahl, and Ove Wolfgang. 2004. Liberalizing the Energy Markets of Western Europe – A Computable Equilibrium Model Approach. *Applied Economics* 36(19): 2137–2149.

Baruya, Paul. 2007. *Supply Costs for Internationally Traded Coal*. London: IEA Clean Coal Centre.

Bayer, Arne. 2012. Investment dynamics and future production costs of coal. In *Long-Term Costs and Reserves of Coal, Oil, and Natural Gas*. Berlin: DIW.

Carl, Jeremy, Varun Rai, and David G. Victor. 2008. Energy and India's Foreign Policy. In *Program on Energy and Sustainable Development*. Working Paper #75.

Cottle, Richard W., Jong-Shi Pang, and Richard E. Stone. 1992. The Linear Complementarity Problem. Edited by W. Rheinboldt, *Computer Science and Scientific Computing*. Boston: Academic Press, Inc.

Crocker, Geoff, and Alex Kovalchuk. 2008. *Prospects for Coal and Clean Coal Technologies in Russia*. London: IEA Clean Coal Centre.

Dahl, Carol A. 1993. *A Survey of Energy Demand Elasticities in Support of the Development of the NEMS*. Golden: Colorado School of Mines.

DERA. 2012: *Energy Study 2012: Reserves, Resources and Availability of Energy Resources*. The Federal Institute for Geosciences and Natural Resources (BGR) on behalf of the German Mineral Resources Agency (DERA). Hanover, Germany.

Eberhard, Anton. 2011. The Future of South African Coal: Market, Investment, and Policy Challenges. In *Program on Energy and Sustainable Development*. Stanford University. Working Paper #100.

EC. 2008. European Energy and Transport: Trends to 2030 – Update 2007. Brussels: European Commission.

EFA. 2008. General Scheme for the Positioning of Power Plants till 2020. Energy Forecasting Agency. Available in Russian only at http://e-apbe.ru/scheme/gs.doc.

Egging, Ruud, Steven A. Gabriel, Franziska Holz, and Jifang Zhuang. 2008. A Complementarity Model for the European Natural Gas Market. *Energy Policy* 36(7): 2385–2414.

Egging, Ruud, Franziska Holz, and Steven A. Gabriel. 2010. The World Gas Model: A Multi-Period Mixed Complementarity Model for the Global Natural Gas Model. *Energy* 35(10): 4016–4029.

EIA. 2008. Annual Coal Report 2007. Available at http://www.eia.doe.gov/cneaf/coal/page/acr/acr_sum.html.

EIA. 2010. *Annual Coal Distribution Report*. Washington, DC: Energy Information Administration.

EIA. 2011. *Rail Coal Transportation Rates to the Electric Power Sector 2001–2008*. Washington, D.C.: Energy Information Administration.

EIA. 2012. *Energy Information Administration. Annual Coal Report, table 32*. Available from http://www.eia.gov/coal/annual/.

Ellerman, A. Denny. 1995. The World Price of Coal. *Energy Policy* 23(6): 499–506.

EMF. 1978. *Coal in Transition: 1980–2000*, vol. 1 and 2. Stanford University: Energy Modelling Forum.

Energy Watch Group. 2007. Coal: Resources and Future Production. EWG-Paper #1/07. Available from http://energywatchgroup.org/.

Graham, Paul, Sally Thorpe, and Lindsay Hogan. 1999. Non-competitive Market Behavior in the International Coking Coal Market. *Energy Economics* 21: 195–212.

GSI. 2010. *Inventory of Geological Resource of Indian Coal.* Available at http://www.cmpdi.co.in/docfiles/coalreserve_010410.pdf, Geological Survey of India.

Haftendorn, Clemens, and Franziska Holz. 2010. Modeling and Analysis of the International Steam Coal Trade. *The Energy Journal* 31(4): 205–229.

Haftendorn, Clemens, Franziska Holz, and Christian von Hirschhausen. 2010. Coalmod-World: A Model to Assess International Coal Markets until 2030, Program on Energy and Sustainable Development, Working Paper 96, Stanford University.

Haftendorn, Clemens, Franziska Holz, and Christian von Hirschhausen. 2012. The End of Cheap Coal? A Techno-Economic Analysis until 2030 Using the COALMOD-World Model. *Fuel* 102: 305–325.

Haftendorn, Clemens, Franziska Holz, Claudia Kemfert, and Christian von Hirschhausen. 2013. Global steam coal markets until 2030: perspectives on production, trade and consumption under increasing carbon constraints. In *Handbook on Energy and Climate Change,* edited by R. Fouquet. Cheltenham: Edward Elgar Publishing Ltd., 103–125.

Hartley, Peter, and Kenneth B. Medlock. 2006. The Baker Institute World Gas Trade Model. In *Natural Gas and Geopolitics: From 1970 to 2040,* edited by D. G. Victor, A. M. Jaffe, and M. H. Hayes. New York: Cambridge University Press, 357–406.

Hirschhausen, Christian von, Johannes Herold, and Pao-Yu Oei. 2012. How a "Low Carbon" Innovation Can Fail–Tales from a "Lost Decade" for Carbon Capture, Transport, and Sequestration (CCTS). *Economics of Energy and Environmental Policy* 1(2): 115–123.

Hotelling, Harold. 1931. The Economics of Exhaustible Resources. *Journal of Political Economy* 39(2): 137–175.

Huppmann, Daniel, and Franziska Holz. 2012. Crude Oil Market Power – A Shift in Recent Years? *The Energy Journal* 33(4): 1–22.

IEA. 2008. *World Energy Outlook 2008.* Paris: OECD/International Energy Agency.

IEA. 2010. *Coal Information 2010.* Paris: OECD/International Energy Agency.

IEA. 2010. *World Energy Outlook 2010.* Paris: OECD/International Energy Agency.

IEA. 2011. *Coal. Medium Term Market Report.* Paris: OECD/International Energy Agency.

IEA. 2012a. *Coal. Medium Term Market Report.* Paris: OECD/International Energy Agency.

IEA. 2012b. *Coal Information 2012.* Paris: OECD/International Energy Agency.

IEA. 2012c. *Energy Statistics of Non-OECD Countries 2012.* Paris: OECD/International Energy Agency.

IEA. 2012d. *World Energy Outlook 2012* Paris: OECD/International Energy Agency.

IEA. 2013. *Coal Information 2013*. Paris: OECD/International Energy Agency.
Indian Bureau of Mines. 2012. *Indian Minerals Yearbook 2011*. Nagpur: Government of India, Ministry of Mines, Indian Bureau of Mines.
Indiastat. 2012. Mines and Minerals Statistics. Available from http://www.indiastat.com. Accessed September 19, 2012.
Kolstad, Charles D., and David S. Abbey. 1984. The Effect of Market Conduct on International Steam Coal Trade. *European Economic Review* 24: 39–59.
Li, Raymond, Roselyne Joyeux, and Ronald D. Ripple. 2010. International Steam Coal Market Integration. *The Energy Journal* 31(3): 181–203.
Lise, Wietze, Benjamin F. Hobbs, and Frits van Oostvoorn. 2008. Natural Gas Corridors between the EU and Its Main Suppliers: Simulation Results with the Dynamic GASTALE Model. *Energy Policy* 36(6): 1890–1906.
Liu, Gang. 2004. Estimating Energy Demand Elasticities for OECD Countries. A Dynamic Panel Approach. In *Discusssion Papers Statistics Norway*.
Minchener, Andrew. 2007. *Coal Supply Challenges for China*. London: IEA Clean Coal Centre.
Minchener, Andrew. 2009. *Future Coal Supply Prospects*. London: IEA Clean Coal Centre.
Mou, Dunguo, and Zhi Li. 2012. A Spatial Analysis of China's Coal Flow. *Energy Policy* 48: 358–368.
NBSC. 2007. *China Energy Statistical Yearbook 2007*. China Statistics Press. Beijing: National Bureau of Statistics of China.
NSWDPI. 2009. *Summary of NSW coal statistics*. New South Wales Department of Primary Industries. Available from http://www.resources.nsw.gov.au/resources/coal.
Paulus, Moritz, and Johannes Trüby. 2011. Coal Lumps vs. Electrons: How Do Chinese Bulk Energy Transport Decisions Affect the Global Steam Coal Market? *Energy Economics* 33 (6): 1127–37.
Platts. 2009. Methodology and Specification Guide – Coal. Available from www.platts.com.
QLDDME. 2009. *Coal Statistics*. Queensland Department of Mines and Energy. Available from http://mines.industry.qld.gov.au/.
Rademacher, Maggi. 2008. Development and Perspectives on Supply and Demand in the Global Hard Coal Market. *Zeitschrift für Energiewirtschaft* 32(2): 67–87.
Ritschel, Wolfgang, and Hans-Wilhelm Schiffer. 2007. *World Market for Hard Coal, 2007 Edition*. Essen: RWE Power.
Rogner, H.-H. 1997. An Assessment of World Hydrocarbon Resources. *Annual Review of Energy and the Environment* 22: 217–262.
Rogner, H.-H., et al. 2012. Energy Resources and Potentials. In *Global Energy Assessment – Toward a Sustainable Future*, edited by Thomas B. Johansson, Nebojsa Nakicenovic, Anand Patwardhan, and Luis Gomez-Echeverri. Cambridge and New York: Cambridge University Press, and Laxenburg, Austria: International Institute for Applied Systems Analysis, 423–512.
Shapiro, Jeremy F., and David E. White. 1982. A Hybrid Decomposition Method for Integrating Coal Supply and Demand Models. *Operations Research* 30(5): 887–906.

Tewalt, Susan J., Harvey E. Belkin, John R. SanFilipo, Matthew D. Merrill, Curtis A. Palmer, Peter D. Warwick, Alexander W. Karlsen, Robert B. Finkelman, and Andy J. Park, 2010, Chemical Analyses in the World Coal Quality Inventory, version 1: US Geological Survey Open-File Report 2010–1196. Available at http://pubs.usgs.gov/of/2010/1196/pdf/of2010-1196.pdf.

Trüby, Johannes. 2013. Strategic Behaviour in International Metallurgical Coal Markets. *Energy Economics* 36: 147–157.

Trüby, Johannes, and Moritz Paulus. 2012. Market Structure Scenarios in International Steam Coal Trade. *The Energy Journal* 33(3): 91–123.

VDKI. 2008. *Jahresbericht 2008. Fakten und Trends 2007/2008*. Hamburg: Verein der Kohlenimporteure e.V.

VDKI. 2011. *Jahresbericht 2011. Fakten und Trends 2010/2011*. Hamburg: Verein der Kohlenimporteure e.V.

VDKI. 2012. *Jahresbericht 2012. Fakten und Trends 2011/2012*. Hamburg: Verein der Kohlenimporteure e.V.

VDKI. 2013. *Jahresbericht 2013. Fakten und Trends 2012/2013*. Hamburg: Verein der Kohlenimporteure e.V.

Warell, Linda. 2006. Market Integration in the International Coal Industry: A Cointegration Approach. *Energy Journal* 27(1): 99–118.

Zaklan, Aleksandar, Astrid Cullmann, Anne Neumann, and Christian von Hirschhausen. 2012. The Globalization of Steam Coal Markets and the Role of Logistics: An Empirical Analysis. *Energy Economics* 34: 105–116.

Part IV

The potential of technology to reconcile coal and climate

11 New technologies to the rescue?

A review of three game-changing coal technologies and their implications for Australia's black coal industry

Bart Lucarelli

11.1 Introduction

Over the past decade, the challenge of reducing greenhouse gas (GHG) emissions from coal-fired power plants has emerged as an existential threat to the global coal industry, at least in theory. With regard to Australia, the issue of coal industry sustainability took on a more urgent tone in July 2012 when the Australian Government passed into law its Clean Energy Act of 2012, which included a policy known as the Carbon Pricing Mechanism (CPM) for taxing CO_2 emissions by large domestic emitters of CO_2. But scarcely two years later, on July 17, 2014, the Australian Senate voted to repeal the CPM elements of the Clean Energy Act.[1]

The repeal of the CPM left Australia without a mechanism for pricing CO_2 emissions, and the pitched political battles over the CPM (see Chapter 6) have highlighted the tensions around coal and climate change that still exist in the country. As a developed economy, Australia arguably is in a better position to afford the high cost of regulating GHG emissions than most emerging economies of Asia. On the other hand, coal exports

This chapter was the most challenging one for the author to prepare, but at the same time it was the most fun. One of the main reasons it was fun was the many helpful editing comments and insights provided by reviewers of this chapter. In this regard, the author wishes to express his sincere thanks once again to Mark Thurber, Associate Director of PESD, for his detailed reviews of earlier versions of this chapter and his many helpful suggestions on how to "tease out" from the three technology case studies the commercialization lessons learned, and the possible pathways available for two of the three technologies that have not yet achieved widespread commercial application. He also wishes to express his deep gratitude to Michael Friederich, coal exploration consultant with Coal Discovery, for taking so much of his valuable time to review earlier drafts and for sharing his many insights on CBM. Last but not least, the author thanks (a) Tim Moore, a CBM exploration expert with Cipher Resources, for reviewing and commenting on the CBM section of this chapter and providing helpful data; (b) Alan du Mee, a non-Executive Director at Genex Pty Ltd, for participating in a phone interview covering a number of critical policy issues raised in this chapter; (c) Chris Mallett, Technical Director of Carbon Energy, for accompanying him on a visit to CE's demonstration UCG plant at Bloodworth Creek, Queensland; and (d) David Mathew for providing useful charts and tables from his library of CBM presentations.

and coal-fired power contribute significantly to Australia's economy, creating important constituencies opposed to the carbon tax. Industry concerns coupled with a sometimes ham-handed rollout of the CPM program turned it into a political liability and ultimately contributed to its repeal (see Chapter 6).

The demonstrated ability of Australia's black coal industry to implement advanced, productivity enhancing technologies, would seem to make Australia a logical candidate to lead in the development and implementation of technologies for reducing the GHG footprint of coal use. And indeed, Australia has led to varying degrees in the development of three such technologies – coal bed methane (CBM) extraction, underground coal gasification (UCG), and CO_2 capture and storage (CCS) – which are assessed in detail in this chapter.[2]

These three technologies, which are at different stages of development, face different regulatory regimes, and have different cost structures and economic drivers in the absence of GHG emission regulation. At the present time, CBM extraction is by far the most economically attractive on its own merits, and as a result it has developed into a significant success story in Australia over the past decade. Its positive impacts on greenhouse gas emissions, while potentially significant due to the rapid spread of the technology, are also incremental in nature – the result of displacing coal used for power generation with gas.

Widespread application of UCG would represent a more radical technological advance. Similar to CBM extraction, it has been pursued due to its potential to access currently uneconomic coal deposits.[3] The main difference is its ability to extract much more of the energy locked within the in situ coal deposits than CBM extraction can ever hope to achieve. It also generates considerable amounts of CO_2 during the production process, which needs to be addressed at the UCG production site. Climate change concerns have not yet been a major driver of UCG development, although it could in theory have a significant positive effect on Australia's GHG emissions if coupled with CCS.

CCS technologies have made substantial technical progress over the past decade but are still the least attractive of the three technologies from a cost-competitiveness perspective. The commercial deployment of CCS technologies is still wholly dependent on the Australian Government and its State Governments implementing aggressive policies for reducing GHG emissions, including the offer of dedicated CCS fiscal and tax incentives, which have not yet materialized in a significant way in Australia (or elsewhere).

The remainder of this chapter consists of four sections. The first three sections assess Australia's progress in developing and commercially deploying CBM, UCG, and CCS technologies, respectively. Section 11.2 describes Queensland's success in implementing CBM extraction technologies and the reasons for that success. It also discusses the limits to expansion that Australian CBM producers are currently facing and the implications of those limits to the future of CBM extraction in Australia. Section 11.3 reviews Australia's mixed experiences to date with the implementation of UCG. Section 11.4 briefly reviews the three main CO_2 capture technologies – post-combustion, pre-combustion, and oxy-fuel combustion[4] – and then assesses the progress of CCS development to date in Australia and elsewhere. Section 11.5 concludes with an overall assessment of the different technologies and their potential contribution under different future scenarios for GHG emissions reductions.

11.2 Coal bed methane extraction

11.2.1 Technology basics

Coal bed methane (CBM)[5] refers to the methane component of the gas adsorbed onto the coal surface within the pore space of coal seams.[6] (Methane is the principal component of the gas; carbon dioxide and nitrogen are the usual minor components.) Geological survey methods and production techniques for identifying and developing a CBM resource differ from those for a conventional natural gas resource, which has formed in structural or stratigraphic traps, resulting in a gas that is ready to flow when drilled and can be readily produced by conventional gas production techniques.

In contrast, CBM is contained within the pore space of coal seams, where it is adsorbed. To extract the adsorbed gas, the coal seams must first be depressurized by pumping out the formation water through a well borehole. The CBM will then flow out from the pores of the coal matrix into the fractures of the coal seam and from there to the lowest pressure area (located near the borehole), and then finally up the borehole to the surface. In some cases the coal seam may be hydraulically fractured to increase the fracture permeability and the gas flow rate. The technologies used for CBM extraction, including fracking,[7] have all been commercially proven for CBM extraction in Australia (Friedrich, 2014).

Two types of drilling technologies are currently being applied in Australia to produce CBM: (1) traditional vertical wells, which rely on simple, low-cost coal exploration drilling rigs; and (2) combination vertical – horizontal directionally drilled (HDD) wells (also referred to as surface to in-seam [SIS] wells). Figure 11.1 shows a basic CBM configuration, including an optional horizontal (lateral) well. Whether vertical-only wells

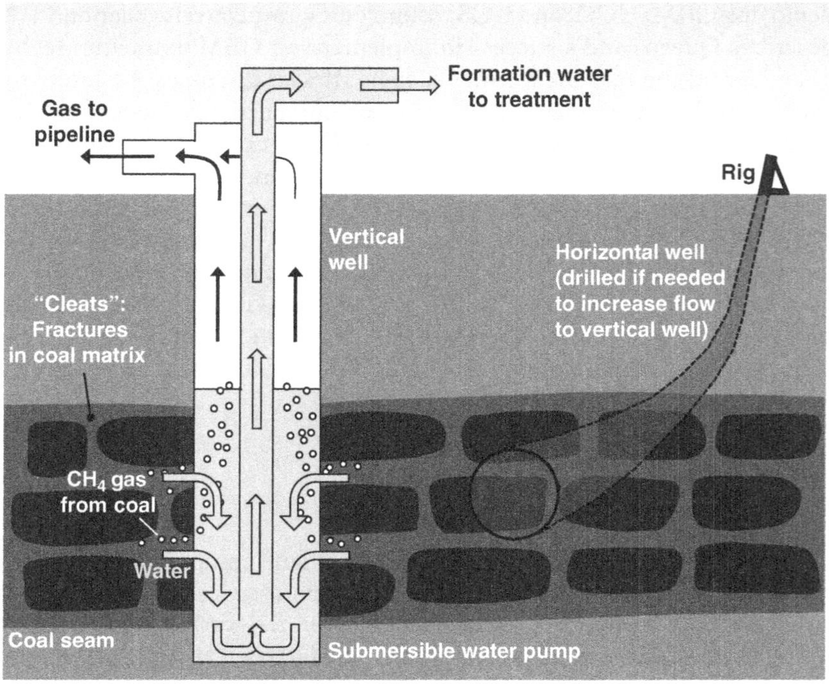

Figure 11.1 Simplified schematic of coal bed methane (CBM) extraction.

or combination vertical-HDD wells are installed within a specific CBM development area is largely determined by a life-cycle cost analysis weighing the comparative costs and revenues of the two technical approaches.[8]

Every CBM project follows six distinct development stages: (1) basin assessment; (2) exploration; (3) appraisal; (4) drilling of production wells; (5) collecting and treating formation water; and (6) CBM extraction. Following is a summary of these main stages.[9]

11.2.1.1 Basin assessment

Before selecting an area for drilling, a CBM developer first completes an initial assessment of a large area of a coal basin, using historical data from boreholes drilled previously for coal exploration and/or conventional oil and gas exploration, seismic data (if available from public and other third party sources), and field reconnaissance. The collected information is used to estimate the seam thickness, coal thermal maturity, coal quality and type, seam depth, structural conditions (dips, faulting), and seam continuity. The developer then relates its findings to other basins and by analogy obtains an initial understanding of (i) the potential "gas in place" and (ii) whether the CBM assessment area has sufficient CBM "prospectivity" to continue to the higher cost exploration stages that would follow. At this stage of development, revenues from potential gas markets and the costs of surface gas infrastructure are also considered (Friederich, 2014).

11.2.1.2 Exploration

If exploration proceeds, the CBM developer drills into the target coal seam a number of widely spaced exploration wells and tests the resulting coal core samples to provide (i) an initial estimate of gas volumes, gas quality and water quality, and (ii) the location of potential development areas, especially highly productive gas bearing areas, known as "sweet spots" or "fairways." During the exploration phase, the CBM developer also targets potential coal seams and quantifies their resource properties (thickness, permeability, coal seam continuity, gas content per tonne of coal, gas saturation, and gas quality) as well as the quality of formation water.

11.2.1.3 Appraisal

The CBM developer next completes a pilot program to assess the commercial viability of an area. A small group of up to five closely spaced wells (500 to 1,500 m apart) is drilled to just below the target coal seam. A submersible pump is installed in one of these wells to pump out the formation water inside and above the seam from the well. The reduction in water pressure then allows the gas to desorb from the coal. These tests are used to estimate reservoir permeability, determine the completion methodology, and establish gas flow profiles (Friedrich, 2014; Moore, 2007).

11.2.1.4 Drilling of production wells

After completing exploration activities, the CBM developer drills large diameter, vertical wells until the main coal seam(s) is (are) reached. If the development is based on vertical wells only, the CBM developer cements the vertical well casing in place and starts removing formation water from the coal seams (see water removal in Figure 11.1).

If the CBM development is based on combination vertical-HDD wells, the developer drills and cements in place the vertical well casing and then drills horizontal wells (also referred to as "laterals") into the coal seams using a flexible drilling tool, which consists of series of hinged drill pipes and an instrumented drill bit that can be progressively manipulated from a vertical to a horizontal alignment. The flexible drilling tool can either be lowered to the bottom of the vertical well, from which point laterals can be drilled into the coal seams, or used to drill a separate in-seam surface well using a "slant drilling rig" (Barker, 2012).

A specially trained drill operator known as a "navigator" manipulates the HDD well drilling tool into the coal seam at the correct angle and drills one or more laterals to the planned length. The number of horizontal wells drilled depends on the number of coal seams that can feasibly be exploited in a given development area. HDD well technology has been perfected to the point where the drill bit can follow a precise path along a coal seam with a thickness as narrow as 2 m for a distance of up to 1,000 m at depths to as much as 1,000 m (Barker, 2012).

11.2.1.5 Collection and treatment of formation water

During the process of drilling each vertical and, if applicable, horizontal well, formation water (water trapped inside the coal seams and immediately surrounding it) drains into a cavity that is created below the vertical well's casing by the injection of compressed air prior to well completion. The formation water, which, in the case of Queensland, is very high in dissolved solids, is then pumped through a sealed pipe within the vertical well shaft (annulus) and collected at ground level, where it is treated and then either used as irrigation or potable water or disposed of in accordance with environmental regulations.[10]

The partially dewatered CBM wells are known as production wells. The number of production wells installed in a CBM development area depends on the size and geological complexity of the area under development. In the Surat Basin of Queensland, production wells have been drilled to depths of 200–600 m with vertical well grid spacing of 750–1,000 m. (CBM developments with well depths shallower than 200 m are generally infeasible because most of the methane has already been lost to the atmosphere due to insufficient water pressure in the formation.) It is technically feasible to drill wells deeper than 600 m; however, the added cost of drilling a deeper well and the decreased permeability of the CBM with depth due to increased pressures makes vertical wells below 1,000 m generally uneconomic in Queensland at this point in time.

11.2.1.6 CBM extraction

Once a sufficient amount of formation water has been drained from the production wells, the gas trapped inside the matrix of the coal seams desorbs, flows through the cleat system of the coal seam and into the vertical well annulus. It then rises through the annulus to the surface, where excess water is removed from the gas. The gas is then compressed before being fed into a high pressure natural gas pipeline or used on-site for power generation or as a compressed fuel for captive truck fleets.

CBM production wells, which produce gas at lower rates than conventional gas wells, are typically characterized by the initial production of large volumes of formation water. Within about one year, coal seam pressures are reduced and CBM begins to flow. Then within five to six years of first gas production, CBM production starts to decline and in-fill wells need to be drilled if CBM is to keep flowing at a steady rate. Figures 11.2 and 11.3 show production profiles for two specific CBM production wells located in Queensland – one inside the Bowen basin and the second within the Surat basin. Figure 11.3 shows a significant resurgence of output after a steady decline in gas production during year five, which is an indication that the well had been "worked over" to clean out accumulated fines and other materials that have built up in the well over time.

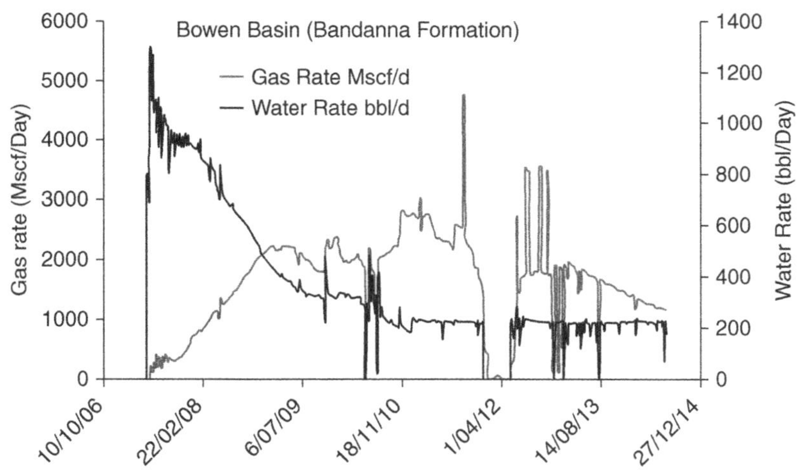

Figure 11.2 Seven-year CBM production profile for an actual CBM production well in the Bowen Basin. Note: Time intervals where gas and water production rates are at zero indicate that the well was shut-in for either maintenance purposes or due to equipment failure. Source: CBM exploration expert who wishes to remain anonymous.

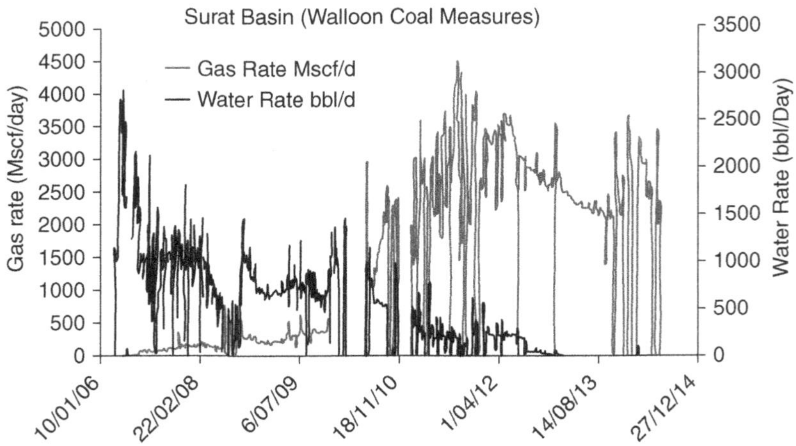

Figure 11.3 Seven-year CBM production profile for an actual CBM production well in the Surat Basin. Note: Time intervals where gas and water production rates are at zero indicate that the well was shut-in for either maintenance purposes or due to equipment failure. Source: CBM exploration expert who wishes to remain anonymous.

11.2.2 Advantages and disadvantages of CBM

From an environmental perspective, the big advantage of CBM (and natural gas in general) is that it has a much lower carbon content than coal and can be burned in highly efficient, combined cycle gas turbine (CCGT) power plants. The net result is that electricity can be generated in a CCGT power plant with half the CO_2 emissions per MWh of electricity generated in a coal-fired power plant. (Because methane has a much higher global warming potential than CO_2, achieving the maximum climate change benefits of using gas requires making sure that gas extraction, transport, and use do not result in venting additional methane to the atmosphere. Mirroring the debate in the US around fugitive methane from shale gas production, there has been an active discussion in Australia about whether CBM production results in significant incremental release of methane to the atmosphere.[11])

To take a specific case, Queensland produced 249 BCF of CBM in 2012. If one assumes that 100% of the CBM produced in 2012 was used to produce electricity, which, in turn, displaced electricity that would have been produced in coal-fired power plants, then Queensland (and all of Australia) achieved a reduction in its CO_2 emissions of approximately 9 million tonnes (Mt) attributable to CBM.[12] (A global analysis would need to account for whether and how the displaced coal was used elsewhere.) This GHG emission reduction may not seem that significant relative to Australia's net CO_2e emissions of 559 million tonnes per annum (Mtpa) in 2012.[13] However, it is somewhat material in comparison with the GHG emissions reduction targets that Australia has committed itself to achieve (as a function of emissions reductions by other nations) by 2020.

As a minimum 2020 target, the Australian Government has agreed to reduce its GHG emissions by 5% of year 2000 GHG emissions, or 28 Mtpa of CO_2e (year 2000 CO_2e emissions were 555 Mtpa). If the amount of CBM produced in Queensland in 2012 is used to displace Australian coal used in domestic power plants, this resource alone could account for up to 25% of Australia's minimum 5% GHG emissions reduction. The Australian Government has also agreed to meet more aggressive GHG emission targets, ranging from 15% to 25% of year 2000 GHG emissions, if other developed countries commit to substantially reduce their GHG emissions. Assuming other developed countries make binding commitments to such reductions, Australia's GHG reductions targets would then range from 179 Mtpa to 270 Mtpa. Under the more aggressive GHG emission reduction targets, CBM usage in Queensland at 2012 CBM production levels will account for at most 3% to 5% of Australia's more aggressive GHG emission reduction targets.[14]

However, the GHG reductions provided by CBM at this early stage in the industry's development show the promise of even greater reductions that Australia can achieve if CBM output continues to grow at double digit annual rates and a significant portion is used domestically to displace domestic coal use in power plants. From a climate change mitigation perspective, the "CBM revolution" in Australia could have a parallel effect in displacing coal to the one the "shale gas revolution" has had in the United States.

CBM development in Australia parallels shale gas development in the US in some other ways as well. Like shale gas, CBM has been driven by basic economics (as opposed to, for example, carbon regulation) and has been enabled by technological progress. CBM is largely extracted from coal seams that are not of a high enough quality or lie below "economic basement," which is the depth below which it is no longer economically feasible to extract the coal using either open-cut or underground mining methods.[15] In this regard, CBM development extends the potential recoverable energy from the coal seams contained within a specific land area.[16]

As discussed in Section 11.2.3, high natural gas prices through July 2014 – especially in the broader Asia Pacific region – offered the promise of high profitability if CBM resources could be developed. Technological progress in areas like horizontal drilling, advanced seismic imaging, and to a lesser extent hydraulic fracturing, then helped make this development possible (see Section 11.2.4).

Also like shale gas extraction in the US, CBM production in Australia has engendered controversy over its environmental impacts. In some cases there are legitimate environmental issues that need to be managed; in others environmental impacts have probably been sensationalized and overblown. The development of CBM production wells actually leaves a very small impact on the surrounding countryside. A typical CBM production well in the Surat Basin area of Queensland might achieve a peak production of 1 to 2 million ft^3 per day of CBM and operate from a single well pad covering an area of 650,000 m^2 (160 acres) (Moore, 2014). Compared to the land impacts of open-cut mining, CBM has relatively benign impacts on land use. However, it does create a fair share of land impacts in the form of thousands upon thousands of vertical well pads, access roads that extend into development areas and numerous pipelines for transporting the gas out of the CBM development areas.

While hydraulic fracturing has seen relatively modest application in CBM, some environmental NGOs have claimed that fracking fluids are polluting groundwater supplies – although these concerns have been largely debunked by the scientific community. Fracking fluids consist of 99.5% water and sand and 0.5% chemicals, and the governments of NSW and Queensland have set strict standards concerning allowable fracking chemicals. In particular, both the Queensland and NSW Governments have banned the use of hazardous Volatile Organic Compounds (VOCs), such as benzene, toluene, ethylbenzene, and xylenes (BTEX), as fracking chemicals.[17] Given that any fracking operations are conducted at depths greater than 170 m and given that well casings are sealed and cemented in place prior to injecting any fracking fluids, environmental protection agencies of the Queensland and NSW Governments have for the most part concluded that fracking, with adequate environmental safeguards and standards in place and applied, is unlikely to lead to the pollution of subsurface water sources.[18]

On the other hand, the storage and disposal of formation water, which is collected from each production well, is a genuine environmental concern that requires careful management and regulatory oversight. Due to the high concentrations of dissolved solids in the formation water, it could pollute surface water and even near-by, subsurface groundwater if it is not properly handled.[19] Removing the water cap from above the coal seams may also depress aquifers over a large area and affect groundwater flows. In Section 11.2.5 we will discuss how formation water management has been regulated.

11.2.3 History and current status of CBM development in Australia

CBM exploration in Queensland began in the Bowen Basin in the 1970s, with limited success. Exploration activities were restarted in 1990, with a focus on high rank coals in the Bowen Basin.[20] However, with the successful production of CBM from low rank coals in the Powder River Basin in the United States, the search quickly widened starting in 2000 to include lower rank coals, especially the sub-bituminous to high volatile bituminous Jurassic coals of the Surat Basin in Queensland.

Between 2000 and 2005, CBM exploration accelerated in both the Bowen and Surat basins, with programs by newly formed entrepreneurial companies that specialized in CBM such as CH4, Arrow Energy, and Queensland Gas Company. Exploration later began in the Gunnedah Basin of NSW and other areas as understanding of exploration techniques and basin prospectivity increased.

The renewed exploration led to initial commercial success in the Bowen Basin, with gas production from the Moranbah area starting in 2005. This success with CBM in the Bowen Basin was enabled by advancements in HDD well technology. Because Bowen Basin coal seams had high gas content but generally low permeability, horizontal drilling was necessary to increase gas flows from each well and make development economic (Friederich, 2014).

The first CBM pilot wells in the Surat Basin were drilled in 2002, and commercial production started in 2006. Surat Basin coals were younger and of lower rank than those in the Bowen Basin. They had lower gas content but contained thick seams with higher permeability. CBM developers soon discovered that, for Surat Basin CBM deposits, they could rely on simple vertical wells and avoid the cost of drilling HDD wells or conducting costly and environmentally sensitive fracking operations (Friederich, 2014). The result was lower drilling and completion costs.

A major trend in Queensland has been the rapid growth in 2P reserves located in the Surat Basin and the decreasing importance of the Bowen Basin as a source of CBM reserves (see Figure 11.4). In June 2005, the Bowen Basin accounted for 82% of Queensland's 2P CBM reserves. By June 2013, the reserve picture had flipped such that the Surat Basin accounted for 77% of 2P reserves while the Bowen Basin contained only 23% of Australia's 2P CBM reserves.

Production trends have reflected this shift in the location of reserves from the Bowen to the Surat Basin (Figure 11.5). In 2005, 97% of Queensland's CBM was produced from CBM fields located in the Bowen Basin. By 2012, a more balanced production profile had been

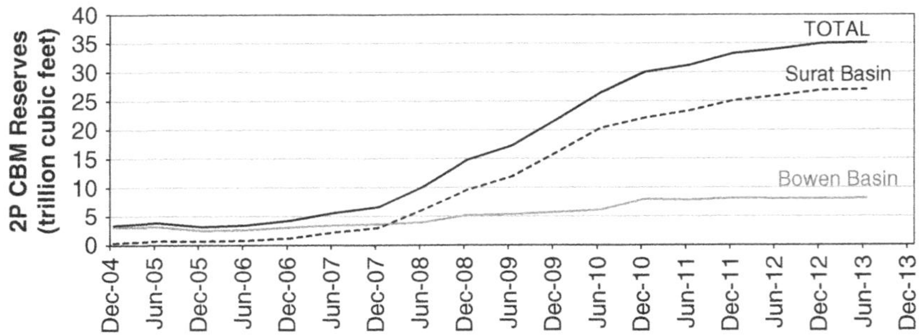

Figure 11.4 Queensland's 2P (proved and probable) CBM reserves. Data source: Government of Queensland, Department of Natural Resources and Mines, 2014.

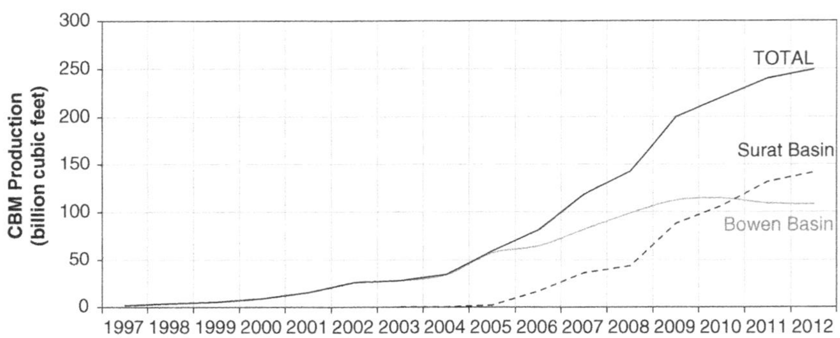

Figure 11.5 Queensland coal bed methane production by basin, 1997–2012. Data source: Government of Queensland, Department of Natural Resources and Mines, 2014.

achieved, with 43% of CBM production coming from the Bowen fields and 57% from the Surat Basin.

Overall, the growth in both reserves and production in Queensland has been breathtaking. Queensland's proved and probable (2P) reserves[21] had grown by end of June 2013 to 35.1 TCF (Figure 11.4), from very low levels as of January 2004.[22] Queensland's CBM production showed corresponding rapid growth, outstripping conventional gas production in 2007 (Figure 11.6). In 2004, CBM production was 34.4 BCF, accounting for only 17% of Queensland's total gas production (CBM + conventional gas), with conventional gas accounting for the remaining 83%. By 2012, Queensland's CBM industry was producing 249 BCF of CBM, or 89% of Queensland's total gas production.

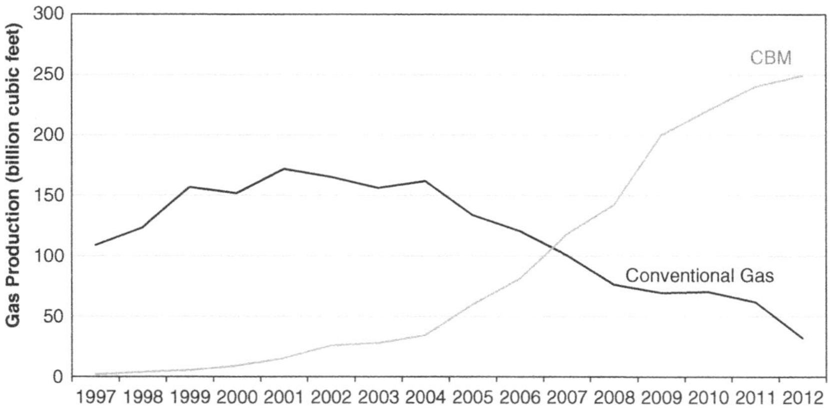

Figure 11.6 Queensland gas production from CBM and conventional sources. Data source: Government of Queensland, Department of Natural Resources and Mines, 2014.

In 2012, the government of Australia provided estimates of Australia's CBM resource potential alongside those for three other sources of natural gas: conventional, tight, and shale gas (Australian Government, 2012a). Australia's economic demonstrated resources (EDR) of CBM as of January 2012 were 33 TCF, relative to 103 TCF for conventional gas (Table 11.1).[23] However, more speculative estimates of resource potentials (inferred and potential undiscovered resources) show CBM as the country's largest gas resource category, and if the highly speculative category known as "potential resource" is considered, CBM supplies are more than three times larger than conventional gas (Table 11.1).

11.2.3.1 The birth of Queensland LNG Industry

As a result of the rapid growth in their 2P CBM reserves, four of Australia's largest CBM developers – Santos, British Gas, Origin Energy and Arrow Energy – soon found domestic markets to be insufficient to support their growth targets. Between 2009 and 2011, three of them – Santos, British Gas, and Origin Energy – partnered with international petroleum and chemical companies (Petronas, Total, Kogas, Petronas, ConocoPhillips, Sinopec, CNOOC, and Tokyo Gas) to develop three independent LNG terminals at the port of Gladstone (Table 11.2). The three terminals are at an advanced stage of construction and require first gas starting in either late 2014 or 2015. They should be operating close to their nameplate capacities by 2017, when they are

Table 11.1 *Australia's total gas resources from all sources, 2012*

	Conventional gas		Coal bed methane		Tight gas		Shale gas		Total gas	
Resource category	PJ	TCF	PJ	TCF	PJ	TCF	PJ	TCF	PJ	TCF
EDR	113,400	103	35,905	33	-	-	-	-	149,305	136
SDR	59,600	54	65,529	60	-	-	2,200	2	127,329	116
Inferred	11,000	10	122,020	111	22,052	20	-	-	155,072	141
Identified resources (EDR + SDR + inferred)	**184,000**	**167**	**223,454**	**204**	**22,052**	**20**	**2,200**	**2**	**431,706**	**393**
Potential in ground resource	n/a	n/a	258,888	235	n/a	n/a	435,600	396	694,488	631
Total resouces (identified + potential)	**184,000**	**167**	**482,342**	**439**	**22,052**	**20**	**437,800**	**398**	**1,126,194**	**1,024**

Notes: Conventional gas identified resources as of January 2011, CBM identified resources as of January 2012 with CBM 2P Reserves = EDR and 2C resources as SDR.
Source: Australian government, 2012a.

Table 11.2 *LNG terminals under construction at port of Gladstone*

Name	Owners	Nameplate capacity (Mtpa LNG)	Projected first LNG	Gas requirement (PJ/yr)	2013 capital cost est. (billion A$)
1. Gladstone LNG (GLNG)	• Santos • Petronas • Total • KOGAS	7.8	2015	468	A$18.0
2. Australia Pacific LNG (APLNG)	• Origin Energy • Conoco-Phillips • Sinopec	9.0	2015	540	A$24.7
3. Queensland Curtis LNG (QCLNG)	• British Gas • CNOOC • Tokyo Gas	8.5	Q4 2014	510	A$19.8

Sources: Intelligent Energy Systems (2013) for all data except 2013 capital costs, which were obtained from Deloitte Access Economics (2014).

estimated to require approximately 1,500 PJ (1.26 TCF) per year of CBM (Intelligent Energy Systems, 2013).

At the time of writing there was an active debate under way as to whether actual CBM production would be sufficient to fill these new LNG terminals coming on line, with particular risk in the gas supply to the Santos GLNG facility.[24] Given the lack of experience with CBM production wells in Queensland, coal industry analysts, and even Santos, are now arguing that CBM developers should have either joint ventured on fewer terminals or spaced their construction out over a longer time period.[25] If gas shortages occur that cannot be solved by procuring CBM from Arrow Energy and other third-party CBM producers, the only other viable CBM procurement option open to owners of the three LNG plants will be to procure conventional gas, which will possibly disrupt gas supplies to retail and power customers in Queensland and NSW (*Sydney Morning Herald*, 2012). This diversion of domestic and industrial supplies to the LNG plants will also lead to sizeable increases in the price of gas feedstock to the Santos LNG plant and reduce the margins earned on its LNG exports.

As a result of the three Queensland CBM-LNG projects, the price of CBM in Queensland has increased substantially over the past eight years. At the start of the CBM revolution in Queensland (2004–05), the price of CBM (ex-gas field) ranged between A$2.00 and A$2.50 per GJ (WestSide Corporation Limited, 2006). These prices reflected initial uncertainty about the reliability of CBM as a primary fuel source, which required CBM producers to offer steep discounts.[26] IES, an Australian economic consultancy, estimated in 2013 that CBM prices ex-field in Queensland ranged from A$4.40 to A$5.60 per GJ, with future prices (2015–2020) likely to be determined on a net-back basis using the delivered price of LNG into Japan[27] as the starting point and then deducting the cost of gasifying the CBM and transporting it as LNG to Japan (estimated at A$6 per GJ) and the pipeline charges from the gas-field to the LNG terminal at Gladstone. IES modelled LNG netback prices of about A$11 per GJ in 2014; A$7.40 to A $10.90 per GJ in 2017; and A$10 to A$13 per GJ in 2023 (Intelligent Energy Systems, 2013, p. 78). If LNG prices into Japan return to their high March 2014 A$ levels while CBM production costs increase only moderately between 2015 and 2023, one can see the huge upside that Queensland's CBM developers hoped to earn by converting their surplus CBM into LNG. This explains why three of Queensland's largest CBM developers rushed into their Gladstone LNG projects despite the many market uncertainties – the potential profits were simply too large to ignore (Intelligent Energy Systems, 2013).

Due to overbuilding of LNG capacity, there may be some shortages of CBM between 2015 and 2020. But if a large enough differential exists between the costs of CBM production and the net back price for Queensland CBM, it is this author's view that Queensland's CBM developers will bring additional reserves into production and close any CBM supply gap. At today's low LNG prices, this scenario appears to be in serious doubt. But one must remember that serious doubts also existed about the recovery of oil, LNG, and coal prices during the global financial crisis of 2008–09.

The successful development of the CBM industry in Queensland has led not only to a shift in the coal industry's mix of saleable fuel products but also to a shift in industry stakeholders. Prior to 2005, Australia's coal industry was dominated by traditional coal mining companies. Today, the coal industry's stakeholders are both traditional coal miners and a growing number of oil and gas producers who specialize in CBM development. The shift in stakeholders may seem insignificant so far, but as CBM and other advanced technologies are increasingly applied to the coal resource sector, the industry may experience a major diversification of equity stakeholders. It is worth remembering that, just ten to fifteen years ago, many of today's large CBM players were small start-ups focused on a non-existent market. Arrow Energy is one such enterprise that expanded rapidly as the CBM market grew. Its growth trajectory is described below to illustrate the kind of industrial transformation wrought by the expansion of CBM.[28]

11.2.3.2 The case of Arrow Energy

Arrow Energy originally went public in August 2000 with an IPO that netted a mere A$5 million on shares valued at A$0.20 each. By August 2009 the company had a market capitalization of A$2.4 billion. In August 2010, Shell and Petrochina acquired Arrow for A$3.5 billion. As of June 2013, Arrow Energy was Australia's largest holder of coal bed methane acreage, with interests in 41,500 km^2 of resource area, and it accounted for 20% of Queensland's total (CBM + conventional) gas production (Arrow Energy, 2014).

What has made Arrow's growth even more impressive is that it did not achieve its first CBM production (and sale) until January 2006, when it brought its Kogan North CBM field into commercial operation. In April 2006, Arrow merged with CH4, which brought CH4's Moranbah field into Arrow's stable of CBM projects. In February 2007, the Tipton West CBM operation started supplying about 6 petajoules per year (PJ/year) to the existing Braemar 1 Power Station. Its output was later increased to 9.5 PJ/year, where it was projected by Arrow Energy to remain for the next fifteen years.

Despite the worldwide financial crisis, FY 2009 turned out to be a year of acquisitions for Arrow Energy. It exercised an option to buy out its partner on the Braemar 2 Power Station, a 450 MW CCGT plant located about 110 km from its Tipton West CBM field, bringing its equity share in the Braemar station and associated 110 km high pressure gas pipeline up to 75%. It then exercised an option to buy out the remaining 25% of ERM Power's equity in the Braemar plant and pipeline and today owns 600 "equity" MWs of power plant in Queensland + pipeline assets. In 2009, Arrow Energy also acquired CBM acreage from a number of small CBM players.

Table 11.3 shows Arrow Energy's extraordinary growth in both reserves and gas output from 2006 through June 2010 (the last year when such detailed figures were published by the company).[29] In less than two years, Arrow Energy increased its gross 2P reserves by 755%, from 719 PJ at the end of June 2007 to 6,150 PJ by the end of December 2009. It also expanded its CBM production from 10.4 PJ in FY 2006–07 to 38.8 PJ in FY 2009–10, a 275% increase over a three-year period. Production and 2P reserves would have been at even higher levels if Queensland's domestic gas market had been large enough to absorb increased production.

Table 11.3 *Growth in Arrow Energy's CBM resources, reserves, and production, 2005–2009*

	FY 2005–06		FY 2006–07		FY 2007–08		FY 2008–09		FY 2009–10	
	FY 2005–06	June 20, 2006	FY 2006–07	June 30, 2007	FY 2007–08	June 30, 2008	FY 2008–09	June 30, 2009	FY 2009–10	Dec. 30, 2009
Reserves (PJ)										
1P		25		86		427		703		808
2P		166		719		2,247		4,092		6,150
3P		2,148		2,760		5,084		9,312		11,042
Production (PJ)	1.0		10.4		28.6		32.4		38.8 (annualized result)*	
Producing wells		50		243		281		n/a		n/a

* Production estimate for FY 2009–10 = two times the half year result. This is a very conservative estimate, but even if it did come to pass, it would still have represented a 20% increase over the previous FY.

Reserves designations: 1P = proved; 2P = proved + probable; 3P = proved + probable + possible

Sources: Arrow Energy, "Annual Reports for FY 2006 through 2008" and "Half Year Financial Statements FY 2010 (issued on 17 February 2010)" for FY 2009–10.

As a result of domestic market constraints, Arrow Energy announced in September 2010 that it would develop a 2 × 4.5 Mtpa LNG facility on Curtis Island within the port of Gladstone and supply surplus gas from its Bowen and Surat basin fields to that plant. Arrow Energy designed the plant for an eventual capacity of 18 Mtpa of LNG (4 × 4.5 Mtpa), with a planned commercial operation date of 2017 for the first LNG train and 2018 for the second train (Coffey Environments). Due to cost overruns and weakening market conditions, the news media are currently speculating that Shell and Petrochina may not give Arrow Energy final approval to proceed with this project (Chambers, 2012). Instead, Arrow Energy's CBM reserves may be sold to one of the other Gladstone LNG projects that are nearing completion, as discussed previously in this section. (The economic competitiveness of LNG from CBM is discussed under Section 11.2.3.1.)

11.2.4 Technology factors affecting CBM development

Advances in HDD well technology and increases in general exploration and drilling know-how played a key role in enabling the early development of Australia's CBM industry. Many of these drilling technology advances could be characterized as evolutionary improvements in the technical expertise required for developing CBM resources. Specific advances in know-how included (i) the improved ability to interpret seismic and core sample data and to determine where the "CBM fairways" are located and their likely productivity, and (ii) greater expertise in drilling both vertical and HDD wells that accrued from the drilling of multiple wells within specific CBM production basins.

Equipment-specific advances that contributed to the growth in Australia's CBM production included:

(a) improved down-hole drilling motors (known as mud motors because they are driven by the hydraulic force of the drilling mud being pumped through the drill pipe and through the motor) which allow the bit to be rotated at the bottom of the hole while the drill pipe remains stationary; and
(b) improved "in-hole" measurement devices that allow directional data to be sent back by telemetry to help guide the drill bit and the drilling motor electronically.

The end result was that between 2000 and 2004, both vertical and in-seam HDD wells were being drilled more quickly and accurately than just a few years earlier. This allowed Australia's CBM producers to bring CBM reservoirs into production very quickly once a firm gas market had been identified and committed to contract.

With the shift in CBM production from the Bowen Basin to the Surat Basin, the application of advanced drilling technology for CBM extraction has actually declined overall. However, this author feels the application of HDD well technology is likely to stage a comeback once the initial CBM reserves of the Surat Basin have been depleted using vertical collection wells only. At that point, these accumulated advances in HDD well technology and know-how will then enable enhanced gas recovery.

In addition to improvements in drilling technology, the development of Queensland's CBM resources was supported to a limited extent by the application of hydraulic fracturing (fracking) technology.[30] With the major improvements in HDD well technology over the past five years and the high permeability of Surat Basin CBM deposits, Queensland's CBM developers have only needed to apply fracking in a small percentage of their CBM production wells. For example, Arrow Energy claimed that, as of July 2013, it had only fracked 4% of its wells (Arrow Energy, 2013). Other sources place the overall percentage of Queensland wells fracked at around 10% (Arrow Energy, 2013). Nonetheless, there are still instances where fracking is used in conjunction with HDD well technology to increase the permeability of coal seams and allow more efficient recovery of CBM.[31]

11.2.5 Regulatory factors affecting CBM development

The main regulatory issues confronted by the governments of NSW and Queensland have involved how to (a) reduce commercial risk for pioneering CBM projects by ensuring that sufficient domestic gas demand would materialize; (b) resolve overlapping claims by holders of mineral (coal) and petroleum development licenses to the same tenement; and (c) ensure sound environmental stewardship, especially in avoiding groundwater pollution.[32]

Overall, the Queensland government has established a regulatory regime that is much more supportive of CBM development than that in NSW. In particular, Queensland has proactively reduced risk for developers by ensuring domestic demand for natural gas, and it has wrestled conscientiously, and in the end fairly effectively, with the challenge of how to resolve overlapping claims. The difference in regulatory environment is a key factor that helps explain why Queensland's CBM industry has seen spectacular growth in both 2P reserves and production while its counterpart in NSW has remained moribund. The trials and tribulations of CBM developers in NSW are discussed in some detail in Section 11.2.5.3 of this chapter.

In the following section of this chapter, the author discusses regulatory actions, particularly on the part of the government of Queensland, to incentivize CBM development, resolve competing claims, and protect the environment.

11.2.5.1 Ensuring a market for CBM

In 2000, the government of Queensland laid out a scheme whereby electricity retailers and large electricity consumers would be required to source 13% of their electricity from power plants fired on natural gas. At that time, gas-fired plants accounted for only 2.4% of Queensland's electricity production. The "gas scheme" requirement became operational in 2005. The requirement for the share of gas-fired power was increased to 15% of total retail sales in 2010 and was slated to increase to 18% of total electricity sales by 2020.

However, by 2012, largely due to the explosive growth of CBM production, the amount of electricity sourced from gas-fired power plants had already increased to 20% of total retail sales. With the passage of the Carbon Pricing Mechanism in July 2012 (see Chapter 6), the Queensland Government felt that the Gas Scheme was no longer needed and discontinued it in 2013 (Queensland Government, 2014b). But in the early years of the CBM industry's ramp up, the Queensland Gas Scheme provided an important base of support to the industry and allowed it to grow by creating a ready market to be tapped at sufficiently attractive prices.

11.2.5.2 Resolving overlapping claims

Since the 1970s, disputes relating to overlapping claims to the same resource tenements – either between two coal miners or between a coal miner and an oil and gas company – would occasionally arise. Until 2007, however, these disputes never became sufficiently widespread to receive government attention.[33] Prior to 2007, overlapping claims disputes were largely handled bilaterally, with the two disputing parties reaching an agreement on how to settle their conflicting claims.[34] If an agreement could not be reached, the issue could be referred to the responsible State minister, who under a 2004 Amendment to the Queensland MRA was empowered, in theory, to make a "preference" ruling on behalf of the warring parties on the basis of the "broader public interest" (Oliver, 2014).

In practice, however, the "preference decision" approach did not work. Between 2004 and January 2014, Queensland's "responsible" ministers did not issue a single preference ruling related to overlapping claims between coal miners and CBM developers, probably due to the political fall-out that might ensue from making such a decision. In any case the non-preferred party, unless agreed by it prior to the minister issuing his decision, would retain the right to ask Queensland's courts to conduct an administrative review of the minister's preference decision (Oliver, 2014).

By 2007, the CBM industry had reached a level of development where disputes related to overlapping claims were growing in number and could no longer be ignored.[35] Existing dispute resolution procedures were criticized by CBM developers and coal miners alike for their "failure to provide a system where competing overlapping production tenures can co-exist and proceed to development in a manner that maximizes resource production with minimal delay" (Oliver, 2014). The overlapping claims problem was threatening to become a significant constraint on resource development.

In January 2011, the government of Queensland issued a draft bill named "Mines and Petroleum Legislation Amendment Bill 2011" ("Draft Bill") whose stated purpose was to "ensure that the framework for overlapping coal and petroleum tenure adequately supported the establishment of Queensland's CSG [coal seam gas, a synonym for CBM] to LNG industry and optimized the use of the State's coal and CSG resources" (Queensland Resources Council, May 2012, p. 7).

However, stakeholders from the mining and CBM industries found the Draft Bill to be inadequate, stating that it did not achieve the right balance between (i) equitable treatment for coal mining companies, who in most cases held Mineral Development Licenses (MDLs) or Mining Licenses (MLs) that were issued prior to the issuance of an Authority to Prospect (ATP) or Petroleum License (PL), and (ii) consideration of the "greater public interest," which was shorthand for support for the rapidly growing CBM industry, which was by this point in the advanced stages of developing LNG export terminals at the port of Gladstone (Queensland Resources Council, May 2012, p. 8). These terminals were expected to become huge revenue earners for the government of Queensland and the source of thousands of new jobs.[36]

Taking their concerns about the Draft Bill as a starting point, CBM developers and coal mining companies worked cooperatively through the Queensland Resources Council (QRC)[37] to prepare a 104-page report setting out detailed procedures and guidelines for addressing overlapping claims between CBM developers and coal miners under two mining regimes: (1) open-cut mining and (2) underground mining. The QRC report, which was issued to the Queensland government in May 2012 as an alternative to the government's Draft Bill, provided a consensus view of the CBM and coal mining industries on how to address overlapping claims disputes between CBM and coal mining companies (Queensland Resources Council, 2012).

After two additional years of extensive consultations with industry stakeholders, the government of Queensland submitted for industry comment a revised Draft Bill that largely reflected the dispute resolution principles enunciated in the QRC report. (Until this Amended Draft Bill is passed into law, the Queensland government's previous procedures and regulations for resolving overlapping claim disputes are being followed, which largely rely on the threat of either party vetoing the other's development plan or the responsible minister making a preference decision if the two parties are unable to reach agreement on their own.) The basic principles in the Amended Draft Bill include the following:

- Only existing exploration tenements and future production tenements will be affected by the Bill unless the parties to existing production tenements agree to be bound by the new dispute resolution terms of the Draft Bill.
- Disputes among parties who are operating under existing coal mining and petroleum production licenses will be resolved under the 2004 dispute resolution procedures, which effectively give veto power to the party which received its license first, unless the responsible minister finds the political courage to make a preference ruling in the public interest.
- For all other parties to an overlapping claim dispute, the following dispute resolution procedures shall allow: a CBM producer to be granted the right to develop CBM resources within a disputed area subject to various time constraints they will need to follow when scheduling CBM extraction activities, and a coal mining company to obtain a right of way to develop the coal resources underlying the area previously allocated under a PL by giving the CBM producer ten years' prior written notice of the company's decision to start coal mining activities. For those areas over which a coal mining company already holds an ML, the coal miner must give the CBM producer operating under a PL a minimum of eighteen months prior notice of the date when coal mining activities will commence.

There are many other details captured in either the Draft Bill or under the QRC report that this author will not attempt to summarize in this chapter, except to note that a great deal of thought and effort has been invested by the government of Queensland into the preparation of the 2014 Draft Bill and its approach to public consultations. Both efforts are additional evidence of the proactive approach taken by the government of

Queensland to regulatory matters affecting its coal mining and CBM industries.

11.2.5.3 NSW opposition to CBM

The NSW government has not as yet shown a similar level of conscientiousness and effectiveness in developing predictable frameworks to manage licensing and expedite CBM development. This has caused viable CBM projects to languish in NSW.

During 2013, Metgasco and Dart Energy, two publicly listed CBM developers with projects in NSW, either mothballed or canceled a number of their NSW CBM projects and said they would only restart their CBM development work in NSW when the NSW government provides greater regulatory certainty (*The Australian*, 2013; Metgasco, 2013a).

In March 2013, Metgasco decided to "suspend its Clarence Moreton Basin field operations because of uncertainty of NSW government support and regulations" and informed its shareholders and the general public that "field operations and subsequent development and production will recommence when the regulatory environment is judged to be more favorable" (Metgasco, 2014b).

In May 2014, Metgasco had its license to drill a CBM exploration well in the North Rivers area of NSW, known as the Rosella project, canceled by NSW's Resources Minister, Anthony Roberts, who cited in his decision speculative information about possible corruption in the awarding of the drilling license (Winestock, 2014).

The minister also referred the matter to NSW's Independent Commission Against Corruption (ICAC), which promptly responded to the Minister and Metgasco that it "will not be proceeding with an inquiry into the Company" because "there was no information in the [Minister's] referral to ICAC that was indicative of corrupt conduct" (Metgasco, 2014c).

In June 2014, after numerous attempts to get the NSW government to reverse its decision on Rosella, Metgasco filed a lawsuit against the NSW government with the NSW Supreme Court (Metgasco, 2014a). In its court filing, Metgasco requested that the Court overturn the suspension order of NSW's Resources Minister and claimed damages from the NSW government. As of February 2015, the case is still being litigated in the NSW Supreme Court.[38]

Dart Energy's experience with the NSW government was just as bad as Metgasco's. In response to NSW government opposition to CBM projects, Nick Davies, Dart Energy's chairman until its takeover by IGas

Energy Plc. (IGas) in September 2014, was scathing in his criticism of the NSW government, stating that the board of Dart was "extremely disappointed with the uncertainty created by recent NSW and Federal government decisions in relation to CSG development in Australia. The consequence is that investment is leaving the country, field operations are being suspended, Australian jobs are being lost, and the impending energy crisis in New South Wales is not being addressed, and indeed, will only get worse" (*The Australian*, 2013).

Partly due to its disappointing experiences developing CBM projects in NSW, Dart Energy lost investor confidence and, in May 2014, agreed to be acquired by IGas, a UK-based CBM developer. As Dart's new owner, IGas informed its shareholders that Dart Energy's eight NSW Petroleum Exploration Licenses MDLs would continue under "care and maintenance," which will "preserve [these] licenses pending regulatory clarity and certainty" in NSW, a sad ending to what could have been an extension of the Queensland CBM success story (IGas Energy Plc., 2014).

As a direct result of this weak government support on regulatory matters, NSW has far fewer CBM wells drilled than Queensland. According to one estimate, NSW had only 10% as many CBM wells as Queensland as of September 2011 (SBS, 2015). Today, that differential is certainly much greater.

Moreover, as of January 2015, NSW had only two significant CBM projects that were still in commercial production: (1) the Coonarah CBM project, located near Narrabri and owned by Santos, which supplies gas to the Wilga Park Power Station under a ten-year agreement with Country Energy; and (2) the AGL Camden CBM project, which supplies about 5% of NSW's gas requirements from CBM deposits in NSW. The Camden project is located in southwest Sydney and has been operating since 2001.

Santos, however, is seeking NSW government approval to develop a large CBM project at Narrabri, which it has named the Narrabri CBM Gas Project. It is located in northwest NSW and can reportedly supply 50% of NSW's gas needs once it is in full production. Santos has filed an EIS with the NSW government, and discussions with the NSW government and NSW stakeholders have started (Santos, 2014). If the Narrabri project is approved, CBM proponents still active in NSW can take some scant comfort in knowing that at least one developer of CBM projects has had its project approved by the NSW government. If, however, the Narrabri project is rejected, this outcome will effectively be the final nail in the NSW CBM coffin.

11.2.5.4 Proper treatment of formation water

As discussed in Section 11.2.2, proper management of CBM formation water is crucial in order to avoid negative impacts on groundwater. The governments of Queensland and NSW both require CBM developers to carefully monitor any temporary formation water containment ponds and arrange for the water to be treated to a standard that will allow it to be used for either potable or agricultural purposes. Until 2009, water produced during CBM production in Queensland was simply pumped into large evaporation ponds, but in 2009 the government of Queensland issued a set of rules and procedures that put a stop to this practice (Queensland government, 2009a).

It was the very success of the CBM industry that created an impetus for action in this area. CBM companies produced 12.5 billion liters of formation water in 2007 (already a significant quantity) and 16.8 billion liters from July 2012 to June 2013.[39] With growth in CBM production (not including quantities needed to support the LNG plants being built at the port of Gladstone), the government of Queensland forecast that the volume of formation water produced is likely to reach 25 billion liters per year and remain at that level for the next twenty-five years. It estimated that 25 km^2 of land would be required to hold this estimated 25 billion liters of CBM formation water if it is held in 2 m deep evaporation ponds[40] (Queensland government, 2009a). Moreover, if CBM producers are successful in developing their LNG plants at Gladstone, it is possible that associated water production could increase to 100 GL per year, which would require a minimum of 100 km^2 of land area to be dedicated to evaporation ponds over the next thirty years. It will also produce over time 7.5 Mt of dry salts from the saline effluent (Queensland government, 2009a).

These concerns about the environmental impacts from the disposal of formation water were what prompted the government of Queensland to stop allowing the use of evaporation ponds to store CBM-formation water. If the water is not re-injected into the ground in an environmentally sound manner or cannot be put to some beneficial use, CBM producers must treat the associated CBM water at their expense and properly dispose of saline effluent and dry salts. CBM producers with existing evaporation ponds were also required to remediate these ponds by May 2012.

Recently a number of CBM production companies have commenced operating water treatment plants to treat the water to a standard that would allow discharge into streams and rivers. In a growing number of cases, CBM companies have initiated reverse osmosis water treatment projects that allow the water to be used as a source of domestic water supply and/or as cooling water for power stations.

11.3 Underground coal gasification

11.3.1 Technology basics

Underground coal gasification (UCG) is a process for converting coal in situ into a synthetic gas (syngas) through partial oxidation. In its simplest form, the process involves first drilling into a coal seam a vertical well known as an injection well, which is used to ignite the coal seam, and then following up with two horizontal wells: (1) an inlet well, which allows an oxidant (either air or pure oxygen) to be continuously injected into the coal seam once it has been ignited, and (2) an outlet or production well, which is used for extracting the syngas produced (Figure 11.7). The partial oxidation that occurs in the coal cavity leads to in-seam temperatures ranging from 900°C to 1200°C. The process yields a syngas consisting mainly of H_2 and CO, with small quantities of CO_2, CH_4, and a very small amount of H_2S.

The extracted syngas is collected above ground, where it is stripped of particulates, tars and, if required, CO_2 and other impurities. The low BTU gas is then dried and can be sold as either a power plant fuel or a feedstock to the chemical industry for the production of ammonia, methanol, and other chemicals.

As with above ground gasification systems, UCG processes that rely on pure oxygen as the oxidant will produce highly concentrated streams of CO_2 without NO_x, which will allow CO_2 to be captured more economically than would otherwise be the case.

11.3.2 Advantages and disadvantages of UCG

UCG has several attractions when compared to traditional coal mining:

- It can use stranded coal – that is, deposits located below "economic basement" – as a fuel source.
- It achieves much greater energy removal rates (80%–90% of the energy contained in the in situ coal) than open-cut mining (60%) – and even greater energy removal rates when compared with CBM.
- It leaves ash, unburned carbon, and other pollutants in the ground.
- It eliminates the cost of mining the coal and then investing in an above ground gasifier and water shift reactor.

Similar to an integrated gas combined cycle (IGCC) plant, the UCG process produces a more concentrated stream of CO_2 as a by-product

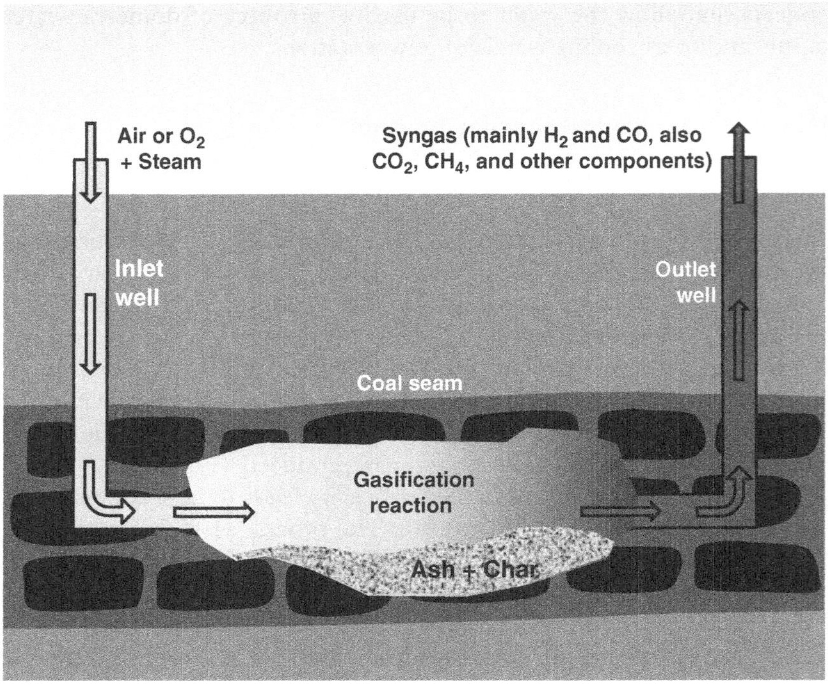

Figure 11.7 Simplified schematic of underground coal gasification (UCG).

than is found in the flue gas of a standard PC boiler power plant (40%–50% CO_2 for an IGCC plant vs. 15% for a PC boiler plant), which in theory should be much less expensive to "capture" from the syngas stream using CCS technology.

On the negative side of the ledger, the UCG process produces BTEX and other volatile organic compounds (VOCs) that can cause damage to the nervous systems of humans and are known carcinogens.[41] If BTEX and other hazardous compounds produced during a UCG operation were to migrate into surrounding aquifers and groundwater, the impact would be disastrous for the UCG industry and the surrounding communities. (Later in this chapter, we shall report on environmental violations that occurred at two of the three UCG pilot projects in Queensland, which resulted in BTEX release into surrounding groundwater, and the impacts that these incidents have had on the further implementation of UCG technology in Australia.) As a result, detailed geological assessments and site planning must be completed prior to the development of any prospective UCG site. A key part of this assessment is determining the condition of the caprock overlying the coal seams. The caprock must be impermeable and without significant fracturing to ensure that hazardous compounds produced during the UCG process do not migrate into surrounding aquifers and groundwater supplies.

More generally, a challenge to the nascent UCG industry is that underground gasification processes inherently cannot be reliably controlled like above ground gasification processes, and the challenge is compounded by the fact that every geological formation is different. The variability of operating conditions within the coal cavity is what allows the production of the VOCs already discussed. The fact that every formation is different would also complicate efforts to reduce costs through standardization and scale if UCG were to become widespread.

Finally, the flip side of UCG's potential compatibility with CCS is the fact that the gasification does produce significant amounts of CO_2, meaning that CCS would effectively be required under any serious climate policy regime. As we will discuss in Section 11.4, any CCS process is likely to add significant cost, even though capturing the more concentrated and purer CO_2 stream from the gasification process should allow for cheaper CO_2 capture than would otherwise be the case.

11.3.3 History and current status of UCG development in Australia

11.3.3.1 Early history of UCG in Australia

UCG became a subject of interest in Australia in the late 1970s after the second oil shock of 1978.[42] A number of Australian university professors started to track development of UCG technology in the United States. The leading proponent of UCG technology through the 1980s was a professor from the University of New South Wales named Ian Smith. Professor Smith pushed for government support of UCG technology in the late 1970s. He was rewarded with contracts from the governments of NSW and South Australia to study the feasibility of developing a UCG demonstration program in those two states.

In 1984, Smith issued his report to the governments of New South Wales and South Australia, titled: "In Situ Gasification of Coal for Australia." He concluded in the report that UCG was a proven technology and that its successful application in Australia could lead to the development of a cost-competitive synthetic fuels industry in Australia and a technology that could be exported worldwide. Smith took his UCG advocacy a step further by recommending that the government of South

Australia utilize UCG technology at the existing Leigh Creek coal mine to produce gas to fire a power plant. He concluded that the cost of producing power from UCG-derived synthetic gas was economically competitive against the price of power from a conventional coal-fired power plant.

However, at the time Smith issued his report, world oil prices were starting their steep, decade long decline, and all interest in the UCG concept disappeared. Both state governments refused to fund further research into UCG technology and, from 1984 until around 1996, UCG was largely relegated to the back burner in Australia.

In 1996, oil prices started to increase again, and Australia's first UCG development company, Linc Energy, was formed by Len Walker, one of Ian Smith's students who worked on the Leigh Creek feasibility study in 1984.[43] Between 1996 and 2007, three UCG development companies were formed in Queensland, which led to the development of three UCG pilot projects. The companies are Linc Energy (1996), Carbon Energy (2004), and Cougar Energy (2007). Their development philosophies and business growth strategies have wide differences, as do the outlooks for their UCG processes and corporate survival. One unifying thread is that actual progress in developing UCG has been halting. The following section provides brief case studies of the three companies and the status of their UCG development efforts.

11.3.3.2 Australian companies developing UCG

11.3.3.2.1 Linc Energy In 1996, Walker convinced investors to fund the establishment of Linc Energy, of which he served as CEO until 2002. During his six-year tenure, he established an association with Ergo Energy Technologies of Montreal, at that time one of the world's UCG technology leaders. The two companies obtained private and government funding to conduct a test burn using ERGO's UCG technology at Linc Energy's test site in the Surat Basin, near the town of Chinchilla.[44]

The test burn, which was successfully conducted over a three-year period (1999–2002), was the first extended UCG demonstration outside of the Yerostigaz UCG plant in Uzbekistan. Linc Energy gasified 35,000 tonnes of coal without violating any of its environmental clearances. Although the test burn was a technical success, Linc Energy was unable to obtain additional funding for taking the project to commercial scale because of the low oil prices that prevailed between 1999 and 2002 due to the Asian financial crisis. It decided to decommission its test facility in 2002, and in that same year, Walker left Linc Energy to pursue other interests.

In 2004, Peter Bond, a self-styled coal industry entrepreneur, acquired a 70% controlling interest in Linc Energy and took over as its CEO. Within two years, he made a number of big moves that pushed Linc Energy back into the forefront of UCG commercialization. Major actions during Bond's tenure included (a) taking Linc Energy public in May 2006 with the issuance of 88 million shares at A$0.25 per share, generating A$22 million (Linc Energy, 2006); (b) terminating Linc Energy's association with Ergo Exergy; (c) buying a controlling 74% interest (now at 91%) in Yerostigaz, the joint stock company located in Angren, Uzbekistan, which operates Uzbekistan's only UCG plant,[45] and using Yerostigaz as Linc Energy's UCG technology arm; (d) restarting Linc Energy's UCG test facility at Chinchilla and conducting additional trial burns; and (e) changing the company's downstream focus from the production of power from UCG syngas to the production of diesel using the Fischer Tropsch process. This last activity included the construction of a small 20-barrel-per-day (bbl/day) syngas-to-diesel pilot plant at the Chinchilla test site.

Between 2008 and 2013, Linc Energy acquired producing coal and oil and gas assets that allowed it to generate operating cash and to advertise itself as "more than just a UCG company."[46] By 2013, Linc Energy, in addition to its UCG activities, was recognized in equity markets as a diversified energy development company active in the areas of (i) conventional and unconventional oil and gas, (ii) traditional, open-cut coal resource exploration and development, and (iii) downstream activities related to the use of syngas produced from UGC.

Linc Energy also pursued UCG development abroad. It entered into an agreement with Exxaro of South Africa (RSA) in May 2013 to develop commercial scale UCG plants in sub-Saharan Africa, with one such project currently underway in RSA.[47] Linc Energy also has early stage UCG project scoping and reconnaissance studies underway in Poland, China, Vietnam, Indonesia, and Botswana. Finally, it has signed agreements with LLC YakutMinerals of Russia and DTEK Oil and Gas of the Ukraine to conduct feasibility studies for developing commercial scale UCG plants in those two countries (Linc Energy, 2014).

In 2013, Linc Energy decided to delist from the Australian Stock Exchange (ASX) and relist on the Singapore Stock Exchange (SGX) in hopes of "unlocking the value of its oil, gas and coal assets and its underground coal gasification (UCG) technology" (Business AsiaOne, 2013). The IPO, which was successfully completed on December 17, 2013, raised S$47.9 million of new equity funding for Linc Energy. As of December 19, 2013, Linc Energy had a capitalized value of S$595 million ($A520.5 million).

Of the three UCG pioneers, Linc Energy is unquestionably the most financially sound, although the collapse of oil and gas prices starting in August 2014 and the prolonged three-year decline in coal prices have left Linc Energy in a weaker financial position today than it was in at the end of FY 2013 (June 30, 2013).[48]

If serious questions exist about Linc Energy's continued role in the UCG industry, they have to do with its UCG technology. The company only publishes generic information about its UCG technology in its annual reports or in its Annual Corporate presentations (Linc Energy, 2013b). This makes it difficult to independently assess whether Linc Energy's UCG technology is a unique, environmentally sound UCG process.

Moreover, in June 2013, Linc Energy decided to exit its UCG demonstration site at Chinchilla due to "regulatory uncertainty in Queensland over the last several years" and its view that the government of Queensland was not treating the CBM industry fairly (Australian Mining, 2013).

Its reputation as a UCG pioneer took a further blow in April 2014 when Linc Energy was notified by Queensland's Department of Environment and Heritage Protection (DEHP) that it was facing criminal charges for "causing serious environmental harm" at its UCG test site in Chinchilla (Bajkowski, 2014). The charges that may be filed against Linc Energy by the DEHP were the result of a nine-month investigation by the DEHP. One of the charges is related to fracturing of the cap rock overlying the coal seam.

The DEHP may serve four charges on Linc Energy due to this alleged incident, with each offense reportedly carrying a maximum penalty of A$2.2 million (Solomons, 2014). As of January 2015, neither the government of Queensland nor Linc Energy had published an update on the court case. The only publicly available information that the author could find was an August 2014 court ruling in Linc Energy's favor, which stated that evidence taken by DEHP from Linc Energy offices had been unlawfully taken by DEHP (Supreme Court of Queensland, 2014). However, it is significant that these charges were made during the tenure of the Queensland's Newman government, which generally favored the coal industry.

The ongoing court case against Linc Energy, coupled with regulatory actions by the Queensland government against fellow UCG pioneers Carbon Energy and Cougar Energy (discussed in the following sections), does not bode well for the commercial application of UCG in Australia. It is difficult to independently assess the merits of these regulatory actions, but there is certainly a general sense that the atmosphere is unfavorable for UCG projects in Australia.

11.3.3.2.2 Carbon Energy Carbon Energy was listed on the ASX on June 23, 2008. Between 2004 and 2007, Carbon Energy operated under the name "Coal Gas Corporation" as a 50:50 JV between (i) Metex Resources Ltd, a mining company that specialized in the development of gold and uranium deposits, and (ii) CSIRO, the Australian government research organization that held the rights and title to special UCG technology as well as various licenses to operate that technology.

In 2008, METEX bought out CSIRO's shares in Coal Gas Corporation for A$2.4 million in cash and 111 million shares of Metex Resources Ltd, which, at the time, gave CSIRO a 27% equity stake in Metex Resources Ltd. Shortly thereafter, Metex Resources Ltd. renamed itself Carbon Energy Ltd. (Metex Resources Ltd., 2008). Today, Carbon Energy is a publicly traded company listed on the ASX with a diversified mix of shareholders ranging from passive financial investors to industrial concerns with a commercial interest in developing downstream businesses in association with Carbon Energy.

Carbon Energy is in many ways the polar opposite of Linc Energy, which has transformed itself into a diversified, vertically integrated entity. Carbon Energy is a small start-up company that is focused on a unique, state-of-the-art UCG technology, with end uses for the UCG syngas left up to companies specializing in those downstream areas. It has relied on financial markets for its growth (and more recently its survival), but it seems to lack the financial acumen and opportunistic drive of Linc Energy when it comes to acquiring and divesting assets. It is therefore not surprising that these two companies are facing very different opportunities and challenges as of January 2015.

Carbon Energy's UCG technology is based on a UCG production system known as Controlled Retraction Injection Point (CRIP), which the US Department of Energy successfully demonstrated at its Rocky Mountain test facility in Colorado in 1988, and which Lawrence Livermore National Laboratory in California continues to develop to this day (US Clean Air Task Force, 2009).[49]

Carbon Energy's patented technology, which Carbon Energy trademarked in 2011 as keyseam®, differs significantly from other UCG systems (Carbon Energy, 2015b). Specific differences include:

(a) injection of oxygen rather than air into the gasification chamber or cavity, producing higher levels of methane and lower NO_x;
(b) use of horizontal directional drilling to position the inlet and outlet pipelines into the coal seams to be gasified;

(c) adoption of the CRIP system to achieve continuous retraction of the inlet and outlet pipes during the gasification process; and
(d) application of advanced geological models and simulation tools for designing the UCG modules and then monitoring their performance.

In summary, Carbon Energy asserts that its keyseam® process, when compared to traditional UCG technology, allows finer control of the in situ gasification process and as a result produces a better quality gas with better control of any environmental impacts (Carbon Energy, 2015b).

Between December 2008 and April 2009, Carbon Energy conducted an initial 100-day test burn at its Bloodwood Creek demonstration site in the Surat Basin area of Queensland.[50] The trial was performed using a module sized to generate 1 PJ (petajoule) per year of syngas with a three-year module life, which was sufficient to fire a 20 MW combined-cycle gas-turbine power plant. According to the company, the 100-day field trial demonstrated that the Carbon Energy CRIP process could produce syngas of consistent quality on a sustainable basis without causing unacceptable environmental impacts (Carbon Energy, 2008).

All seemed to be proceeding according to plan for Carbon Energy until June 2010, when Queensland's environmental regulator – at that time the Department of Environment and Mineral Resources (DERM), later to become the DEHP – discovered that Carbon Energy was aware that it had violated its Environmental Authority (EA) three times between May and August 2009 but had not reported these violations to DERM until much later.[51]

In response to these environmental violations, DERM required Carbon Energy to provide a detailed environmental evaluation of the impacts that these spills had on its Bloodwood Creek site. Carbon Energy submitted the report to DERM in August 2010 and proceeded to remediate the site by pumping out any polluted water contained in containment ponds and removing any affected soils. Water sampling results completed by Carbon Energy at the time confirmed "that no ongoing environmental harm has occurred as a result."

Before allowing Carbon Energy to restart its Bloodwood Creek facility, however, DERM required Carbon Energy to demonstrate to it that the Bloodwood Creek test facility could be operated without creating further violations of Carbon Energy's EA. As part of its settlement with DERM, Carbon Energy agreed to adopt enhanced underground monitoring of water quality and to adopt improved risk management practices (Carbon Energy, 2013, p. 18; and *The Chronicle*, 2011).

In November 2010, DERM accepted Carbon Energy's environmental report as addressing its site remediation requirements, and in February

2011, it approved Carbon Energy's application to start UCG Panel 2, which allowed Carbon Energy the opportunity to complete its UCG demonstration project at Bloodwood Creek (Carbon Energy, 2013, p. 18).

Despite Carbon Energy's apparently good faith efforts to remediate its site, on July 12, 2011 DERM "advised the Company of charges in relation to spillage of process water in 2009" (Carbon Energy, 2012).[52] In the end, Carbon Energy negotiated a settlement with DERM on December 5, 2012 that did not require it to plead guilty in Court to any violations of its EA (Carbon Energy, 2013).[53]

While all of this regulatory activity was underway, Carbon Energy continued to develop its keyseam process, and during March 2012 it declared its UCG demonstration project at Bloodwood Creek a success. However, given its unsettled regulatory issues, it also announced that, based on a strategic review of its business model, it had decided to reduce costs at the Bloodwood Creek site to a minimum "until a full-scale commercial agreement with a partner was reached and *regulatory certainty achieved* [emphasis by author]" (Carbon Energy, 2015a).

In October 2012, Carbon Energy ceased all UCG activities at the Bloodwood Creek UCG trial site, and in January 2014 it commenced drilling into panel 2 to collect cavity water and core samples to provide environmental data that would facilitate the decommissioning of panels 1 and 2 at the Bloodwood Creek site.

The future of Carbon Energy is now uncertain, partly because of this unfortunate environmental incident, but largely due to the ongoing depressed environment for coal investments of any type in Australia. For the past two years, Carbon Energy has been surviving on an austerity budget with very limited resources. Carbon Energy's available cash and cash equivalents had declined to A$1.7 million as of June 30, 2013 from A$6.5 million as of July 1, 2012 (Carbon Energy, 2013, p. 59). Its financial situation remained weak throughout FY 2014, with its shares trading in the range of A$0.01 and A$0.06 per share, a precipitous decline from its peak of $0.83 per share on September 18, 2009, creating doubts as to whether Carbon Energy could continue as a "going concern" (Carbon Energy, 2014a).

Carbon Energy did manage to obtain funding during the second half of of the 2014 calendar year from a variety of sources including a A$3.7 million R&D rebate from the Australian Tax Office and net proceeds from an October share placement, which provided Carbon Energy with A$2.2 million of new funding. However, as of December 31, 2014, Carbon Energy still only had A$3.5 million in cash and cash equivalents, with a forecasted expenditure of A$1.6 million during the period January 1, 2015 to March 31, 2015 (Carbon Energy, 2014a, appendix 5b). To survive, Carbon Energy has had to implement strict cost control measures, including cutting the salaries of senior management as well as the number of junior staff (Carbon Energy, 2014a, p. 7).

Carbon Energy is hoping to attract partners for its planned Blue Gum and Bloodwood Creek 2 projects in the Surat Basin, with Blue Gum focused on producing pipeline-quality gas from the syngas produced at a yet to be permitted UCG plant in the Surat Basin Area. It has also been trying to sell its non-core coal assets to third-party investors as a means of raising additional cash (Carbon Energy, 2014a, pp. 3–4). However, given the parlous state of the coal industry worldwide, Carbon Energy is facing difficulties in selling off its non-core coal assets[54] and finding JV equity partners for its Blue Gum and Bloodwood Creek UCG projects. Its difficulties are compounded by the reluctance of the Queensland government to allow the commercial development of UCG projects due to environmental concerns.[55] In any case, new UCG projects in Queensland will take at least three years to get permitted, and it is uncertain whether it will be successful in obtaining the necessary clearances for a second UCG project.

Needless to say, the financial outlook for Carbon Energy as of January 2015 is bleak. At the price range for its shares in 2014 (A$0.02–A$0.06/share), Carbon Energy will find it exceedingly difficult to tap public equity markets for sufficient additional funding to remain a going concern.

The company acknowledged these financial uncertainties in its 2014 Annual Report (Carbon Energy, 2014b) when it stated:

> The ability of the Company and the Consolidated Group to continue as going concerns and fund the development and commercialisation of its keyseam UCG technology is dependent upon the receipt of agreed technology and service fees, the receipt of the R&D cash incentive to be lodged with the ATO and the ability to secure additional funds from a range of opportunities. (p. 56)

It went on to state that "should the Company and the Consolidated Group not receive the forecast cash inflows and additional funding referred to above *there are material uncertainties as to whether the Company and the Consolidated Group will be able to continue as going concerns*" (emphasis added).

The recent election in Queensland resulted in the Australian Labor Party returning to power, which is another bit of bad news for Carbon Energy, as the previous National-Liberal government was much more sympathetic to energy projects in general.

11.3.3.2.3 Cougar Energy Cougar Energy Limited was a publicly listed Australian company that traded on the ASX from 2007 until May 2013, when it was restructured and renamed "Moreton Resources Ltd." The original founder of Linc Energy, Len Walker, who served as Managing Director of Cougar during this time, helped shift Cougar Energy's business away from traditional minerals exploration and development toward UCG. The UCG technology applied by Cougar Energy was a variant of the old Soviet technology that was applied by Linc Energy during Walker's tenure there. Cougar Energy entered into a licensing agreement with Ergo Exergy Technologies of Canada to apply Ergo Exergy's UCG technology to all UCG projects that both Cougar Energy Ltd. and its subsidiary, Cougar Energy (UK) Ltd, which was formed to develop international UCG projects, chose to develop.[56]

Within a year of being appointed MD of Cougar Energy, Walker brought a UCG test burn project to an advanced stage of development at its Kingaroy site, which is located around 210 km east of Brisbane in Queensland. According to Cougar Energy, its site contained a JORC-compliant resource of 73 Mt, with two primary seams at depths ranging from 130 to 300 m and having thicknesses of 5–17 m. Cougar Energy deemed this coal resource sufficient to meet the fuel requirements of a 400 MW, combined cycle gas turbine (CCGT) power plant for thirty years.

On April 30, 2008, the Queensland Environmental Protection Agency (EPA) granted Cougar Energy an "environmental authority" (EA) to conduct a test burn of 20,000 tonnes of coal in situ at its Kingaroy site without requiring Cougar Energy to submit an EIS. It also did not require that Cougar Energy commit to work under approved environmental monitoring and operating conditions laid down in its government-approved Groundwater Assessment and Impact Study (Queensland Ombudsman, 2012).

The Cougar test burn project commenced on March 15, 2010, scarcely three days after the Queensland DERM, which was by then the regulatory successor to the EPA and responsible for managing water pollution aspects of mineral development projects under Queensland's Environmental Protection Act (1994), received Cougar Energy's draft Groundwater Assessment and Impact Study (Queensland Ombudsman, 2012). Cougar Energy produced its first syngas on March 16, 2010 (Queensland Ombudsman, 2012). Shortly after commencing its test burn, Cougar Energy publicly announced its plans to increase the project size so that it could produce sufficient syngas to fuel a gas-fired combined cycle plant with a first stage capacity of 200 MW (Cougar Energy Limited, 2010).

Unfortunately for Cougar Energy and the entire UCG industry, the Kingaroy project did not proceed as planned. On March 20, 2010,

Cougar Energy had to shut down the Kingaroy pilot plant due to "blockages in both the injection well and the production well" (Cougar Energy Limited, 2010). Three months later, on June 30, 2010, Cougar Energy informed DERM that it had detected minute levels of benzene (2 parts per billion) in two groundwater samples taken on two separate occasions in March 2010 (Cougar Energy Limited, 2010).

From that point onward, DERM commenced "breach of environmental authority" proceedings against Cougar Energy, and on July 8, 2011, DERM ordered Cougar Energy to shut down its UCG test burn facility at Kingaroy. On July 17, 2010, DERM informed Cougar Energy that it needed to conduct a full environmental evaluation of the Kingaroy site and that it would face legal action for violating the terms of its EA, especially because Cougar Energy had failed to notify DERM in a timely manner after it had detected breaches of environmental standards for the project (*The Australian*, 2011).

Cougar Energy for its part claimed that DERM had made a decision based on flawed conclusions and contested the DERM statement that it had caused any environmental harm to groundwater supplies in the area. On October 17, 2011, Cougar Energy launched an unprecedented civil suit against DERM for "negligence and breach of statutory duty." It also filed a complaint with the Queensland Ombudsman (Queensland Ombudsman, 2012).

It would take another year for the Cougar-DERM dispute to be resolved. In the end, all parties had egg on their faces. The Ombudsman, which issued its report in September 2012, was highly critical of the weak environmental oversight provided by the EPA and its successors, DERM and DEHP. In particular, the Ombudsman felt Queensland's environmental regulatory agencies did not create adequate standards and procedures for managing the Kingaroy project, especially given that the Cougar Energy UCG project and other UCG test projects were based on novel technology that had a high level of uncertainty concerning impacts on water quality. The EPA, DERM, and DEHP ended up looking incompetent.

However, before the Queensland Ombudsman issued its report, Cougar Energy's Board of Directors decided on July 24, 2012, to replace Len Walker as Cougar Energy's Managing Director and CEO, with Rob Neill, a UGG and coal industry veteran, named as his replacement. It also decided to seek a settlement with the government of Queensland over the Kingaroy water pollution issue. Approximately one year later, on July 30, 2013, Cougar Energy agreed to drop its lawsuit against DERM (by then DEHP), and in September 2013 it accepted a fine of A$75,000 as a settlement for violating the terms of its EA.

Rob Neill lasted as CEO and Managing Director of Cougar Energy until September 10, 2013, when he resigned his executive positions. Shortly thereafter, at an extraordinary shareholders meeting held on September 19, 2013, Cougar Energy shareholders approved a board proposal to change the company's name to Moreton Resources and reposition the company away from being a UCG development company and toward becoming a traditional coal resource development company.

As of December 31, 2014, Moreton Resources was still completing its process of decommissioning its UCG panel at the Kingaroy site. It has confirmed that the Kingaroy site will no longer be considered as a site for a UCG project. It will instead be developed as a conventional coal mining site (Moreton Resources Ltd., 2014). Similar to Carbon Energy, it is surviving on limited resources, with cash and cash equivalents equal to A$6 million at the end of 2014.

11.3.4 Technology and regulatory factors affecting UCG development

As in the case of CBM, the government of Queensland has played a more proactive role than the government of NSW in addressing important regulatory issues such as overlapping claims and environmental protection.[57] However, the Queensland government's regulatory performance has been significantly less supportive and proactive than in the case of CBM. There are several likely explanations. First, Queensland's environmental protection agencies found themselves in a severe state of flux between 2007 and 2013; they were reorganized at least three times during this period, with associated transfers of environmental protection responsibilities.

Perhaps as a result of this unstable regulatory environment, Queensland's EPA and its successors, DERM and DEHP, were severely handicapped in their attempts to establish and impose adequate environmental protection safeguards on UCG developers prior to and during their UCG test burns. During the test burns, successors to the EPA may not have had sufficient time to assume ownership of their new environmental protection responsibilities and, as a result, did not provide sufficient oversight and monitoring of the test burns and any impacts on groundwater quality surrounding the UCG test burn sites. More stringent regulation *at the outset* could have helped provide more certainty for UCG developers and perhaps helped them avoid making the serious missteps that (along with adverse market conditions) have now thrown all UCG efforts in Australia into doubt.

Second, UCG is simply a more complex, less mature, and arguably more risky technology than CBM at its current stage of development. Regulators always face the challenge of how to balance caution against the tolerance for experimentation that can allow a new industry to develop. In the case of UCG, a retrospective analysis of environmental regulations suggests that maybe it was too permissive in the beginning, perhaps swinging to overly punitive once some problems were discovered. Unlike processes such as CBM or shale gas extraction, the near-term economic prize from UCG in the form of royalties and employment may simply not be large enough to push regulators to allow UCG to move forward.

Whatever the reasons for the environmental lapses of early UCG projects, their impacts on the future development of UCG in Queensland, and perhaps in the rest of Australia, have been disastrous. It is unlikely that new pilot UCG projects will be funded by private investors in Queensland and equally unlikely that new pilot UCG projects will be approved by the Queensland government over the remainder of this decade. There will need to be a prolonged cooling off period before renewed interest in UCG is shown by private investors and the Queensland government.

The issue of overlapping claim disputes between UCG and CBM developers was also handled by the Queensland government in a manner that was unfavorable to further UCG development. The CBM industry, having reached full commercial development, already held Petroleum Licenses to the most promising areas for UCG development. In some cases, coal mining companies held Mining Development Licenses to promising UCG development areas. Given the unproven nature of UCG technology, the government of Queensland was unwilling to provide UCG developers with the same level of regulatory certainty that it was willing to provide to coal miners and CBM developers.[58]

Given the unexpected environmental violations of the three UCG pilot projects, perhaps the Queensland government was wise in retrospect to proceed with caution on the overlapping UCG-CBM claims issue. Nevertheless, the lack of resolution of such disputes has effectively put UCG companies in a state of development limbo. Unlike coal mining and CBM, UCG and CBM projects are, for the most part, mutually exclusive. In order for a CBM project to proceed, the water cap immediately above the coal seam and the water contained within the matrix of the coal seam must be removed so that the seam can be depressurized and the CBM trapped within the coal matrix will desorb and flow out of the well. However, once the water cap and formation water are removed, the basis for an efficient UCG process is destroyed.

For a UCG process to proceed efficiently and in an environmentally sound manner, the coal seams and the conversion cavity must remain at a minimum hydrostatic pressure. This minimum pressure is necessary to ensure the controlled growth of the UCG cavity and to minimize the risk that pollutants and contaminants produced during the UCG process will escape from the UCG cavity and pollute nearby aquifers and other groundwater sources. Because CBM production wells operate below this minimum hydrostatic pressure, an area is no longer suitable for syngas production using the UCG process once CBM production commences.[59] (Water in the coal matrix also supports the in situ conversion gasification process by allowing a water shift conversion process to occur, which creates favorable conditions for the production of additional methane and hydrogen.)

In June 2013, an independent scientific panel (ISP) appointed by the government of Queensland issued its report on the underground coal gasification pilot trials by the three companies. The report contained three overarching recommendations and eight specific recommendations. The eight specific recommendations were mostly related to management of the environmental performance of future UCG projects, pilot or otherwise. The report focused heavily on the deficiencies and failings of the three pilot plant proponents to follow industry standard measures for controlling environmental incidents from petrochemical plants.

Its overarching recommendations suggested that (1) Carbon Energy and Linc Energy should be allowed to continue with their pilot projects but only for the purpose of successfully demonstrating that "the self-cleaning cavity approach advocated by these two project proponents for decommissioning their pilot projects is environmentally safe" (Moran, Da Costa, and Cuff, 2013); (2) the government and each project proponent should develop "a planning and action process" to demonstrate decommissioning, especially the self-cleaning cavity process;[60] and (3) until an environmentally sound decommissioning process has been proven, the government should not allow any commercial facilities to commence operations. The advice provided by the ISP in its report was accepted by the Queensland government in principle. If used as the basis for implementation of future UCG projects, the ISP's recommendations should help prevent the types of environmental incidents that occurred during the three pilot projects. But it is unclear when any future UCG projects in Australia will actually take place.

11.4 Carbon capture and storage

"Carbon capture and storage (CCS)" is a blanket term for technologies that capture CO_2 at the point where a fossil fuel is used and then inject the CO_2 underground where it will not escape to the atmosphere. Because

CCS requires a massive upfront capital expenditure, has not been fully demonstrated at commercial scale, and has as its only function the disposal of CO_2, it has no prospect for entering into widespread commercial operation until a significant penalty (carbon tax or price) is imposed on coal consumers for releasing CO_2 to the atmosphere and/or a high value mass market for CO_2 is created.[61] In this respect, it is very different from CBM and UCG, which are potentially profitable (in the case of CBM, demonstrably so) even in the absence of any new price or other type of supporting regulation. (As discussed in Section 11.2, the Queensland government did provide some incentives for natural gas use that helped support the nascent CBM market, but this was a far smaller regulatory intervention than would be required to make CCS attractive.) The development of CCS technology will therefore proceed on the basis of very different technical, environmental, and commercial considerations than CBM and UCG technologies.

11.4.1 Technology basics

A CCS system consists of three parts:

- CO_2 capture, which involves the removal of the CO_2 contained in the flue gas of a coal-fired power plant as well as its liquefaction via compression and cooling;
- CO_2 transportation, which is mostly focused on pipeline transport but could also include ship, rail, and truck transport; and
- CO_2 storage, which involves injecting the liquid CO_2 into long-term underground storage.

Of the three parts, the capture component imposes the greatest cost penalty on electric power. When implemented as part of a new coal-fired power plant it increases the life-cycle cost of electricity from the new power plant by 80%–90% (Folger, 2013). It also creates a significant adverse impact on power plant performance. As a result of its importance to the commercial success of any CCS technology, the capture component of any CCS system has received the most financial support and attention from government agencies and is discussed in greater detail than the CO_2 transport and storage components.

11.4.1.1 CO_2 capture

The capture of CO_2 contained in coal can occur either before the coal is burned in a boiler (pre-combustion CO_2 capture) or after it is burned (post-combustion CO_2 capture). A third capture technology, oxy-fuel combustion, combines features of both pre- and post-combustion capture.

11.4.1.1.1 Post-combustion capture of CO_2 The post-combustion process involves the combustion of coal in the presence of air and then the absorption of the CO_2 from the resulting flue gas via a chemical solvent extraction process (IEA, 2013). In this process, flue gases are passed through an absorption column, also referred to as a "CO_2 scrubber," in which the solvent chemically binds with the CO_2, removing it from the gas stream.[62] After CO_2 absorption, the remaining flue gases consist primarily of water vapor and a climatically inactive form of nitrogen. The CO_2-rich solvent is recovered from the CO_2 scrubber in solid form and heated, using steam from the power plant's steam turbine, to release nearly pure CO_2.[63]

Heating the solvent to release the CO_2 requires a lot of steam, which can be taken from either the power plant's steam turbine or a stand-alone steam turbine, and results in a 15%–25% reduction in the power plant's net dependable capacity and a 30% reduction in its net efficiency (relative to a plant without CCS).[64] If air is used as the oxidant for combustion, the CO_2 content of the flue gas being treated in the solvent extraction vessel is very dilute – around 15% CO_2 by volume. This low CO_2 concentration requires a very large solvent extraction plant to be built, adding to the high costs of the solvent extraction process.

On the positive side, post-combustion solvent extraction systems can be retrofitted onto an existing coal-fired power plant as well as included as part of a new plant design. For a retrofit to be practical, however, the plant site must have sufficient unused space for the CO_2 capture plant, which can equal the land requirement of the power island (boiler, turbine and generator, and ancillary systems). There also need to be substantial modifications to the power plant's anciliary systems to support the solvent plant that will be used for removing CO_2 from the flue gas stream.

Although the CO_2 capture process is an extension of the current technical concepts for removing nitrogen oxides (NO_x), sulfur oxides (SO_x), and other power plant emissions from a power plant's flue gases, the large volumes of CO_2 produced and the dilute nature of the CO_2 in the flue gas stream means that the capital cost and space requirements for a CO_2 capture plant are much higher than for NO_x and SO_x removal systems.

11.4.1.1.2 Pre-combustion capture of CO_2 The pre-combustion process involves the removal of the CO_2 before the combustion of the final energy product. The first step in the pre-combustion process is to convert the original solid fuel into a synthesis gas (syngas) via gasification, with the resulting syngas consisting of CO_2, carbon monoxide (CO), hydrogen (H_2), and water. The syngas is then cleaned of particulates and other impurities that might cause problems in the subsequent CO_2 removal step. Next, the CO is reacted with steam in a shift reactor to produce more CO_2 and H_2. After the shift reaction, the gas consists primarily of CO_2 and H_2 with a small amount of methane (CH_4). The CO_2 is then separated from H_2 and other flue gas constituents using either a chemical or physical solvent absorption process.

Because the CO_2 is in a highly concentrated form, the size of the solvent absorption plant and compressors can be much reduced. The H_2 can be sold as fuel for power generation or as feedstock for making chemicals such as ammonia and methanol.

The leading pre-combustion method currently being applied is called IGCC with Carbon Capture. In the IGCC process, the syngas is purified, stripped of its CO_2, and fed into a CCGT plant to generate power.[65] Proponents of the IGCC process claim that it will allow CO_2 to be recovered from coal in a relatively simple and cost-efficient way when compared against alternative carbon capture technologies.

Most of the process equipment required for pre-combustion CO_2 removal, such as downdraft gasifiers and gas clean-up equipment, is commercially proven and widely available from credible equipment suppliers. The IGCC process is already used in the petrochemical, refinery, and power industries.

11.4.1.1.3 The oxy-fuel combustion process The oxy-fuel combustion process is a very promising but, at this point, commercially unproven technology for CO_2 capture. It involves combusting coal in an oxygen-rich atmosphere inside a specially designed boiler. The resulting flue gas has a very high concentration of CO_2, which can, in theory, be more economically separated from other flue gas components than with post- and even pre-combustion capture systems. The oxy-fuel combustion process includes the following components: (a) an air separation unit (ASU) that separates oxygen from air, leaving a product that is 85% oxygen and around 15% nitrogen plus trace amounts of argon and water; (b) a specially designed boiler and auxiliary unit that can safely combust coal in an oxygen rich boiler atmosphere; and (c) a CO_2 purification unit (CPU) for drying and compressing CO_2 prior to its liquefaction.

The process begins with the combustion of coal and air in a specially designed boiler capable of firing either 100% air or a mixture of air and pure oxygen (O_2). Once steady state operating conditions have been achieved on 100% air, the plant operator switches the plant to the oxy-fuel combustion mode, and coal is then combusted in the presence of a concentrated stream of oxygen and CO_2 recycled from the CO_2 capture plant. The resulting flue gas has a very high CO_2 content but a relatively low NO_x content, since the elemental nitrogen in the air from which NO_x emissions are largely formed has been mostly removed during the production of oxygen in the air-separation plant. That leaves only the nitrogen in the coal and the small amounts in the O_2-nitrogen mixture and air as sources of nitrogen for NO_x formation.[66] Any NO_x by-products that are still formed via the oxy-fuel combustion process can be controlled with standard low NO_x burners and should not require the installation of SCR to manage NO_x emissions.

After the coal is combusted in the high-oxygen furnace environment, the resulting flue gas is cleaned of ash and other particulates, water, and sulphur. After compression, the purified flue gas ranges from 80% to 90% CO_2, with water vapor, argon, oxygen, and NO_x as the remaining impurities. The flue gas is then condensed, removing the water vapor and creating an almost pure CO_2 stream. The CO_2 can then be compressed, dried and further purified before being transported to a storage site. The chemical solvent extraction process is avoided completely. Since CO_2 converts into a liquid at a pressure of 70 bar, power plants relying on the oxy-fuel combustion process (or the solvent extraction method) for CO_2 capture still need to rely on compressors to convert the gaseous CO_2 into a liquid. The compression process is very energy intensive, as is the process of separating oxygen from air in an air separation plant.

The primary challenges facing developers of the oxy-fuel combustion process are (1) increasing the efficiency of air separation plants;[67] (2) reducing the cost of physical solvents; and (3) demonstrating the effectiveness and environmental soundness of injecting CO_2 streams that have less than 99% purity into underground storage areas. Nonetheless, oxy-fuel combustion is currently viewed by the CCS community as the strongest contender to become the least-cost CO_2 capture system, especially for use in large new (or re-powered) coal-fired plants.

Other advanced capture systems remain under development, including (1) a chilled ammonia process for removing CO_2 from a flue gas stream; (2) oxygen chemical looping (as an enhancement to the oxy-fuel process); (3) ion transport membranes and other advanced membranes for higher efficiency ASU plant and CO_2 capture (Aldous et al., 2013).

11.4.1.2 CO_2 transport

The transport of CO_2 can be by pipeline, ship, rail, and even truck, but due to its low (or even negative) value, it is difficult to justify its transport over long distances by anything other than a pipeline. Pipeline transport of CO_2 is a proven technology and does not pose any significant technical or safety risks, although substantial investments in pipelines may be required if CO_2 is generated far from suitable storage sites. Large volumes of CO_2 are already safely transported over long distances in high-pressure pipelines in the US. The pipelines extend over 2,500 km, with the transported CO_2 used to enhance oil production in mature oil fields. Although CO_2 transport is considered a commercially proven technology, additional research and development activity is needed in two areas: (i) planning and implementing pipelines and CO_2 transport hubs capable of accepting CO_2 streams with CO_2 concentrations as low as 95%; and (ii) developing large scale sea transport vessels for CO_2.

11.4.1.3 Geological storage of CO_2

Underground storage of CO_2 involves the injection of liquid CO_2 into subsurface geological formations that have the capacity and geological conditions to keep the injected CO_2 locked away for hundreds, if not thousands, of years. Underground injection of CO_2 has been successfully performed by oil companies since the 1970s as a means of enhancing oil production from declining oil wells. But these early projects did not involve long-term storage. The focus was on increasing oil production, with a certain percentage of the injected CO_2 rising to the surface along with the oil. Over time, more of the injected CO_2 will escape from the depleted wells.

Researchers have expressed a high level of confidence that indefinite geological storage of CO_2 is feasible for non-EOR underground storage projects. They cite as one of the strongest bits of evidence the fact that CO_2 is found in many parts of the world, trapped underground between the pore spaces of sedimentary rock in the same way that oil and natural gas have been stored for millions of years until man started their extraction. The idea behind geological storage is to replicate the conditions that have allowed these natural CO_2 accumulations underground.

Two alternatives have received the greatest attention as potential underground storage sites for CO_2 produced from fossil fuel power plants. They are (1) depleted oil and gas reservoirs, which have proven their capability to hold oil and gas over millions of years and in some cases have already been injected with CO_2 for purposes of enhancing oil and gas recovery; and (2) deep saline underground aquifers. These aquifers are filled with salt water and located a minimum of 800 m underground. In many cases, they are capped by impermeable rocks which prevent the injected CO_2 from escaping.

The natural mechanisms for sequestering CO_2 indefinitely include a combination of physical and geochemical processes. Before storing CO_2

in any geological formation, it must first be compressed to a dense "super-critical" fluid with a density that occupies less than one per cent of the space it would occupy in a gaseous state. Then natural processes would be expected to accomplish the long-term underground storage of the CO_2 emissions generated by power plants.[68]

A significant amount of research is needed to empirically confirm the theory of long-term storage of CO_2 at identified storage sites. The required research ranges from laboratory and bench scale research studies to large commercial scale demonstrations of specific CO_2 storage sites (Aldous et al., 2013). Significant effort is needed to develop models and data sets for testing the flows of CO_2 within a storage area and to predict leakage rates from specific sites. Sophisticated computer models and standards will be needed even to determine what constitutes "CO_2 leakage" from a storage site. There is a related need for improved technologies and methods for monitoring and verifying the amount of leakage from a CO_2 storage site over time (Aldous et al., 2013). The research program that has been proposed for CO_2 storage by the CCS scientific community is impressive in terms of its comprehensive scope. However, it also highlights the high level of uncertainty that still exists concerning the efficacy of long-term CO_2 storage projects under development and in implementation. Unpleasant surprises may emerge that make the theory of long-term CO_2 storage seem less compelling in practice.

11.4.2 Advantages and disadvantages of CCS

The advantage of CCS to Australia's coal industry is simple and significant: it is the one technology that could allow coal to be burned with minimal emissions of CO_2 to the atmosphere. Other improvements in coal conversion processes – for example in the efficiency of conversion into usable energy – can reduce CO_2 emissions per kWh to an extent but will not be able to prevent coal plants from remaining very significant sources of GHGs. For this reason, CCS options that can be retrofitted into existing coal-fired power plants are viewed by the IEA and most industry analysts as one of the essential technologies to adopt if the GHG emission reduction target proposed by the United Nation's Intergovernmental Panel on Climate Change (IPCC) for 2050 is to be met.[69] If one accepts that coal will remain a significant fuel for some time and that climate change mitigation is important, some form of CCS would seem to be required.

The biggest disadvantage of CCS is cost, coupled with a lack of experience at scale. A 2011 engineering cost study conducted by WorleyParsons for the GCCSI estimated that post-combustion CCS would add 53%–76% to the levelized cost of electricity (LCOE) when incorporated into a new coal-fired power plant (WorleyParsons Schlumberger, 2011). (CCS using pre-combustion capture was only projected to add 37% to the cost of a new IGCC plant, but the total LCOE was comparable to other coal plants with CCS due to the higher cost of IGCC itself.) The WorleyParsons analysis assumes a 10%–15% degradation in output due to the use of CCS. It also assumes a reduction in net plant efficiency of 30% for post-combustion CCS and 22% for pre-combustion CCS. It estimates capital and O&M costs to be almost double what they would be for a power plant without CCS (WorleyParsons Schlumberger, 2011).[70]

A second study completed in 2011 by staff from the Melbourne-based Global CCS Institute examined the comparative costs of coal-fired power plants equipped with CCS against other low carbon alternatives such as geothermal, hydropower, onshore wind, nuclear, offshore wind and solar.[71] The study noted that, based on 2011 prices, CCS was significantly more expensive than geothermal, hydropower, onshore wind, and nuclear but less expensive than offshore wind and solar. The report's authors noted that the lower-cost alternatives to CCS are limited by (i) resource constraints, (ii) availability of suitable sites (conventional geothermal, hydropower, and onshore wind), (iii) perceived safety issues and associated regulatory risk (nuclear), or (iv) the challenge of managing intermittency (onshore wind). In summary, the institute's authors concluded in 2011 that, despite the high cost of CCS, it is still one of the cheaper GHG reduction options to implement and should form part of any country's slate of technology options for reducing GHG emissions (Abellera and Short, 2011).[72]

11.4.3 History and current status of CCS in Australia

Three government-funded organizations are leading the CCS research, development, and demonstration programs in Australia. They are:

- The Commonwealth Scientific and Industrial Research Organization (CSIRO), which is Australia's national science agency.[73] Initially its focus was on research into improved agriculture methods and crop selection, but over time it has broadened its research mandate to include all sectors of Australia's economy. Energy, environment, and mining are today very important components of CSIRO's overall research program. With respect to CCS, CSIRO conducts research in collaboration with local and foreign universities and industry into (i) improved post-combustion capture systems,[74] (ii) understanding of the chemistry and basic science of coal gasification with a focus on determining the suitability and performance of Australian coals as feed stocks for coal gasification systems, and (iii) the performance of CO_2 storage sites.[75]

- The Cooperative Research Centre for Greenhouse Gas Technologies (CO2CRC), which is part of the Australian Government's broader Cooperative Research Centre's program, an organization that has existed since 1990. The CO2CRC program was established in 2003 as a not-for-profit, limited liability public company. It is a joint venture between foreign and domestic industry, universities, and government agencies. It also has a commercial entity known as CO2CRC Technologies Pty Ltd. (CO2Tech), which holds exclusive rights to patents acquired through its various collaborative projects. CSIRO is one of CO2CRC's research collaborators.
- The Global CCS Institute (GCCSI), which was established in 2008 with Australian government funding of A$100 million. The GCCSI has its headquarters in Melbourne, Australia, and serves as the global research and advocacy arm of the CCS industry. Although it is connected to the CCS industry, GCCSI is generally recognized for its unbiased, fact-based reporting of CCS activities and issues worldwide.[76]

These three research, development, and information-sharing organizations have been in the forefront of all CCS activity in Australia. Initially, CCS projects in Australia received their funding through the government's National Low Emissions Coal Initiative, an A$500 million program started in 2005 during the John Howard led government. Government financial support for CCS was later expanded through the Rudd Administration's A$5.1 billion Clean Energy Initiative's CCS Flagships Program. In 2010, the CCS Flagships Program had funding that would have been sufficient to support up to four commercial-scale power plant projects with a combined capacity of 1000 MW and a number of smaller industrial CO_2 removal projects.

The Australian government seemingly launched itself to the forefront of global CCS project development with this initial flourish of institution building and its announcements of CCS project funding. Proposals for funding under the CCS Flagships Program were solicited in 2009, and four projects were short-listed in December 2009 for funding awards:

- Wandoan IGCC Plant, a 400 MW project proposed by GE and Stanwell Corporation using GE IGCC technology and pre-combustion CO_2 removal with the MEA solvent process. The project proponents planned to achieve commercial operations by 2015, with 2.5 Mtpa of CO_2 slated to be captured and stored.
- Zerogen, another IGCC plant (530 MW) proposed by the Queensland government and various private companies. It was expected to cost A$4.3 billion and capture and store 2 Mtpa of CO_2. The estimated start of commercial operations was unknown.

- Collie South West Hub, a multi-user industrial capture facility using post- and pre-combustion technologies to capture up to 2.5 Mtpa of CO_2 with commercial operations starting in 2015. The project was proposed for the southern Perth Basin in western Australia.
- Carbon Net, another multi-user project designed to capture and transport between 4 and 10 Mtpa of CO_2 from power plants located in the state of Victoria. The project was proposed by the state of Victoria, with commercial operation scheduled to begin sometime between 2015 and 2019.

However, the enthusiasm of the Australian government for large-scale CCS demonstration projects started to wane around 2010. The government of Australia was supposed to announce final project awards under its Flagships Program by August 2010 but delayed its final announcements until 2012. During 2011, massive flooding hit Queensland and other states. The Australian government, which estimated the cost of rebuilding damaged infrastructure at A$5.6 billion, decided that two-thirds of that amount should be funded through spending cuts and spending deferrals across a number of government programs. In January 2011, the government proposed a combination of reductions and deferrals to the CCS Flagships Program and the GCCSI of A$250 million, but those reductions and deferrals were almost doubled to $420.9 million in May 2011.

Even some project proponents started to lose interest in CCS once they became fully aware of the capital costs and technical uncertainties of their proposed carbon capture (CC) technologies. Of the four short-listed CCS Flagship projects, the two Queensland-based projects – Wandoan and Zero-Gen – were canceled between 2010 and 2011. Only two projects ended up being funded under the CCS Flagships Program: the Collie Multi-User Hub project and the Carbon Net project in Victoria. The Collie CO_2 Hub project is facing some opposition from farmers on environmental grounds, largely due to concerns about possible damage to groundwater aquifers amid reports that a suitable saline aquifer with an impermeable cap rock has yet to be found for storing the captured CO_2 (Vivian, 2014).

11.4.4 *Can learning and new technology make CCS economical in the future?*

Proponents of CCS technologies do not deny that CCS as practiced today is very costly. They reply with three counterarguments. First, post-combustion CCS technology, albeit costly, is still cheaper than most other GHG emission reduction options. Without it, the cost of achieving reductions in GHG emissions in line with the UN IPCC 2050 target would be as much as 40% higher (Abellera and Short, 2011). Second, high-value, beneficial uses for CO_2 (in addition to Enhanced Oil Recovery) may eventually reach commercial feasibility and, by generating substantial revenues, pay for a portion of the CCS project and allow it to be commercially deployed at a much reduced cost. Third, through learning and the application of advanced (third-generation CC) technologies the cost of CCS will be progressively reduced over time to more "acceptable" levels.

With respect to the first counter-argument, the verdict is in: governments, industry, and a majority of the general Australian public are not willing to pay the high price of implementing CCS on a wide scale using existing commercially available CC technology. Indeed, a cautious approach is being followed in all developed countries thus far, which involves the implementation of a few commercial-scale CC demonstration plants and/or a few CO_2 transport and sequestration projects but not to the extent that their power sectors will be heavily exposed to existing CC technology. The hope is that more efficient and less costly CC technologies will eventually be ready for commercial deployment in ten to twenty years. In the interim, governments and industry are focused on not falling into a "technology trap" by investing in expensive and inefficient CC plants based on first-generation technologies. It must also be noted that these comparative economic results were based on 2011 prices for the various technologies. Given the substantial cost reductions for PV solar panels and on-shore wind turbines over the past four years, the cost comparison reported by GCCSI may no longer be valid.

Concerning the possibility of high value, beneficial uses for CO_2, no other reuse technologies apart from EOR have proven themselves to be ready for commercial deployment on a wide scale. Beneficial reuse has proven to be a disappointment, with most technologies requiring CO_2 to be available for free (EOR being the exception) and in some cases at negative cost in order for beneficial reuse options to be commercially feasible. Beneficial reuse, including EOR, therefore is only likely to serve a small niche role in making CCS affordable.

Finally, there is the combined prospect of cost reductions through (i) the process of learning,[77] and (ii) the application of advanced or third-generation CC technologies.[78] With respect to cost reductions through learning, researchers distinguish between the costs of deploying "first-of-a-kind" (FOAK) CC plants and "nth-of-a-kind" (NOAK) CC plants. The cost of FOAK plants is typically based on first-generation CC technology and current EPC prices while the cost of NOAK plants is based on second-generation CC technologies such as enhanced heat recovery for solvent regeneration, improved solvents, and improved efficiencies of cryogenic, air-separation units, and even more due to EPC and O&M cost reductions resulting from "learning."

The magnitude of projected cost savings through learning and the application of second-generation CC technologies varies dramatically. A number of academic and policy researchers have suggested very large reductions in the cost of CO_2 abatement for NOAK plants ranging from 25% to 70%.[79]

Engineering companies, which are the ones tasked with the building of CC plants, take a much more cautious view. For example, WorleyParsons Schlumberger (2011), as part of its engineering cost study of CCS systems for GCCSI, forecast that NOAK CC plants will achieve very modest reductions of 3% in both the cost of CO_2 abatement and the life-cycle cost of electricity relative to those same costs for FOAK CC.

Finally, with respect to the ongoing efforts to develop and eventually deploy third-generation CC technologies, one can find reason for some optimism. Between 2007 and 2013, significant progress has been made in testing, at the bench and pilot scale level, a number of third-generation CC technologies, which, if successfully deployed at commercial scale, can result in dramatic reductions in life-cycle CO_2 abatement costs for CC systems.[80]

Third-generation CC technologies applicable to pre- and post-combustion CC systems include (i) improved liquid solvents such as advanced amines, potassium carbonate, advanced mixtures, and ionic liquids; (ii) solid adsorbents such as supported amines, carbon based adsorbents, and sodium carbonate; and (iii) polymeric, amine-doped, and bio-mimetic-based membranes (Aldous, 2013). These technologies are at early stages of development, ranging from laboratory to pilot scale, and issues related to energy requirements for solvent regeneration, and environmental (air emission) impacts resulting from the regeneration and degeneration of solvents are still issues limiting their commercial application.

With respect to oxy-fuel and pre-combustion systems, two generic technologies are cited in the technical literature as showing considerable promise. They are (i) ion transport membrane technology (ITM) using a type of ceramic material that demonstrates a high oxygen flux (high rates of oxygen transport) and is highly selective for oxygen (Repasky, Stein, Armstrong, 2013); and (ii) chemical looping combustion, which is a two-stage (sequential) chemical process that converts fossil fuel into a stream of H_2, CO_2, and H_2O without the production of any NO_x (Aldous, 2013).

ITM technology, which has reached the pilot scale stage of development in the United States, has the potential to achieve significant reductions in the capital and operating cost of air-separation units (ASUs), which represent a major portion of the capital and operating costs of both oxygen blown pre-combustion plants and oxy-fuel combustion plants. A study completed in 2013 by Air Products for the US DOE compared the technical and economic performance of cryogenic and ITM oxygen plants for two cases: (i) an IGCC power plant with carbon capture, and (ii) an oxy-fuel combustion power plant with carbon capture. The study indicated that ITM technology could lead to a reduction in the levelized cost of electricity of 5% to 10% in the IGCC example and 5% to 20% in the oxy-fuel combustion example with higher cost savings resulting from a lowering of the percentage of CO_2 emissions that are captured (Repasky, Stein, and Armstrong, 2013).

Chemical looping combustion (CLC) refers to a novel flameless, low NO_x combustion process. The CLC process takes place in two sequential steps: (1) an oxidation step where an oxygen carrier (typically a metal) reacts with air to generate oxygen in the form of a metal oxide; and (2) a reduction step in which the metal oxide reacts with a fossil fuel to produce heat (for generating steam in a boiler), CO_2, and H_2O. CLC, when applied in a power plant, thus produces as its end products electricity and a pure mixture of CO_2 and H_2O. The H_2O is then condensed out of the $CO_2 - H_2O$ mixture, making the pure, high pressure CO_2 stream sequestration-ready (Aldous et al., 2013; Folger, 2013). CLC has the potential to make carbon capture a much simpler and cheaper process but it is still at an early stage of development with unresolved technical issues such as materials handling and the selection of suitable oxygen carrier still being debated in the scientific community (Aldous et al., 2013; Folger, 2013).

Delaying the widespread implementation of CCS until third-generation CC technologies are commercially deployed may, in the end, turn out to be a self-defeating option from a climate change perspective. The two promising technologies mentioned above have, at most, advanced to the pilot stage of development. Certain key components of each technology are still at the conceptual design stage for which detailed designs currently do not exist. In the case of ITM, Air Products has suggested 2020 as the earliest date for its commercial deployment of large-scale ITM systems suitable for large-scale IGCC or oxy-fuel combustion plants (Repasky, Stein, Armstrong, 2013). However, as with all novel technologies still at the pilot scale stage of development, relying on project proponents to suggest commercial deployment schedules usually results in disappointment and lost credibility when these schedules prove to be overly optimistic. It is more likely that commercial deployment of ITM oxygen separation technology at power plant scale will not occur prior to 2025, with 2030 as an even more realistic year for its commercial deployment. Chemical combustion looping faces an even more distant commercial deployment date than ITM, with 2030 as the earliest year of deployment and 2040 probably the more realistic date – if it should ever achieve commercial deployment.

Finally, despite the advantages demonstrated through conceptual design studies, it is uncertain whether any of these promising designs and even pilot-scale demonstration projects will result in the advancement of these promising technologies to the stage of commercial deployment. In short, waiting for commercial deployments of third-generation CC technologies may be self-defeating by preventing any chance of achieving a meaningful reduction in GHG emissions prior to 2050. This, then, is the cost-performance dilemma that Australia and all other countries face with their CCS implementation programs – do they implement known technologies that are hugely costly and grossly inefficient or do they wait for less costly technologies to emerge? This is a difficult choice for which no easy answer exists.

11.5 Summary and conclusions

11.5.1 Assessing advanced coal technologies

Australia's black coal industry has a long history of applying state-of-the-art mining and exploration technologies and making them work in the Australian context. However, today's technology challenge – reduction of GHG emissions – is a much more difficult (if not intractable) challenge in comparison with the more localized issues that faced the coal industry in the 1950s.

The relative advancement of CBM, UCG, and CCS in Australia thus far says a lot about the prospects for these technologies to be deployed in other locations in the near future. CBM extraction technology is today a technically proven and economically attractive technology for application in areas having suitable geological conditions, available markets, and access to supporting pipeline infrastructure. UCG has significant commercial potential but remains as yet an unfulfilled promise. The particular environmental stumbles in Queensland by the proponents of the three UCG pilot projects have at least temporarily spoiled UCG's prospects throughout the country, meaning that further technology advancement may need to take place outside of Australia for the moment. The trajectory of CCS thus far has been characterized by initial optimism that has not been followed up by real action, mirroring a pattern observed elsewhere in the world. While CCS development (at least to a more advanced stage of technological understanding) would seem to be a necessary part of serious climate action, governments including Australia's seem to have lost their stomachs for supporting CCS once the costs become real.

11.5.1.1 Assessing CBM

As of February 2015, the outlook in Australia for two of the three advanced coal technologies reviewed in this chapter were less than promising. CBM extraction technology was the one exception – a commercially proven advanced coal technology that has already contributed to a significant reduction in GHG emissions from Queensland's power sector. The technology has proven so successful that it has created favorable conditions for the development of an entirely new LNG export industry in Queensland.

The CBM case shows the importance of scale to industry growth. Scale effects can be positive or negative. In the case of CBM, the main adverse effect of scale has been the production of significant quantities of formation water, which has appreciably increased CBM production costs and also created an environmental risk that was not considered in the early days of the CBM industry in Queensland.

Yet, the impacts of scale have been mostly positive. In addition to catalyzing an LNG export industry in Queensland, the development of a large and mature CBM industry has improved drillers' knowledge about fairways and drilling conditions while creating market conditions that allow the hiring of a critical mass of highly trained drillers and personnel for maintaining and operating CBM production wells.

Indeed, the CBM industry has risen from nothing in a very short time. In this respect, except for a much more limited use of fracking, the growth in Australia's CBM extraction industry has some parallels to the shale gas boom in the United States. The case of Arrow Energy also bears a striking resemblance to the Australian experiences of Utah International during the 1970s, when this single company, which was previously a mining contractor, spearheaded Queensland's transformation from a very small black coal producer into a powerhouse that overtook NSW as Australia's leading producer and exporter of black coal. The former Utah assets in Queensland have since been acquired by BHP Billiton, Australia's largest miner, paralleling Shell's buy-out of Arrow Energy from its original shareholders.[81]

The CBM industry story has come to be about more than just producing CBM. Arrow Energy and a number of its competitors are engaged in expansion programs that will transform them into diversified energy companies with interests in pipelines, gas-fired power plants, and LNG facilities. The CBM industry in Queensland has reached a scale where, with further exploration and resource development and competitive domestic pricing of CBM, it could easily provide for the fuel requirements of any new power plants that might be built in Queensland over the next twenty years.

Australia's CBM resource base is likely to expand rather than contract over time and appears to be sufficient to support domestic gas requirements, although this statement must be qualified somewhat by the need for domestic gas prices to remain close to the high "net-back" LNG prices that prevailed up until May 2014 for this prediction to come true. Another encouraging trend is that Queensland's CBM developers have found ways to reduce the cost and time required for well completion. If well completion costs continue to decline into the foreseeable future, Queensland is likely to experience further increases in 2P reserves and annual CBM production once LNG prices recover from the very low levels experienced between October 2014 and February 2015. [82]

On the other hand, due to the early stage of development for CBM extraction in Queensland, a high level of uncertainty exists about the life cycle production profiles for wells located in both the Surat and Bowen basins. There is the possibility that expected levels of peak CBM production and the duration of that production may be less than expected, which may require CBM developers to drill more "make-up" wells than expected, leading to higher costs of production over time.

It is also possible over the long term that CBM developers will need to exploit deeper coal seams than are being exploited today. This prospect raises the risk of significant increases in the cost of CBM over time.[83]

Nonetheless, on balance, the development of the CBM industry in Queensland over a relatively short period (2004–2014) has proven to be a spectacularly successful application of an advanced coal technology. In lieu of huge (and unlikely to occur) reductions in the capital and operating costs of CCS systems, CBM is probably the only significant coal technology option that can lead to a significant and sustainable reduction in Australia's GHG emissions.[84]

CBM could have been an even greater success story for Australia if the government of NSW had shown a similar level of political support for its nascent CBM extraction industry as Queensland. Thus far, it has not, and CBM in NSW must be considered a failure as of this point in time due to regulatory uncertainty and the lack of a supportive political environment.

11.5.1.2 Assessing UCG

The second promising technology, and one in which Australia already has a substantial technological lead, is underground coal gasification (UCG).

Proponents argue that UCG, once it is proven at commercial scale, will allow syngas to be produced at significantly lower prices than achievable with an above ground coal gasification plant. The syngas stream would have a very high concentration of CO_2 when compared to a coal-fired power plant operating with a pulverized coal (PC) boiler, which should allow the use of more energy efficient and lower cost CC processes such as physical solvent CO_2 capture and perhaps even advanced membrane extraction of CO_2. The prospect for lower cost CO_2 capture is one of the tantalizing possibilities of UCG.

For UCG to eventually be a successful commercial technology, however, improvements appear necessary in two supporting technologies:

(a) *Carbon capture and storage*: In situ gasification produces the same level of CO_2 as a coal-fired power plant but at much higher concentrations in the gas stream. The higher CO_2 concentrations will create favorable conditions for reducing the cost of CO_2 capture, but at today's levels of CCS technology development these costs will still be formidable. Improved technologies for capturing the CO_2 produced during the UCG process are urgently required if any syngas derived fuels and chemicals are to prove economically competitive options against CBM and conventional gas.

(b) *Air separation plants*: Oxygen-blown UCG has a number of advantages over the use of air. It allows a higher concentration of CO_2, greater production of H_2 and CH_4, and less CO. It also minimizes the formation of NO_x during the gasification process. However, air separation plants are capital intensive and expensive to operate. If significant improvements can be made to the efficiency of such plants, and in particular to the compressors used to extract oxygen from the air or through the application of novel ITM and chemical looping combustion technologies, these breakthroughs could tip the balance toward UCG processes that use oxygen instead of air to produce their energy products.[85]

Unfortunately, UCG technology is now at an impasse in Australia due to a series of environmental incidents experienced at the three UCG pilot plants in Queensland. These incidents can be linked to poor site planning and inadequate environmental controls – although the regulator itself bears some responsibility for failing to provide clear rules and oversight. The environmental violations by the three UCG pioneers – Linc Energy, Carbon Energy, and Cougar Energy (now Moreton Resources) – have created serious doubts on the part of the Queensland government and the general public about the ability of private companies to implement UCG in an environmentally sound manner. After what may have been an excessively permissive (and unclear) approach at the outset,

Queensland government regulators have recently shown scepticism toward UCG (in contrast to the proactive support provided to CBM), contributing to UCG's development impasse. Part of the difficulty may be that the environmental impacts of a UCG project are inherently more complex and site-specific than for a CBM extraction project.

There are certainly legitimate unanswered questions about UCG's environmental impacts, in particular (i) pollution of subsurface water tables with BTEX, (ii) above ground release of dangerous VOC emissions, and (iii) subsidence of the ground above the UCG site, which will only be settled through extended demonstrations and test burns. The findings and recommendations of Queensland's ISP in its report related to the pilot UCG test burns, if fully implemented, should go a long way toward addressing these legitimate environmental concerns about UCG.

However, it is unclear when equity investors will start to support the UCG industry again and also when UCG developers might actually move forward on new UCG projects. As a result of what UCG developers call "regulatory uncertainty," only Carbon Energy, with its keyseam® process, is still pursuing UCG development in Queensland, and then just barely due to its weakened financial position. Linc Energy has withdrawn completely from Queensland and is focused on developing UCG projects in Asia, Africa, and Eastern Europe, while Cougar Energy has withdrawn from the UCG space entirely and rebranded itself as a traditional coal resource developer with a new name, Moreton Resources.

If carbon pricing is ever reinstated in Australia, commercial-scale UCG projects will be significantly burdened by the costs of CO_2 emissions. On the other hand, if there were to be significant advances in CCS technology (and reductions in its cost), UCG might emerge as an attractive way to extract energy from coal seams while producing a more easily captured CO_2 stream.

11.5.1.3 Conflicts between CBM and UCG

Additional challenges are presented by the fact that CBM and UCG are, for the most part, mutually exclusive technologies. To produce CBM, the cap of water that holds the CBM under pressure within the matrix of the coal seams needs to be removed. For UCG to work, however, it needs the same water to remain in place.

Once CBM producers pump out the formation water and the water cap, UCG developers will consider the surrounding coal resource to be permanently unusable as a UCG site. Since CBM developers have already obtained petroleum licenses to develop CBM on the most prospective areas of the Bowen, Surat, and Cooper basins, UCG developers will find it difficult to develop their industry at the same pace as CBM developers. The successful commercial application of UCG technology in Australia is unlikely to happen without the development of regulations by the governments of Queensland and NSW that encourage, if not require, "peaceful coexistence" between CBM and UCG proponents.

On the positive side, both UCG and CBM producers are able to produce their energy products without limiting the ability of traditional coal miners to extract coal resources amenable to open-cut mining. Energy products from both UCG and CBM can be extracted from coal resources at depths that are uneconomic to mine via even underground mining methods, which should minimize, if not exclude entirely, any concerns that either technology might displace the production of black coal resources. In the case of CBM projects applied to coal deposits above economic basement for underground mining, the extraction of CBM from the coal seams actually enhances the value of those resources to traditional underground coal miners by completing the process of degassing the coal seams at little or no cost to the coal mining company.

11.5.1.4 Assessing CCS

In 2009, Australia appeared to be one of the early leaders in developing and implementing technology for carbon capture and storage. However, the initial enthusiasm seems to have waned, and few concrete demonstration projects were moving forward at the time of writing. The Global CCS Institute, which was established in Melbourne, does continue to play a constructive role in research on CCS and its potential contribution to climate mitigation, as do other Australian research institutions with an interest in CCS (e.g., CSIRO and CO2CRC).

In this respect, Australia is not too different from other developed countries, such as the United States, which set out ambitious plans for CCS development that were scaled back once real investment was required.[86] What is being observed in Australia and elsewhere is that the massive costs of CCS mean the technology will struggle to move forward (even in the demonstration phase) until the Australian government imposes a significant price on GHG emissions (as well as continuing and perhaps expanding its support for CCS-related RD&D activities). As already mentioned, that commitment is not there today in Australia or in other countries.

To have a major impact on CO_2 emissions prior to 2050, CCS technology will need to be retrofitted onto existing coal-fired power plants with widespread implementation by 2020. Otherwise, governments that choose to encourage CCS as a major plank of their CO_2 reduction programs will need to consider the more costly option of retiring aging coal-fired power plants early and then installing CCS as part of a complete re-powering exercise. But at present, retrofitting a CCS system onto an existing baseload, coal-fired power plant creates very large drops in power plant efficiency. Retrofits also involve significant additional capital costs for the CCS plant (beyond the capital costs of installing CCS on a new power plant) due to the need to modify the power plant to accommodate the CO_2 capture system, which will, in most cases, experience sub-optimal performance due to site limitations and being bolted onto an already built facility.

Advanced CCS technologies remain quite immature, but even proven technologies, such as MEA solvent extraction, face formidable technical and cost challenges scaling up to handle the huge volumes of CO_2 generated by coal-fired power plants of 600 MW and more.

All CO_2 capture systems are likely to pose significant operating risks for power plants due to their need to rely on new and exotic power plant auxiliary systems for carbon capture. Even if one assumes that technology and cost issues are resolved at the individual plant level, significant technical and economic issues are likely to emerge as Australia and other countries attempt to implement these systems on a wide scale. Constraints, for example, could take the form of solvent, equipment, and skilled personnel shortages that lead to higher than expected life-cycle costs for power plants that incorporate carbon capture systems, even after CCS systems have been fully demonstrated at commercial scale. Queensland's LNG facilities at Gladstone provide a telling example of how the widespread adoption of any large-scale technology – mature or not – can lead to a major inflation of project costs.

Many of the technology options being explored today will struggle to achieve commercial deployment even by 2030 due to either (i) their high first costs and limited economies of scale or (ii) their unproven operating efficiencies. There is no guarantee that a single one of the third-generation carbon capture options discussed in this chapter will prove technically feasible at scale and, even if scale-up is successful, that it will be able to accomplish CO_2 capture at an acceptable cost.

Countries that implement CCS at scale could also see significant increases in the price of electricity due to the high cost of installing and operating CCS systems.[87] In Australia, for example, this might translate into significant increases in the price of power in states that rely heavily on coal-fired power plants for electricity generation, such as Victoria, South Australia, NSW, and Queensland.[88]

Table 11.4 provides a "reality check" for those who may still believe that CCS might offer a medium-term option for reducing CO_2 emissions

Table 11.4 *The hope vs. reality of CCS*

The hope	The reality
Cost of CCS can be defrayed by putting CO_2 to productive/beneficial uses such as enhanced oil recovery (EOR), producing biomass (microalgae), and making cement.	Markets that reuse CO_2 are limited in size (EOR), rely on unproven technologies, and require low or zero cost CO_2.
3rd generation CC technologies will become commercially available by 2020 and will lead to a significant reduction in CCS costs and improve efficiency of carbon capture.	Most 3rd generation CC technologies are still in the laboratory and pilot test phases. The most promising technologies are unlikely to be commercially available until 2030, if at all.
Up to 90% CO_2 can be captured from existing power plants.	90% CO_2 capture as a retrofit technology comes at a very high price, and using today's post-combustion MEA technology, the retrofitted CO_2 extraction plant would increase the price of power by at least 60%.
Captured CO_2 can be safely sequestered for hundreds, if not thousands, of years in saline formations and depleted oil wells.	Distance of such reservoirs from power plant sites will limit applications. No long-term studies to confirm that CO_2 will be permanently and safely sequestered.
Transport and injection technology is well understood and commercial.	Private companies unlikely to take long-term sequestration risk without some form of government-backed indemnity.

from new and existing power plants. It separates the hopes being offered under the pro-CCS banner from the reality of where the components of CCS systems stand today in the technology development process and will likely continue to stand in 2020, by which time the widespread commercial deployment of CCS is needed to meet UN IPCC GHG emission reduction targets for 2050.

11.5.2 Scenarios for the future application of advanced coal technologies

Even though GHG emissions reduction programs are, for the most part, not yet in place in Asia, the climate change debate that has transpired over the past few years has created a sense of inevitability about their eventual implementation, in some cases creating the same effects as if they had actually been implemented. A number of companies that were planning to build new coal-fired power plants in Australia and the US have either cancelled their plans to do so or have indefinitely delayed these plans due to the uncertainty created by the GHG debate.

In response to these uncertainties, the coal and power industries are looking for technology to come to the rescue. Given the economic stakes for Australia, either Australia's black coal industry or new entities, such as CBM extraction companies and UCG production companies, might plausibly be at the forefront in finding innovative solutions to the issue of GHG emissions. The level of technical innovation and entrepreneurial energy that has been injected by Australian companies into CBM and UCG development is impressive and provides some cause for optimism. Indeed, independent of any future carbon policy in Australia, CBM has significant momentum and will continue to move forward. UCG's moribund state in Australia at present is related to correctable environmental issues not connected with climate change.

Further progress on CCS, on the other hand, will be contingent on both significantly higher levels of government support – including both financial support and enhanced policy efforts to cut carbon emissions – and technology breakthroughs that lead to a dramatic lowering of the life-cycle costs of CCS. At this point in time, the Australian government is unlikely to be providing significant financial and policy support to CCS. Moreover, given the excessive costs of first generation CCS technology, one must question whether using public funds to support an expanded commercial-scale, demonstration program for CCS is a prudent use of public funds.

As a result of the many political, economic, and technical uncertainties confronting GHG reduction programs in Australia and the Asia Pacific region, it is valuable to consider multiple scenarios for Australia's black coal industry (broadly defined in this chapter to include new production stakeholders such as CBM and UCG developers as well as traditional coal miners).

For example, if Australia's new coalition government (or a successor government) does not reinstate carbon pricing, but Asia's main coal consuming countries decide to implement strong GHG reduction programs, one can easily imagine Australia's black coal industry diversifying into underground (and perhaps above ground too) coal gasification, from which a syngas based liquid fuel and chemicals industry could emerge. The industry has already successfully diversified into production of natural gas via CBM extraction technologies, with a good share of the credit for this diversification being the large unmet demand for LNG in Asia. Apart from the still-unsettled environmental protection issues resulting from three UCG pilot projects in Queensland, which remain significant but resolvable issues, there is no other major reason why UCG could not be the next big thing for Australia's black coal industry.

If an outcome favoring UCG does come to pass, Australia's black coal industry will significantly diversify its product (and export) mix from mainly solid fuels to a diverse palette of solid, liquid, and gaseous fuels, with solid coal used as a feedstock or primary source for producing liquid and gaseous fuels.

Alternatively, Australia may reinstate carbon pricing, and major Asian coal consuming countries may also choose to implement GHG reduction programs with significant bite – for example if new carbon rules promulgated by the US Environmental Protection Agency inspire more aggressive action on climate policy in Asia.[89] Admittedly, the prospect for very aggressive GHG action in Asian countries and Australia seems remote at present, but it is nevertheless an important case to consider. Australia's black coal industry under this scenario would still diversify into liquid and gaseous fuels, but its prospects would additionally depend on the economic feasibility of implementing CCS facilities at existing and new coal-fired power plants throughout Asia. Even under this scenario, the relative prices of steam coal and LNG will remain important drivers of the scenario's outcome.

The economic feasibility of CCS will depend on complex factors including (a) the technical performance and reliability of CCS systems working in tandem with coal-fired power plants, (b) US$ per GJ price differentials for steam coal and natural gas (pipeline gas or LNG), and (c) the level of CAPEX and OPEX for CCS and either "coal-to-gas" or "coal-to-liquids" plants. Affordability would also be a key concern for a number of Asia's main coal consuming nations in deciding what GHG emissions reduction direction to follow, and the pace at which to move in a given direction.

Under all plausible technology development scenarios, CCS technologies are likely to remain expensive and their application at new coal-fired power plants will lead to a reduction in demand for new coal-fired power plants. Their application at existing power plants would lead to a reduced "net back price" being paid to Australia's steam coal producers. Even under optimistic CCS implementation scenarios, the size of Australia's coal export markets is likely to decline over time due to the higher costs of power resulting from the application of CCS technology. Price will trump all, and, if CCS is widely implemented throughout Asia, Asian consumers will eventually reduce their demand for Australian coal. On the other hand, the application of CCS systems by Asian power utilities and independent power producers would at least allow coal exporters to continue serving their traditional solid coal export markets in some capacity under an aggressive Australia/Asia GHG reduction scenario.

The timing and structure of the coal industry transformation under such a scenario would be determined by the relative progress in the commercial deployment of CBM extraction, UCG, and CCS technologies. CBM extraction has been widely implemented in the state of Queensland since 2004 and has added a new business dimension to the steam coal industry in that State. UCG has been demonstrated at the pilot and pre-commercial stages but is facing obstacles in Australia due to environmental compliance issues that still need to be resolved. CCS had not yet been demonstrated in Australia at commercial scale at the time of writing. And yet it cannot be denied that CCS and UGC – either in tandem or individually – have the theoretical potential to transform Australia's black coal industry in the same manner that mechanization and advancements in coal exploration techniques transformed it during the period 1950–1980 (see Chapter 6) and CBM extraction technologies are transforming it today.

There will, of course, be winners and losers, with those companies adhering to the old methods of extracting and using coal going the way of other declining industries in Australia's past, such as the town gas industry and ships and railway locomotives that ran on coal in the early 1950s (see Chapter 6). The choices may appear stark, but Australia's black coal industry has been here before and has shown a remarkable capacity to adapt. What remains to be seen is whether Australia can repeat the technology adaptation miracle of the past sixty years in a brave new world of carbon emission limits. For now, one must declare the technology rescue story for Australia's black coal industry as a work in progress with its final chapter yet to be written.

Notes

1. The repeal was made in the form of a series of amendments to the Clean Energy Act, which removed the obligation of liable parties to pay the CPM past July 1, 2014. The legislation that eventually passed the House and Senate was an amended version of "Exposure Draft – Clean Energy Legislation (Carbon Tax Repeal) (Jobs and Competitiveness Program) Act 2013" as issued by the Australian government, Department of Environment.
2. CBM extraction technology releases and captures clean-burning and less carbon-intensive methane from deep coal seams; UCG gasifies coal underground, which in theory could lead to more cost-effective capture of CO_2 as well as local pollutants. CCS encompasses a variety of methods of capturing CO_2 at different stages of the coal conversion process and preventing its release into the atmosphere.
3. Although CBM projects occasionally access CBM at depths that are amenable to underground mining, removal of CBM before the start of mining in most

cases will add value to these unmined coal resources by completing the degassing portion of the underground mining project at little or no cost to the coal miner. In the case of Queensland, the coal seams identified for CBM extraction have lower CV contents and other unfavorable characteristics, such as thin seams with partings that make these deposits undesirable for underground mining.

4. The following generic types of CO_2 capture systems are considered: (1) the post-combustion system is assumed to rely on MEA solvent extraction technology for capturing the CO_2; (2) pre-combustion capture is assumed to be an Integrated Gasification Combined Cycle (IGCC) plant equipped with a solvent extraction plant; and (3) oxy-fuel combustion is assumed to be an oxy-fuel combustion power plant with the direct capture of high purity CO_2 via compression of the oxy-fuel flue gas.
5. In Australia, coal bed methane (CBM) is also referred to as coal seam gas (CSG), which is a broader definition more in line with that of the petroleum industry, as methane is just one of the components of the coal-derived gas. However, for purposes of maintaining consistency with terminology used in the US and elsewhere, the author uses CBM throughout this chapter when referring to this coal-derived fuel.
6. The potential size of a particular CBM source depends mainly on: (a) thickness, depth, and continuity of the coal seam reservoir; (b) gas content, expressed in cubic meters of methane per tonne of coal with the gas content influenced by the thermal maturity of the coal as estimated from its vitrinite reflectance with bright vitrinite-rich coals typically being "gassier"; (c) gas quality, expressed as the percent methane component by volume of the gas. The deliverability of gas from a CBM reservoir – an important indicator of the commercial potential of a CBM resource – is dependent on two factors: (1) the gas saturation or its holding capacity, which is the ratio of the volume of gas present in the coal compared to the theoretical maximum volume of gas that could be contained in the coal (if the saturation or holding capacity is high, gas flows much more easily when the seam is dewatered); and (2) permeability, which is mainly dependent on the type and extent of fractures or cleats in the coal seams (Friederich, 2014).
7. In Australia, fracking is spelled "fraccing," but the author has chosen to use the more common spelling "fracking."
8. Drilling and maintaining a combination HDD-vertical well is more expensive than for a simple vertical CBM production well. Before deciding on which drilling technology option to pursue, the CBM developer will complete a life-cycle cost analysis of the two options by comparing on a discounted cash flow basis the internal rates of return of the two development options. Specifically the CBM developer will compare the additional capital and operating costs and the net revenues that will result from a combination HDD-vertical production well as opposed to a simple vertical well. The additional capital and O&M costs will be a function of (a) depth of coal seam(s), (b) various coal seam properties such as permeability and gas content (m3/tonne), (c) the number of HDD laterals that will be drilled off of the main vertical well, and (d) the length of each lateral. Additional revenues will be the additional CBM revenue earned over the

lifetime of a vertical-HDD well. The discounted cash flow analysis will also consider the time value of increasing gas production in the early years of a well's life resulting from the application of HDD well technology. The option that generates the highest internal rate of return (IRR) on equity invested will be the one selected for a specific CBM development area.

9. There is also a seventh phase known as decommissioning, which is assumed to happen at the end of the CBM extraction stage and involves cementing closed the well openings.
10. Until 2008, CBM producers were allowed to store the formation water extracted from their production wells in open-air lagoons. Since 2008, the governments of Queensland and NSW have required CBM developers to build water treatment facilities to recycle their formation water to a standard that will allow it to be used for at least crop irrigation. The formation water is treated to a level where the main residue for disposal is concentrated brine, which is the current disposal concern. It is uncertain how the large quantities of brine will be safely disposed of, but burial in an environmental secure manner seems to be the most economical option (Moore, 2014).
11. CBM is a GHG with anywhere from twenty-eight to eight-six times the global warming potential of CO_2 (depending on time period of analysis [twenty to a hundred years]) in the atmosphere. Some detractors of CBM development in Australia have argued that "methane seeps" from around the CBM wells are creating significant methane releases into the atmosphere. Natural gas wells will always have some methane leaks – while in production and also after abandonment or decommissioning. According to the US EPA, methane leaks at operating US gas wells range from 1% to 2% of a well's annual gas production. Gas leaks also occur when a well is shut down or decommissioned, but this issue remains unexplored (Johnson, 2014). The significance of those leaks and seeps remains an ongoing topic of debate in the science community; assertions of much higher losses than estimated by the US EPA have not been proven at this time. CSIRO is conducting a detailed geotechnical investigation into the causes of methane seeps in the Surat Basin with natural processes being one of the possible causes (see Sullivan, 2013). For completeness, it should be mentioned that coal mines – both open-cut and underground mines – also generate fugitive methane emissions that add to the GHG emissions of coal mines beyond coal's carbon content.
12. Queensland produced 249.2 BCF of CBM in 2012. If used to produce electricity, CBM-fueled power plants would have produced 11.9 million tonnes (Mt) of CO_2. Coal-fired power plants would have produced 20.9 Mt of CO_2, resulting in a net savings of 9 Mt.
13. Latest government estimates based on the UNFCCC standard were obtained from http://www.environment.gov.au/system/files/resources/6b894230-f15f-4a69-a50c-5577fecc8bc2/files/national-inventory-report-2012-vol1.pdf, p. 40, which indicates 559 Mtpa CO_2e emissions including emissions resulting from "land-use, land-use changes and forestry." If one excludes this emissions category, Australia's net GHG emissions drop to 544 Mtpa.
14. See: http://www.climatechange.gov.au/sites/climatechange/files/files/climate-change/national-targets.pdf

15. As mentioned in note 3, CBM extraction in Queensland has not always targeted coal seams below the technical basement for underground mining. However, for many of those shallow seam CBM projects, the coals have not been of a quality or the tenements were of a nature that did not justify an underground coal mine. Where the coal was of suitable quality to justify an underground mine, CBM extraction added value by allowing degassing of the coal seams at little or no cost to the coal miner.
16. In Queensland, CBM has in some instances been extracted from seams at depths that are economical to develop as underground mines, raising issues of competing claims for the same resource, which is covered toward the close of this section.
17. Allowed chemicals range from (i) innocuous substances such as guar gum, a food thickening agent, and acetic acid or vinegar, to (ii) substances toxic if used in high concentration. Examples of toxic substances that can be found in fracking fluid are: (a) sodium hypochlorite, which is used in pool chlorine; (b) sodium hydroxide, which is lye or caustic soda; (c) ammonium persulphate, which is used in hair bleach; (d) ethanol; and (e) orange oil. Even these identified toxic chemicals are only toxic in high concentrations. As a part of a fracking fluid, they are used in such highly diluted amounts that the risk of groundwater pollution is considered very low by the environmental regulators of both Queensland and NSW.
18. See, for example, a recent report issued by NSW's Chief Scientist and Engineer, titled "Independent Review of Coal Seam Gas Activities in NSW: Managing environmental and human health risks from CSG activities," September 2014. In her cover letter to the report, the NSW's chief scientist states, "The management of potential risks associated with CSG, as with other industries, can be done with effective controls and regulation. This includes engineering solutions involving high levels of industry professionalism, monitoring and modeling, and comprehensive risk assessments." Available at (http://www.chiefscientist.nsw.gov.au/__data/assets/pdf_file/0006/56922/140930-Final-Managing-Environmental-and-Human-Health-Risks.pdf).
19. Formation water collected from CBM wells also may contain concentrated levels of mercury, arsenic, cadmium, and uranium, which may have leached from the soil and then become concentrated in the formation water.
20. These Bowen Basin coals are of Permian age and are mainly of high volatile to medium volatile bituminous rank.
21. The Queensland government defines 2P reserves as proved + probable reserves.
22. Australia's first stand-alone, commercial CBM project commenced in December 1996 at the Dawson Valley project (then owned by Conoco), adjoining BHP's Moura coal mine. Prior to the commercial production of CBM in 1997, the mining industry was focused on innovative ways to remove methane from operating underground mines. Known as coal mine methane (CMM) removal projects, three became operational in 1996 at BHP's Moura, Tower underground mines in Queensland, and Appin in NSW, with the extracted CMM used to fuel specially designed, on-site gas-fired

generator sets. Later CMM projects include those of Xstrata Coal in the Hunter Valley of NSW and at Oakey Creek in the Bowen Basin of Queensland (2006); and Anglo Coal at Grasstree in the Bowen Basin (2008 and 2012) (see Anglo American, 2014).

23. According to Geoscience Australia, the EDR of CBM is roughly equivalent to proved and probable petroleum reserves. For coal, EDR is roughly equivalent to measured and indicated coal resources as per JORC.
24. Starting in late 2012, anecdotal evidence emerged suggesting that production growth might not be on pace to meet the gas requirements of at least one and possibly two of the LNG plants at Gladstone, with one reputable market analyst claiming that huge uncertainty exists around the security of contracted CBM supplies to the Santos GLNG project between 2015 and 2019. (See *Sydney Morning Herald*, 2012; Kebede, 2013; Validakis, 2014; Chambers, 2014.) Geoscience Australia has also jumped into the debate about whether the LNG terminals at Gladstone will suffer a gas shortage by stating that although "[2013] booked coal seam gas reserves exceed current LNG contract requirements," there is a risk that planned CBM production wells might not meet their expected production rates of 400,000–500,000 ft^3 per day. Moreover, a leading CBM expert who wishes to remain anonymous is of the view that Queensland's 2P CBM reserves are unlikely to increase at significant growth rates from 2014 onward.
25. Building three LNG terminals in the same general location at once has also caused the significant construction cost overruns for each terminal ranging from 15% to 35% (*Sydney Morning Herald*, 2012).
26. Between 2004 and 2006, oil prices were also at much lower levels than they are today, which contributed to the low price of CBM.
27. In March 2014, the Platts Japanese Korean Marker (Platts JKM), which is a survey estimate of the spot price of LNG sold into the Japanese and Korean markets, was US$18.40 per GJ. As of February 2015, the Platts JKM had plunged by 49% to US$9.40 per GJ. In A$ terms, the decrease has not been as serious due to the depreciating value of the A$ relative to the US$. The A$ values for spot LNG sales into Japan and Korea were A$19.95 and A$12.74 per GJ for March 2014 and February 2015 respectively, a drop of 36%.
28. Detailed information on CBM reserves and production was not available from the public filings of other large CBM producers such as Origin Energy, Queensland Gas, and Santos. But the story would probably be the same – extraordinary levels of growth over a relatively short period of time.
29. After Shell and Petrochina acquired Arrow, Arrow stopped releasing detailed information on reserves and production.
30. Arrow Energy (2013), "Fracking: What You Need to Know, 12 July 2013." Fracking is the process of pumping large volumes of fracking fluids at high pressure down the vertical shaft of the production well and into the area surrounding the coal seams of a CBM development area. The process creates fractures in the coal seams and surrounding subsurface areas of "100's of meters around the wells."
31. Although fracking is not used extensively in the Surat Basin today, in the future CBM developers may turn to fracking in order to access the low

permeability resources at depths below 800 m, which contain large amounts of gas-in place (2–5 TCF) and will require fracking to achieve required gas flows (Moore, 2014).

32. In this section of the chapter, the author focuses on the actions of the Queensland government to address overlapping claims issues that exist between coal mining companies and CBM developers who have licenses over the same geographic area. In the section on UCG, the author discusses the rationale of the Queensland government for not offering UCG developers a similar level of Queensland government support, which has left UCG developers feeling abandoned.
33. For this section of Chapter 11, "overlapping claims" refers to instances where a CBM developer and a coal mining company each hold licenses to develop a mineral or petroleum resource over the same tenement. In the case of the CBM developer, its license would be issued under the Petroleum and Gas (Production and Safety) Act of 2004 (Qld. P&G Act) and the coal miner under Queensland's Mineral Resources Act of 1989 (Qld MRA).
34. One exception to the previously mentioned procedure was a special dispute between the BHP-Mitsui Alliance (BMA) and the Cherwill Creek Coal Company, which, due to BMA's particular licensing arrangements, needed to be resolved by the Queensland Parliament through an amendment to the Qld. MRA. However, this was a very special case with BMA, owner and operator of the Peak Downs coking coal mine, being regulated under the Central Queensland Coal (CQC) Associates Agreement that was entered into by the Queensland government and CQC Associates on January 1, 1969. BMA held rights to locate infrastructure (power facilities, warehouses, and stockpiles) on a portion of the land allocated to it under the CQCA Agreement but did not hold the development rights to the coal underneath that land. Cherwill Creek in 1989 acquired rights to develop the coal underneath that land via the Qld. MRA. The amendment bill ("The Mineral Resources [Peak Downs Mine] Amendment Bill 2008") was passed into law in 2008 and effectively settled the overlapping claims dispute in favor of BMA (Office of the Queensland Parliamentary Counsel, 2008).
35. Overlapping claims disputes after 2007 were almost exclusively between Queensland's CBM developers and traditional coal mining companies. These issues could have been settled internally if the coal miners had developed the coal bed methane deposits within their license areas – either alone or in partnership with a specialized CBM development company. From a public interest perspective, if the coal mining industry actively developed CBM resources within their tenements by obtaining a Petroleum Development License, the outcome might have still been suboptimal from a state government perspective. Coal miners, having a narrow and biased business strategy, might have delayed or declined funding to the CBM side of their business, imposing long delays in implementing CBM projects.
36. The Draft Bill only considered overlapping claims between CBM developers and coal miners even though disputes were also starting to break out between CBM developers and UCG developers. Overlapping claim disputes between CBM and UCG developers will be discussed under section 11.3 on UCG.

37. The QRC is a resource industry advocacy group funded by the mining and petroleum industries active in Queensland.
38. Metgasco believed that a Court decision might be announced by March 2015.
39. Against this 34% increase in formation water, CBM production increased 150%. The leveling off in the production of formation water is most likely a reflection of the reduced use of fracking and the maturing of the industry from one in which wells were mostly being dewatered to a time when a significant number of CBM wells are in production while new wells are being developed.
40. This figure does not include an allowance for land required to manage and maintain the ponds, nor does it allow for land required to store runoff from seasonal rainstorms.
41. These VOCs include benzene, toluene, ethylbenzene, and xylene (BTEX), as well as phenols and polycyclic aromatic hydrocarbons (PAH).
42. The earliest commercial scale application of UCG, but under non-commercial conditions, was the Yerostigaz UCG plant in Uzbekistan, which started operating in 1960. For almost fifty years the Uzbekistan Yerostigaz project has been the largest pre-commercial demonstration of UCG technology in the world.
43. Between 1989 and 1996, Walker completed a series of self-funded UCG-to-power studies, which thoroughly convinced him of the technical feasibility and economic potential for UCG in Australia.
44. Ergo Exergy's technology was a variation on the theme of the Yerostigaz technology applied in Uzbekistan.
45. Yerostigaz was formed during the Soviet era to operate the country's only UCG facility. Yerostigaz's UCG plant was commissioned in 1961 and within four years was producing 1.4 billion m^3 of syngas, which was supplied to the Angren Power Station. It currently is producing around 30 million m^3 of UCG syngas per month or 300 million m^3/yr, assuming the plant operates at an 84% capacity factor.
46. In late 2008, Linc Energy acquired SAPEX, a South Australian oil and gas exploration company. Linc Energy bought SAPEX primarily for its Mineral (Coal) Licenses, which covered 24.6 million acres located in the Arckaringa, Cooper, Eromanga, Walloway, and Stansbury Basins of South Australia (SA) (Linc Energy, 2013a). At the time of its acquisition by Linc Energy, SAPEX was reportedly developing a UCG project around its tenements in the Adelaide area of SA. SAPEX also held a Petroleum License for a large area (69,000 km^2 [17 million acres]) within the Arckaringa Basin that was expected to contain a large, unconventional (shale) oil and gas resource (Linc Energy 2013a; Tulli, 2014).

    Between 2008 and 2013, Linc Energy acquired a number of high quality coal tenements in the Surat and Galilee basins. In August 2010, Linc Energy sold one of its Galilee coal assets, the Carmichael coal tenement, to Adani of India for A$500 million in cash and a royalty of A$2.00 per tonne indexed to the Brisbane CPI for the next twenty years. Linc Energy has subsequently used cash from the Carmichael sale to buy producing coal assets in Australia and to fund the continued development of (i) new conventional oil and gas projects in the US (Alaska, Wyoming, Texas, and Louisiana) and (ii) unconventional shale oil and gas resources in South Australia (Arckaringa Basin).

47. Under terms of its agreement with Exxaro, Linc Energy will receive a total license fee of A$30 million, of which A$23 million has already been received (Linc Energy, 2014).
48. As of July 31, 2013, Linc Energy incurred a net loss of A$25 million and had A$124 million in cash and cash equivalents (Linc Energy, 2013a). For FY 2014, Linc Energy incurred a loss of A$239 and as of June 30, 2014 had cash and cash equivalents of A$49 million. For the second quarter of FY 2015 (October–December 2014), Linc Energy reported a net loss before taxes of A$165 million, mostly due to impairments resulting from the collapse of energy prices, but still had a healthy balance of S$104 million in cash and cash equivalents (Linc Energy, 2015).
49. "In the 'CRIP' process, directional drilling is used to create a channel connecting the production well to the injection well. A gasification cavity forms at the end of the injection well in the horizontal section of the coal seam. Once the coal in the cavity area is expended, the injection point is withdrawn (usually by burning a section of the liner) and a new gasification cavity is initiated" (US Clean Air Task Force, September 2009, p. 5, note 3).
50. Carbon Energy holds mining leases in the Surat Basin at a location called Bloodwood Creek, which is located 55 km west of Dalby in southeast Queensland. The Bloodwood Creek site reportedly contained at the time of the test 218 Mt of JORC-compliant indicated resources and 450 Mt of inferred resources. The UCG facility that was used for the initial test burns was based upon the US DOE experience at its Rocky Mountain-1 (RM-1) demonstration facility during the 1980s. Carbon Energy later claimed that the improvements it made to the US DOE CRIP system after 2009 had moved its technology "from an experimental stage to commercial reality."
51. The incidents were the spilling of process water from the UCG plant onto surrounding land and then allowing the spilled water to be transported from the site.
52. It was reported at the time that DERM would pursue damages in the courts against Carbon Energy with a maximum penalty of A$125,000 for each incident plus a second maximum charge of A$50,000 for failing to notify DERM of any violations of its EA in a timely manner.
53. It is not known if Carbon Energy might have paid a fine to the DERM as part of any out of court settlement and if so the amount of those fines.
54. The company's other plan for survival is to divest all of its coal assets in Queensland (except for Bloodwood Creek and Blue Gum) to raise cash to keep itself in business.
55. Carbon Energy has long been trying to jointly develop its Bloodwood Creek site with a partner having expertise in the chemical industry. In 2008, Carbon Energy executed a Memorandum of Understanding ("MOU") with Incitec Pivot Ltd. ("IPL"), a manufacturer of ammonia and ammonia-based derivatives. The MOU provided for both companies to form a joint venture company upon the successful conclusion of Carbon Energy's field trial at its Bloodwood Creek site. The MOU provided IPL with exclusive global rights to Carbon Energy's UCG technology for the manufacture of ammonia and ammonia-derived products. Incitec still owns 75.6 million Carbon Energy

shares, making it the fourth largest Carbon Energy shareholder, but given the lack of political and regulatory support within Queensland for UCG at this time, it appears unlikely that Incitec Pivot will be building an ammonia plant at the Bloodwood Creek site any time soon. The Blue Gum project, which is also located in the Surat Basin, will produce pipeline-quality gas from the syngas produced at a yet to be permitted UCG plant in the Surat Basin Area. It is a great concept but will take years to bring to market.

56. During the time that Len Walker was CEO of Linc Energy, he developed a close working relationship with Ergo and Blinderman, which was discontinued by Peter Bond, new owner of Linc Energy, once Walker left Linc Energy in 2002.
57. In the case of UCG, this point about NSW is moot given that UCG developers were not active in NSW at all. It is possible that the NSW government's reputation for offering scant support and regulatory certainty to demonstrations of novel coal conversion technologies led to this outcome.
58. In 2009, the government of Queensland issued a brief policy document, which provided its position (and actions) related to UCG development in Queensland (Queensland Government, 2009b). In short, it provided developers of the three UCG pilot projects (Linc Energy, Carbon Energy, and Cougar Energy) with "an opportunity to demonstrate the technical, environmental and commercial viability" of their technologies by awarding them the rights to any underground coal resources "notwithstanding that a petroleum tenure may overlap part or all of the relevant MDL (Mineral Development License)" (Queensland government, 2009b, p. 3). However, the policy document goes on to state that no other UCG pilot projects will "be allowed on public interest grounds" until the completion of the pilot test burns and the submission of a report by "Government" to describe the outcomes of the pilot projects and to recommend whether the industry was viable or not on technical, environmental, and commercial considerations (Queensland government, 2009b, p. 3).
59. It has been suggested that formation water from the CBM process could be injected into the coal seams during CBM well decommissioning. However the technical feasibility and cost implications of re-injecting formation water into CBM-depleted coal seams are unknowns at this time (Moran, Da Costa, and Cuff, 2013).
60. "Self-cleaning cavity" is ISP shorthand for UCG proponent claims that UCG panels, once depleted of their resource coal, can be decommissioned without any environmental remediation. Both Linc Energy and Carbon Energy claimed that elevated temperatures in the UCG cavity will cause any water in the cavity to vaporize and rise through the outlet well to the surface, where it can be treated. This process, which presumably will involve injecting water into the cavity and removing the vapor with pollutants for treatment at the surface, will continue until the UCG process has been fully extinguished. At that time, assuming that remaining pollutants in the cavity are at acceptable levels, the inlet and outlet wells will be sealed and the site considered "decommissioned" under terms of their EA. ISP noted that the clean cavity approach to decommissioning remains an unproven concept (Moran, da Costa, and Cuff, 2013).

61. Enhanced oil recovery (EOR) is one example of a high value beneficial reuse market, but it cannot be described as a mass market. Given the quantities of CO_2 that need to be disposed of in an environmentally sound manner, EOR is, at best, a significant niche market for CO_2.
62. The most commonly used solvent is monoethanoamine (MEA), an amine-type solvent, which has been used for decades in the gas processing and petrochemicals industries for removing CO_2 from petrochemical feed stocks. Some researchers are studying the use of other solvents such as chilled ammonia in place of MEA. The MEA technology has been proven at medium scale for those industries where the value of the final gas product is very high. Applying this technology at a scale sufficient to remove 90% of CO_2 emissions from the flue gas of a large coal-fired power plant having a capacity of 600 MW or more is still considered unproven technology.
63. The recovered solvent is recycled back into the solvent extraction plant and used to extract CO_2 from new flue gas streams. About 5% to 10% of the solvent is lost in the extraction and regeneration process and needs to be replaced with each extraction pass.
64. Assumes that MEA is used as the chemical solvent. Alternative "physical" solvent extraction processes are available from companies in the oil, petrochemical, and industrial gases industries such as (1) Linde/Lurgi's Rectisol process, (2) Lurgi's Purisol process, (3) Union Carbide's Selexol process, and (4) Shell's Sulfinol process. However, these physical solvent processes are considered much more expensive than MEA when used in a CO_2 capture plant located at a coal-fired power plant site.
65. Pre-combustion CO_2 removal processes have thus far focused on "above ground" pre-combustion processing. However, the pre-combustion concept can also be applied to underground coal gasification projects, whereby the coal is gasified in situ with the coal seams and surrounding geological structures and hydrological conditions acting as the gasifier and supporting an "in situ" water shift reaction. UCG processes would theoretically avoid much of the capital cost associated with an above ground IGCC system and would also avoid the cost of mining the coal. Finally, the UCG option allows the production of syngas (and CO_2) from deep, sub-economic coal seams that would not be mined in any case. The UCG option was discussed as a stand-alone technology, but in cases where carbon taxes are being imposed, UCG could be combined with an above ground carbon capture plant based on physical or chemical solvent extraction.
66. Maintaining a small amount of NO_2 in the inlet oxygen stream along with the injection of a portion of the captured CO_2 (less than 15%) into the inlet oxygen stream is necessary to prevent an explosion in the boiler that might occur if the coal was combusted in a very high purity oxygen atmosphere.
67. Air separation is a very energy intensive process, which reduces overall power plant efficiency. Technical improvements are currently being explored to reduce costs and improve efficiency, but this exercise remains a work in progress with no guarantee of success.

68. A number of natural trapping mechanisms have been identified and are being assessed by researchers for efficacy (CO_2 Capture Project, 2015):
    - *Structural trapping*: CO_2 can be "structurally" trapped within an underground reservoir that contains an impermeable caprock above the reservoir, which prevents the CO_2 from moving upwards. The injected CO_2 will rise within the reservoir until it reaches the caprock, where it will be permanently trapped within the reservoir.
    - *Residual trapping*: As the injected CO_2 migrates up through the geological storage site towards the caprock, some becomes trapped inside the pore spaces of the reservoir rock, a process that is similar to water becoming absorbed in a sponge.
    - *Dissolution trapping*: The injected CO_2 will over time dissolve in the water that exists in the geological formation, such as the saltwater contained in a deep saline aquifer or the water that fills the pore spaces in the reservoir rock.
    - *Mineral trapping*: CO_2, over a long period of time, will eventually react with the minerals present in the reservoir and bind to those minerals and form carbonates, similar to those found in natural limestone.
69. The IPCC in its 4th Assessment Report on Climate Change states that a maximum atmospheric concentration of 450 ppm of CO_2 equivalent emissions will need to be achieved for a maximum global temperature increase of 2° C to be achieved (see Intergovernmental Panel on Climate Change, 2007).
70. The cost comparisons are based on carbon capture (CC) plants designed to remove between 90% and 100% of the CO_2 generated. The study did not explore the cost implications of reducing the percentage of CO_2 captured in order to bring down capital costs. This option, if considered, would likely lead to a higher cost per tonne of CO_2 captured but might be worthy of consideration at this early stage of development for carbon capture technology.
71. See Abellera and Short (2011, p. 5) for cost comparisons in US$ per MWh.
72. The cost comparison was made using prices for PV solar panels and onshore wind turbines in 2010/2011. Given the significant reductions in capital costs and improvements in efficiency for PV solar panels and to a lesser extent to onshore wind turbines, these results may not be valid using 2015 prices for solar, wind, and CCS technologies.
73. CSIRO was known as the Council for Scientific and Industrial Research from 1926 until 1949 when it was reconstituted by the Australian government into an independent statutory body.
74. With respect to post-combustion capture (PCC) technology, CSIRO has collaborated with others in the implementation of pilot scale PCC systems at five operating coal-fired power plants, three in Australia and two in China. It has also developed and tested new "designer amines, ionic liquids, honeycomb monolithic carbon fibre composites [HMCFC], mixed matrix membranes and enzyme CO_2 capture technologies." Finally, it has patented a technical process that uses UV radiation to eliminate nitrosamines and organic nitrates, two known carcinogens that are produced during MEA solvent process for capturing CO_2.

75. CSIRO's CO_2 storage research has focused on the development of computer models to predict the "fate" of CO_2 stored underground and methods for (i) monitoring the impacts of stored CO_2 on surrounding groundwater resources, and (ii) identifying CO_2 leaks from specific underground storage sites.
76. GCSSI provides nonmembers with free online access to its growing library of technical reports and databases on CCS (www.globalccsinstitute.com). Starting in 2015, GCCSI will become more reliant on member fees for its outreach services. It remains to be seen whether nonmembers will continue to enjoy free access to its online library of reports or will be required to subscribe to its report and database service.
77. In its simplest form, cost reduction through learning involves discovering ways to reduce capital and operating costs of new CCS plants by constructing and operating multiple CCS facilities.
78. Advanced, third-generation CCS technology is almost totally with reference to the carbon capture component of CCS systems. Current CCS technology refers to current technology as applied today, second-generation refers to technologies that policy makers and engineering companies believe will be commercially deployed between 2020 and 2030, and third-generation technology refers to advanced, game-changer technologies that will be ready for commercial deployment after 2030.
79. See Al-Juaied and Whitmore (2009), who estimate a whopping 70% reduction in the cost of CO_2 abatement for NOAK CCS plants, and Folger (2013), who estimates a more moderate cost savings of 25%.
80. Excellent reviews of these advanced technologies can be found in Aldous (2013) and Folger (2013).
81. One difference, however, is that it is unlikely that coal miners will swoop in at this late stage and scoop up CBM assets from their current oil and gas company owners. Traditional coal miners are expected to continue their "spectator" roles while oil and gas E&P companies, such as British Gas, Shell Oil Company, and Santos, continue to dominate the CBM landscape, either directly or through acquired affiliates such as Arrow Energy and Origin Energy.
82. Except for the period 2015–2020, CBM resources should be adequate to meet domestic and export requirements through 2030 based on current estimates. There are still large undeveloped CBM resources available at typical CBM depths within the Surat and Bowen basins, plus there has been a renewed push to develop conventional gas and CBM resources located in the Cooper Basin, which is already linked by pipeline to the Port of Gladstone (Friederich, 2014).
83. The author does not subscribe to the view that Queensland's CBM industry has entered into a long-term, rising cost phase of development, which assumes that Queensland has already developed its lowest cost CBM reserves, and new technological advances will not be forthcoming.
84. CBM can also be used as a replacement for diesel fuel in passenger and coal trains, which would lead to an even further reduction in Australia's carbon footprint. CBM, therefore, is rightly viewed as a game-changing technology that is not only leading to a redefinition of what constitutes the black coal

industry of Australia but is also becoming an icon of Australia's contribution to global GHG emission reduction.

85. See Repasky, Stein, and Armstrong (2013) for advanced oxygen separation technologies that may change the economics of pre-combustion and oxy-fuel combustion CCS and improve the economics of UCG syngas production.
86. In January 2015, the US DOE announced the suspension, and most likely the permanent cancellation, of its FutureGen project, which had been under development since 2003. The DOE Secretary cited two reasons for the project's suspension: (1) the inability of the project company to spend its allocated funding within the time period allowed under federal spending regulations and (2) its inability to raise the required share of private sector funding (http://will.illinois.edu/news/story/department-of-energy-suspends-futuregen).
87. In North America, huge cost overruns of over 100% at Southern Power's Kemper IGCC-CCS plant in Mississippi led to the utility having its credit rating downgraded by S&P and the firing of the top executives of Southern's Mississippi Power affiliate, which was responsible for managing plant construction. The plant is also one year behind its completion schedule (MIT, 2014). Saskatchewan Power has also had significant cost overruns for its 110 MW Boundary Dam CCS project, with that project's completion also delayed by one year. This project was a retrofit of a single, small unit using proven MEA solvent extraction technology, but it still experienced a one-year completion delay and a C$100 million budget overrun (Global CCS Institute, 2014).
88. Because CCS leads to a significant reduction in coal plant efficiency, widespread implementation of CCS technologies throughout Asia could also lead to a significant increase in the consumption of coal, which could lead to intermittent coal shortages (see Chapter 12).
89. Two other business as usual (BAU) scenarios are possible:
    - *BAU scenario 1*: Australia does not reinstate its CPM and major coal-consuming countries of Asia do not succumb to international pressures to implement carbon pricing schemes or prescriptive policies that restrict the building of new coal-fired power plants or speed up the retirement of existing coal-fired power plants.
    - *BAU scenario 2*: Australia reinstates its CPM, but major coal consuming countries of Asia do not implement carbon emission pricing schemes or other significant GHG emission reduction programs.

    Both of these BAU scenarios have been discussed in Chapter 6 and, therefore, will not be considered in this chapter because the conclusion is that Australia's black coal producers will continue with the current mix of solid fuel and CBM-to-LNG products, and overall production of solid fuel should not be severely affected by government GHG regulations.

References

Abellera, Chester, and Christopher Short (2011). "The Costs of CCS and Other Low Carbon Technologies." *Global CCS Institute*. Available at http://cdn.globalccsinstitute.com/sites/default/files/publications/24202/costs-ccs-and-other-low-carbon-technologies.pdf (accessed September 8, 2014).

Aldous, R. et al. (2013). "CSLF Technology Assessment, CCS Technology Development: Gaps, Opportunities and Research Fronts." Cooperative Research Centre for Greenhouse Gas Technologies, Canberra, Australia, CO2CRC Publication Number RPT13-4571.

Al-Juaied, Mohammed and Adam Whitmore (2009, July). "Realistic Costs of Carbon Capture." *Belfer Center*. Available at http://belfercenter.ksg.harvard.edu/publication/19185/realistic_costs_of_carbon_capture.html (accessed September 09, 2014).

Anglo American (2014). "Moranbah North Mine and Capcoal Underground Reduce Greenhouse Gas Emissions Development." *Anglo American Australia*. Available at http://australia.angloamerican.com/our-difference/case-studies/greenhouse-gas-emissions (accessed February 18, 2015).

Arrow Energy (2014). "Arrow Energy – Go Further." Available at http://www.arrowenergy.com.au/ (accessed September 08, 2014).

Arrow Energy (2013). "Fracking : What You Need to Know, 12 July 2013." Available at http://www.arrowenergy.com.au/ (accessed September 08, 2014)

The Australian (2013, April 2), "Dart Energy Slashes costs amid CSG changes." Available at http://www.theaustralian.com.au/archive/business/dart-energy-slashes-costs-amid-csg-changes/story-e6frg9ef-1226610812450.

The Australian (2011, July 8). "Coal Gas Company Banned in Queensland for Contaminating Groundwater." *The Australian*. Available at http://www.theaustralian.com.au/business/mining-energy/coal-gas-company-banned-in-queensland-for-contaminating-groundwater/story-e6frg9df-1226090511694 (accessed September 8, 2014).

Australian government (2014). Department of Climate Change and Energy Efficiency (undated). "Fact Sheet: Australian Emissions Reduction Targets." Available at http://www.climatechange.gov.au/sites/climatechange/files/files/climate-change/national-targets.pdf (accessed October 19, 2014).

Australian government (2012a). *Australia Gas Resource Assessment, 2012*. Available at http://www.ga.gov.au/webtemp/image_cache/GA21116.pdf (accessed September 8, 2014).

Australian government (2012b). "National Inventory Report 2012 (Vol 1) : The Australian Government Submission to the United Nations Framework Convention on Climate Change-Australian National Greenhouse Accounts, April 2014." Available at http://www.environment.gov.au/system/files/resources/6b894230-f15f-4a69-a50c-5577fecc8bc2/files/national-inventory-report-2012-vol1.pdf, p. 40 (accessed October 19, 2014).

Australian Mining (Vicky Valiadakis). (2013, November 7). "Linc Energy dumps coal gasification project." Available at http://www.miningaustralia.com.au/news/linc-energy-dumps-coal-gasification-project (accessed February 20, 2015).

Bajkowski, Julian (2014, April 15). "Queensland Government Hits Underground Coal Gasification Player Linc Energy with Environmental Damage Charges – Government News." *Government News*. Available at http://www.governmentnews.com.au/2014/04/queensland-government-hits-underground-coal-gasification-player-linc-energy-environmental-damage-charges/ (accessed September 08, 2014).

Barker, Geoff (2012, June 27). Proceedings of 5th Annual CBM & Conventional Gas Conference Bali Indonesia http://riscadvisory.com/presentation-by-geoff-barker-cbm-geology-well-design/ (accessed February 18, 2015).
Carbon Energy (2015a). "Bloodwood Creek." Available at http://www.carbonenergy.com.au/irm/content/bloodwood-creek-australia.aspx?RID=221h (accessed February 20, 2015).
Carbon Energy (2015b). "Keyseam Difference." Available at http://www.carbonenergy.com.au/irm/content/keyseam-difference.aspx?RID=285 (accessed February 20, 2015).
Carbon Energy (2014a). "Quarterly Report, December 2014." Available at http://www.carbonenergy.com.au/IRM/Company/ShowPage.aspx/PDFs/2168-10000000/QuarterlyActivitiesReportAppendix5B (accessed March 2, 2015).
Carbon Energy (2014b). "2014 Annual Report." Available at http://www.carbonenergy.com.au/IRM/Company/ShowPage.aspx/PDFs/2050-10000000/CarbonEnergy2014AnnualReport.
Carbon Energy (2013). "2013 Annual Report" (Director's Report: Board Information, p. 18). Available at http://www.carbonenergy.com.au/irm/content/annualreport/12board-information.pdf (accessed February 21, 2015).
Carbon Energy (2008). "Half Year Financial Report" (December 2008, p. 2). Available at http://www.carbonenergy.com.au/irm/archive/half-year-reports.aspx?RID=178 (accessed February 21, 2015).
Chambers, Matt (2014, January 25). "Gladstone Pays High Price for Its New Industry." *The Australian*. Available at http://www.theaustralian.com.au/business/mining-energy/gladstone-pays-high-price-for-its-new-industry/story-e6frg9df-1226810038403?nk=953cf9e6e30c82f365ac1c11d6ca3f0b (accessed September 8, 2014).
Chambers, Matt (2012, November 20). "Shell Rethinks Arrow LNG as Costs Blow Out." *The Australian*. Available at http://www.theaustralian.com.au/business/mining-energy/shell-rethinks-arrow-lng-as-costs-blow-out/story-e6frg9df-1226519911937 (accessed September 8, 2014).
The Chronicle (2011, June 12). "Carbon Energy Faces Environmental Charges." *Toowoomba Chronicle*. Available at http://www.thechronicle.com.au/news/carbon-energy-faces-environmental-charges/903560/ (accessed September 08, 2014).
CO_2 Capture Project (2015). "CO_2 Storage." Available at http://www.co2captureproject.org/co2)_storage.html (accessed February 21, 2015).
Coffey Environments. *Environmental Impact Study – Arrow LNG Plant*. Available at http://www.arrowenergy.com.au/__data/assets/pdf_file/0018/2367/0120–20Introduction.pdf (accessed September 8, 2014).
Cougar Energy Limited (2010, March 16). "Cougar to Ramp Up UCG Process Down Under." *Downstream Today*. Available at http://www.downstreamtoday.com/news/article.aspx?a_id=21673&AspxAutoDetectCookieSupport=1 (accessed September 08, 2014).
Dart Energy (2013), *Annual Report 2013*. Available at http://www.dartgas.com/content/Document/AnnualReport_2013_WEB.PDF (accessed September 8, 2014).

Deloitte Access Economics (2014, July). "Gas Market Transformations – Economic Consequences for the Manufacturing Sector." Available at http://pdf.aigroup.asn.au/Deloitte%20Gas%20Market%20Transformations%20-%20Manufacturing%20Impacts%20Report%20-%20web%20final%20-%20July%2014%202014.pdf (accessed September 9, 2014).

Folger, Peter (2013, October 21). *Carbon Capture: A Technology Assessment*. Congressional Research Service Report for US Congress.

Friederich, Michael (2014). E-mail message to author. July 19, 2014.

Geoscience Australia (2009/2012/2013). "Australia's Identified Mineral Resources." Available at http://www.ga.gov.au/image_cache/GA16805.pdf (accessed September 25, 2014).

Global CCS Institute (2014, August 7). "Boundary Dam Integrated Carbon Capture and Sequestration Demonstration Project." *Global CCS Institute*. Available at http://www.globalccsinstitute.com/project/boundary-dam-integrated-carbon-capture-and-sequestration-demonstration-project (accessed September 09, 2014).

IGas Energy Plc. (2014), "IGas Energy to Acquire Dart Energy" (Board presentation dated May 9, 2014, informing IGas Energy shareholders and potential investors of its intention to acquire Dart Energy and status of Dart Energy's NSW 8 PELs. May 9, 2014, p. 18). Available at http://www.igasplc.com/media/10154/recommended-acquisition-of-dart-energy-limited.pdf (accessed February 19, 2015).

Intelligent Energy Systems (2013, November 28). *Study on the Australian Domestic Gas Market*. Department of Industry, and Bureau of Resources and Energy Economics. Available at http://www.industry.gov.au/Energy/EnergyMarkets/Documents/IESStudyontheAustralianDomesticGasMarket.pdf (accessed September 8, 2014).

Intergovernmental Panel on Climate Change (2007). "IPCC Fourth Assessment Report (AR4)." *Intergovernmental Panel on Climate Change*. Available at http://www.ipcc.ch/publications_and_data/publications_ipcc_fourth_assessment_report_synthesis_report.htm (accessed September 08, 2014).

International Energy Agency [IEA] (2013). *Technology Roadmap: Carbon Capture and Storage*. Paris: International Energy Agency. Available at http://www.iea.org/publications/freepublications/publication/technologyroadmapcarboncaptureandstorage.pdf (accessed September 8, 2014).

International Energy Agency (2008). *CO_2 Capture and Storage: A Key Carbon Abatement Option*. Available at http://www.iea.org/publications/freepublications/publication/CCS_2008.pdf (accessed February 18, 2015).

Johnson, Jeff (2014). "Methane's Role in Climate Change." *Chemical and Engineering News*, July 7.

Kebede, Rebekah (2013, December 04). "UPDATE 1-Santos' Gladstone LNG Faces Additional Well Costs." *Reuters*. Available at http://uk.reuters.com/article/2013/12/04/australia-santos-outlook-idUKL4N0JJ0EV20131204 (accessed September 08, 2014).

Linc Energy Ltd. (2015). Unaudited Second Quarter and Six Months to Dec 2014: Financial Statements Announcement, January 30, 2015, p. 3.

Linc Energy Ltd. (2014). *Corporate Presentation 2014*. Report. 2014.

Linc Energy Ltd. (2013a, November 28). *Annual Report 2013*. Available at http://www.lincenergy.com/data/annual_reports/flipbook/2013/2013AnnualReport.html (accessed September 8, 2014).

Linc Energy Ltd. (2013b). *Corporate Presentation 2013*. Report. 2013. (http://www.lincenergy.com/data/asxpdf/ASX-LNC-532.pdf) (accessed February 19, 2015)

Linc Energy Ltd. (2006). *2006 Prospectus for the Issue of 88,000 Shares at an Issue Price of $0.25 per Share to Raise $22,000,000*. Report no. ACN 076 157 045. Available at http://www.lincenergy.com/data/prospectus/LNC-Prospectus-2006.pdf (accessed September 8, 2014).

Metex Resources Limited (2008). "March 2008 Quarterly Report." (http://newsstore.smh.com.au/apps/previewDocument.ac?docID=GCA00834424MEE&f=pdf (accessed February 20, 2015).

Metgasco (2015). "Quarterly Activities Report for Quarter ending December 30, 2014 (issued January 30, 2015) http://www.metgasco.com.au/sites/metgasco.com.au/files/2015-01-30-Quarterly-activities-report.pdf (accessed February 19, 2015).

Metgasco (2014a). "2014 Annual Report." Available at http://www.metgasco.com.au/sites/metgasco.com.au/files/2014_annual_report_0.pdf (accessed on February 19, 2015).

Metgasco (2014b, June). "About Metgasco." *Metgasco* (single page posting on Metgasco website). Available at http://www.metgasco.com.au/information/about-metgasco (accessed February 19, 2015).

Metgasco (2014c, June). "ICAC: Metgasco Has No Case to Answer." *Metgasco* (single page posting on the Metgasco website). Available at http://www.metgasco.com.au/asx-announcements/icac-metgasco-has-no-case-answer (accessed February 19, 2015).

MIT (2014, May 19). "Kemper County IGCC Fact Sheet: Carbon Dioxide Capture and Storage Project." *MIT*. Available at http://sequestration.mit.edu/tools/projects/kemper.html (accessed September 09, 2014).

Moore, T. A. (2007, July). "Exploration and Development of a Low Rank Biogenically-Derived Coalbed Methane Prospect, Huntly Coalfield, New Zealand." Lecture, CBM Workshop, Bali.

Moore, Tim (2014). E-mail message to author. July 31, 2014.

Moran, Chris, Joe Da Costa, and Chris Cuff (2013, June). "Independent Scientific Panel Report on Underground Coal Gasification Pilot Trials." Available at http://www.fraw.org.uk/files/extreme/derm_2013.pdf (accessed September 8, 2014).

Moreton Resources Ltd. (2014). "Quarterly Report 31 December 2014." Available at http://www.moretonresources.com.au/wp-content/sharelink/20150129-quarterly-activities-report-87758337599988541.pdf (accessed February 22, 2015).

NSW government (2014, September). "Independent Review of Coal Seam Gas Activities in NSW: Managing Environmental and Human Health Risks from CSG Activities." Available at http://www.chiefscientist.nsw.gov.au/coal-seam-gas-review/final-report-september-2014 (accessed February 19, 2015).

Office of the Queensland Parliamentary Counsel (2008). "Mineral Resources (Peak Downs Mine) Amendment Bill 2008." 1–9. Available at https://www.legislation.qld.gov.au/Bills/52PDF/2008/MinResPDMAB08Exp.pdf (accessed September 8, 2014).

Oliver, Rhett (2014, January 28). "Changes Imminent to Queensland Overlapping Mining Tenure Framework." *CBP Lawyers*. Available at http://www.cbp.com.au/Publications/Changes-imminent-to-Queensland-overlapping-mining- (accessed September 08, 2014).

Queensland government (2014a, July 1). "Production and Reserves Statistics." *Queensland*. Available at http://mines.industry.qld.gov.au/mining/production-reserves-statistics.htm (accessed September 09, 2014).

Queensland government (2014b, July 7). "Queensland Gas Scheme." *Queensland*. Available at http://www.business.qld.gov.au/industry/energy/gas/queensland-gas-scheme (accessed September 8, 2014).

Queensland government (2009a, May 2009). "Management of Water Produced from Coal Seam Gas Production: Discussion Paper." Available at http://www.impulsehydro.com/cmsAdmin/uploads/coal-seam-gas-water-discussion-paper_May_2009_001.pdf (accessed October 24, 2014).

Queensland government (2009b, February 18). *Underground Coal Gasification Policy*. Brisbane: *Queensland. Dept of Employment, Economic Development and Innovation*. Available at http://mines.industry.qld.gov.au/assets/mines/ucg_policy_february_2009.pdf (accessed September 8, 2014).

Queensland Ombudsman (2012, September). "The Underground Coal Gasification Report." *Queensland Ombudsman*. Available at http://www.parliament.qld.gov.au/documents/tableOffice/TabledPapers/2012/5412T1124.pdf (accessed September 8, 2014).

Queensland Resources Council (2012, May). "Maximising Utilisation of Coal and Coal Seam Gas Resources – A New Approach to Overlapping Tenure in Queensland." *Queensland Resources Council*. Available at https://mines.industry.qld.gov.au/assets/mines-pdf/overlapping-tenures-final-report-may-2012.pdf (accessed September 8, 2014).

Repasky, John A., VanEric E. Stein, and Phillip A. Armstrong (2013, April 29). "Ceramic and Coal: ITM Oxygen for Power Generation with Reduced CO_2-Emissions, Detailed Engineering Study Results." In *Ceramic and Coal: ITM Oxygen for Power Generation with Reduced CO_2-Emissions*. Proceedings of AIChE Spring National Meeting. Available at https://aiche.confex.com/aiche/s09/techprogram/P145728.HTM (accessed September 09, 2014).

Santos (2014). "The Narrabri Gas Project." Available at http://www.santos.com/library/FACT-SHEET-Narrabri-Gas.pdf (accessed February 19, 2015).

SBS (2013, September 3). "Factbox: CSG in Australia." Available at http://www.sbs.com.au/news/article/2013/03/12/factbox-csg-australia (accessed February 23, 2015).

Solomons, Mark, and Mark Willacy (2014, April 16). "Landholders Claim Secrecy over Experimental Coal Gasification Plant." *ABC*. Available at http://www.abc.net.au/news/2014-04-16/landholders-claim-secrecy-over-experimental-coal-gasification/5395626 (accessed September 08, 2014).

Straits Times (2013, October 5). "Linc Energy Plans ASX Delisting to Join SGX." *Straits Times*. Available at http://business.asiaone.com/news/linc-energy-plans-asx-delisting-join-sgx#sthash.qiY74KJv (accessed February 20, 2015).

Sullivan, Rachel (2013, November 11). "Measuring Methane Seeps in the Surat Basin." *ECOS Magazine*. Available at http://www.ecosmagazine.com/?paper=EC13258 (accessed September 08, 2014).

Supreme Court of Queensland (2014, August 1). "Linc Energy Ltd v Chief Executive Administering the Environmental Protection Act 1994 & Anor [2014] QSC 172," p. 1. Available at http://www.queenslandreports.com.au/docs/db_keydecisions/QSC14-1721.pdf (accessed February 20, 2015).

Sydney Morning Herald (2012, June 28). "Santos Slumps on LNG Cost Blow-out." *Sydney Morning Herald*. Available at http://www.smh.com.au/business/mining-and-resources/santos-slumps-on-lng-cost-blowout-20120628-213uh.html (accessed September 08, 2014).

Tullii (2014, January 8). "Arckaringa Basin May Hold 95 Billion BOE, LNCGY." *Tullii's Instablog*. Available at http://seekingalpha.com/instablog/3957081-tullii/2561141-arckaringa-basin-may-hold-95-billion-boe-lncgy (accessed September 8, 2014).

US Clean Air Task Force (2009, September). "Coal Without Carbon – An Investment Plan for Federal Action." *US Clean Air Task Force*. Available at http://www.catf.us/resources/publications/files/Coal_Without_Carbon.pdf (accessed September 8, 2014).

Validakis, Vicky (2014, January 17). "Doubts Raised over Santos CSG Well Capacity to Feed GLNG Plant." *Australia Mining*. Available at http://www.miningaustralia.com.au/news/doubts-raised-over-santos-csg-well-capacity-to-fee (accessed September 8, 2014).

Vivian, Geoff (2014, January 31). "Prospective Carbon Capture Site Lacks Ceiling." *Phys Org*. Available at http://phys.org/news/2014-01-prospective-carbon-capture-site-lacks.html (accessed September 08, 2014).

WestSide Corporation Limited (2006, November). *Prospectus*. Report no. ABN 74 117 145 516. Available at http://www.westsidecorporation.com/Portals/0/Westside%20Files/PDFs/WSA%206169%20Prospectus%20V13%20271106%20%5B1%5D.pdf (accessed September 8, 2014).

Winestock, Gregg (2014, June 3). "ICAC clears Metgasco of Minister's Slur." *Financial Review*. Available at http://www.afr.com/p/national/icac_clears_metgasco_of_minister_hRToDr2K70as3rKKHJueGM (accessed September 08, 2014).

WorleyParsons Schlumberger (2011). *Economic Assessment of Carbon Capture and Storage Technologies*. Available at http://cdn.globalccsinstitute.com/sites/default/files/publications/12786/economic-assessment-carbon-capture-and-storage-technologies-2011-update.pdf (accessed September 8, 2014).

12 The real drivers of carbon capture and storage in China

Richard K. Morse, Varun Rai, and Gang He

12.1 Introduction

The capture and permanent storage of CO_2 emissions from coal combustion may well be imperative to reconcile the world's trajectory of coal use with stability of the global climate. Coal is the world's fastest growing fossil fuel, and coal combustion is the largest single source of anthropogenic CO_2 emissions (see Chapter 1). This trend presents a forceful case for the development and wide dissemination of technologies that can decouple coal consumption from CO_2 emissions, and the leading candidate technology to do this is carbon capture and storage (CCS). Indeed, IEA climate mitigation scenarios call for CCS worldwide to provide overall emissions reductions of 5–10 billion tonnes (Bt) of CO_2 annually ($MTCO_2$/yr) by 2050 (14%–19% of the total greenhouse gas emissions reductions in IEA model scenarios). Because China's coal-fired power sector is the world's largest, the IEA states that CCS in China will have to supply 20%–25% of total emissions reductions, and over 60% of those reductions will need to be obtained from coal-fired power plants (IEA 2008).

At the same time, the deployment of CCS is nowhere near the implied requirement to stay on pace with IEA projections, and a broadly applicable commercial model for CCS is yet to be demonstrated. The majority of existing projects have required favorable niche circumstances (for example, government subsidies or special funding) or niche applications (often the opportunity for captured CO_2 to be used in enhanced hydrocarbons recovery) to make high costs recoverable. Very little progress has

The authors would like thank Yu Yuefeng of Shanghai Jiao Tong University, Zheng Song of GreenGen, Zhang Hong of China Coal Industry Association, Li Hongjun of China Coal Information Institute, Zhang Chi of BP China, Wang Wanxing of Energy Foundation China Sustainable Energy Program, Peng Suping of China University of Mining and Technology, Yang Xianfeng of China Coal Transportation and Distribution Association, and Mao Yushi for their support and assistance with our research. The opinions expressed here are purely our own and the authors are responsible for all content. Thanks also to David Victor, John Bistline, and Jeremy Platt for their detailed comments on earlier drafts of the chapter. Finally, this chapter benefited significantly from the comments and insights of Mark Thurber.

been made in applying these CCS technologies to coal-fired power plants in any significant way (Rai et al. 2008; MIT 2014). As of 2014, there were only a few very-small-scale pilot projects in operation that applied CCS to coal power plants, with two larger scale coal plant projects (the Kemper project in Mississippi and the Boundary Dam project in Saskatchewan) scheduled to come on line within a year (MIT 2014). The key factors that have so far deterred investments in CCS projects include technological uncertainty, especially regarding performance of capture technologies at scale; high costs; regulatory uncertainty including over liability for sequestered CO_2; and lack of clear carbon policy that could provide a revenue stream for CO_2 capture.

China simultaneously presents the most challenging and most critical test for CCS deployment at scale. Its coal sector is the world's largest, and the rapid industrialization of China is inexorably tied to the same process that fueled the West's development – burning coal. However, the stark reality to be explored in this paper is that China's incentives for keeping on the forefront of CCS technology learning do not translate into incentives to massively deploy CCS in power plant applications as IEA scenarios would have it. In fact, fundamental and interrelated Chinese interests – in energy security, economic growth and development, and macroeconomic stability – directly argue against large-scale implementation of CCS in China unless such an implementation can be almost entirely supported by outside funding. In this chapter, we consider how these core Chinese goals play out in the specific context of the country's coal and power markets, and we use this analysis to draw conclusions about the path of CCS implementation in China's energy sector. Finally, we consider possible leverage points for international climate policy to spur the development of CCS in China.

We make the following four principal observations.

First, the primary driver of pioneer CCS projects in China is the strategic development of its energy security agenda, with particular emphasis on diversity of energy supply, reliable and cheap electricity, and the development of domestic intellectual property for energy technologies. Unfortunately, many analysts who are rushing to declare that CCS has arrived in China are mistaking these motives for an enthusiastic embrace of CCS for purposes of large-scale CO_2 emissions reductions. A crucial distinction must then be made: while these energy security drivers are likely to foster the development of China's CCS demonstration efforts, they do not translate into incentives to deploy CCS at scale. (In fact, as we argue in the following paragraphs, CCS at scale will place a heavy burden on China's coal supply chain and is more likely to harm China's energy security than to help it.) The only manner in which CCS would concretely serve energy security needs would be if China were subject to a stringent greenhouse gas reduction regime at some future time, in which case CCS could facilitate the use of domestically available coal (for either power or for transport using coal-to-liquids technology). However, we argue that such a scenario is likely many years from being a reality, mainly because China has no core interest in agreeing to and enforcing such a regime on itself until it has developed economically to the point where a large domestic constituency values climate change action as much as rapid economic growth. But this aspect of energy security does explain why China remains interested in developing proprietary CCS technology in the demonstration phase. The possibility that some future scenario might leave China dependent on CCS in order to use domestic coal provides motivation for the country to develop strong indigenous technology *capability* in this area (which might additionally lead to commercial opportunities down the road) so that it can control its energy destiny if CCS ever becomes a necessity. Thus, technology demonstration initiatives like the GreenGen plant at Tianjin will continue to be popular, but these should not be confused with enthusiasm for wide deployment of CCS in China soon.

Second, deployment of CCS at any serious scale in China (as many climate models imply) could in fact have detrimental effects on China's energy security by stretching to the breaking point a coal supply chain that is already overburdened at many critical points. As noted earlier, a widely overlooked point about large-scale implementation of CCS in China is that it would almost assuredly result in significantly more coal use due to the parasitic load from CCS (typically estimated at 20%–30%, although improved technology in the future could bring this number down). China's coal supply chain has some structural issues – among them the mismatch between the location of reserves and points of energy consumption that requires a massive and already overburdened transportation infrastructure (see Chapters 2 and 3 and the Appendix) – that make it difficult to rapidly (and cost-effectively) expand supply even though reserves are plentiful. If every coal plant on which CCS is installed will require 20% more fuel on average,[1] energy planners will need to expand coal supply and the supply chain that much more to accommodate the technology. Moreover, because significant investment in the bottlenecks in coal supply will be required, the full costs of CCS at scale in China when one takes into consideration the costs of additional rail, port, and shipping infrastructure could be significantly higher than the already daunting figures for CO_2 capture, transport, and injection.

Third, the Chinese government's non-negotiable desire for cheap power to fuel economic development and keep inflation in check precludes cost recovery through higher electricity prices for CCS at scale.

This constraint will make it even more difficult for CCS for power to develop and scale up in China. In 2008, this structural problem was illustrated by the inability of Chinese generators to incorporate the full cost of their primary input, coal, into the electricity price (see Chapters 2 and 3). A huge run-up in coal prices caused massive losses among Chinese generators. If China's power sector is structurally unable to accommodate fluctuations in fuel prices, it is difficult to see how it can support the massive investments that would be required for CCS. Special government subsidies may be able to fund isolated CCS projects, but without power markets that allow for cost recovery, rapid scaling of the technology in China is likely to prove elusive.

Fourth, in the absence of viable financial models for domestic funding of CCS, international finance will be crucial to the realization of CCS at scale in China. A handful of internationally funded marquee projects are in progress. But international policy mechanisms that have been used previously for engaging the developing world, such as the UN's Clean Development Mechanism, are not well matched to the scale of the CCS challenge and will need to be reshaped if they are to stimulate CCS investment at a scale that will meaningfully impact global CO_2 emissions. A further complication of international finance strategies may be political: countries may be disinclined to provide the substantial funding required for CCS when the associated jobs and technology learning accrue to an economic competitor like China.

These observations have important implications for emissions mitigation efforts. First, the climate policy community needs to understand in a more nuanced way the motivations behind Chinese activities in CCS, which are largely driven by factors other than an interest in CO_2 mitigation, in order to design effective policy. Efforts to engage China in joint technology development on CCS will likely find a receptive audience in the country, as will international funding models. But the assumption of inherent Chinese motivation for wide CCS deployment will likely lead to disappointment because actual results fall far short of outside projections. If such deployment is deemed necessary for the stabilization of the global climate, concerned nations will likely need to play a central role in picking up the (sizable) bill. Second, the climate modeling community needs to move beyond understanding CCS costs in terms of just capture, transport, and storage. Any realistic analysis must capture the full costs of CCS – including through the entire coal-to-power value chain – in order to really understand the potential for CCS uptake in China.

This chapter is split into three sections. Section 12.2 explores the underlying drivers for initial CCS activities in China. Section 12.3 examines some serious impediments to CCS that have thus far received scant attention. We argue these issues will be critical for any serious evaluation of CCS at scale in China going forward. Finally, Section 12.4 considers how to promote deployment of CCS in China, emphasizing the importance of well-designed financial and strategic incentives.

12.2 The strategic drivers of initial CCS efforts in China

China's involvement in CCS projects should be understood in the context of its overarching energy-security agenda. That agenda is complex, but at its heart are security of fuel supply, availability of cheap and reliable electricity supply, and access to key energy technologies. Assessing how CCS could help contribute to longer-term Chinese energy-security goals is the primary motivation behind China's CCS efforts so far.[2] The early progress of CCS in China is being largely driven by considerations other than climate change.

China's CCS efforts began with a series of central-government policies supporting CCS research and development. In late 2005, the Ministry of Science and Technology signed a CCS memorandum that initiated government-sponsored CCS research. In 2006, the National Medium and Long-term Science and Technology Development Plan (2006–2020)[3] formally designated CCS a "cutting-edge technology" in pursuit of clean and high-efficiency use of coal (a broader high-priority energy target in China). A 2007 government document (Ministry of Science and Technology et al. 2007) established that CCS efforts will focus on RD&D, capacity building, and demonstration of CCS technologies. China's State High-Tech Development Plan, the 863 program,[4] later allocated RMB 300 million (USD 43 million) to CCS technology development from 2008 to 2010 and further formalized China's pursuit of the technology (IEA 2008). While these policies support important CCS research in China, they are not geared towards CCS deployment.

A few CCS projects that span a range of technologies and commercial models (see Table 12.1) have been developed for the Chinese market.[5] Broadly, CCS could be disseminated at scale in China in three contexts: industrial applications of CCS, for example, at steel and cement plants; fuel transformations, for example, coal to synfuels, especially coal to liquids (CTL); or CCS at power plants, either pre- or post-combustion.[6] Of these, the last two options are generally considered to have better scale-up opportunities and to be better suited for mitigation of China's growing emissions: the IEA projects that in 2050 nearly 25% of CCS in China will be fuel-transformation-related CCS, while another 60% will be power-generation based CCS (IEA 2008). Accordingly, in our analysis we focus only on these two major applications of CCS.

Table 12.1 *Pioneer CCS pilot projects in China*

CCS project	Technology	Partners	Financial arrangement	CO_2 capture	Status
Huaneng Beijing Thermal Power*	Post-combustion	• Huaneng • Australia CSIRO	USD 4 million research project by CSIRO	3,000 tonnes annually; CO_2 sold to the food industry	Operational since 2008
Huaneng Shanghai Shidongkou Power Plant	Post-combustion	• Huaneng • Xi'an Thermal Power Research Institute • Shanghai Electric	RMB 150 million (about USD 21.4 million)	120,000 tons captured annually; CO_2 sold for food processing and welding	Started in 2008; operational since 2010
GreenGen Corporation	IGCC** pre-combustion de-carbonization; gasification or partial oxidation shift plus CO_2 separation	• Huaneng plus other 7 state-owned energy companies: China Datang Group, China Huadian Corporation, China Guodian Corporation, China Power Investment Corporation, Shenhua Group, State Development & Investment Corporation, China Coal Group • Peabody Energy	Registered capital: RMB 300 million (about USD 44 million); Huaneng has 51%, and other 7 in the group have 7% each; total investment will reach RMB 7 billion	Planned for 60,000–100,000 tonnes capture and storage annually	Started in 2009; 200MW IGCC operational on Dec. 12, 2012; CCS and EOR expected by 2020

Shenhua CTL	Coal to synfuels (direct coal liquefaction)	• Shenhua Group • Sasol • West Virginia University	USD 1.4 billion; support from nat'l government's 12th Five-Year Science and Technology Plan	100,000 tonnes annually captured and stored in nearby deep saline aquifer	CTL operational; 100,000 tonnes storage demonstrated in 2011; total storage capacity at existing site est. at 300,000 tonnes
Shengli oil field CCS + EOR	Post-combustion CCS; enhanced oil recovery (EOR)	• China Petrochemical Corporation • Shengli Oil Field	Support from nat'l government's 11th Five-Year Science and Technology Plan	40,000 tonnes annually captured and used for EOR	Operational whole procedure since 2010.
Jilin Oil Field CCS + EOR	CCS + EOR	• Ministry of Science and Technology • PetroChina • Jilin Oil Field	Support from Ministry of Science and Technology and Petro China	100,000 tonnes annually captured and used for EOR	Operational since 2007. Total storage capacity about 800,000 to 1,000,000 tons

* Huaneng had reportedly planned to launch its second pilot carbon capture project in Shanghai at the end of 2009, but the plan was held back by high costs (*Reuters* 2009).

** "IGCC" stands for Integrated Gasification Combined Cycle. Other active IGCC projects with studies of CO_2 capture in China include the Dongguan IGCC project, the Huaidian Banshan IGCC project, and the China Power Investment Corporation IGCC project.

Sources: General project information from company websites and ACCA21 (2011). Project status updates obtained through author interviews.

The first major CCS projects in China – Shenhua's CTL project in Ordos, Inner Mongolia and the GreenGen IGCC plant in Tianjin – have progressed relatively rapidly because they explore technologies with implications for China's long term security of energy supply.[7] CTL could provide China with alternatives to importing oil, and the GreenGen IGCC plant tests a coal combustion technology that would reduce coal demand by increasing efficiency while also reducing local air pollution. (The CCS phase of GreenGen is slated to come later and remains somewhat of an afterthought in comparison with the IGCC development work.) In the case of post-combustion technologies that do not have potential benefits for fuel security, China has been slower to undertake major projects and is eager to spread risks across international partnerships. In the remainder of this section we focus in particular on the Shenhua CTL and GreenGen projects and how these climate-friendly efforts align with Chinese non-climate objectives.

12.2.1 Shenhua CTL

Shenhua's CTL project is the first major CCS project in China, but it is not driven primarily by climate concerns. The strategic logic of coal-to-liquids technology is all about replacing oil imports.[8] Integrating CCS into CTL processes would further boost security of oil supply by providing high purity CO_2 streams with almost no additional capture cost (the largest cost of CCS in power generation) that could be pumped into declining oil reservoirs to yield previously unrecoverable oil supplies in a process known as enhanced oil recovery (EOR). To prove the relevant coal-to-synfuels and CO_2 EOR technologies, Shenhua's Inner Mongolia project targeted the production of 1 million tonnes (Mt) of liquid transport fuel annually. The 3.6 Mt of CO_2 that the project is eventually expected to generate annually will be used for EOR in nearby declining hydrocarbon reservoirs, yielding new oil and gas revenue streams that will bolster the project economics (Ren 2008). As shown in Table 12.1, Shenhua's CTL project is now operational, albeit at a much lesser scale than what is ultimately envisioned. One hundred thousand tonnes of CO_2 were stored in a demonstration of storage capability; the site in which this initial CO_2 was injected has a total storage capacity estimated at 300,000 tonnes.

Because coal is China's largest domestic energy resource, CTL serves as a hedge against oil import dependence. Indeed, deployment of CTL in China, as in other places, is closely linked to oil prices: in 2008 when the oil price crested at $147, China had 12 CTL projects in the pipeline. Many projects have been put on hold since oil prices crashed, but in addition to the Shenhua project, two other CTL plants are being developed by the Yitai Group and the Lu'an Group.[9]

But CCS for CTL in China has limited relevance to global CO_2 mitigation goals, for a number of reasons. First, CTL is not currently a major source of emissions in China, and its growth is much less certain than the growth of coal-fired power. Second, CCS on a new CTL plant essentially results in a transport fuel that is roughly net equivalent in carbon emissions to oil; it only creates a significant net benefit for climate relative to the very high net emissions of CTL without CCS. Third, the simplicity of CO_2 capture from a CTL process means that technology learning from such applications is not transferable to the largest source of CO_2 in China – coal combustion at power plants. Relative to post-combustion CCS, where the capture process represents over 50% of CCS costs, or IGCC-CCS where the costs of IGCC alone have thus far proven commercially prohibitive, capturing CO_2 from a CTL process is much cheaper and more straightforward.[10] Indeed, removing most of the CO_2 from the gaseous stream after coal gasification is necessary to improve the yield of liquid fuels during the synthesis step of the Indirect Liquefaction process. In the Direct Liquefaction process used by the Shenhua CTL project, a pure stream of CO_2 also becomes available (with no capture equipment) during the production of hydrogen, which is used in the hydrocracking step to break down the structure of coal. Therefore, while CTL is certainly a response to dependence on the global oil market for energy imports, CTL-CCS does not address the major challenge of reducing emissions from the largest source of CO_2 in China – coal combustion at power plants.

In addition, CTL-CCS projects are only likely to be pursued where EOR opportunities exist. CTL projects are very capital intensive in their own right, which means that project owners look for stable revenue streams over long time horizons. Adding carbon storage increases capital requirements and adds further financial risk to these projects. These risks are not likely to be justified absent additional revenue streams from incremental oil recovery.[11]

12.2.2 GreenGen IGCC

CCS technology integrated with IGCC,[12] as envisioned by GreenGen, is the major focus of state-supported CCS for power plants in China. The main advantages of CCS with IGCC are the high combustion efficiencies of IGCC and the relative ease of CO_2 capture compared with post-combustion CO_2 capture (where the CO_2 partial pressure is much lower and hence capture is more complicated and expensive). While no

IGCC project has yet been deployed, the central government's 863[13] and 973[14] R&D programs favor developing CCS in concert with IGGC.

The drivers for making IGCC power plants a state priority in China are again rooted in energy security concerns (there are also co-benefits in the form of reduced local pollution and synergies with chemicals production).[15] First, China's own internal energy policy provides strong incentives for developing IGCC as an efficient coal combustion technology.[16] China's national energy-efficiency policy depends heavily on generating electricity from coal more efficiently. In addition, long-term planning for the reliability of electricity supply necessarily emphasizes using coal efficiently, as remaining domestic resources are finite and production and associated transport costs will increase as the best and cheapest coal reserves are exhausted over time.

Second, developing intellectual property and technological capability in IGCC will save China from dependence on foreign power plant manufacturers in the future;[17] the concern is that foreign technology is expensive and may be subject to political restrictions on availability, as is already the case for some technology. For this reason, China has directed Xi'an Thermal Power Research Institute to develop a domestic gasifier and turbine. China has pursued a similar strategy for existing state-of-the-art coal combustion technologies through an aggressive national program for developing domestic supercritical and ultra-supercritical coal-combustion technology.[18] An additional benefit for China of developing indigenous technology is the commercial opportunity to export it to other countries. The appeal of domestic IP and future opportunities for export are perhaps the strongest additional motives for China to investigate carbon capture technologies.

The potential direct benefit to Chinese energy security through higher efficiency and the development of potentially lucrative domestic IP help explain why the Chinese government is putting its own capital behind GreenGen, the world's leading IGCC-CCS project. The financial risk of GreenGen is significantly borne by the government through eight state-owned companies (see Table 12.1).[19] While similar projects elsewhere (for example, FutureGen[20] in the US) have struggled to get started due to large capital requirements and uncertain benefits that deter first-movers, full-fledged government support through state-owned companies makes GreenGen's implementation highly likely. The information that GreenGen provides the Chinese government about the costs of IGCC and associated CO_2 capture will be crucial as China prepares the road-map of its power generation build-out beyond 2020.

GreenGen is centered around a 400 MW IGCC coal plant in Tianjin. The project plans to incorporate polygeneration (the flexibility to produce a combination of electricity and chemical feed stocks) and CCS. The project was designed to take place in three stages:[21]

- *Stage one (2005–2010)*: 2000 t/d gasifier, 250 MW IGCC operation, hydrogen production and CO_2 separation at pilot scale;
- *Stage two (2010–2015)*: 3000 t/d gasifier, 400 MW IGCC, hydrogen production and CO_2 separation at 100 MW scale;
- *Stage three (2015–2020)*: 3000 t/d gasifier, 400 MW IGCC with full scale CO_2 capture.

An important observation about GreenGen's development sequence is that CCS remains for the moment distinctly an afterthought relative to the more strategically important goal of developing IGCC. The initial 200 MW of IGCC capacity came on line at the end of 2012, but the design of any carbon capture system is still very much on the drawing board, and carbon storage is even further from a reality, with no CO_2 storage sites having been selected at this point.

Looking beyond GreenGen, integrating CCS into IGCC plants looks to be the most promising route for deployment of CCS for power in China, because it both addresses emissions in the power sector and is well aligned with China's own incentives to increase efficiency and reduce local pollution, but several key barriers remain. First, IGCC itself has to be proven both technologically and commercially in the Chinese market. The push for IGCC in China envisions gaining efficiencies and additional revenue streams by integrating multiple processes – power, process heat, and chemicals – in order to make the technology cost competitive (Liu et al. 2008). (At the same time, parallel advancements in ultra-supercritical coal-combustion technology are raising the efficiency levels that IGCC needs to compete with to be the leading "clean coal" option.) Second, the commercial and technological integration of CCS into IGCC needs to be proven.

12.2.3 Other CCS projects and international partnerships

China is leveraging international support for developing CCS wherever possible, but especially in cases in which domestic benefits are less clear. It is not a coincidence that the projects with fewest direct benefits to China are the ones that involve international financing and technology sharing models. These projects have tended to progress at a slower rate than those inherently motivated by China's security interests.

Several international CCS research efforts were launched in China. Australia is a partner in Huaneng's Beijing post-combustion capture demonstration plant. The EU showed a burst of interest in the late 2000s in supporting the development of CCS in China. In late 2007, an

EU-China MOU called for "research of near zero emissions coal power generation technology through carbon capture and storage" (Council of the European Union 2007). That agreement spawned the Near Zero Emissions Coal project between China and the UK, which aimed to research and develop an IGCC polygeneration CCS demonstration project by 2014, but this project has since languished (MIT 2014). In May 2008, Japan and China announced a cooperative project to capture CO_2 from a Chinese coal-fired power plant and inject it into a Chinese oil field for EOR. While all of these projects certainly represent useful research efforts, they have moved slowly (or stalled completely) and do not represent a level of investment comparable to the Shenhua CTL and GreenGen IGCC projects, suggesting again that finance and progress will be closely correlated to China's energy security interests. Those interests are not well aligned with installing post-combustion capture on the biggest source of emissions in China – the existing fleet of conventional coal plants.

12.3 Obstacles to wider deployment of CCS in China

As described in the previous section, China's prioritization of CCS technology demonstration projects provides a good window into the inherent motivations behind the country's efforts. But the fact that a few projects are being enthusiastically pursued because they line up with longer-term energy security goals should not be taken to indicate that wider deployment of CCS in China is likely in the near future, particularly in the power sector. Deployment of CCS at scale has not yet been successfully demonstrated in any country, due to obstacles including high costs, uncertain storage conditions and capacity, and the absence of suitable regulatory frameworks. Beyond these universal barriers, though, implementation of CCS in the Chinese power sector presents two special impediments that have not been fully considered in most analyses: the structure of the power market and the supply chain for coal. These formidable obstacles need to be understood in order to gauge the true potential of CCS in China.

12.3.1 The structure of China's power market constrains CCS cost recovery

Depending on the technology, adding CCS is estimated to increase the cost of generating coal-fired power by 40%–80% (IPCC 2005). IGCC with CCS is estimated to increase electricity cost by 40%–60% (on top of the cost of electricity from IGCC), while post-combustion CCS will increase electricity costs by 60%–80% (on top of the cost of electricity from subcritical or supercritical coal combustion without CCS). No matter what technology goes forward as the leading contender for deployment of CCS, the costs will be daunting. In the European and US markets, policy makers need to consider how to cover these costs through some combination of carbon market incentives, subsidies, and increased end-user electricity prices. The Chinese power sector, on the other hand, is not market-oriented, and regulation of the energy system is largely driven by the political priorities of the central government. As we will explore in the remainder of this section, the resulting complex organization and inflexibility of Chinese energy markets makes it uniquely difficult to sustainably recover the costs of CCS. Chinese energy markets in general exist in various states of reform and liberalization; for the purposes of CCS, the most relevant markets are the coal market and the power market, where the relative imbalance of reform creates serious problems (see Chapters 2 and 3).

China now has mature markets for coal that are increasingly exposed to international prices. As a market-based coal sector has evolved, it has not been immune to price volatility. In 2008 alone, the price of coal at the leading port, Qinhuangdao, mirrored international price volatility and fluctuated from RMB 580 to RMB 950,[22] reaching historical highs before crashing again.

The power market, however, is not market-oriented. China's central planning apparatus, the National Development and Reform Commission (NDRC), keeps tight control over electricity prices in China in order to meet its larger socio-political agenda. Reform and liberalization of China's power markets remains a distant prospect.

The conflict that arises between unevenly reformed coal and power sectors illustrates a key problem for CCS in the Chinese energy system: the power market cannot internalize increased costs. This means that the Chinese power market is designed in a way that makes it nearly impossible for it to deploy a commercially viable CCS model on its own.[23] In 2008, much of the Chinese power market could not even bear the cost fluctuations of its primary input – coal. Coal currently is the largest cost for power generators in China, and when the coal price rises in response to supply shortages or other market events, the power sector cannot proportionally increase electricity prices in response.[24] In 2008, coal prices reached new heights due to transport constraints and the inability of the Chinese coal sector to keep up with booming demand. The price spike could not trump politics, and power generators were forced to operate at massive losses – it is reported that Chinese power companies lost an estimated RMB 70 billion (RMB

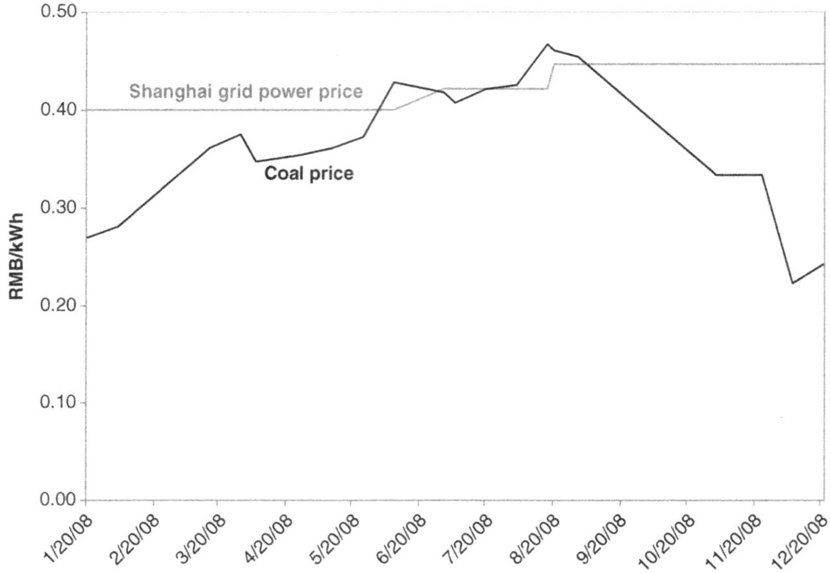

Figure 12.1 Coal and power prices in China in 2008. The power price is the Shanghai grid price as reported by the Shanghai Development and Reform Commission. The coal price is the Qinhuangdao price from Reuters. Note: Coal cost is expressed in terms of kWh and reflects the average efficiency of coal plants in China, assumed to be 2,775 kcal/kWh. Numbers are indicative for the national market and may exclude some additional costs of coal and power generation.

40 billion for the Big Five only) in 2008.[25] China's state-owned power generators are often forced by the government to operate at losses when independent power producers simply shut down. Figure 12.1 illustrates how tight control of power prices affects the profitability of coal fired generators (for example, between April and September 2008) due to the inability of generators to pass on basic costs. Where the price of coal is above the price of power, as was the case in much of 2008, the costs of coal alone (even excluding all other relevant costs) cannot be recovered by power generators. If the structure of Chinese power markets means that generators can barely support the cost of coal, supporting the cost of CCS is currently unthinkable absent major power sector reforms.

12.3.2 *Major CCS deployment would create new stress on the coal supply chain*

Many analysts and climate modelers have failed to consider that CCS costs are in fact not limited to the power sector – they would extend to the entire coal value chain. There is a significant reduction in the efficiency of power generation when CO_2 capture is added, as capture systems consume a lot of power: adding CO_2 capture reduces power generation efficiency by 20%–30% (or 10–15 percentage points).[26] CCS thus requires an increase in coal consumption of around 20%–25% to produce the same amount of electricity. We estimate that CCS at scale in China as prescribed by the IEA Blue scenario would demand approximately 200–300 Mt per year of additional coal production.[27] This level of incremental coal production above baseline targets would be accompanied by further stress on the coal supply chain that could itself threaten the paramount objective of maintaining cheap and reliable electricity. Infrastructure in the coal supply chain already struggles to keep pace with the growth in coal demand to fuel the economy; wide deployment of CCS would greatly exacerbate these strains. Beyond the obvious additional cost increases for generators, ramping up coal production to these levels would require new mining capacity, rail infrastructure, port expansions, and shipping capacity – investments on a massive scale. The appendix at the end of this chapter considers in detail the costs throughout the coal supply chain that could be associated with a 200–300 Mt per year expansion in coal production and transport capacity to support the CCS deployment demands of the IEA Blue scenario. Costs would be well in excess of RMB 100 million (USD 15 billion). If one were to include other environmental externalities of coal production and consumption, these numbers would be even higher.[28]

The costs of CCS, therefore, cannot be understood as limited to the power sector and instead permeate China's entire coal-based energy infrastructure. Because industry and infrastructure development is normally arranged in a five-year plan system, major infrastructure projects first are proposed, approved, and funded in the plan and then deployed into construction. Thus, ensuring the availability of coal supply for CCS at scale entails coordination of the entire energy industry and economic development system.

As China's central planners confront China's energy challenges, CCS investments have to be understood not only in terms of their own system-wide costs but also in terms of opportunity costs in foregone alternatives. China is experiencing a boom in clean-energy development. Wind power grew from 0.76 GW in 2004 to 13.24 GW in 2008. The Chinese government has a plan to reach 30GW of wind in 2020, which requires more than RMB 1 trillion (USD 146 billion) in additional investment. Deployment of solar is planned to reach 10GW, and an estimated RMB 300 billion (USD 43 billion) is required to achieve that target. Nuclear is on track to account for 5% of the energy mix in 2020, and RMB 750 billion (USD 109 billion) is required. CCS would likely come at the expense of some of these investments, a sober reality that is sure to weigh heavily on the minds of China's central planners[29] – especially given that these other alternatives enhance China's diversity of energy supply whereas CCS does not.

12.4 The path forward: paying for CCS in China

As we have illustrated, in their current form China's energy markets will be incapable of handling the high costs of CCS. Deployment of CCS in China, even at moderate scales, will thus require serious policy interventions and strong financial support. This leaves two broad options for financing CCS at scale: direct financial support from the Chinese government or international financing mechanisms, likely through climate treaties or direct engagements.

The first option, domestic financial support for deploying CCS at scale, is highly unlikely given China's current energy strategy. China currently has few strategic or financial incentives to deploy CCS beyond the demonstration phase. While China is increasingly stressing sustainable development, this has not meant CO_2 mitigation thus far, and economic development is still China's top priority. Spending on CO_2 mitigation through CCS is a diversion from that goal. Opportunity costs are high (see Section 12.3.2), and CCS will not measure up against other development and economic priorities. Further, investment in environmental controls at power plants is focused not on CO_2 mitigation but on abatement of local pollutants from coal combustion, such as SO_2 and NO_X, that directly impact China's citizens on a daily basis. Controlling these pollutants also requires large investments, and China is moving ahead with the planning and deployment of these technologies. Support for CCS RD&D will continue because it aligns with longer-term Chinese objectives of energy security, but as the locus of domestic priorities lies elsewhere, new and much stronger incentives will be required for any type of CCS deployment at scale in the Chinese power market.

For these reasons, international financial support will be needed if CCS deployment in China at scale is to become a reality in the near to medium term. Financial mechanisms under international climate agreements could conceivably be used to help support CCS projects in China. It was contemplated to include CCS in the Clean Development Mechanism (CDM) under the UN's Kyoto Protocol. Under such an arrangement firms would have earned the market price of Certified Emissions Reductions (CERs) for every ton of CO_2 abated by a CCS project. The CDM would have provided price levels of roughly €8–€23 per tonne (USD 11–USD 32) of CO_2 for CCS investors.[30] While meaningful, these price levels are not sufficient when the marginal cost of abatement for CCS is on the order of $50 to $100 per tonne of CO_2 (even without considering additional obstacles discussed previously) (IPCC 2005; IEA 2008).[31] Another issue is that, for capital-intensive projects like CCS, stability of revenue streams is critically important for the project's viability. Paying for the additional cost of CCS projects solely with a carbon offset mechanism like CDM would have proved problematic on this front because of carbon price volatility.[32] Thus, carbon offsets alone will not enable the widespread deployment of CCS. Funding CCS in China will require policy makers to rethink the design and application of international mechanisms for financing CO_2 mitigation.

International negotiations for a post-2012 climate agreement have focused on potential successors to the CDM. The EU has proposed "no lose" sectoral crediting mechanisms that, if properly designed, could be better equipped to stimulate CCS investment. As currently envisioned for the power sector, the UN would establish a benchmark (likely based on emissions intensity), and provide carbon credits to sectors that perform better than the baseline.[33] CCS, which theoretically allows a very low emissions intensity, could be a potential beneficiary of such a policy. However, potential policies along these lines are still being shaped and are not yet clearly defined in such a way as to provide credible support for CCS.

Irrespective of the financial mechanism, those who argue that CCS in China must be a part of any climate solution need to soberly evaluate the scale of investment required. In the Chinese context, investments in CCS that would yield about 1.2 Gt/yr CO_2 emissions reductions by 2050 (IEA's ACT scenario) amount to a cumulative investment of USD 300–400 billion between 2030 and 2050, or an annual average investment of USD 25–30 billion. The total amount is even higher for the IEA's BLUE scenario for CCS in China, at USD 450–600 billion. These estimates exclude operating costs; including these can increase total costs by a further 50%–100% (IEA 2008).[34]

Thus, if deploying CCS in China is considered an imperative for climate policy, policy makers will need to take note of the staggering scale of investment required and calibrate climate policy accordingly. If the scale of the offset mechanisms in the Kyoto Protocol and the US's stalled-out climate legislation are any indicators of the likelihood and political viability of such a massive climate-based wealth transfer from the developed to the developing world, the precedent for funding CCS at scale through climate policy is not promising. If, for example, CCS were to be included as a CDM-like offset methodology in an international treaty as well as in a US offset scheme like that proposed in the ill-fated Waxman-Markey bill, and 25% of these offset markets were dedicated to CCS funding, the total value of these funds would be around USD 7 billion annually.[35] This optimistic scenario still amounts to about 25% of the implied requirement for covering CCS in China (USD 25–30 billion annually), as described prevoiusly. Purely apart from the financial capacity of the developed world to support CCS in China, there is the question of the political viability of a program that could be seen by OECD (Organization for Economic Cooperation and Development) voters as steering technology and jobs on a massive scale to an economic competitor.

Climate finance will certainly be critical for getting CCS off the ground in China, but given the political difficulties, policy strategies for international engagement will need to include additional types of CO_2 emissions reduction incentives. Cost-sharing agreements that are aligned with China's own incentives are likely to be the most tenable path forward in the near term. CCS is still critical for the abatement of emissions from coal, but in the face of daunting costs and risks, investors and policy makers should not ignore the full portfolio of options in the coal-fired power sector. Increasing the efficiency of Chinese coal combustion may represent a much greater opportunity for CO_2 mitigation in the near term. Because it naturally aligns with energy security incentives as discussed previously, improvement of combustion efficiency is already a top priority of Chinese policy and spending. Increasing the average efficiency of the Chinese coal fleet from its average of 32% in 2009 to the EU's average of 38% – a shift that would be possible utilizing existing proven technologies – could represent a 20% reduction in emissions from China's power sector.[36] A critical step in this direction will indeed be to prove the viability of high-efficiency IGCC power generation as the GreenGen project hopes to do – an effort that could also build a foundation for later deployment of CCS.

12.5 Conclusions

The expansion of coal-fired power in China made the country the world's largest coal market and leading emitter of GHGs. As many analysts, policy makers, and academics have realized, any serious strategy for avoiding the worst consequences of climate change must therefore urgently confront emissions from Chinese coal combustion. Wide deployment of CCS technology is required to address this challenge. But without a careful assessment that takes into account the structure of Chinese energy markets and China's own priorities, the true potential for deployment of CCS in China will be miscalculated and climate policy will not address the coal challenge in a pragmatic way.

China is already undertaking CCS projects, both on its own and in international partnerships. But not all CCS projects are created equal, and the projects that are most viable support long-term priorities of the Chinese government: diversity of energy supply, domestic sourcing of energy, cheap and reliable electricity, and control of intellectual property for key technologies. Absent major changes in Chinese climate policy, it is these priorities that will continue to provide the driving logic for CCS in China. While these priorities have led to support of projects like Shenhua CTL and GreenGen, they do not automatically translate into support for wider CCS deployment.

Policy makers and observers need to understand that not all CCS projects are well matched to major sources of emissions in China, and as such China's CCS efforts offer disparate degrees of promise for CO_2 mitigation. CTL-CCS is likely to be the major source of sequestered CO_2 in China in the near term because China's incentives to add CCS to CTL projects for EOR are high and the risks are comparatively low. But the real climate benefits of CCS with CTL are minor, as CTL is not a major source of emissions in China. The key to reducing China's emissions from coal is applying CCS at power plants, where China has already made IGCC a priority and is charging ahead with GreenGen. While IGCC-CCS is likely to have the greatest chance of success among CCS for power options in China, deploying CCS in this context would still involve the formidable task of reinventing how power and chemicals are made in China.

Effective climate change policy requires both the vigorous promotion and careful calculation of CCS's role in Chinese power generation. Carbon finance must be applied to deploying CCS at scale in China. As the world attempts to craft a new international policy architecture for mitigating climate change, international offset policy (and potential US offset standards) needs to create methodologies that directly address CCS funding at scale. The more closely these policies are aligned with China's own incentives and the unique context of its coal and power markets, the better chance they have of realizing the optimal role for CCS in global climate efforts.

Appendix Costs associated with increasing Chinese coal production capacity by 300 Mt to support CCS demands of IEA Blue scenario

Increased costs would begin at the mine site. The average cost of extracting one ton of coal from underground in Shanxi was about RMB 115 (USD 17) in 2008 by modern mechanical systems, though it can as low as RMB 50 (USD 7) per tonne by hand.[37] The best resources have already been extensively exploited, and the remaining coal is increasingly expensive to mine. Ramping up production by 300 Mt would require 340,000 additional workers and RMB 19 billion (USD 2.8 billion) more investment in the fixed assets in the coal industry.[38]

Much of China's coal production is far from the coastal centers of industry that consume it, which requires a massive coal transportation infrastructure. Typical coal supply routes illustrate the complexity and challenge of moving nearly 3 Bt of coal per year. For instance, in Shanxi Province, where about one-fourth of China's coal is produced, most of the coal from the local mine is first transported by trucks to the local rail loading station. From that station the coal is transported 653 km by the Da-Qin rail line to the major port Qinhuangdao on the eastern coast. Qinhuangdao serves as a major trading hub that supplies industrial China with coal via maritime shipping supply routes to the south. This elaborate transport system of coal is used because the rail lines are so overburdened that it is cheaper to rail coal to Qinhuangdao and then ship it all the way to southern China on a boat than to send it by rail directly. The main railway line transporting coal from Shanxi to Qinhuangdao had already reached 340 Mt of coal carrying capacity in 2008. Within the next few years it might reach 400 Mt at most with restructuring of the lines and upgrading of the capital stock. A second rail line with 200 Mt annual capacity from Lüliang in Shanxi to Qingdao port is under discussion, and about RMB 100 billion (USD 14.6 billion) is needed to build such a new line. If one assumes 300 Mt of extra coal for CCS use, this means that it needs the equivalent of another Da-Qin rail line to transport added coal, which entails on the order of RMB 133 billion (USD 19.4 billion) for rail alone given the scale of investment by the Lüliang line.

If the rail-to-maritime routes are maintained, increased transport capacity would mean an expansion of the shipping fleet and domestic ports. About 500 Mt of coal were moved by maritime trade in the Chinese market in 2010. Even moving an additional 150 Mt would require expanding the dry shipping fleet by around 1,500 Panamax vessels and expanding port capacity by one-third of existing capacity. The Caofeidian port near Tangshan illustrates the scale of costs associated with additional port capacity. Construction of the port started in 2006 and targeted 50 Mt in annual coal handling capacity in the first stage of development. It is estimated that RMB 15 billion (USD 2.2 billion) in investment is required.[39]

Notes

1. Calculated using "Efficiency Loss" figures on p. 65 of IEA (2008).
2. The IEA has done some analysis about CCS development in the context of China's clean coal development; see IEA (2009).
3. See Ministry of Science and Technology (2009), Article 5: "Advanced Energy Technology."
4. The name "863" comes from the fact that the program was created in the third month of 1986. The program is intended to stimulate the development of advanced technologies in a wide range of fields for the purpose of ending China's financial dependency on foreign technologies.
5. We argue that risk and the ability to mitigate it are leading indicators of the viability of CCS models in energy markets.
6. These models can largely be understood based on technology. Our goal here, however, is to broaden the analysis to encapsulate wider-ranging commercial conditions.
7. Huaneng Beijing Thermal Power plant is not a major CCS effort because its capture and sequestration volumes amount to 3,000 tonnes per year, which is used for carbonating beverages.
8. As Chinese economic growth has become increasingly contingent upon oil imports – in 2008, China imported 179 Mt of crude oil (49.8% of its requirement) at a cost of about USD 19 billion – China's interest in CTL as an energy security contingency plan has grown despite its high cost. China has 114.5 Bt (13.5% of world total reserves) of coal reserves, but only 2.1 Bt (1.3% of world total) of oil reserves and 66.54 trillion m^3 (1.1% of world total) of gas reserves.
9. Yitai and Lu'an both planned test plants that would produce 160,000 tons of liquid fuel annually and use indirect coal liquefaction technology developed domestically by the Institute of Coal Chemistry in Shanxi.
10. The CO_2 streams available during coal-gasification processes are at an order of magnitude higher partial pressure than those in the flue-gas exhaust of power plants. This not only reduces the size and complexity of the required capture systems, but it also permits the use of commercially available and relatively cheap capture technologies (IPCC 2005).
11. Volatility of oil prices hampers CTL projects, which are more attractive where oil prices remain high over the long run. Given that carbon storage with CTL is favored only when EOR opportunities are available, volatile or low oil prices doubly hamper carbon storage based on CTL.
12. In IGCC, steam and oxygen are used to gasify coal into syngas, a mixture of carbon monoxide (CO) and hydrogen (H_2). After removal of impurities (H_2S, COS, etc.), the syngas is passed through a water-gas shift reactor

that generates CO_2, which is then captured in the next stage. The remaining syngas is combusted in a gas turbine to generate electricity. The hot flue-gas stream of this stage is used to generate steam that is used in a steam cycle to generate more electricity (hence "combined cycle").

13. The national high technology research development program, named for starting year and month.
14. The national key fundamental research development program, named for starting year and month.
15. Most analysts predict that CCS integrated with IGCC would reduce prohibitive capture costs, and thus is the most viable path forward for integrating CCS into coal-fired plants, despite the fact that IGCC technology itself is largely commercially unproven. IGCC could also have significant environmental benefits when compared to conventional coal combustion, reducing emissions of NO_x, SO_2, and mercury.
16. The average efficiency of China's coal plants is 32%–34%, and IGCC might reach 40%–45% efficiency.
17. China has historically relied on foreign power-plant technology. Today, many foreign firms operate JVs with Chinese companies to produce equipment inside of China. The Chinese government has emphasized the pursuit of domestic intellectual property as a strategy for weaning dependence on foreign firms.
18. In 1995–2000, NDRC initiated the "Development of 600MW supercritical coal-fired power unit" project. In 2001–2005, MOST's 863 project of "Ultra-supercritical coal-fired power generation technology" further advanced the domestic RD&D efforts.
19. Peabody Energy, a US coal company, is also seeking participation in GreenGen.
20. Two of FutureGen's major backers, American Electric Power and Southern Company, backed out of the project in June 2009 just weeks after the Obama administration confirmed renewed funding for the project.
21. From GreenGen's company website.
22. Historical data from China Coal Transportation and Distribution Association, with average heat content of 5,800 kcal/kg.
23. There is precedent for the NDRC allowing power plants with more advanced environmental controls to charge slightly higher tariffs. It allows a price increase of 1.5 cents/kWh for the installation of FGD. However, these increases are not well matched to the costs of CCS. Although special power tariffs may be allowed for demonstration plants, CCS deployed beyond the demonstration phase would entail a significant increase in power prices. In the absence of cost recovery by pass-through in the power price, it is likely significantly larger direct subsidies would be required.
24. China tried to use a coal-electricity price linkage system to incorporate the coal cost into the power tariff. The system was proposed in 2004 and designed such that when coal prices increased by 5% in six months, the power tariff would adjust accordingly. In May 2005, NDRC adjusted the power tariff with a RMB 2.52 cents increase, followed by another adjustment in May 2006. The system was only used twice since and failed to track the dramatic rise in coal prices in 2008.

25. China Electricity Council, Office of Statistics and Information. Report by officer Xue Jing.
26. The loss in efficiency depends on the combustion technology used. For example, the efficiency loss is higher for post-combustion capture than for IGCC. See IPCC (2005) and IEA (2008).
27. IEA Blue scenario projections of CCS in China are for 2 Gt CO_2 reductions by 2050, over 60% (or 1.2 Gt) of which come from the electricity sector. Our calculations show that achieving that scale will require 375 GW (gross) CCS-equipped coal capacity. Assumptions made for that calculation were: 50% gross efficiency; 5% auxiliary consumption; 10% efficiency loss due to capture; that is, 35% (50 – 5 – 10) net efficiency with capture, 85% capture rate, 0.8 plant load factor. Given that China's ~600 GW of coal capacity in 2009 consumed 1.2 Bt of coal per year, an additional 25% for 375 GW (due to decreased efficiency) would represent an additional 190 Mt of coal.
28. In *The True Cost of Coal*, Mao et al. (2008) report that when the environmental costs of coal are considered, along with the impact of price distortion caused by existing regulations, the total external costs of coal reached RMB 1,745 billion in 2007, equal to 7.1% of China's GDP for that year.
29. Su Wei, director general of the climate change unit at the NDRC, made this point in an interview, stating: "Carbon capture and storage, particularly for China, is not one of the priorities – the cost is an issue. If we spent the same money for CCS on energy efficiency and the development of renewables, it would generate larger climate-change benefits" (Morales and van Loon 2009).
30. Based on historical data from the Reuters CER index. The Dec '08 and Dec '09 CER contracts have fluctuated between €7.48 and €22.85 (at the time of publication). Higher CER prices are possible and depend on a number of factors, including market design and supply and demand.
31. See IEA Clean Coal Centre (2008) for further arguments that a CDM-like offset program is insufficient to stimulate CCS investments.

    As discussed at several points earlier in the chapter, the actual marginal cost will vary with the application and the commercial and regulatory context for that technology. For example, if China builds IGCC/Poly generation for other reasons, then the marginal cost for CCS is much lower. On the other hand, if it does Greenfield power generation projects where IGCC with CCS must compete with conventional coal, then the marginal abatement cost is totally different (much higher).
32. Hedging carbon price volatility would be possible with the existence of liquid, long-term derivatives markets for carbon. Thus far carbon policy has not been stable enough to provide these markets.
33. There were a number of proposals about the design and administration of a sectoral crediting system. For further information, see Meckling and Chung (2009), Schmidt et al. (2008), and Baron et al. (2009).
34. The IEA BLUE and ACT scenarios roughly call for 2 Gt and 1.2 Gt CO_2 per year reductions in China through CCS by 2050.
35. The following assumptions are used to generate this calculation, which is purely indicative: (1) average annual CER issuance would be 306 million,

based on UNFCC estimates at http://cdm.unfccc.int/Statistics/Registration/AmountOfReductRegisteredProjPieChart.html; (2) a hypothetical US cap-and-trade scheme would maximize usage of international offsets at 1.5 billion per year; (3) CER price is €15; (4) US offset price is the average of EPA's high and low scenarios in 2015: USD 15; and (5) euro-to-dollar exchange rate is 1.413.

36. Indicative calculation based on emissions factors in Baruya (2008).
37. From interviews with local firms and industry experts.
38. Calculated with data from *China Coal Industry Yearbook 2006*, based on inputs associated with new production that year. Fixed assets include facilities and equipment. In 2006, coal production was 2.33 Bt, the total workforce in the coal industry was 2,657,230, and fixed assets investment was RMB 147.9 billion. Using linear growth as an indicative calculation, 300 million tons of additional coal requires about 340,000 additional workers and RMB 19 billion more investment.
39. From Caofeidian Port's website.

References

ACCA21 (2011, September). *Carbon Capture, Utilization and Storage: Technology Development in China*. Edited by the Administrative Centre for China's Agenda 21, Ministry of Science and Technology. Available at http://www.acca21.org.cn/gest/etc/CCUS_en.pdf (accessed September 29, 2014).

Baron, Richard, Barbara Buchner, and Jane Ellis (2009). *Sectoral Approaches and the Carbon Market*. Paris: International Energy Agency. Available at http://www.oecd.org/environment/cc/42875080.pdf (accessed September 29, 2014).

Baruya, Paul (2008). *Competitiveness of Coal-fired Power Generation*. London: IEA Clean Coal Centre.

China Coal Industry Yearbook 2006 (2007). Coal Information Research Institute.

Council of the European Union (2007). "10th China-EU Summit, Beijing, 28 November 2007, Joint Statement." Available at http://www.consilium.europa.eu/ueDocs/cms_Data/docs/pressData/en/er/97355.pdf (accessed September 29, 2014).

IEA (2009). *Cleaner Coal in China*. Paris: International Energy Agency.

IEA (2008). CO_2 Capture and Storage: A Key Carbon Abatement Option. Paris: International Energy Agency.

IEA Clean Coal Centre (2008). "*Economic instruments and clean coal technologies*." Available at http://www.iea-coal.org.uk/documents/82109/7073/Economic-instruments-and-clean-coal-technologies (accessed September 29, 2014).

IPCC (2005). *IPCC Special Report on Carbon Dioxide Capture and Storage*. New York: Cambridge University Press.

Liu Hengwei, et al. (2008). "Strategic thinking on IGCC development in China." *Energy Policy* 36(1): 1–11.

Mao Yushi, et al. (2008). *The True Cost of Coal*. Available at http://www.greenpeace.org/eastasia/pagefiles/301168/the-true-cost-of-coal.pdf (accessed September 29, 2014).

Meckling, Jonas O. and Gu Yoon Chung (2009). "Sectoral Approaches to International Climate Policy: A Typology and Political Analysis." Harvard Kennedy School Discussion Paper 2009–02, January. Available at http://belfercenter.ksg.harvard.edu/files/2009_Meckling_Chung_Sectoral_Approaches.pdf (accessed September 29, 2014).

Ministry of Science and Technology (2009). Website of the Ministry of Science and Technology of the People's Republic of China. http://www.most.gov.cn/kjgh/ (accessed September 29, 2014).

Ministry of Science and Technology et al. (2007, June). "China's Scientific & Technological Actions on Climate Change." Available at http://www.ccchina.gov.cn/WebSite/CCChina/UpFile/File199.pdf (accessed September 29, 2014).

MIT (2014). "CCS Project Database." Available at https://sequestration.mit.edu/tools/projects/ (accessed September 29, 2014).

Morales, Alex and Jeremy van Loon (2009, August 6). "China Balks at Global Warming-Gas Capture Costs." *Bloomberg*. Available at http://www.bloomberg.com/apps/news?pid=newsarchive&sid=awnmd15dPMVU (accessed March 13, 2015).

Rai, Varun, Ngai-Chi Chung, Mark C. Thurber, and David G. Victor (2008, November). "PESD Carbon Storage Project Database." Program on Energy and Sustainable Development Working Paper #76. Available at http://pesd.fsi.stanford.edu/publications/carbon_storage_database (accessed September 29, 2014).

Ren, Xiangkun (2008, May 29). "Shenhua CTL (Clean Coal) Projects and CO_2 Sequestration." Presentation. Second U.S.-China Emissions Control Science & Technology Symposium in Hangzhou, China. Available at http://www.netl.doe.gov/publications/proceedings/08/CO2E/PDF/session%201/2008.5.Hangzhou%20CO2%20forum-EN.pdf (accessed September 29, 2014).

Reuters (2009, May 25). "China's Huaneng pushes carbon capture but costs bite." Available at http://www.reuters.com/article/2009/05/25/us-china-huaneng-carbon-capture-idUSTRE54O15Y20090525 (accessed September 29, 2014).

Schmidt, Jake, Ned Helme, Jin Lee, and Mark Houdashelt (2008). "Sector-based approach to the post-2012 climate change policy architecture." *Climate Policy* 8(5): 494–515.

Sheppard, Lynn (2009). "Communication on financing CCS and other clean carbon technologies in emerging and developing countries." European Commission 2009. Press release available at http://europa.eu/rapid/press-release_MEMO-09-295_en.htm?locale=en (accessed March 13, 2015).

Part V

Conclusions and implications

13 Major factors affecting the production, trade, and environmental impact of coal

Mark C. Thurber and Richard K. Morse

13.1 Introduction

Coal seems to present the world with a stark conflict between economic and environmental objectives. On the one hand, major developing countries in Asia see no obvious path to continued growth and development without the energy that coal provides (IEA, 2013). On the other, climate change mitigation objectives cannot be met if coal use continues on the trend that it has been following, with absolute growth exceeding any other energy source over the past decade. The tension between these two facts is a compelling reason to take a closer look at the factors that will shape coal's future.

Because the future growth in coal consumption will come from Asia, we focused in particular on the domestic and international markets that serve this region. We devoted special attention to China, whose scale of coal production and consumption dwarfs that of any other country. We also considered India, which is potentially the biggest source of future coal demand from the international market (as well as a large producer itself), along with three key suppliers of coal on the international market: Australia, Indonesia, and South Africa.

The studies presented in this book suggest that the amount of coal that is produced and consumed around the world, the observed patterns of international trade, and the environmental impacts of coal consumption depend importantly on three broad factors: (1) the geology and geography of coal-producing regions; (2) the policy environment for coal production and, critically, transport; and (3) the policy environment for coal consumption. These factors influence coal supply, demand, and, critically, price. (Factors influencing the price of alternatives to coal, such as natural gas, nuclear, hydropower, and intermittent renewables, also have an important effect on coal demand and price but are beyond the scope of this book.[1]) In the remainder of this concluding chapter, we discuss each of these categories in turn and consider the policy-relevant lessons we can learn from the ensemble of studies in the book. These include the following.

First, geography can be as important as geology when it comes to coal. Indonesia has vaulted to the top of global coal exporters in part because its inland waterways enabled it to build out transport infrastructure in modular fashion, bypassing some of the rail and port constraints that have beset other producers. Proximity to major East Asian markets also gave Indonesia an advantage relative to Queensland, Australia, which in turn had a transport distance advantage over New South Wales. Some coal deposits, such as those in the US Powder River Basin or China's Xinjiang province, are cheap to extract but far from major markets and largely stranded for now due to a lack of adequate transport infrastructure.

Second, coal industries flourish when governments act to reduce the risk of investment in mines and transport infrastructure and struggle when governments, intentionally or otherwise, act in a way that increases risk. The major exporting countries of South Africa, Australia, and Indonesia all saw their coal industries blossom following periods in which their respective governments were very focused on providing a low-risk investment climate. In recent years, the coal industries of all three countries have seen a significant worsening of their sectors' investment climates due to regulatory uncertainties created by their respective governments. In some cases, they may be victims of their own success, as their growth over many years has reduced government focus on industry development in favor of other goals such as income distribution, environmental protection, and increasing government revenues from their now prosperous coal industries. More broadly, volatile international prices and serious uncertainty around climate change policy are injecting uncertainty and risk into investment decisions.

Third, where governments demonstrate political will and policy focus, "regulatory workarounds" can sidestep institutional challenges – such as entrenched state-owned companies, ill-suited legal frameworks, or a challenging political economy – that would otherwise stifle coal industry development. The governments of China and India both deployed regulatory workarounds – including the establishment of "hybrid markets" having both controlled price and market price elements – aimed at bypassing bottlenecks associated with state ownership of coal mines and rail networks. On balance, China's efforts have been significantly more successful, although it is still struggling to find a perfect solution to the "coal-power conflict." Both Australia and Indonesia used direct government-to-business agreements to encourage coal sector investment in the initial stages of developing their respective coal sectors.

Fourth, exports and domestic use of coal should be viewed as complementary rather than in conflict. The coal industries of South Africa, Australia, and Indonesia all relied on export demand, most notably from Japan, to grow into the major players they are today. Domestic demand alone would not have yielded the large and dependable revenue streams needed to support major investment in these industries. A scarcity mentality that leads to policy restrictions on exports – as was observed in South Africa between 1950 and 1970 – reduces the incentive for the industry to invest and ultimately discourages the investments in mines and transport that make coal cheap and available for domestic users.

We studied two countries with very large production and domestic consumption: China and India. Of these, it is probably no coincidence that the country with the healthier coal industry and better ability to keep up with domestic demand, China, also exports some coal (and ships appreciable quantities of coal by sea from the north to the south of the country); India's uniquely high-ash coal makes exports infeasible.

Fifth, climate concerns are injecting additional uncertainty into coal production, trade, and use at the margins, but barring much more significant climate policy action they seem unlikely to put a major dent in Asia's coal consumption over the next several decades. Among the countries we studied, Australia was the most aggressive in implementing domestic climate policies, but it repealed its carbon tax in 2014 after only two years in operation. Even so, carbon policy uncertainty probably remains a deterrent to the construction of new coal-fired power plants in Australia. Carbon pricing in other major consuming markets in Asia has thus far been absent or too modest to significantly change patterns of coal use.[2] No major producing country has as yet proposed taxing the carbon content of the coal it exports. Plans to build coal terminals in the US Pacific Northwest – which could have an important impact on the global trade – were facing significant opposition on climate grounds at the time of writing, but their economic prospects look tenuous anyway at today's depressed coal prices.

Most major Asian consumers would like in principle to reduce both greenhouse gas emissions and local air pollution by burning less coal, but there are still few affordable alternatives to coal for much of Asia. Japan and Korea remain very interested in coal imports as a way of shoring up their energy security, and past experience suggests that the notional caps on coal output in China's five-year plans do not necessarily bite if the economy demands more coal.

Sixth, technologies that could reconcile coal use with climate mitigation goals – notably carbon capture and storage (CCS) – are *not* on track to play a major role because supporting policies (including, critically, carbon pricing) have thus far remained too weak even to propel demonstration projects at scale. Australia, as a high-income country with an economically important coal industry, should have a natural motivation to push forward CCS development. Indeed, the country established the Global CCS Institute (GCCSI) in 2009 to do just this. However, this early enthusiasm for CCS development in Australia has not translated since then into timely progress in implementation of pre-commercial demonstration projects that could advance learning and drive down costs.

China would be a high-impact location to scale CCS because of the pace at which it continues to build new coal-fired power plants. However, the country's focus on energy security and commercial advantage may incentivize some niche projects but not widespread deployment of CCS.

In the final section of this concluding chapter, we will speculate about possible future scenarios for worldwide coal production, trade, and consumption. We will focus in particular on how climate policy may affect coal markets. The studies in this book highlight the fact that this global outlook is strongly dependent on the local situations in key producing and consuming countries.

13.2 Geology and geography

The geology of coal deposits and the geography of where they are located critically shape patterns of production, trade, and use. To start with, the cost of extraction is strongly tied to geology. Coal deposits near the surface may be amenable to open-cut (also known as open-pit or opencast) mining, which tends to be significantly cheaper than underground mining – especially with the application of modern technology for removing overburden and moving coal in bulk. Indonesia's industry was propelled forward by its access to large deposits that could be produced with open-cut methods (see Chapter 7); Queensland overtook New South Wales as a producer based in large part on its ability to develop open-cut mines (see Chapter 6); US coal producers hope that open-cut mines in the Powder River Basin can form the basis of a competitive export industry, assuming new ports can be built (see Chapter 8); and China's coal future may involve tapping large coal deposits in Xinjiang through open-cut mines if water and transportation issues can be resolved (see Chapter 3).

The characteristics of the coal itself are also important. For example, a key characteristic of India's coal, and challenge for its industry, is that its coal has an extremely high ash content which makes it unsuitable for export (see Chapter 4). It is instead burned in power plants in India that are tailored to its high-ash character. As we discuss further in Section 13.3.3, the existence of a viable export market for India's coal industry over the course of its development might arguably have improved its profitability and general health, and in turn allowed it to more effectively meet India's domestic demand. By contrast, many countries with high-value coking coals built their early export industries around these resources before shifting to a greater focus on steam coal. For example, strong demand from Japan for coking coal helped catalyze the early-stage development of the coal export industries of both South Africa (see Chapter 5) and Australia (see Chapter 6) .

We have seen throughout this volume that transportation is a key link in the coal value chain, which makes geography as important as geology in determining which coal resources are produced. Shipping coal over long distances by rail can be an expensive proposition, in part because rail systems are rarely competitive and are frequently state-owned monopolies, as in China, India, South Africa, and, up until 2000, Australia (Queensland and NSW).

Fan (2009) has estimated that costs associated with rail logistics account for 30% to 60% of the cost of coal for power companies in China; such costs can include both official rail rates and "unofficial" payments to expedite the shipment with the railway monopoly (see Chapter 3). The cost of transportation poses a major obstacle to developing the massive coal resources in far western Xinjiang province. And even if rail transportation is economically feasible in theory, the institutional challenges of building and operating these railways could be one argument in favor of a "coal-by-wire" approach in which power plants are built near the mines and electricity rather than coal is transported east.

Other countries have seen their coal exports crimped by a lack of port capacity at the limited number of geographically favorable locations where deep-draft bulk carriers can dock. For example, Australia's Port of Newcastle became notorious for long vessel queues from 2004–2011, with several other Australian ports also affected, although port capacity added since 2010 has improved the situation (see Chapter 6). In South Africa, by contrast, port capacity expansions have left the capacity of state-owned railway Transnet as the major bottleneck for exports (see Chapter 5). In the face of sharply declining domestic demand, US producers would like to be able to ship more coal to Asia from the Power River Basin in Montana and Wyoming, but they were constrained at the time of writing by the lack of coal port capacity at the nearest feasible shipping points in Oregon and Washington (see Chapter 8).[3]

Due to favorable geography, Indonesia's coal industry has been able to almost entirely sidestep the port and rail challenges that have bedeviled other countries, which partly explains its meteoric rise to the position of world's largest coal exporter. Many of the nation's operating coal mines are located within a reasonable distance of navigable waterways, and their coal can be transported by barge directly to customers in Indonesia (or even as far away as the Philippines) or to floating transshipment facilities in calm waters offshore of Kalimantan and Sumatra, from which it is shipped onward to customers throughout Asia and to a limited extent in Europe and even the United States (see Chapter 7). New transshipment facilities can be built and installed at already approved anchorage areas in eighteen months or less at modest upfront costs relative to land-based ports. The ability to deploy transshipment facilities offshore of Kalimantan and Sumatra has enabled Indonesian coal producers to increase their export volumes seamlessly over the past two decades.

Indonesia also, of course, has the key geographical advantage of being closer to major East Asian coal consumers than Australia (not to mention even more distant alternatives such as South Africa and Colombia), which has enabled it to compete in part on shipping cost (see Chapter 6). Within Australia, Queensland has benefited from being closer to Asian markets than New South Wales but still suffers from a transport cost disadvantage relative to Indonesian coal exporters into east and south Asia and relative to South African coal exporters into south Asia.

13.3 The policy environment for coal production and transportation

While geology and geography set the basic context and potential for a country's coal industry, government laws and policies and, as importantly, the implementing rules and regulations in support of those laws and policies, play a critical role in determining the degree to which a country achieves its potential in coal production, and on what time frame. For example, India and China both have the coal resources to meet their domestic demand, but features of the policy, legal, and institutional environments in these respective countries have caused China's coal production to better keep pace with its consumption than India's in recent years. In fact, back in the 1970s, China's central planning policies came to be perceived by China's policy makers as a serious drag on coal production and in turn electricity generation and economic growth. China's government was able to put in place a series of reforms over time that opened the industry – previously dominated by large, central-government-owned mines – to smaller players, created stronger pricing incentives for production, and ultimately had a dramatic positive impact on output (see Chapter 2). (In recent years, China has reversed course to an extent and tried to close small mines as a strategy for boosting efficiency and improving safety.) At the time of writing, Indian Prime Minister Narendra Modi was reportedly contemplating reformist policy moves, including a break up of state-owned coal behemoth Coal India Limited (CIL), in an effort to boost coal output and avoid constraining electricity generation (Das, 2014).

Certain features of the policy and market environment for coal were common among the countries we studied. Notably, governments rarely seem to structure rail markets in a way that creates significant competition. Several of the countries we studied have state-owned railway monopolies (China, India, and South Africa), and where railways are private (Australia and the United States), there is still only limited competition.

Other policy and market structure variables vary widely across our cases. Countries fall on a spectrum with regard to their central planning versus market orientation, with China and India having the strongest history of central planning and Australia and the United States traditionally most inclined to rely on markets. The countries we studied also vary in their use of state-owned enterprises. In China, India, South Africa, and Indonesia, state-owned power companies are the main domestic customers for coal. Only India relies principally on a state-owned enterprise (SOE) – Coal India Limited – to produce coal, although China does lean on one very important SOE (Shenhua) in its coal sector.

13.3.1 Reducing risk or increasing it

As illustrated by the preceding comparison of China and India, a government's specific policy choices clearly matter – China, for example, found that it had to reduce SOE dominance in its coal sector in order to increase output. At the same time, one striking observation from the cases in our book is that countries can succeed in coal with policy, legal, and market frameworks that differ widely. Moreover, many countries have developed their coal industries quite effectively at some points in their history and rather unsuccessfully at others. The key determinant of success seems to be whether government regulations and other actions increase the risk of investment in coal mining and transportation or decrease it.

Typically, the governments in the countries we studied made effective coal sector development a priority only after the status quo had become obviously problematic. Australia, for example, asserted tighter regulatory control over the sector in the post-WWII period, after its war effort had been seriously hampered by labor problems and production shortfalls in the coal industry (see Chapter 6). After Soeharto took over power in Indonesia in the latter half of the 1960s, the imperative of stabilizing a faltering economy helped motivate his moves to improve the climate for foreign investment generally and to develop ironclad legal frameworks with foreign coal mining companies in particular. These took the form of

Coal Contracts of Work (CCOW) – legally binding agreements between foreign mining companies and the Government of Indonesia. These CCOW protected the foreign investors from changes in general laws until their expiration. (The most important of these CCOW, which are known as 1st Generation CCOW, and which laid the groundwork for Indonesia's later success in coal, are set to expire between 2019 and 2025 (see Chapter 7).)

As mentioned earlier, China opened its coal sector to local coal mines and introduced market pricing for some coal only after exclusive reliance on central state-owned coalmines selling at planned prices started to threaten economic growth (see Chapter 2). Because there can be a significant time lag from exploration through production and transport of coal, policy and regulatory measures to decrease investment risk may not boost output until years later. (And conversely, the effects on output of increasing investment risk in an actively producing coal industry may not be felt right away.)

Another common pattern was for countries to relax their focus on coal sector development after their industries had become successful. Australia's Commonwealth Government, for example, focused under the Rudd and Gillard administrations (2007–2013) on increasing government royalties from mineral extraction and reducing greenhouse gas emissions, in the process increasing the regulatory uncertainty and perceived investment risk in coal (see Chapter 6). In the cases of South Africa and Indonesia, the relaxation of the previous focus on coal sector development coincided with a shift to less authoritarian but at times more nationalistic governments, leading in some cases to a kind of populist resource nationalism. South Africa's post-apartheid government has prioritized diversification of ownership in the coal sector but sometimes failed to respond adequately to threats to the sector's overall health (see Chapter 5). Indonesia's government is likewise far more representative in the post-Soeharto era but also more dispersed among diverse constituencies that have aspirations ranging from resource nationalism ("kick out the foreign investors") to obtaining greater benefits from mining companies in the form of increased salaries, worker benefits, and taxes and royalties. This development has generally worsened the climate for making mining sector investments in Indonesia's coal sector (see Chapter 7). The point is not that such changes in governance and goals are bad, merely that, when they reduce investor returns and increase investors' perception of risk and uncertainty, they have the potential to reduce coal output over the longer term.

Governments may shift priorities from coal industry expansion to other goals and then back again, depending on which needs seem most urgent at a given moment. This has the potential to create a cyclical pattern in the industry. China encouraged small mines starting in the late 1970s to increase coal output, sought to close them down in the late 1990s because of their low efficiencies and poor safety records, increased dependence on them again amid the high economic growth in the 2000s, and more recently has refocused on closing them down (see Appendix, this volume). After focusing for some time on broadening ownership, South Africa's government may be shifting some focus back to integrated planning of mine, rail, and port development in the face of coal export shortfalls relative to potential, although the extent of the shift remains to be seen (see Chapter 5).

Investors in coal mining and coal transport increasingly have to contend with certain kinds of risk that were not present a decade ago, notably the threat of climate change regulation (which will be discussed further in Section 13.4 of this chapter) and volatile global prices for coal (see Chapter 6). Moreover, environmental groups that oppose coal on climate change grounds are actively trying to become "risk enhancers" through activism in local stakeholder processes such as environmental impact reviews. These groups have realized that they can deter investment in new coal-related projects through the threat of increased cost, delay, and uncertainty even if they are unable to get governments to expressly prohibit these projects (see Chapter 8). Of course, these tactics are mainly effective in democratic societies with open stakeholder processes and are unlikely to, for example, play a major part in curbing China's appetite for coal.

They also come with several risks. For one, they are not always guaranteed to produce the environmental ends they seek to achieve given the complexity of the global energy market (see Chapter 8). In addition, the strategy of fighting coal in local arenas which are ill-suited to address global climate change risks causing confusion, polarization, and ultimately a broader degradation in the investment climate that could spill over into non-coal-related areas. The chaos that could result from widespread local adjudication of climate disputes is one strong argument for implementing sensible carbon pricing at the national level.

13.3.2 The importance of policy and regulatory workarounds

Governments may be motivated to grow their coal industries but find that legacy aspects of the policy and regulatory environments as well as market arrangements are not conducive to this. In China and India, centrally planned pricing of coal and state-owned rail monopolies have historically posed obstacles to expanding domestic coal production. However, it has proven difficult for these two governments to raise coal prices to market levels once entrenched interests expect heavy discounts or to confront state rail monopolies that have a lot of political clout.

The governments of Australia's coal-rich states of New South Wales and Queensland faced the challenge of attracting investment in coal mines and transport infrastructure in the context of laws that were not suited to the task and could not easily be changed (see Chapter 6). Indonesia had the problem of how to convince investors that they would be protected in the country's relatively undeveloped legal and regulatory environment (see Chapter 7).

In all of these cases, governments sought policy workarounds that would let them achieve their goals given their particular policy and political constraints. China and India tackled their legacies of central planning in broadly similar ways, but with differing results. Below-market coal pricing meant that both governments were forced to come up with schemes to allocate coal to customers and then provide rail capacity to ship it. Both governments convened stakeholders along the coal value chain – China at an annual "coal conference" (see Chapter 3) and India on a "linkages committee" (see Chapter 4) – and attempted to facilitate a functional allocation of coal. However, as is typical with such non-market frameworks, both approaches have faced issues of corruption and shortages (see Thurber and Chang [2011] for analogues to this situation from the case of natural gas).

Both countries eventually pursued a similar solution to this allocation problem: allowing a portion of coal produced to be sold at market rather than planned prices. China has largely transitioned to market pricing of coal, whereas India's efforts remain more of a work in progress. China started allowing some coal to be sold at market prices in the early 1980s. The fraction of market-priced coal was increased in the 1990s, and since 2002 the government has no longer systematically controlled any coal prices (see Chapters 2 and 3). (When not held back by excessively powerful energy-consuming interests, such "hybrid markets" with both planned and market price tiers can develop a momentum of their own, incentivizing more and more production under the "market price" tier to the point where this segment can entirely subsume the planned part of the market.) India began an "e-auction" system in the early 2000s that offers a certain fraction of coal at market prices, but the vast majority of coal is still sold through the linkage system (in part due to the power of the state electricity boards buying the coal), which is one factor holding back expansion of coal production (see Chapter 4).

China and India have also pursued conceptually similar but practically different workarounds aimed at loosening the grip of large SOEs in mining and rail transport. Again, China's efforts started earlier and have yielded more results to this point. Most importantly, China opened up coal production to local state mines and town-and-village-enterprise (TVE) mines starting in the 1980s (see Chapter 2). It also created a large, vertically integrated company, Shenhua, with operations that spanned the value chain from mining through railways to power generation. The government hopes that the creation of vertically integrated "coal-power bases" modeled after Shenhua may circumvent railway constraints and solve the "coal-power conflict" – diverse producers selling coal at liberalized prices to large state-owned power companies obliged to sell power at fixed prices – in various other locations as well (see Chapter 3). However, most of these additional coal-power bases remain works in progress with uncertain long-term impacts.

India, for its part, has allowed "captive" miners into the coal sector – private players (for example, in the power and/or steel industry) that build entire value chains from mining through transport to end use consumption (see Chapter 4). India's encouragement of captive mining is somewhat analogous to China's promotion of vertical integration, such as the integration of Chinese state power company Huaneng upstream into coal mining (see Chapter 3). While the captive mining framework does allow some Indian coal to be produced that otherwise would not be, it is not nearly as scalable, for example, as China's opening of its own coal sector to local players starting back in the 1980s. One challenge is that captive miners must be able to actually consume the coal themselves and have the financial and management capability to build and operate an integrated supply chain. In addition, captive mining does not solve basic issues around land rights acquisition in India that significantly slow the pace of new mine development. For now, CIL still performs the vast majority of coal extraction.

Australia and Indonesia did not have the same legacies of significant coal production by SOEs under a system of central planning. Their challenge, instead, revolved around how to incentivize private (and often foreign) companies to make the large investments needed to grow the coal industry. Lacking suitable laws on coal licensing, the state government of Queensland in Australia developed in the 1970s a direct government-to-business agreement known as the Central Queensland Coal Associates (CQCA) Agreement Act. The CQCA codified not only exploration, development, and production terms but also fiscal and land use terms in a way that provided investors with legal protections in the event that subsequent changes to mining and general tax laws would adversely affect their economic positions (see Chapter 6). Indonesia also needed a way to make its commitments credible in the face of an immature and evolving legal system, which it did through "coal contracts of work" whose provisions superseded other laws – both past and future – under Indonesia's constitution (see Chapter 7). In fact, these contracts were patterned after the CQCA in Queensland. In the cases of both Australia and Indonesia, governments that were focused on creating the conditions for their coal sectors to succeed were able to find skillful policy workarounds compatible with existing legal and policy structures.

13.3.3 Policies on exports versus domestic use of coal

A number of countries that we studied made efforts at different times to restrict coal exports so that the resource was more available for domestic use, following a kind of scarcity logic. This proved to be shortsighted and counterproductive. As we see in this book, the real limitation on production of coal (or any other energy commodity[4]) is not the sheer quantity in the ground but rather the amount that can be profitably extracted and transported to customers given available technologies, infrastructure, and selling prices. In many countries, export revenues have in fact underpinned major investments in coal mines, transportation, and technology development that enabled these industries to extract more resources more efficiently – and in the process to better serve their domestic markets.

Concerns about domestic resource availability led South Africa to restrict exports and control coal prices between 1950 and 1970, but the result was anemic growth in coal production, with limited technological progress and poor financial performance. Performance improved and growth accelerated in the 1970s under a more favorable policy environment that included support for exports (see Chapter 5).

Indonesia's government initially aimed its nascent coal industry at providing fuel for planned domestic power plants that would replace oil-fired generators, but it did not initially attract much foreign investment in part because of the limited size of that market (see Chapter 7). The industry only took off after the government allowed exports and dropped the "import substitution" policy that had viewed coal narrowly as a substitute for foreign oil. In the face of a big current push to expand coal-fired electricity generation capacity in Indonesia (the "Fast Track Coal-fired Power Plant Program" of state-owned power company PLN), the government has put in place a new policy aimed at ensuring that coal is available for domestic buyers: the domestic market obligation (DMO).

To date the DMO seems to have been implemented in a fairly mild way, with limited impacts on coal producers and their export customers. But precisely how the policy will be implemented over the longer term remains unclear, and any steps that significantly disrupt the relationship between coal producers and their foreign customers could have serious negative ramifications for the health of Indonesia's coal industry. In Indonesia's case, domestic and export demand should instead be seen as quite complementary, with the possibility that new PLN demand for coal for domestic power stations could in fact help spur the development of Sumatra's vast low rank coal resources (see Chapter 7).

Being connected to the international market can provide coal industries with both a more diverse customer base through exports and a spur to greater efficiency through competition from imports. Though it became a net importer of coal in 2009, China continues to export, and as of late 2013 China's government was reportedly mulling the possibility of reducing barriers to exports to help producers in the face of domestic coal oversupply (*Reuters*, 2013). When coal prices are higher, on the other hand, domestic power companies are able to import more coal to keep input prices down and make domestic coal producers compete more aggressively (see Chapter 9).

In contrast, Coal India Limited is exposed to neither the rigors of direct competition from international coals, given the largely prescribed system through which it sells coal domestically at low prices, nor the upside incentive that could come from being able to export, since India's high-ash coal is unexportable (see Chapter 4). Over the long run, this state of affairs has likely had a negative effect on CIL's performance.

Facing low natural gas prices and an adverse regulatory climate, steam coal producers in the United States, and especially PRB producers, desperately need to be able to export more of their coal to Asia in order to thrive going forward (see Chapter 8). While there are no blanket policy prohibitions on such exports, there were many NGO players at the time of writing that were trying to increase the investment risks of the required new port developments rather than decrease them.[5]

13.4 Coal use and climate change policy

While many policy factors affect the attractiveness of coal relative to alternatives in end-use applications, the biggest shadow now looming over coal markets is the threat of more significant policy action around the world to combat climate change. In the introduction to this book, we asked whether climate change policy is likely to significantly curtail coal production and use, whether the structure and robustness of actual coal markets will instead derail progress toward climate mitigation goals, or whether technology will actually be able to reconcile coal and the climate. In the remainder of this section, we consider, first, what we can learn from how climate policy has affected coal markets so far and, second, the progress to date toward viable technological solutions for reducing the carbon dioxide emissions from burning coal.

13.4.1 Climate policy and the threat of climate policy

The impact on coal of greenhouse gas (GHG) emissions reduction policies actually in force around the world has thus far been muted, largely because only a few jurisdictions have such policies and they are mostly weak. We now consider a few such policies, which include carbon pricing mechanisms (a carbon tax or cap-and-trade system), emissions standards, restrictions on financing for coal, and notional caps on future coal consumption.

Australia's carbon tax – recently repealed at the time of writing – had a negative effect on domestic demand for coal within Australia and caused for the two years of its implementation a modest increase in retail electricity prices, although the time period for analyzing these impacts (July 2012–June 2014) was too short to draw any definitive conclusions. Such effects over time would presumably have led to an appreciable reduction in coal sales at home but, assuming the future carbon tax was maintained at or below the 2014 level, would have had only a marginal impact on the competitiveness of Australian coal in Asia's export markets. But even these modest potential impacts on exports now seem to be moot given the repeal of Australia's carbon tax and the collapse of the shaky political coalition that was behind the carbon tax in the first place (see Chapter 6).

Europe's Emissions Trading System (ETS) is probably causing only limited shifts in the electricity dispatch order from coal to natural gas given the low carbon prices observed at present. And to help compensate for the early retirement of nuclear capacity, Germany seems to be leaning on coal-fired power as its default source of new baseload capacity (see Chapter 8).

In the United States, proposed new source performance standards regulating CO_2 emissions under the Clean Air Act (EPA, 2014a) effectively dictate that no new coal power plants can be brought on line without carbon capture and storage (CCS) technology, which for now is cost prohibitive. The immediate effect of these rules will be limited, as new coal plants were already unable to compete with natural gas units in the US due to low gas prices. However, these regulations could have the effect of locking in this state of affairs and making sure that coal without CCS does not re-emerge as a viable generation option in the US even if natural gas prices go up in the future.

A newer and more ambitious Clean Air Act rule aims to reduce CO_2 emissions from existing power plants by creating state targets for emissions intensity reductions that could be achieved in part by closing existing coal plants, increasing coal fleet efficiency, and preferentially dispatching natural gas units over coal ones (EPA, 2014b). If it holds up through expected legal challenges, this rule could further increase the momentum away from coal in the United States.

At the same time, tighter restrictions on burning coal in the US may make some additional coal available in global markets (see Chapter 8). Understanding the exact magnitude of this kind of effect will require further development of global models of the coal trade such as the one described in Chapter 10, as well as a better understanding of the dynamics of substitution between coal and alternative fuels like natural gas.

The Obama administration's 2013 climate plan included a push to end public financing of new coal power plants abroad, which has picked up support from some European countries (Adragna, 2014) and multilateral agencies such as the Asian Development Bank and the World Bank. This initiative includes exceptions for the world's poorest countries (for which it mandates that the most efficient coal technologies be used) or for coal plants that incorporate CCS. International financing packages for South Africa's new Medupi and Kusile power plants (see Chapter 5) that were put together around 2010 and 2011 attracted significant controversy in the US and elsewhere (Goldenberg, 2010; Natter, 2014). But the South African case illustrates that keeping the electricity flowing in a developing country with critical energy shortages is still likely to take priority over climate issues, even while South Africa has conceded the eventual need for its participation in climate mitigation efforts (see Chapter 5).

China hoped to cap its coal output at around 3.8 billion tonnes per year during its 12th Five Year Plan. Although this cap indicates policy makers' recognition that massive coal consumption has negative ramifications for China, prominently including its effect on local environmental quality, there are reasons to doubt that such caps actually constrain consumption if economic trends create a sufficient demand for coal. First, there are large "gray" markets for coal in China that are not officially reported, and, second, planning targets for coal in the previous Five Year Plan were significantly exceeded (see Appendix, this volume).

At this point, the *threat* of future climate policy around the world, coupled with climate-related local actions by environmental NGOs, is probably affecting coal markets as much as the actual policies and laws that are in place, although still far less than would be required to halt the overall global growth in coal use (and associated CO_2 emissions).

The uncertainty that has swirled around Australia's climate regime ever since Kevin Rudd proposed pricing carbon in 2008 has probably had some effect – along with many other factors – in deterring investment in the country's coal industry (see Chapter 6). The recent repeal of the country's carbon tax may reduce some direct burdens on the industry, but it arguably does not eliminate this uncertainty and may even increase it, as there remain significant elements of Australian society that are pushing for action on climate. The US still lacks coherent federal climate policy for the moment, but environmental groups are capitalizing on general climate-related concern to fight coal-related projects on the local level, in the process sowing uncertainty among investors (see Chapter 8).

However, the fact remains that the largest markets for coal in Asia – both developing and developed countries – do not at this time appear very likely to implement aggressive climate policies. Japan has a small carbon tax on fossil fuels, and Korea launched a cap-and-trade market in early 2015, but both countries still seem to view coal as the most reliable and affordable source of baseload power, especially in light of stumbles in their respective nuclear industries and high natural gas prices in Asia (Iwata, 2014; Han and Humber, 2012). Developing countries have, quite understandably, been unwilling to constrain energy access and crimp the economic growth that will raise the living standards of their populations. In a scenario that assumes that all countries which have already announced climate policies follow through on them, the IEA projects coal use growing between 2011 and 2035 by 384 million tonnes in China, 507 million tonnes in India, and 269 million tonnes in ASEAN countries (IEA, 2013). Indian climate change expert Rajendra Pachauri has clearly expressed his view of the impossibility of India's not using coal given its limited alternatives and the fact that it still has 400 million citizens without basic lighting (see Chapter 4).

13.4.2 The potential of technology to reconcile coal use and climate change

The two studies in Part IV of this volume suggest that technologies to significantly reduce CO_2 emissions from coal combustion do not at present have the policy support that would enable them to play an important role in reducing greenhouse gas emissions. Australia, which has both a large coal industry and some inclination as a developed country to advance climate change mitigation, would appear to be among the most promising locations for "clean coal technology" development. Indeed, the establishment in 2009 of an Australia-based international organization, the Global Carbon Capture and Storage Institute (GCCSI), seemed to reflect seriousness, optimism, and policy momentum around developing carbon capture and storage (CCS) as a viable option for reducing GHG emissions from coal at scale.

However, most of the actual CCS demonstration projects that were on the drawing board at that time have since been canceled by their project proponents or have advanced very slowly (see Chapter 11). Without the looming threat of a strong carbon price or sufficient political support for generous funding of expensive CCS demonstrations, there is simply little motivating force to propel CCS technology forward.

Another technology – coal bed methane (CBM) extraction, which has allowed a reduction in Australia's GHG emissions by substituting CBM for coal – has succeeded in Queensland because it is profitable and because the government of Queensland provided strong policy and regulatory support for its implementation (see Chapter 11). Developers of CBM projects in NSW have been less fortunate. In NSW, the government has not provided the same level of regulatory support, and the CBM extraction industry is now in serious decline, with investors and developers abandoning NSW for other areas having more attractive CBM regulatory regimes.

Underground coal gasification (UCG), which could ultimately integrate with CCS, is struggling due to challenges of both environmental compliance and government policy support in both Queensland and NSW. But even in the case of UCG, the government of Queensland has provided stronger regulatory support than the government of NSW, which has made a difference in outcomes (see Chapter 11). In Queensland, there has at least been proof of UCG technology at the pre-commercial stage of development and tentative agreement within the government of Queensland on a pathway for implementing commercial scale UCG projects in a manner that will ensure protection of the local environment (see Chapter 11).

China, with world-leading CO_2 emissions attributable mainly to coal combustion, seems another logical breeding ground for clean coal advances. A few CCS projects in China have made some progress, but the nature of this progress reveals that the government's top priority is energy security rather than climate, with technology development and future commercial advantage being a secondary but also important motivator (see Chapter 12). Technologies that improve efficiency of coal use can make China's energy reserves go farther and have therefore been prioritized relatively more. This category includes ultra-supercritical and integrated coal gasification combined cycle (IGCC) technology. Technologies that would enable replacement of imported fuels, such as coal-to-gas and coal-to-liquid conversion, are also given significant attention. By contrast, post-combustion CCS, which is expensive and would further stress the coal supply chain because its energy penalty means more coal input is required, has moved slowly and attracted interest mainly as a technology development project rather than a near-term prospect for implementation at scale. Barring very significant funding support from foreign countries (which would be a politically difficult proposition in these countries), it seems unlikely that pre- or post-combustion CCS will be widely disseminated in China (see Chapter 12). However, collaborations with foreign companies that want to test and develop CCS technologies in China – in the process enhancing China's technological capability – might be a viable and mutually beneficial way to advance the state of the technology.

The core problem is that coal is popular in large part due to its low cost, while technologies to capture and sequester CO_2 remain staggeringly expensive. Until the world becomes serious enough about climate change to put in place stable frameworks that price carbon at a high enough level, it is unlikely that such technology will play a major role in the power sector.[6]

13.5 Alternative futures for coal

Whether we like it or not, coal will continue to be a major energy source around the world for a long time. Too many countries simply have too few affordable alternatives. That being said, the studies in this book hint at different possible ways that coal production, trade, and use could evolve, and the factors that will shape this evolution. In this final section we speculate on how possible futures for coal and the coal trade are affected by policy.

13.5.1 The future of the international coal trade

As of the middle of 2014, coal was relatively plentiful on the international market, and prices were far down from their 2008 and 2011–2012 highs. Whether plentiful coal will continue to be available at low prices on the international market over the longer term is less predictable and will depend heavily on the kinds of policy factors that we discussed in the country case studies of Part II.

For example, India's demand for imported coal is a major wildcard. Despite the country's ample reserves, India's coal production has not kept pace with demand due to a murky land rights regime, insecurity in major coal producing areas, and the bureaucratic encumbrances of state producer Coal India Limited (see Chapter 4). The country has increasingly turned to coal imports to fill the gap (see Figure 1.6 in Chapter 1). If India continues to grow its economy without resolving the issues that have held back coal production, it could absorb an increasing amount of internationally traded coal.

China's demand for coal imports is of a significantly different character. Although the increasing imports observed in Figure 1.8 in Chapter 1 might suggest that China's coal demand is outstripping its production capacity, the reality is different. In 2012, China consumed close to 3 billion tonnes of steam coal, far more than any other country (IEA, 2014). (The United States was in a distant second place, consuming a bit less than 800 million tonnes.) China's imports, while large enough to significantly influence the international coal trade, are small relative to its own consumption. Rather than importing because of fundamental production capacity limits, China's power producers import to arbitrage between international and domestic prices (see Chapter 9). For this reason, the increase in imports shown from 2009 through 2012 does not necessarily presage ever-growing Chinese demand in the international market.

There are uncertainties on the supply side as well. Indonesia has become the world's leading coal exporter by a significant margin when measured on raw tonnes exported (see Figure 1.8 in Chapter 1),[7] but in recent years the government has begun to intervene in more and more aspects of the coal market, and there is some concern that this trend, if it continues, could turn the country into a less reliable exporter (see Chapter 7). South Africa has been hampered by rail constraints and, at times, an uncoordinated approach to coal sector policy on the part of the government (see Chapter 5). The vast reserves of coal in the US Powder River Basin could likely provide significant, economically competitive supply to Asia (see Chapter 10), but only if stakeholder processes in the US allow the construction of new ports in the Pacific Northwest (see Chapter 8) – and if premium Newcastle coal prices return to their 2011–2012 levels.

If India's import demand continues to grow, developed countries such as Japan and Germany continue to use coal as their default baseload power alternative, and developing countries in Asia turn to coal in the absence of other affordable choices (Allchin, 2014), the international coal market could become tighter over the long term. Market tightness will only intensify if, for example, suppliers such as Indonesia have less to offer due to a more adverse regulatory environment for coal production and export, and US Powder River Basin coal ends up stranded due to environmental opposition.

13.5.2 Climate policy and coal

Climate change is a profoundly difficult policy problem. Carbon dioxide is an inescapable by-product of hydrocarbon combustion, and combustion is at the heart of a huge range of economically important activities. Mitigating these emissions or switching to alternatives to combustion is possible but in most cases very expensive. Moreover, because the climate change problem is global, policy solutions need to encourage expensive actions by a large number of countries with widely different capacities, resources, and interests, while minimizing "free riding."

We can consider two alternative scenarios for the trajectory of climate policy and what they might mean for coal. In the first, relatively little progress is made on the climate policy front over the next few decades beyond today's halting efforts at carbon pricing in a few locations. Some additional developed countries and subnational jurisdictions put in place relatively modest carbon prices, as exist today in Europe, parts of the United States (California and the US Northeast), and parts of Canada (most notably British Columbia and Québec). Carbon pricing does not spread to developing countries. Developed countries and regions within them continue to put in place politically popular "complementary policies" such as renewables supports, although not at the same level observed in the heyday of Europe's feed-in tariffs due to the huge costs that these entailed (for very little climate mitigation benefit).

In the second scenario, bottom-up climate policy actions around the world – and especially the use of flexible, efficient instruments such as carbon taxes or cap-and-trade systems with cap and floor prices – start to gain more momentum as policy makers grapple with the reality of climate change. Increasingly, jurisdictions learn from observing previous programs and get better at designing carbon pricing mechanisms that are relatively transparent, not open to manipulation, and predictable over time.

As more of these programs come into force, policy makers, industry, and the public realize that well-designed programs do not in fact wreck economies and bankrupt businesses, and that costs of mitigation are reduced significantly once businesses are given the flexibility and incentives to reduce emissions at least cost. In this scenario, developing countries take on carbon pricing too, just at much lower carbon prices than developed ones. This helps shift the global debate beyond whether countries should price carbon to how they should share the effort.

At first glance it might seem that the first scenario of relatively limited carbon policy action is best for the coal industry, but we actually posit that the second is preferable. In either of these scenarios, the coal industry will be needed for a long time. But in the first scenario, it will face growing policy and market uncertainties because of the widening gap between awareness of climate change and its impacts and what is actually being done about it. The state of US climate policy provides a good illustration of the problem.

Asia needs PRB coal, and the US coal industry would be more than happy to provide it. However, it is struggling right now to build coal ports, in part because the absence of coherent climate policy at the federal level encourages climate activism in all kinds of local environmental impact review processes not well equipped for considering the issue. (The federal policy now being developed by the EPA is itself needlessly complex, labyrinthine, and rife with uncertainty for industry, but it is the default alternative until the US Congress steps up to the task of putting in place a straightforward carbon price.) As long as climate policies are weak, the threat of stronger action in the future will continue to cast a long shadow of uncertainty over coal investments.

As the studies in this volume make clear, coal will be with us for many years, whether we like it or not. There are simply too many countries – in particular, Asian countries – that have too few alternatives for powering their economies. At the same time, the environmental toll of burning coal is undeniable. The first step is for developing countries to put in place and enforce well-designed regulations to reduce the local pollution impacts of coal. This is something economically tractable that we know how to do. The global climate change challenge is much more difficult, and yet it is a challenge we must face if we want to preserve a livable world.

Arguably, the best world for the coal industry is one in which governments everywhere focus on reducing the risk of coal extraction and transportation, *and* they all price carbon in a transparent and predictable fashion. In this ideal world, coal producers no longer have to fight battles over climate in every local region where they seek to build new infrastructure. They have long-term certainty about carbon pricing that can be seamlessly incorporated into business plans. (As the case of Australia illustrates, removing a climate policy does not mean removing the risk of future climate policy.) And the majority of coal is burned in the locations with the least ability to afford alternatives, which is mainly in poor and emerging markets in Asia, not in the US, Europe, and OECD Asia.

Notes

1. As illustrated by the decrease in coal use and greenhouse gas emissions in the United States following the "shale gas revolution," the competitiveness of natural gas relative to coal is one of the most important factors shaping patterns of energy use and greenhouse gas emissions around the world.

2. At the time of writing, China was running seven carbon market pilots in designated provinces and cities, Japan had in place a small tax on the carbon content of fossil fuels, and Korea had just launched a cap-and-trade system.
3. Ports in British Columbia already ship some PRB coal to Asia, but at present they are capacity-constrained and/or less economical than the proposed ports in Oregon and Washington.
4. For a vivid illustration of the fallacy of resource scarcity ideology in another context, that of oil, see Stern (2013).
5. Even apart from these environmental challenges, the adverse coal price environment of late has significantly increased investment risk (see Chapter 8).
6. In applications such as UCG, IGCC, or oxy-fuel combustion that produce concentrated streams of CO_2, the cost barriers to carbon capture may be lower but still appear prohibitively expensive for the foreseeable future (see Chapter 11).
7. Because Indonesia's coal exports have increasingly been dominated by low-energy coals, the increase in the country's actual energy exported in the form of coal has not grown as fast as the export of raw tonnes (see Chapter 7).

References

Adragna, Anthony (2014, March 26). "Netherlands joins bid to end financing for coal-fired plants to spur green growth investment." *Bloomberg BNA*. Available at http://www.bna.com/netherlands-joins-bid-b17179889127/ (accessed July 17, 2014).

Allchin, Joseph (2014, March 4). "Bangladesh's coal delusion." *New York Times*. Available at http://nyti.ms/1n63Ylg (accessed March 3, 2015).

Das, Krishna N. (2014, May 21). "Modi exploring breakup of Coal India, opening up sector – sources." *Reuters*. Available at http://in.reuters.com/article/2014/05/21/india-coal-india-idINKBN0E10KB20140521 (accessed July 13, 2014).

EPA (2014a). "Standards of performance for greenhouse gas emissions from new stationary sources: Electric utility generating units." *Federal Register* 79 FR 1429. January 8, 2014.

EPA (2014b). "Carbon pollution emission guidelines from existing stationary sources: Electric utility generating units." *Federal Register* 79 FR 34829. June 18, 2014.

Fan, Bi (2009). "Thoughts on comprehensive solution of coal and power disputes" *(guanyu xitong jiejue meidian maodun de silu)*. Macroeconomic Management 8.

Goldenberg, Suzanne (2010, April 8). "World Bank's $3.75bn coal plant loan defies environmental criticism." *The Guardian*. Available at http://www.theguardian.com/business/2010/apr/09/world-bank-criticised-over-power-station (accessed July 17, 2014).

Han, Sangim and Yuriy Humber (2012, April 13). "Nuclear halt in South Korea seen boosting coal: Energy markets." *Bloomberg*. Available at http://www.bloomberg.com/news/print/2012-04-13/nuclear-halt-in-south-korea-seen-boosting-coal-energy-markets.html (accessed July 17, 2014).

IEA (2013). *World Energy Outlook 2013*. Paris: International Energy Agency.

"IEA Coal Information Statistics," International Energy Agency (IEA), 2014. Available at http://www.oecd-ilibrary.org/energy/data/iea-coal-information-statistics_coal-data-en (accessed July 17, 2014).

Iwata, Mari (2014, March 27). "Japan's answer to Fukushima: Coal power." *Wall Street Journal.*

Natter, Ari (2014, July 14). "Environment groups oppose Ex-Im Bank reauthorization bill over coal policy tweak." *Bloomberg BNA*. Available at http://www.bna.com/environmental-groups-oppose-b17179892305/ (accessed July 17, 2014).

Reuters (2013, October 12). "China mulls lifting coal export tax in 2014: paper." Available at http://www.reuters.com/article/2013/10/12/us-china-coal-idUSBRE99B01X20131012 (accessed March 8, 2015).

Stern, Roger (2013, February). "Oil scarcity ideology in U.S. national security policy, 1909–1980." Program on Energy and Sustainable Development Working Paper #105.

Thurber, Mark C. and Joseph Chang (2011). "The policy tightrope in gas-producing countries: Stimulating domestic demand without discouraging supply." Advance summit paper for the Pacific Energy Summit, Jakarta, Indonesia, February 21–23, 2011.

Part VI

Detailed information on the coal value chain in China

Appendix: A statistical review of coal supply, demand, and transport in China

Kevin Jianjun Tu

A.1 Introduction

Coal currently accounts for about 30% of the world's primary energy consumption, second only to oil. In China, coal is the most dominant fuel, representing almost 70% of the country's primary energy consumption in 2011 (IEA, 2013) and 93% of its proven reserves of fossil fuels (BP, 2010). In 2010, China's reported coal output of 3.24 billion tonnes (Gt) represented nearly half of the world total.[1] China's heavy reliance on indigenous coal, notably for more than 80% of its electricity generation, brings benefits in terms of energy security (NBS, various years-a). The conversion of coal to chemicals, liquid fuels, and synthetic natural gas can allow an even greater reliance on its domestic resources than possible in many other countries. The Chinese government has even encouraged state-owned energy companies to invest in coal mines outside of China to secure external coal supplies that would further enhance China's energy security.

Coal has undoubtedly played a key role in China's rapid economic growth since the Chinese economy was opened to the outside world in 1978; nevertheless, the unprecedented exploitation and utilization of coal has also created enormous environmental and social challenges such as (1) coal mining wastes and land subsidence-related environmental impacts in many areas of the country; (2) severe water shortages in major coal mining regions such as Shanxi, Shaanxi, and Inner Mongolia; (3) mounting safety challenges as more than 250,000 miners have died in China's numerous coal mine accidents since 1949;[2] (4) bottlenecks in China's transport infrastructure (especially rail); (5) deteriorating air quality in many urban centers and adjacent regions; and (6) greenhouse gas (GHG) emissions abatement pressure especially after China overtook the United States as the leader in CO_2 emissions in 2006.[3]

Historically, China has been a net coal exporter because of a large reserve base, low labor costs, attractive export prices, and government tax incentives. In 2003, China's coal exports peaked at 94 million tonnes

(Mt) with net coal exports at 83 Mt. Since then, spiking domestic coal demand coupled with, at times, central government restrictions on coal exports has led to a decrease in exports (NBS various years-b). In 2009, the prolonged dispute on contract thermal coal prices between China's major coal enterprises and power companies and attractively priced coal from the international markets (see Chapter 9) turned the country into a net coal importer.

The world hard coal trade is dominated by seaborne transport (as opposed to overland transport via rail), which accounted for 90% of world coal exports in 2013 (see Figure 1.3 in Chapter 1). Given that the volume of world seaborne coal trade is less than one-third of China's annual coal production, a moderate shift of China's coal import/export balance has the potential to disrupt the international coal trade.

This study examines the industrial organization of the Chinese coal industry. The objectives of this study are to (1) comprehensively review the Chinese coal value chain and (2) identify problems and policy challenges in the Chinese coal industry.

A.1.1 Historical evolution of the Chinese coal industry

While planning is a key characteristic of centralized, communist economies, a plan established for the entire country normally contains detailed economic development guidelines for all regions, and the five year plans of China are a series of such economic development initiatives. As China has transitioned from a planned economy before 1978 to an increasingly market-oriented one in recent years, the name for the 11th Five Year Plan was officially changed to "guideline" instead of "plan." Nevertheless, it is still commonly referred to as a plan.

Since 1949, the governance structure of the Chinese coal industry has witnessed continual change (see Table A.1). The former Ministry of Coal Industry (MCI) has been created and abolished, several times. Since 1978, state-owned coal mines became encumbered by heavy welfare obligations to their bloated workforces and millions of retired workers. Unable to meet the burgeoning demand for domestic coal, Beijing had to encourage private investment in the coal industry. As a result, the share of coal production by township and village enterprises (TVE) grew from 15.4% in 1978 to 46.2% in 1995 (CCII, various years). In 1998, the MCI was finally abolished. With the emergence of the National Energy Administration (NEA) under the National Development and Reform Commission (NDRC) in 2008, the era of the centralized governance for the Chinese coal industry has been permanently ended. In 2010, the creation of the National Energy Commission was yet another attempt

Table A.1 *Chronicle of the Chinese coal industry*

Period	Major event	Average annual production (Mt)	Average annual fatalities	Output % of TVE mines
Recovery Period (1949–1952)	The PRC was founded in 1949. The Administration of the Coal Industry (ACI) was created under the Ministry of Fossil Fuels (MFF).	48.7	530	4.5
1st FYP (1953–1957)	Number of Greenfield mines under construction reached 194, with total capacity of 75.37 Mt. Number of completed mines reached 205, with total capacity of 63.76 Mt. Setting the target of raising annual national coal output to 131 Mt, which was the exact production level in 1957.	98.5	700	4.6
1955	MFF was abolished and the ACI was upgraded as the Ministry of Coal Industry (MCI).	98.3	677	4.8
2nd FYP (1958–1962)	The Great Leap Forward started in 1958 and ended in 1961. As a result, the national coal production target was revised upward to an unrealistic level of 900 Mt. Not surprisingly, overreporting became rampant across the country, and the national economy entered a period of chaos.	306.6	4,120	6.8
1958	15 major coal-producing provinces established regional ACIs.	270.0	2,662	12.8
Adjustment Period (1963–1965)	The coal industry did not recover until 1965 when the output grew again.	221.1	1,261	4.1

Table A.1 *(cont.)*

Period	Major event	Average annual production (Mt)	Average annual fatalities	Output % of TVE mines
3rd FYP (1966–1970)	Setting the target of increasing national coal production capacity by 68.06 Mt, and the coal production actually grew as high as 7.1% annually during this period. Nevertheless, the political turmoil across the country soon forced the military to take over the management responsibility of the coal industry in August 1967.	259.3	1,848	5.9
1966	The beginning of the Cultural Revolution, which resulted in long-lasting and detrimental impacts on the Chinese economy as a whole	251.5	1,478	2.3
1970	The MCI was abolished, and the Ministry of Fuels and Chemical Industry (MFCI) was created.	354.0	2,903	8.9
4th FYP (1971–1975)	Setting the target of raising national coal output to 400–430 Mt by 1975. As the average annual production growth rate was 4.2% during this period, this target was easily met, which indicates that the coal industry actually still functioned well in spite of the political turmoil.	423.0	3,837	11.3
1975	The MCI was recreated.	482.2	4,526	13.1

5th FYP (1976–1980)	Setting the target of establishing 8 large-scale coal production bases.	581.6	5,325	16.2
1976	The end of the Cultural Revolution, and the Chinese economy soon took off thereafter.	483.5	4,826	14.8
1978	The Chinese economy was opened to the outside world.	617.8	5,830	15.4
6th FYP (1981–1985)	The start of the resurgence of the TVE mines in China.	732.8	5,534	25.7
1983	The State Council issued the Notice to Encourage the Private Sector's Investment in the Mining Industry, which marked the beginning of the "let the water flow" policy.	714.5	5,431	23.8
7th FYP (1986–1990)	Setting the target of raising annual national coal production level to 1,000 Mt. With the private sector's rapid development, this target was met in 1990, when national output reached 1,080 Mt.	987.1	7,015	35.7
1988	The Ministry of Energy (MOE) was created, and the MCI was abolished. The creation of three key central level coal enterprises.	979.9	6,751	35.9
8th FYP (1991–1995)	The private sector's market share peaked at the expense of the state-owned enterprises, tax revenue losses, mounting environmental degradation, and a daunting safety record.	1,190.8	5,817	42.1
1993	The MOE and the three central level coal enterprises were all abolished. The MCI was recreated. Since 1993, coal produced by TVE and private mines has been freely sold in the market. But the former State Planning Commission still issued guidance prices for coal supplied to power plants, and negotiators at the national coal-ordering conference simply used these in place of the earlier benchmark prices.	1,150	5,152	42.5

Table A.1 *(cont.)*

Period	Major event	Average annual production (Mt)	Average annual fatalities	Output % of TVE mines
1995	The market share of TVE mines peaked at 46.2%.	1,361.0	6,222	46.2
9th FYP (1996–2000)	In 1997–98, the ongoing national campaign to close small coal mines formally started, with detrimental impacts on statistical collection.	1,373.0	6,049	38.3
1998	The MCI was downgraded as the State Coal Industry Bureau (SCIB) under the State Economic and Trade Commission (SETC), and authority over key state-owned mines was devolved to local governments.	1,332.0	6,304	41.9
10th FYP (2001–2005)	Serious underreporting of national coal production occurred.	1,866.0	6,272	34.6
2001	The SCIB was abolished and most of its functions were handed over to the State Administration of Work Safety (SAWS). The State Administration of Coal Mine Safety (SACMS), under SAWS, oversees safety at coal mines.	1,384	5,670	23.9
2004	At the 2004 coal-ordering conference, the guideline price for utility coal was removed by the NDRC, but power prices remained fixed while coal prices soared.	2,123	6,027	38.0
11th FYP (2006–2010)	The coal production target set for 2010 is 2.6 Gt, which was already exceeded in 2008.	2,882	3,362	37.2
2006	The former Energy Bureau of the NDRC worked with the National Bureau of Statistics (NBS) to revise national coal	2,529	4,746	38.3

	statistics between 1999 and 2004. Shanxi's first resource consolidation campaign: setting minimum coal mine capacity as 0.3 Mt/annum.			
2007	In Jan. 2007, the 11th FYP for coal industry was issued by the NDRC.	2,692	3,786	38.0
2008	The National Energy Administration (NEA) replaced the Energy Bureau of the NDRC.	2,802	3,215	36.6
2009	Shanxi's second resource consolidation: setting minimum coal mine capacity as 0.9 Mt/annum. A prolonged dispute on contracted thermal coal prices between major coal enterprises and utilities. The complete deregulation of coal prices at the central government level.	2,973	2,631	38.2
2010	The creation of the National Energy Commission is another attempt by Beijing for national coordination of China's energy sector. NBS revised coal statistics between 1997 and 2008.	3,413	2,433	35.2
12th FYP (2010–2015)	The coal production target set for 2015 is 3.9 Gt, which is expected to be exceeded.	N/A	N/A	N/A
2013	The State Electricity Regulatory Commission (SERC) was merged into the NEA.	3,645	1,067	N/A

Source: TFSUC (2009).

Coal production statistics in recent years are based on revised figures from the NBS, but the shares of TVE mines are based on original statistics from various sources including CCII, CCTD, CNCA, and SAWS. The only exception is the production data of 2010, which is based on data from CCTD. In comparison, production data from the NBS in 2010 was 3,240 Mt, which is significantly lower than the figure of 3,413 Mt from CCTD. "FYP" stands for Five Year Plan. Sources: Tu (2007), Thomson (2003), CCII (various years), NBS (various years-b, a), IEA (2009a), SETC (2000), SXCOAL, and CCTD.

by Beijing to nationally coordinate China's energy sector. In 2013, the State Electricity Regulatory Commission (SERC) was merged into the NEA, which reflects the Chinese leadership's desire to increase regulatory efficiency within the energy sector. Nevertheless, the management responsibility of the Chinese coal industry is still fragmented among many governmental agencies.

A.1.2 Current state of government policy and regulations

A.1.2.1 Review of the 11th Five Year Plan (FYP)

China's 11th Five Year Guidelines (or Plan), issued in March 2006, set out the objectives and overarching principles of the country's continued modernization drive between 2006 and 2010. Under the overarching guidelines, detailed plans for each sector were prepared by the relevant government departments. The 11th FYP for the coal industry was issued by the NDRC in January 2007. The plan defines major tasks of the coal industry as follows: (1) to optimize distribution of coal production capacity across different mining districts, (2) to adjust and control coal output and resource capacity, (3) to establish thirteen large-scale coal production bases (see Chapter 3), (4) to nurture large-scale coal mining groups, (5) to consolidate and retrofit small- and medium-scale coal mines, (6) to encourage scientific innovations in the coal industry, (7) to improve coal mine safety, and (8) to establish resource-efficient and environmentally friendly mining districts.

The plan devotes an entire chapter to environmental impact assessment, which indicates Beijing's increasing awareness of the ecological degradation caused by coal mining activities. In addition, the NDRC set detailed control targets for coal production, mine capacity development, and small coal mine consolidation for each coal-producing province. Nevertheless, the actual national coal production level in 2008 (2,802 Mt) already exceeded the output control targets for 2010 (2,600 Mt) set by the NDRC, which indicates the difficulty of regulating an increasingly market-oriented industry through command-and-control style plans.

A.1.2.2 Outlook of the 12th Five Year Plan

The essence of the 12th Five Year Plan of the Chinese coal industry is the promotion of orderly, industry-wide harmonious development with due consideration of the ecology and environment. As a result, capacity expansion is expected not to be treated as the sole target of sector development. Instead, operational efficiency should be increased to alleviate coal's environmental impacts, and the safety record needs to be further improved. The experience gained during Shanxi's coal resource consolidation is expected to be introduced to the rest of the country. In other words, small-scale private coal mines in the rest of the country are likely to be either merged with state-owned enterprise (SOE) mines or forcefully shut down. SOE mines will be expanded to improve operational efficiency of the coal industry. Not surprisingly, production restrictions, closure of small coal mines, and sector consolidation are the three main themes of the 12th FYP.

In addition, the Chinese government plans to stabilize coal output in central China, and state investment is expected to be discontinued in regions in east China that are unsuitable for the development of large-scale coal mines. Meanwhile, closure of small coal mines will be implemented in east China.[4] While overcapacity has become a concern for the Chinese coal industry after the 2008/09 economic slowdown, coal production capacity with appropriate safety configurations remains insufficient. As a result, safety standards and enforcement are expected to be further strengthened especially for small coal mines.

A.1.2.3 Review of coal industry policy

The Chinese government has established the rule of law as fundamental to the country's long-term transformation from state control of the economy to a market economy within a state planning framework. The legal and regulatory system for the coal industry has been reformed along these lines. A multiplicity of laws, regulations, and standards currently in force deal with six core aspects of the coal industry: (1) resource administration, (2) safety supervision, (3) environmental protection, (4) industry administration, (5) energy conservation, and (6) emissions control. However, some existing laws, regulations, and standards are considered to be outdated. Given the rapid development of the industry, the mechanisms and capacity required for ensuring effective enforcement of these laws, regulations, and standards are sometimes inadequate, which leads to severe sustainability challenges.

The multiple laws, regulations, and standards listed in Table A.2 aim to cover all aspects of the coal value chain and deal with various environmental issues simultaneously. However, though the coverage of these laws and policies is quite extensive, the ambiguous wording and the lack of supporting economic instruments, specific regulatory standards (e.g., on point source or ambient emissions or technologies), and technical support often make them difficult to be implemented at the local government levels. Other problems with the existing laws,

Table A.2 *Coal industry policy in China*

Type of policy	Title
Laws and regulations	Environmental Protection Law
	Atmospheric Pollution Prevention and Control Law
	Solid Waste Pollution Prevention and Control Law
	Environmental Impact Assessment Law
	Energy Conservation Law
	Circular Economy Promotion Law
	Cleaner Production Promotion Law
	Coal Law
	Mineral Resources Law
	Water and Soil Conservation Law
	Land Administration Law
Administrative orders and standards	Decisions of the State Council on Implementing the Scientific Development Concept to Strengthen Environmental Protection
	Opinions of the State Council on Promoting the Sound Development of the Coal Industry
	Special Regulation on Land Reclamation, Provisional Measures for Environmental Protection in the Coal Industry, Emission Standard for Pollutants from the Coal Industry
	Cleaner Production Standard for the Coal Industry, Emission Standard of Coal Bed Methane/Coal Mine Gas (trial version)
	Guideline of Environmental Impact Assessment for Coal Mining Projects.
Other relevant government policies	11th Five Year Plan on the Prevention of Acid Rain and Sulphur Dioxide Pollution
	11th Five Year Plan on the Development of the Coal Industry, Coal Industry Policy
	Opinions (of the NDRC) on Emission Reductions and Energy Conservation in the Coal Industry
	Opinions (of the NDRC) on Coal Mine Methane Control and Utilization, Technical Policies for Sulphur Dioxide Emission Prevention and Control
	Opinions of the NDRC on the Experiment of Sustainable Development in the Coal Industry in Shanxi
	9th Five Year Plan on Cleaner Coal Technology and Development Outline in 2010
	Preferential Policy on Power Generation Demonstration Projects with Cleaner Coal Technologies

Source: TFSUC (2009).

regulations, and standards include (1) individual policy is often governed by too many agencies (compromising the effective implementation of government policy and resulting in inadequate monitoring) and (2) insufficient support and encouragement for enabling technologies key to the sustainable development of the Chinese coal industry.

A.1.3 Appendix structure

The author presents China's coal resources and reserves in Section A.2, followed by an analysis and projection of China's coal demand to 2030 in Section A.3. Section A.4 details coal transportation by mode. In Section A.5, the analysis focuses on grey coal markets and associated coal statistical distortion. In Section A.6, China's increasing importance in the global climate change regime and the future of the Chinese coal industry are discussed.

A.2 Coal supply

A.2.1 Coal resources and reserves

As illustrated in Figure A.1, 89% of China's gross coal resources are located in twelve provinces (or municipalities or autonomous regions) to the west of the Daxing'anling-Taihangshan-Xuefengshan series of mountain ranges; these provinces include Shanxi, Shaanxi, Inner Mongolia, Ningxia, Gansu, Qinghai, Xinjiang, Sichuan, Chongqing, Guizhou, Yunnan, and Tibet. Twenty provinces (or municipalities or autonomous regions) to the east of the listed mountain ranges account for only 11% of gross coal resources in China.

Similarly, 93.6% of gross coal reserves are located in eighteen provinces to the north of the Kunlunshan-Qinling-Dabieshan series of mountain ranges; these provinces include Beijing, Tianjin, Hebei, Liaoning, Jilin, Heilongjiang, Shandong, Jiangsu, Anhui, Shanghai, Henan, Shanxi, Shaanxi, Inner-Mongolia, Ningxia, Gansu, Qinghai, and Xinjiang. Only 6.4% of gross coal reserves lie in fourteen provinces to the south of these mountain ranges. Not surprisingly, the unbalanced resource distribution dominated by geologic conditions largely determines the pattern of coal transportation in China, which generally follows routes from the north to the south and from the west to the east (TFSUC, 2009).

China's coal resources cover a land area of 600,000 km^2, encompassing 6% of the country's 9.6 million km^2. There are five major coal endowment districts in China: northeast, north China, south China, northwest, and Tibet and Yunnan. The China Geological Survey reported that China's inferred coal resources total 5,555 Gt (MLR, 1999). In 2006, the Ministry of Land and Resources (MLR) reported that China's total coal reserves stood at 1,160 Gt across 7,965 mining

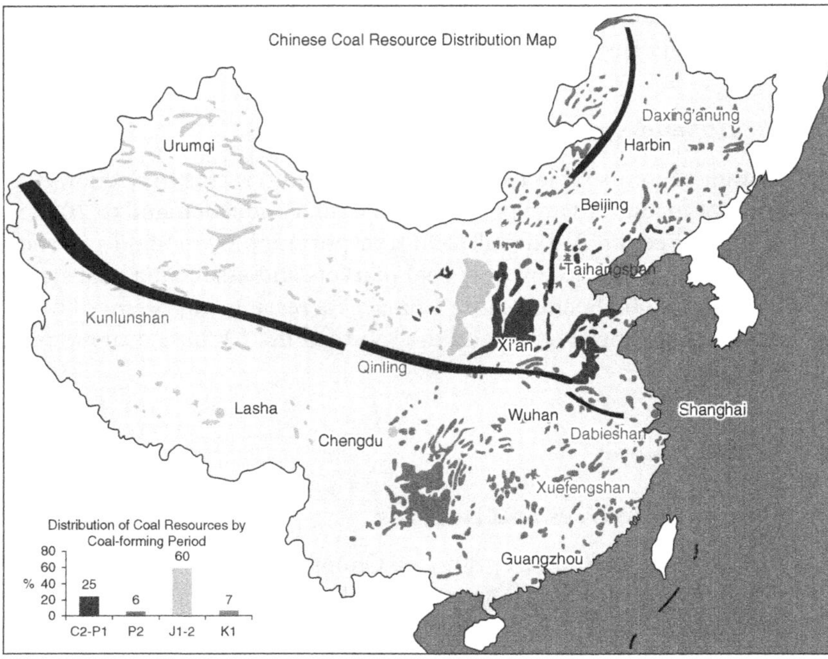

Figure A.1 Chinese coal resource distribution map. Source: Adapted from Ministry of Land and Resources, 1999.

districts, comprising 334 Gt of "basic reserves" and 826 Gt of "prognostic reserves." China's coal resource classification is considered to be incompatible with international standards. BP's statistics have not updated China's proven coal reserves at 114.5 Gt for many years, which is significantly lower than the figure of 182.5 Gt in Table A.3 (Huang et al., 2008, BP 2010).

Figure A.2 shows China's coal resource distribution by type of coal.[5] Lignite accounts for 13.5% of China's coal resource. Inner Mongolia and Yunnan alone possess 76.5% and 13.8% of China's lignite resources, respectively. Low-rank bituminous coals are primarily concentrated in coalfields in the Erdos Basin, Datong, south Liaoning, and Xinjiang. Coking coals are primarily produced in Shanxi, Hebei, Shandong, north Jiangsu, north Anhui, Heilongjiang, Liaoning, Xinjiang, Guizhou, Sichuan, Hunan, and Jiangxi, and Shanxi alone accounts for 55% of China's proven coking coal reserves. Meager coal and anthracite are produced in Shanxi, Guizhou, Yunnan, south

Table A.3 *Coal resources and reserves in China at the end of 2006 (in Gt)*

Planning area (province/ municipality/autonomous region)	No. of mining districts	Proven reserves	Basic reserves*	Prognostic reserves*	Total reserves*
Jing-Jin-Ji (Beijing, Tianjin, Hebei)	277	2.70	7.69	9.57	17.25
Northeast (Liaoning, Jilin, Heilongjiang)	1,208	5.45	14.45	17.57	32.03
East China (Jiangsu, Zhejiang, Anhui, Fujian, Jiangxi, Shandong)	1,440	11.01	25.38	31.17	56.54
Central south + Hainan (Henan, Hubei, Hunan, Guangdong, Guangxi, Hainan)	1,576	8.37	15.79	17.12	32.91
Jin-Shang-Meng-Ning (Shanxi, Shaanxi, Inner Mongolia, Ningxia)	1,320	126.51	220.16	532.15	752.31
Southwest (Chongqing, Sichuan, Guizhou, Yunnan)	1,454	18.25	29.04	62.93	91.96
Xin-Gan-Qing + Tibet (Gansu, Qinghai, Xinjiang, Tibet)	690	10.25	20.98	155.80	176.77
Total Coal	**7,965**	**182.54**	**333.48**	**826.30**	**1,159.78**
Coking Coal		**64.3**	**124.2**	**150.3**	**274.8**

* Resources.
Note: Basic reserves are based on relatively extensive geological exploration activities; they are suitable for resource extraction in the near future or longer term. The proven coal reserves are the portion of basic reserves that can be economically recovered in the near future. Prognostic reserves are primarily based on geological inferring and modeling instead of extensive exploration activities. Total reserves = Basic reserves + Prognostic reserves.
Sources: Huang et al. (2008) and CCTD.

Hunan, west Beijing, Fujian, Jiangxi, Guangxi, Yunnan, and Ningxia (Editorial Board of CCCI, 1999).

A.2.2 Coal production

A.2.2.1 Historical production

There are only thirteen major coal mining districts or coalfields in China that are suitable for surface mining operations, and coal suitable for surface mining represents only 4.1% of China's ensured reserves.[6] In the mid-1990s, surface mining accounted for only about 3% of national coal production (Wu, 1996). This ratio increased slightly to 4% in 2001

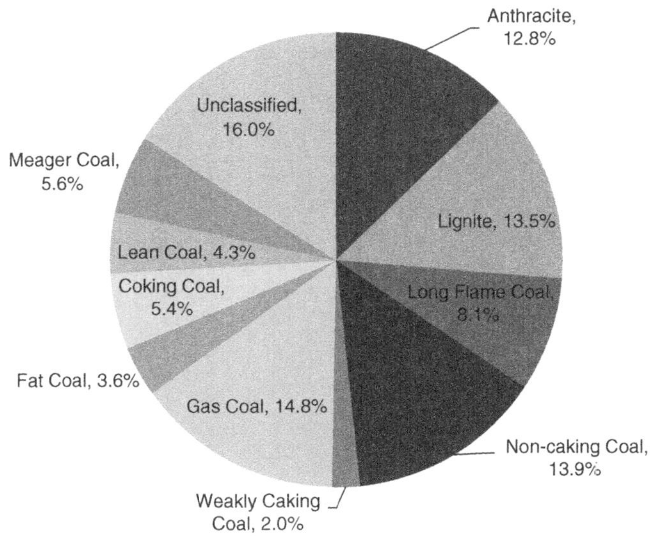

Figure A.2 China's coal resource distribution by type of coal. Source: Editorial Board of CCCI (1999).

(Fan et al., 2003) and about 5% to 8% in recent years. Though the Chinese government encourages the development of large-scale surface mines, coal output from surface mines is unlikely to exceed 10% of national production in the near future due to resource constraints.

As illustrated in Figure A.3, coal has been China's most dominant fossil fuel since the inception of the PRC in 1949, when coal accounted for 99.2% of China's fossil fuel output. In 1978, coal's share of China's total fossil fuel production was lowered to 72.6% due to the rapid development of the oil industry in the previous decade. In 1990, the baseline year of the Kyoto Protocol, coal represented 78% of China's total fossil fuel output. In 2009, coal's share of China's total fossil fuel production rebounded to 85% due to (1) the loss of oil self-reliance in 1993, making energy security a primary concern for Chinese decision makers (greatly benefiting domestically abundant and low-cost coal); and (2) the economic boom in the past three decades (requiring enormous amounts of energy that are difficult to meet by any alternative energy production, transport, and distribution infrastructure other than coal).

Figure A.4 presents China's coal production by type of coal. Anthracite is primarily used in the fertilizer and coal chemical industries in China. In 2009, anthracites accounted for 18.5% of national coal production.

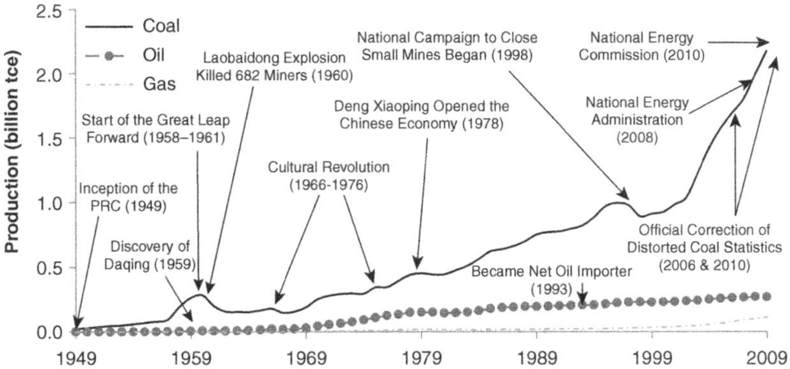

Figure A.3 History of fossil fuel production in China, 1949–2009. Sources: NBS, various years-b and CCTD.

Note: Coal output data between 1999 and 2004 was first revised by the National Bureau of Statistics in 2006. Coal output data between 1997 and 2008 were revised again by the same agency in 2010. Production unit is tce (tonnes of coal equivalent), 1 tce = 29.31 GJ.

Lignite is suitable for electricity generation and coal gasification. Lignite production in China increased rapidly from 2.7% of national coal production in 1990 to 12.3% in 2009. Finally, bituminous and sub-bituminous coals accounted for 69.2% of national coal output in 2009.

A.2.2.2 Coal production bases

Under the 11th FYP, thirteen large coal bases were planned (see Chapter 3). Together these thirteen large coal bases include ninety-eight mining districts as listed in Table A.4. They cover a land area of 103,388 km^2, with rich reserves at 853 Gt of a broad range of coal types, generally of high-quality, good mining conditions, and the ability to support large-scale mining operations and ancillary facilities. Most large--scale Greenfield coal mines in China are expected to be added in these bases. In principle, each coal base allows the existence of only one primary developer. Given that many coal mining enterprises often operate at the same coal base, the emphasis on the single developer concept indicates that resource consolidation will continue to be the overarching policy of the Chinese coal industry in the years to come. In 2010, output at these thirteen coal production bases reached 2.8 Gt, accounting for 87.5% of the national total. During the 12th FYP period, Xinjiang was promoted

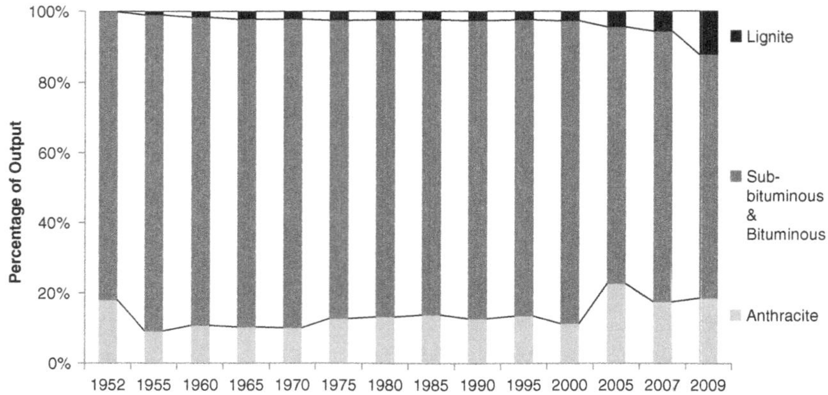

Figure A.4 Coal production by type of coal in China, 1952–2009. Sources: CNCA (2006), CCII, and industrial sources.

from a coal reserve base into the fourteenth coal production base in China.

A.2.3 Industry structure and organization

A.2.3.1 Type of enterprises

There are three types of coal mines in China.

Key SOE mines, once managed by the former MCI, are currently supervised by provincial governments except for Shenhua Group and China Coal Energy Company. This category accounted for half of national coal output in 2009. Coal production and sales of Shenhua Group, the largest state-owned coal mining company in China, reached 328 Mt and 357 Mt in 2009.[7] In comparison, Peabody Energy Corp., the world's largest private-owned coal mining company, produced only 172 Mt of coal in the same year (Peabody, 2009 and US EIA, 2009). Moreover, Shenhua's fatality rate was 0.027 death/Mt of coal in 2009, significantly lower than national average at 0.86 death/Mt of coal in the same year. Not surprisingly, the Chinese government is trying to promote the Shenhua model (see Chapter 3).

The numerous township and village enterprises (TVE) in the Chinese coal industry are mines invested in and controlled by agricultural collective economic organizations or peasants. They have caused many regulatory challenges such as environmental degradation, tax evasion, and mining accidents; their number once exceeded 100,000 in 1991 but has

Table A.4 *List of 13 coal production bases in China*

Planning zone	Coal base	Province	Acreage (km^2) / reserves (Gt)	List of mining districts (no. of mining districts)	Major rail lines
Coal export zone	Shendong	Inner Mongolia, Shanxi, Shaanxi	9,000 / 146.6	Shendong, Wanli, Zhungeer, Baotou, Wuhai, Fugu (6)	Shen-Shuo-Huang Line, Long-Hai Line
	Shaanbei	Shaanxi	14,440 / 109.8	Yunshen, Yuheng (2)	Zhengzhou-Xi'an section of Long-Hai Line, Hou-Xi Line, Xi'an-Nanjing Line, Xi-Kang Line
	Huanglong	Gansu	3,262 / 15.7	Shanchang (inc. Yonglong), Huangling, Xunyao, Tongchuan, Pubai, Chenghe, Hancheng, Huating (8)	Ning-Xi Line
	Jinbei	Shanxi	5,278 / 58.0	Datong, Pingshuo, Shuonan, Xuangang, Hebaopian, Lanxian (6)	Da-Qin Line, Feng-Sha-Da Line, Ji-Tong Line
	Jinzhong	Shanxi	8,372 / 917	Xishan, Dongshan, Fenxi, Huozhou, Liliu, Xiangning, Huodong, Shixi (8)	Shi-Tai Line, Tai-Jiao Line
	Jindong	Shanxi	8,209 / 76.7	Jincheng, Lu'an, Yangquan, Wuxia (4)	Hang-Chang Line, Tai-Jiao Line
	Mengdong (Northeast)	Inner Mongolia, Northeast	11,723 / 68.4	Zalainuoer, Baorixile, Yimin, Dayan, Huolinhe, Pingzhuang, Baiyinhua, Shengli, Fuxin, Tiefa, Shenyang, Fushun, Jixi, Qitaihe, Shuangyashan, Hegang (16)	Bin-Sui Line, Sui-Jia Line
	Ningdong	Ningxia	1,100 / 8.8		Ning-Xi Line

Table A.4 *(cont.)*

Planning zone	Coal base	Province	Acreage (km^2) / reserves (Gt)	List of mining districts (no. of mining districts)	Major rail lines
				Shizuishan, Shitanjing, Lingwu, Yuanyanghu, Hengcheng, Weizhou, Majiatan, Jijiajing, Mengcheng (9)	
Coal import zone	Lianghuai	Anhui	6,863 / 27.1	Huainan, Huaibei (2)	Jing-Hu Line, Huai-Nan Line
	Luxi	Shandong	4,500 / 18.6	Yanzhou, Jining, Xinwen, Zaoteng, Longkou, Zibo, Feicheng, Juye, Huanghebei (9)	Jing-Hu Line, Yan-Shi Line
	Henan	Henan	7,794 / 19.4	Hebi, Jiaozuo, Yima, Zhengzhou, Pingdingshan, Yongxia (6)	Zhengzhou-Guangzhou section of Jing-Guang Line, Luoyang-Shimen section of Jiao-Liu Line
	Jizhong	Hebei	2,556 / 6.5	Fengfeng, Handan, Xingtai, Jingxing, Kailuan, Yuxian, Xuanhuaxiahuayuan, North Xhangjiakou, Pingyuan Large Coal Fields (9)	Shijiazhuang-Guangzhou section of Jing-Guang Line
Self-reliance zone	Yungui	Yunnan, Guizhou	20,191 / 43.5	Panxian, Puxing, Shuicheng, Liuzhi, Zhina, Qianbei, Laochang, Xiaolongtan, Zhaotong, Zhenxiong, Enhong, Junlian, Guxu (13)	Nan-Kun Line, Zhu-Liu Line, Qian-Gui Line

Sources: Acreage and resources are based on Wu and Wang (2006) and http://www.chinacoal.org.cn/coaljjyx/266/18786.aspx [2011–02–11].

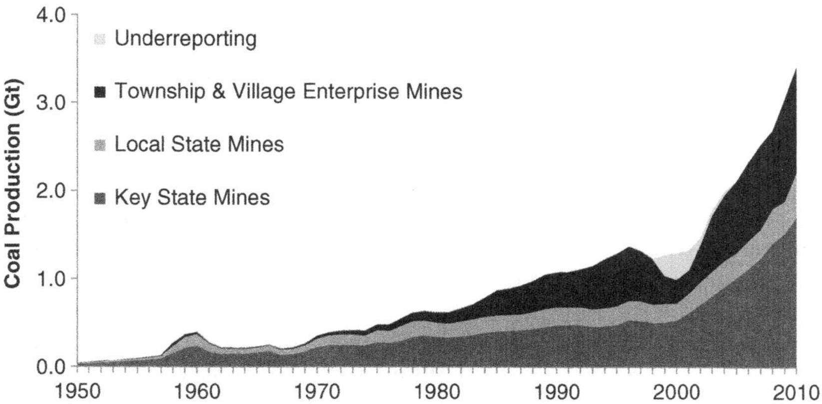

Figure A.5 Coal production by enterprise type in China, 1949–2009. Underreporting represents coal statistical revisions made by the NBS in 2006 and 2010, which has not been disaggregated by the NBS in terms of type of coal enterprises. Sources: Thomson (2003), CCII (various years), SXCOAL, and CCTD.

been reduced significantly thereafter (CCII various years). TVE mines accounted for about 38% of national coal output in 2009.

Finally, local SOE mines are owned and supervised by the local instead of the central government; their market share shows a declining trend in recent years. In 2009, local SOE mines accounted for only about 12% of national coal output.

Due to the Chinese government's continuous policy initiatives to close small coal mines and consolidate the coal industry, the ownership boundary among different types of mines has been blurred by the establishment of joint ventures (either de facto or de jure) between SOE mines and private investors, which is especially applicable to local SOE mines. As a result, the statistical collection and regulations of the Chinese coal industry are expected to gradually shift away from the ownership-focused system towards a capacity-based one, which is evidenced by the fact that reporting of coal production statistics by enterprise type has been discontinued in recent years. The policy preference for large-scale mines with capacity above 0.9 Mt/annum has successfully increased the production share of this category from 30% in 1991 to 46% in 2008. They are expected to become increasingly dominant at the expense of small-scale mines (CCII, various years).

A.2.3.2 Cost comparison of different producers

Table A.5 provides a coal production cost breakdown by type of enterprise in Shanxi. In recent years, production costs of all types of coal enterprises increased rapidly primarily due to the growth of wages, fuel prices, royalties, and other government taxes.

Generally speaking, key SOEs have the highest production costs due to (1) relatively higher technological, environmental, and safety standards maintained by key SOE mines; and (2) a bloated workforce in many key SOE mines inherited from the central planning era. In comparison, TVE mines have the lowest production costs owing to (1) poor technological, environmental, and safety standards at most TVE mines; (2) lowest wage rates; and (3) common practices of underreporting coal output.

Because of TVE mines' lower production costs, they have noticeable competitive advantages against SOE mines, and TVE mines can be profitable with significantly lower retail coal prices than SOE mines. Nevertheless, their lack of access to the national railway network has weakened their ability to market to coal end users in remote regions.

A.2.4 The role of government

Government supervision and administration of the Chinese coal industry is mainly performed through licensing and regulation of market entry, work safety, trading and processing, and other activities in the sector, as illustrated in Figure A.6. According to Article 24 of the Coal Law 1996 and the Rules on the Administration of Coal Mining Licenses, a coal mine must obtain a coal mining license before commencing operation. At present, coal mining licenses are mostly issued by provincial development and reform commissions or coal industry administrations. Agencies responsible for licensing differ from place to place; not every province has a coal industry bureau and other agencies assume responsibility. Issuance of coal mining licenses is restricted to the provincial level; county and city agencies are responsible for only initial or partial examination of applications (IEA, 2009a).

The importance of government regulation in the coal mining industry is also reflected in the taxes and fees imposed on coal enterprises. The production costs of Chinese coal enterprises have increased significantly in recent years due to (1) higher taxes and fees imposed on coal mining enterprises, (2) the promotion of new pricing mechanisms for the acquisition of coal mining rights, (3) the establishment of the mechanism of coal mine environmental recovery deposit (a type of environmental fee), and

Table A.5 *Coal production cost breakdown by type of enterprise in Shanxi (in yuan/tonne)*

Type	Cost	2000	2001	2002	2003	2004	2005	2006	2007
Key SOE mine	Average retail prices	129.26	143.23	160.42	173.57	226.68	296.2	309.58	330.65
	Unit cost and expenditure	132.16	139.74	155.7	168.26	191.96	247.6	257.63	286.88
	Raw coal production costs	69.01	72.38	80.92	85.46	121.9	156.3	159.93	179.31
	Profit	−2.9	3.49	4.72	5.31	34.72	48.6	51.95	43.77
	Taxes	2.62	2.79	3.29	3.43	3.62	6.16	6.93	7.66
Local SOE mine	Average retail prices	61.96	78.80	103.76	109.03	107.30	206.10	195.68	
	Production costs	45.76	62.29	75.71	86.09	74.98	116.82	152.28	
	Energy costs	10.06	10.36	10.35	13.19	10.84	10.58	15.74	
	Other materials	15.70	13.28	14.18	20.48	17.51	20.24	20.24	
	Labor	17.34	26.98	27.88	29.81	30.95	30.31	29.24	
	Investment	2.66	11.67	23.29	22.61	15.68	55.68	87.05	
	Interests	4.56	2.97	2.94	4.25	6.84	11.47	1.67	
	Taxes	6.08	7.97	10.82	13.04	12.98	24.59	26.44	
	Other fees	3.67	5.00	5.94	4.98	10.00	10.00	9.24	
TVE mine	Average retail prices				149.34	192.26	177.89	154.93	
	Production costs				100.03	115.63	94.03	100.15	
	Energy costs				12.24	13.67	12.00	13.23	
	Other materials				25.63	27.15	23.14	20.27	
	Labor				34.32	41.65	40.24	45.50	
	Investment				20.91	24.63	19.56	15.72	
	Interests				6.23	12.73	39.39	4.27	
	Taxes				10.55	15.16	17.53	24.52	
	Other fees				4.07	4.86	4.45	4.64	
	Expenditures for town and villages				1.53	10.76	5.59	15.13	

Sources: Wang and Horii, 2008.

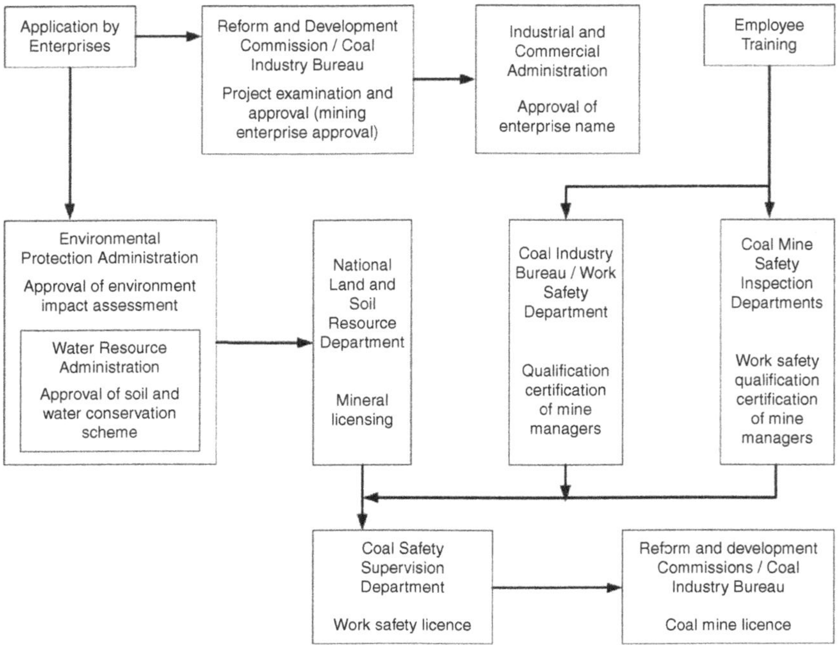

Figure A.6 Coal mine licensing procedures. Source: IEA (2009a).

(4) the improvements in the social insurance system for employees at industrial and commercial enterprises.

A.2.5 Reference coal production trajectory

In the past, China appears to have maintained a short-term planning function whereby detailed projections for coal demand are provided for the coming years, but it is extremely rare to see projections for coal use in each subsector over longer time periods from official sources. Even when such information is available, benchmarking past projections with actual statistics generally indicates very poor track records in terms of providing accurate and meaningful forecasts.

China is struggling to reconcile its centralized planning tendencies with a growing market-oriented economy with "socialist characteristics." One example of central planning is the government's preference for setting coal production control targets. China's coal production control target in 2010 was set as low as 2.6 Gt in the 11th FYP for the coal industry. Over the medium to longer term, the government projections then assumed that coal consumption in 2020 will increase to 2.9 Gt (Minchener, 2007).

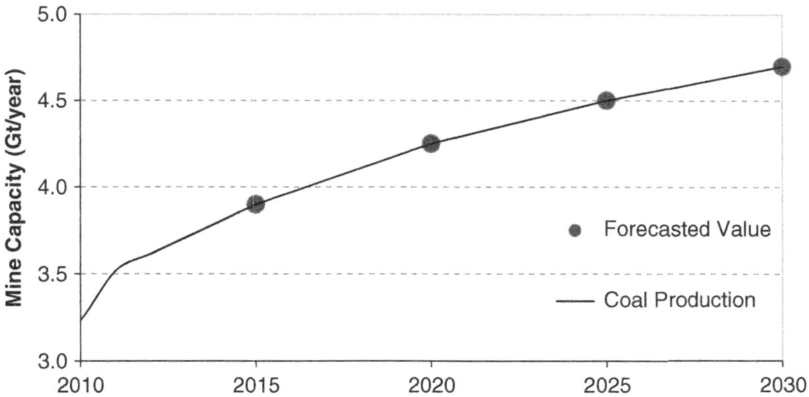

Figure A.7 Reference coal production trajectory in China to 2030. Reference coal production forecast in this study is based on expert opinions solicited during the author's interviews with Chinese academics and officials between 2009 and 2011 and the author's own estimation.

China's coal production control target in 2015 was set at 3.9 Gt in the 12th FYP for the coal industry. Given that China's coal production had already reached 3.87 Gt in 2014, the coal production control target will be exceeded again during the 12th FYP period.

Figure A.7 presents the reference trajectory of China's coal production growth to 2030. Though the Chinese government has a very strong political desire to control the national coal output level to around 3.8 Gt/annum by the end of the 12th FYP period, coal output in China may be significantly higher than the reference trajectory under the following two scenarios: (1) the existence of sizable grey coal markets in China (i.e., the official coal output statistics in recent years have been seriously underreported, which is evidenced by NBS's continuous upward revisions of national energy statistics) and (2) the Chinese government's ambitious planning targets are unable to effectively prevent a domestic coal consumption spike (similar to what happened during the 11th FYP period, when the original 2010 coal production control target of 2.6 Gt/annum was far exceeded later by an actual coal output level of 3.24 to 3.41 Gt/annum).

A.3 Coal demand

A.3.1 Overview of coal consumption in China

China first overtook the United States as the leading coal consuming country in 1986 with an annual consumption representing 21% of the world total. After that, China's coal demand increased by 251%, reaching 3.24 Gt in 2010. In comparison, global coal consumption only grew by 55% during the same period. In 2010, China's coal demand accounted for 48% of the global total and was 20% higher than the combined coal consumption of the remaining top ten coal-producing countries (BP, 2011).

At the inception of the PRC in 1949, coal accounted for more than 90% of China's primary energy consumption. Since then, this ratio gradually decreased to 86.5% in 1965 and 70.7% in 1978 due to the rapid development of the Chinese oil industry. Since the start of China's open-door policy era, coal's share of national primary energy consumption fluctuated over time, reaching 70.4% in 2009 (NBS, various years-b).

Figure A.8 shows China's coal consumption by sector over time. In 1980, electricity, residential, and iron and steel were the three largest coal-consuming sectors in China, accounting for 21%, 19%, and 9.8% of national coal consumption. In 2009, electricity, iron and steel, and non-metal minerals were the three largest coal-consuming sectors, accounting for 49%, 9.0%, and 7.9% of national coal demand. Since 1980, national coal consumption in China increased at 5.6% annually, reaching over 3 Gt in 2009. In comparison, coal consumption for power generation increased more rapidly, at 8.7% annually during the same period, reaching 1.44 Gt in 2009.

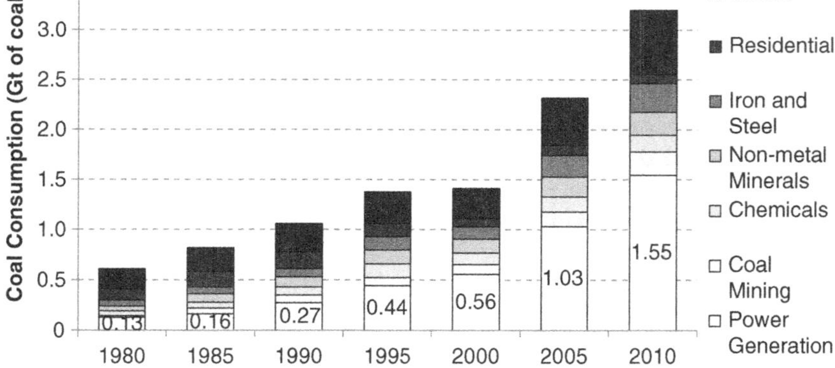

Figure A.8 Coal consumption by sector in China. Sources: NBS (various years-a) and CNCA (2006).

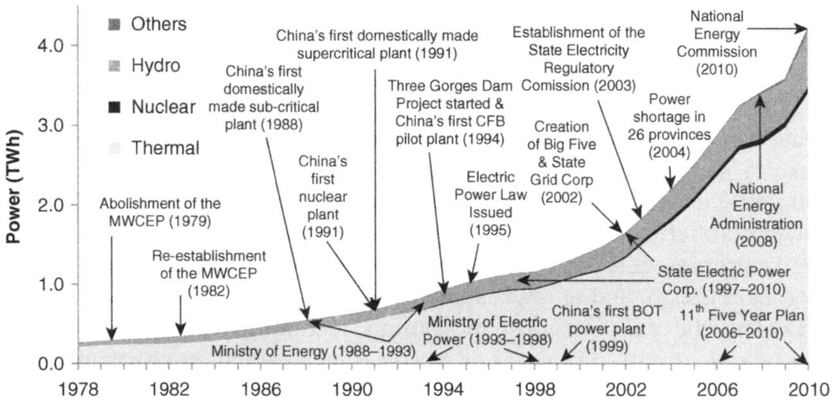

Figure A.9 China's electricity generation by fuel and major events, 1978–2010. Sources: NBS (various years-b) and CEC (various years).

A.3.1.1 Power generation

Since 1978, China's rapid economic development has propelled its power demand to growth of 9% annually. Due to coal's dominance in China's energy sector, the thermal power sector has grown at an almost identical pace as the power industry as a whole. Thermal power's share of national electricity generation has been maintained around 80% between 1978 and 2010 (NBS, various years-a). In the Chinese context, the dominance of coal, the high cost of oil, and lack of available natural gas make coal-fired generation the vast majority of all thermal power (China classifies both coal and natural gas-fired power generation as thermal power).

Figure A.9 presents coal consumed by the Chinese power sector and associated energy efficiency trends over time. Since 1978, coal consumption for power generation in China has increased at 7.6% annually, reaching 1.44 Gt in 2009. In comparison, thermal power increased at more than 8.9% annually during the same period. The lower growth rate of coal consumption can be largely explained by energy efficiency gains achieved by the power industry. For instance, average coal-fired generation efficiency in China has increased from 26% in 1978 to 36% in 2009, the equivalent of 38% improvement (NBS, various years-a). In addition, industrial statistics show that the rate of electricity distribution loss was lowered from 9.64% in 1978 to 8.06% in 1990 and then from 7.81% in 2000 to 6.55% in 2009, the equivalent of a 32% improvement on energy efficiency during the study period (CEC, various years, and NBS). If the Chinese power sector was still operating at the 1978 efficiency level, additional coal amounting to 543 Mt would have been wasted in 2009. As a result, energy conservation is expected to continue to be the top priority in the Chinese electricity industry in years to come.

The average size of new coal-fired units in China is about 500 MW, while increasingly larger units of up to 1.0 GW are commissioned that are close to state-of-the-art in their class. The rapid pace of construction, coupled with closures of small, less-efficient units, means that average power plant efficiency in China is rapidly catching up with that of developed countries. It is estimated that the average efficiency of China's coal-fired fleet is expected to approach 40% by 2030 (IEA, 2009a).

During the past decades, indigenous technological capacity of the Chinese electricity industry has become increasingly sophisticated. In October 1988, China's first domestically made subcritical power generator was commissioned at Pingwei Power Plant in Anhui. In June 1992, China's first 600 MW subcritical generator was commissioned at Shidongkou II Plant in Shanghai. In November 2004, the first indigenous 600 MW supercritical generator started to operate at Qinbei Power Plant in Henan. In fewer than two years, China's first ultra-supercritical generator (1.0 GW) was commissioned at Huaneng Yuhuan Power Plant in Zhejiang.

In 2002, the State Council issued a Circular for the Reform Plan for the Electric Power System, which officially kicked off the "Plant-Grid Separation" reform in the Chinese electricity industry. Under this plan, the generation segment of the former State Power Corp., which included both power plants and electric grids all over mainland China, was divided into the so-called "Big Five" power generation groups: China Guodian Corporation (Guodian), China Datang Corporation (Datang), China Huaneng Group (Huaneng), China Huadian Corporation, and China Power Investment Corporation (CPI).

The transmission and distribution segment of the former State Power Corporation was divided into the State Grid Corporation, headquartered in Beijing with its five regional subsidiaries, and China Southern Power Grid Corporation, headquartered in Guangzhou. In addition, four auxiliary groups were established: (1) China Power Engineering Consulting Group Corporation (CPECGC), (2) China Hydropower Engineering Consulting Corporation (CHECC), (3) Sinohydro Corporation (Sinohydro), and (4) China Gezhouba (Group) Corporation (CGGC).

In the generation segment of the Chinese electricity industry, competition was introduced following the Plant-Grid Separation reform in 2002. By the end of 2008, there were 4,300 electricity production enterprises in China with capacity above 6 MW, and SOE and state-held share enterprises accounted for 90% of the total (SERC, various years).

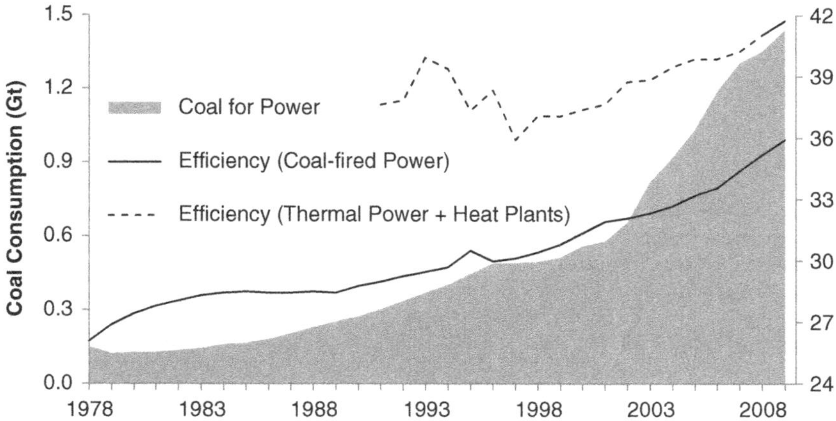

Figure A.10 Coal consumption and efficiency trends at power plants, 1978–2009. Sources: NBS (various years-a), CNCA (2006), Editorial Board of CEIYB (various years), and CEC (various years).

A.3.1.2 Iron and steel industry

The Chinese iron and steel industry has expanded significantly in the past, reflecting a boom in construction and infrastructure development, as well as strong demand for consumer durables. Iron and steel outputs in China were 105 Mt and 95 Mt in 1995. Since then, they have increased by 347% and 425%, reaching 471 Mt and 501 Mt in 2008, respectively. China is currently the world's largest iron and steel producer. In comparison, Japan, the world's second largest producer, produced only 119 Mt of crude steel in 2008. In 1995, coke input to the Chinese iron and steel industry was 65 Mt. Since then, it has increased by 226%, reaching 255 Mt in 2008. In comparison, iron production in China has increased by 347% during the same period. As a result, the coke ratio at key Chinese iron and steel mills has dropped by 27% since 1994, reaching 392 kg of coke per tonne of iron in 2008, which indicates a significant resource efficiency gain (NBS, various years-b, -a; Chinese Society of Metals).

The energy and feedstock intensity improvements witnessed by the Chinese iron and steel industry can be largely explained by (1) commissioning of large-scale state-of-the-art iron and steel mills and retrofitting of the existing ones, (2) higher levels of recycled steel production, and (3) reliance on merchant producers for coke production, since the average capacity of merchant producers has been increased significantly by the

Table A.6 *Energy intensities of key Chinese iron and steel enterprises (in kgce/tonne)*

	1980	1990	2000	2005	2008		
Category	average	average	average	average	worst	average	best
Comprehensive energy intensity for steel production	1,646	1,274	920	876	783	630	557
Comparable energy intensity for steel production	1,285	1,017	781	714	842	610	504
Sintering*	105	80	69	65	91	55	39
Pelletizing*	74	63		40	49	30	17
Coking*	217	185	160	142	335	120	61
Iron Smelting*	556	526	466	456	531	428	364
Converter*	67	38	29	36	38	6	–13
Electric Furnace*	378	303	266	201	177	82	37
Steel Rolling*	266	187	118	89	238	59	27

* Data in 1980 and 1990 are based on all iron and steel enterprises. Sources: Cai (2002) and Wang (2009).

government's mandatory closure of small-scale and backward coking ovens in recent years.

Table A.6 suggests that energy intensities of key Chinese iron and steel enterprises have improved rapidly. Between 1980 and 2008, average comprehensive energy intensity for steel production in China has improved by 62%. In comparison, fuel economy of major iron and steel manufacturing processes generally shows less significant improvements, which suggests that part of the energy efficiency gains for steel production in China is related to an increased level of steel recycling.

Energy efficiency levels of Chinese iron and steel enterprises differ significantly among themselves. While average energy intensity for steel rolling was 59 kilograms of coal equivalent (kgce)/tonne in 2008, the best operators only required half of the aforementioned energy to roll one tonne of steel. In comparison, energy consumed by the least efficient enterprise was more than 300% higher than an average producer. Currently, configurations of the best Chinese producers can be considered to be very close to best available techniques, and Chinese iron and steel enterprises with the worst fuel economy performance lag far below national average standards and thus are the targets for closure, triggered by the government's mandate to improve energy intensity of the Chinese economy (Wang, 2009).

A.3.1.3 Chemicals

To coordinate the development of the Chinese coal chemical industry, the NDRC developed a Draft Plan for Coal Chemical Development in China in 2006. According to this unpublished draft plan, cumulative investment of the Chinese coal chemical industry is expected to reach more than 1.0 trillion yuan between 2006 and 2020, with 50% of expenses on equipment and 10% on technology development. In addition, coal to a liquid (CTL), dimethyl ether (DME) for gasoline substitution, coal-to-olefin, and coal-based methanol are expected to grow to 30 Mt, 20 Mt, 8.0 Mt, and 66 Mt by 2020, respectively (NDRC 2006).

Since the circulation of the draft plan by the NDRC in 2006, the Chinese coal chemical industry has shown strong signs of overheating, with numerous CTL, DME, olefin, and methanol projects announced across the country. Following the worldwide economic slowdown in late 2008, the existing coal-to-methanol capacity in China has already exceeded domestic demand especially when facing competition from gas-derived methanol from overseas markets. In addition, technological immaturity, significant financial risks, and negative environmental impacts of large-scale coal chemical development have raised doubts regarding the overly ambitious targets. As a result, this draft plan was not officially released by the NDRC and the Chinese government has become more cautious about coal chemical development.

Unlike in most parts of the world, coals have long been used as a dominant fuel and feedstock in the Chinese chemical industry due to resource constraints. The aggregate energy consumption in the Chinese chemical industry has increased by 87% since 2000, reaching 346 Mtce in 2009. During the same period, coal and coke inputs have increased by 59% and 86%, reaching 123 Mt and 21 Mt in 2009. As a result, coal and coke's share in the energy mixture of the Chinese chemical industry has been lowered from 48% in 2000 to 42% in 2009 (NBS various years-a). Considering the looming potential of coal-to-liquids (CTL) and other coal chemical development across the country, the Chinese chemical industry's reliance on coal may rebound in the future.

Currently, major end products of the Chinese coal chemical industry can be categorized under five technological routes: (1) conversion to coke, (2) coal gasification, (3) coal liquefaction, (4) coal to chemical products, and (5) coal to calcium carbide. Only coking and coal liquefaction are discussed in the following paragraphs due to their existing scale or potential for growth in China. We focus in particular on coke production and CTL.

A.3.1.3.1 Coke production China's coke consumption spike in recent years is primarily driven by the rapid expansion of the Chinese iron and steel industry. Since 2000, coke consumption for iron and steel manufacturing in China increased by 215%, reaching 272 Mt in 2009. During the same period, the iron and steel industry accounted for 83% of national coke consumption, followed by 8% at the chemical industry (NBS, various years-a).

Compared with increasingly stringent environmental regulations and enforcement in the rest of world, the lax environmental enforcement and cheap labor in China have significantly strengthened Chinese coke producers' competitive advantage in the past. As a result, China's coke production as a percentage of the world total increased rapidly from 20% in 1990 to about 60% in recent years and became increasingly important for international consumers (NBS, various years-a, China Energy Statistical Textbook, and http://www.resource-net.com). Between 2000 and 2007, coke exports from China were maintained at a level around 15 Mt/annum. Because of the strong political desire for resource conservation and environmental protection, coke exports from China declined significantly to 12 Mt in 2008 and then to a mere 0.5 Mt in 2009.

With 49% of China's coking coal proven reserves, Shanxi has long been China's leading coke-producing province. Between 1990 and 2009, Shanxi alone produced 32% of China's cumulative coke output (NBS, various years-b, -a). Figure A.11 shows coke production composition by type of coke ovens in Shanxi. In 1985, backward production capacity[8] accounted for 47% of coke output in Shanxi. Since then, "backward" capacity expanded rapidly and its share of provincial output peaked at 88% in 1995. Following the Chinese government's ban on indigenous coke ovens in 1996, market share of backward capacity declined over time. Nevertheless, the spiking domestic demand coupled with lax enforcement especially at local levels left plenty of room for the survival and booming of backward coke manufacturers. The compulsory phase out of backward capacity in Shanxi was not targeted by the government with vigor until the early 2000s. With continuous governmental pressure for closure, the market share of backward capacity in Shanxi has been reduced to a minimal level.

The Chinese coking industry has followed a similar trend as Shanxi. In 2001, mechanized coking ovens accounted for 72% of the national coke output at 131 Mt. Since then, large-scale producers have expanded their production capacity at a breakneck pace at the expense of backward coke oven operators. In 2008, market share of backward capacity was lowered to about 3% of the national coke output.[9] Because of the government's

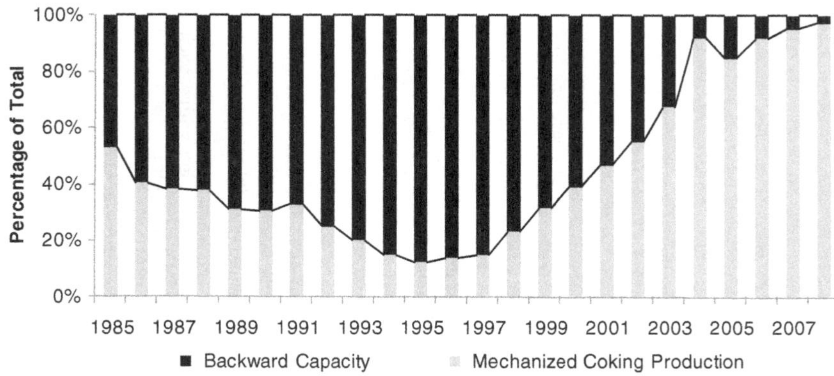

Figure A.11 Coke production composition by type of coke ovens in Shanxi, 1985–2008. Backward capacity includes both indigenous coke ovens and modified-indigenous coke ovens. Source: Shanxi Bureau of Statistics (various years).

continuous closure of backward capacity, the implied resource savings are substantial. Nevertheless, given that the Chinese coking industry still lags far behind international best practice, resource conservation and environmental protection are expected to continue to be the top priority in this sector.

A.3.1.3.2 Coal-to-liquids (CTL) conversion Table A.7 lists major CTL projects in China. Shenhua is the industrial leader in CTL development in China. In August 2004, Shenhua began the installation of the world's first commercial-scale direct coal liquefaction plant in Inner Mongolia. With an annual capacity at 1.08 Mt/annum,[10] the first phase of this project has a price tag of 12.3 billion yuan. A trial operation of the plant in December 2008 achieved continuous operation of 303 hours. As this facility has been demonstrated as technologically viable, Shenhua plans to expand its overall CTL capacity to 20 million tonnes of oil equivalent (Mtoe)/annum in the long run. Nevertheless, falling oil prices since late 2014 have brought significant uncertainties to the future of CTL in China.

There are some important negative aspects of CTL that offset the advantages it appears to bring in terms of energy security enhancement, namely that CTL has (1) intensive water requirements (3 to 4 tonnes of water per tonne of products for direct CTL and 4.5 to 5.5 tonnes of water for indirect CTL); (2) impacts on local air quality; (3) coal-combustion-related carbon emission and process GHG emissions; (4) substantial

Table A.7 *List of major CTL projects in China*

Developer	Location	Capacity (kt/yr)	Investment (billion yuan)	Process	Note
Shenhua	Inner Mongolia	1,080	12.3	Operational since 12/2008	One of the three direct CTL trial plants at phase one (3,240 Mt/annum in total)
Shenhua	Inner Mongolia	180	N/A	Depending on the direct CTL unit	Indirect CTL unit added to the above direct CTL process
Shenhua	Ningxia	4,000	55	Under construction	Indirect CTL: construction started in 09/2013, with targeted commission date in 10/2016.
Lu'an Group	Shanxi	160	3.5	Trial operation in 12/2008	Indirect CTL
Lu'an Group	Shanxi	1,800	23.9	To be commissioned in 2015	Indirect CTL
Yitai Group	Inner Mongolia	160	2.7	Trial operation in 3/2009	The first commercial scale indirect CTL unit in China
Yitai Group	Xinjiang	1,000	19	Approved in 07/2014	Yitai plans to expand its capacity to 5,400 kt/annum in the future
Gun Mining Group	Shaanxi	1,100	20	Under construction without permitting	Phase one: 5 Mt/annum, first production line with capacity of 1.1 Mt/annum; phase two: 10 Mt/annum

Sources: Various Chinese media and industrial sources.

energy efficiency losses in the coal-to-oil conversion process, in contradiction with China's energy intensity abatement policy (3 to 4 tonnes of coal are required for direct CTL and 4.5 to 5.5 tonnes for indirect CTL to produce one tonne of liquid products); and (5) significant financial risks associated with the wide fluctuation of oil prices (8,000 to 1,100 yuan/t of capacity for direct CTL and 1,100 to 1,300 yuan/t of capacity for indirect CTL).[11]

A.3.1.4 Building materials

Building materials in China can be divided into (1) construction materials and products manufacturing; (2) non-metal mineral products; and (3) inorganic and non-metal mineral new materials.

Currently, China ranks number one in the world in terms of cement, flat glass, and ceramics production. In 2006, coal consumption by the Chinese building materials industry reached 197 Mt, which represents 7.7% of national coal consumption and is the equivalent of a 13% growth over the 2005 level (NBS and China Building Materials Network). Manufacturing of non-metal mineral products is the most dominant subsector. In 2006, coal consumption by this subsector accounted for 85% of the sector total. Energy consumption of this subsector fluctuated over time but has still more than doubled since 1995, reaching 269 Mtce in 2009. In 1995, coal accounted for 73% of the subsector total energy consumption, which gradually decreased to 65% in 2009.

Cement kilns are the largest energy-consuming equipment in the Chinese building materials industry. Figure A.12 compares energy intensity levels of various cement kilns. In most cement-producing countries, natural gas and electricity are the primary energy carriers. Due to coal's dominance in China's energy sector, coal and electricity are the primary energy sources consumed in the Chinese cement industry. The dry hollow kiln is the most inefficient type of cement kiln, with energy intensity at about 300 kgce/tonne. In comparison, medium- or large-scale dry process kilns' fuel economy performance can be as low as 150 kgce/tonne. Market share of more efficient kilns such as the dry process type has been expanded rapidly by the government's energy intensity initiatives and mandatory closure of small inefficient kilns.

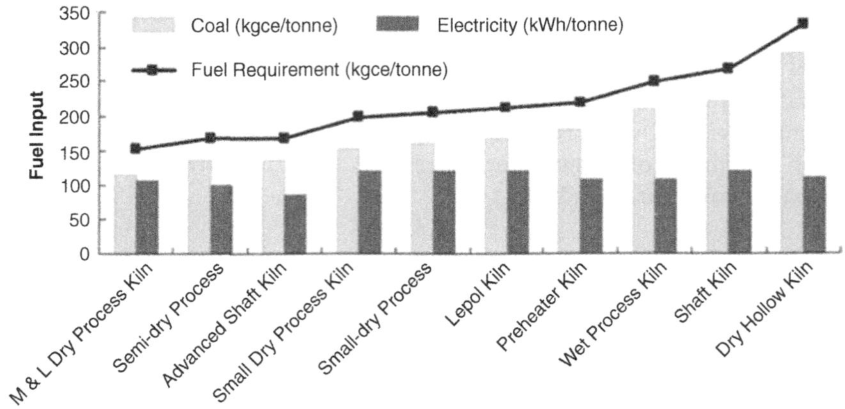

Figure A.12 Coal, electricity, and fuel requirements for different types of cement kilns. Source: China Cement Association.

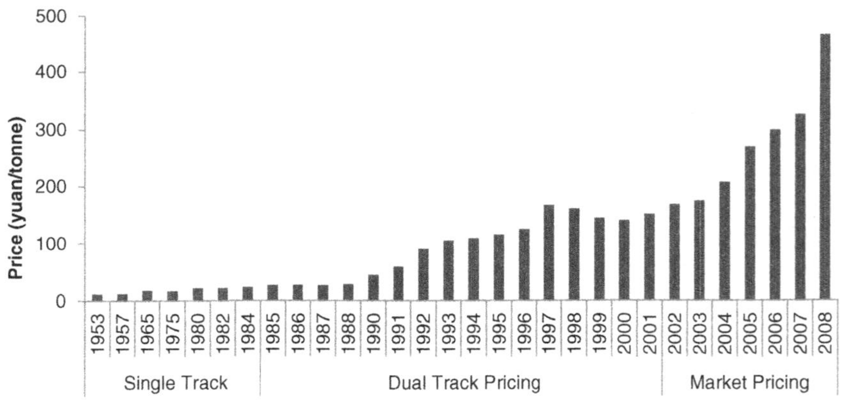

Figure A.13 Average mine mouth sales price for coal produced by key SOEs in China. Sources: Yang et al. (2009), Wang and Pan (2009), and NDRC (2006).

A.3.2 The role of government

The government's influence on retail coal prices is one of the most important determinants of coal end uses in China. Figure A.13 shows the historical development of average mine mouth sales price of coal produced by key state-owned enterprises in China over time.

During the transition from the planned economy, the regulation of coal prices in China underwent several major changes, described in the following paragraphs (for a comprehensive discussion of the evolution of China's regulation of the coal market, see Chapter 2).

Single-track pricing period (1949–1985): At the inception of the PRC, commodity pricing in China was dominated by the theory promoted by the former Soviet Union. According to this approach, part of value-added income in heavy industries should be realized in the light industry or other production sectors, so prices of raw materials including coal were set at arbitrarily low levels. Although coal production costs in China increased over time, coal prices were only adjusted marginally several times between 1949 and 1984; as a result, the Chinese coal mining industry ran a deficit during most of this period.

Dual-track pricing period (1985–2002): After the Chinese economy entered a period of rapid development around 1985, energy demand in the country increased rapidly. To meet the surging coal demand, Beijing had to encourage the development of small coal mines and to deregulate prices for coal produced by TVE mines. In addition, state-owned mines were allowed to sell their above-quota output at market prices. Since 1993, the Chinese government has gradually deregulated coal retail prices in most sectors. Nevertheless, it still imposed "supervised prices" for thermal coal consumed by power plants between 1994 and 2001. During this period, average coal retail prices increased from 26.05 yuan/t in 1985 to 167.81 yuan/t in 2002.

Market-oriented pricing period (2002–present): Following the Southeast Asia Financial Crisis, energy demand in China surged and coal prices rebounded. In 2002, the Chinese government abolished the "supervised thermal coal prices" implemented since 1994. To facilitate the thermal coal contract negotiation for utility use, the state still issued reference coal prices at the Annual Coal Trade Fair. Nevertheless, due to the immaturity of the Chinese coal market, thermal coal contract prices in 2002 and 2003 could be determined only with extensive state intervention. In August 2004, the former State Development Planning Commission had to temporarily intervene in prices of thermal coal for utility use in Henan, Anhui, Shandong, Shanxi, and Shaanxi. Nevertheless, China still encountered a serious coal and power shortage throughout the country. To lower financial pressure on the power sector caused by spiking coal prices, the Chinese government issued the Coal-Electricity Price Linkage Mechanism in the same year (it was infrequently used in subsequent years due to political pressure to keep power prices low). In 2006, the NDRC finally allowed prices of thermal coal for utility use to be fully subject to market pricing (though numerous attempts to cap prices occurred in later years). According to the Guidelines Regarding Improving the Linkage Amongst Coal Production, Transport, and Demand, issued by the NDRC on December 14, 2009, thermal coal contracts of power plants should be directly negotiated between coal enterprises and utilities, without state intervention, thus marking the complete deregulation of coal pricing at the central government level.

A.3.3 Reference trajectory of power generation capacity

Several key trends are likely to drive investment patterns in the Chinese power sector going forward. Given the Chinese government's emphasis on energy conservation and air pollution control (e.g., SO_2 emissions) during the 11th FYP period and the preference for domestically made equipment, capacity of large-scale supercritical and ultra-supercritical (USC) power plants has expanded rapidly in China. This trend is likely to continue. During the 12th FYP period, energy conservation and air pollution regulations have been continuously tightened, and a 17% GHG emissions intensity reduction target was also introduced. As a result, lower carbon fuels including renewables (especially wind), nuclear, and natural gas are planned to grow substantially in the coming decades, as shown in Table A.8.

In this study, the reference scenario of the Chinese power sector to 2030 was simulated by the CIMS-China model,[12] which was first developed by the author between 2001 and 2004 to explore the policy options to alleviate China's energy security, air pollution abatement, and climate change challenges (Tu et al., 2007). China's power consumption projection in Table A.8 is used as the baseline input in the model, and the Chinese government's targets in the 12th FYP period have also been incorporated.

Based on the update of the CIMS-China model,[13] Figure A.14 presents the reference forecast on generation capacity mixture in the Chinese electricity industry by 2030. In 2010, installed capacity of the Chinese electricity industry was 1.0 TW. To meet China's surging power demand, installed power capacity in China is expected to increase about 6% annually, exceeding 3.1 TW in 2030.

Among all the generation technologies, wind power is expected to show the strongest growth in percentage terms, at 16% annually during the study period. In comparison, installed capacity of coal-fired units is expected to expand moderately, at about 5% annually, to meet rising domestic demand and build a sufficient reserve margin in the coming decades.

Table A.8 *Demand and governmental targets for the Chinese electricity industry*

	Category	2005	2010	2015	2020	2025	2030
Power demand forecast*	GDP (trillion 2005 yuan)	18.5	31.4	45.9	64.4	88.7	118.1
	GDP growth rate (%)	11.2	7.9	7.0	6.6	5.9	
	Power consumption (GWh)	2,494	4,192	6,019	8,054	10,605	13,572
	Power consumption growth (%)	10.9	7.5	6.0	5.7	5.1	
	Electricity to GDP Elasticity	0.98	0.95	0.86	0.86	0.86	
Hydro target (GW)	Renewables Plan (2007)	117	190		300		
	12th Five Year Plan**			300			
Nuclear target (GW)	2005–2020 Nuclear Plan (2007)	6.85	12.528	24.968	44.968		
	12th Five Year Plan**			50			
Wind target (GW)	Renewables Plan (2007)	1.266	5		30		
	12th Five Year Plan**			90			
Biomass target (GW)	Renewables Plan (2007)	2	4		24		
	12th Five Year Plan**			N/A			
Solar target (GW)	Renewables Plan (2007)	0.07	0.3		1.8		
	12th Five Year Plan**			5.0			

Note: Cells with grey background indicate either forecast or planned government targets.

* GDP growth rates between 2010 and 2030 are based on Li et al. (2010). Growth rates of power consumption levels by 2020 are based on forecasts by the China Electricity Council. Electricity to GDP elasticity after 2020 is set as the same value as the previous FYP period.

** Preliminary targets revealed by Chinese governmental officials during various conferences and media interviews. Sources: CEC (various years), NDRC (2006), and Li et al. (2010).

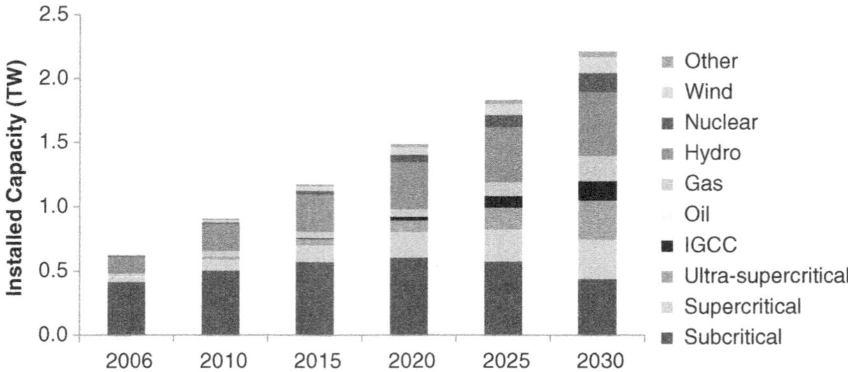

Figure A.14 Reference projection for China's electricity generation capacity mix in 2030. Only grid-connected Greenfield power plants are included in the assessment. Capacity in 2010 is based on data from China Electricity Council.

Despite the Fukushima Daiichi crisis, the nuclear share in China's electricity capacity mixture is expected to increase over time due to strong support by the Chinese government. As a result, its share in total installed capacity is expected to increase from about 1% in 2010 to near 5% in 2030.

Natural gas-fired generation is projected to increase during the study period in order to alleviate air pollution in the urban centers and meet peak load demand. The expected gas consumption growth is projected to be met with pipeline gas imports from central Asia and Russia, LNG imports in coastal China, and exploration and development of domestic gas resources including unconventional plays.

In comparison, at the beginning of the 11th FYP period, China's installed oil generation capacity, at about 10 GW, was planned to be gradually phased out. Nevertheless, to meet energy intensity targets set by the central government, restricting power supply had become a widespread practice at the local government level by the end of 2010. As a result, diesel-fired engines have been installed across the country. If this practice continues in the future, oil-fired power generation is expected to maintain a marginal but still important role in China.

Due to the rapid capacity addition of advanced coal units such as supercritical and ultra-supercritical power plants to China's thermal generation fleet, average coal consumption for power generation in China was simulated to fall from 335 gce/kWh in 2010 to 288

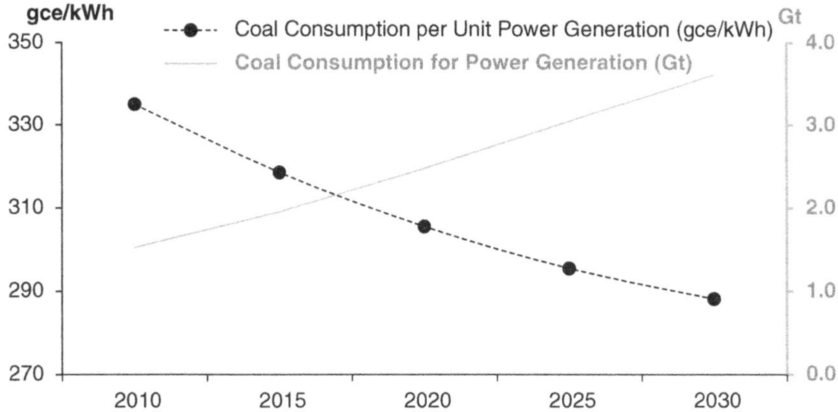

Figure A.15 Coal consumption and efficiency trend in the Chinese power industry. This figure indicates that it is difficult to restrict China's national coal production below 4.5 Gt by 2030; "gce" on the left axis represents "grams of coal equivalent."

gce/kWh in 2030, the equivalent of about a 14% gain in efficiency within two decades. Nevertheless, the expected energy efficiency improvements alone are insufficient to retard rapid growth in coal consumption in the Chinese power sector. Figure A.15 shows that national coal consumption for power generation in China is expected to grow by more than 4% annually, reaching 3.6 Gt in 2030, which suggests that China's national coal output will be difficult to cap below the reference trajectory defined in Figure A.7.

Nevertheless, the aforementioned forecast on coal consumption is highly sensitive to assumptions about China's GDP growth rate in the coming decades. While a lower economic growth trajectory could lead to a significant drop of thermal coal consumption for power production, a rapidly growing Chinese economy would imply faster power demand growth, and thus faster coal demand growth (allowing for some gains in the efficiency of converting coal to electricity). Finally, resource constraints are considered to be one important factor underlying power generation technology composition in the Chinese electricity industry. For instance, gas-fired power capacity in China by 2030 could be significantly lower than the reference trajectory if access to gas resources is impeded by either institutional or technological obstacles in China.

A.4 Coal transport

Rail, waterway (including inland waterway and coastal marine transport), and road are the three major transport modes to move coal in China. The characteristics of coal resource distribution and the long distance between China's major coal-producing and coal-consuming centers dictate that coal must be moved from west to east and from north to south.

Railway is by far the most important coal transport mode in China. The volume of coal and coal products moved by rail increased from 629 Mt in 1990 to 1,327 Mt in 2009, when it accounted for 48% of total cargo tonnage hauled by rail and 44% of national coal production. Water transport, especially coastal shipping from north to south, is the second largest coal transport mode. Coal throughput has increased rapidly from 78 Mt in 1990 to 949 Mt in 2009 at major coastal ports and from 63 Mt in 1990 to 374 Mt in 2009 at major inland ports (MOT, various years; NBS, various years-b). In comparison, road transport has remained a supplemental mode for moving coal within China, since it is used primarily for local distribution of coal.

Since 1985, coal production in China has increased by 250%, reaching 3.24 Gt in 2010. In comparison, growth of coal transport by rail lagged far behind the national coal supply spike. During the same period, coal transport by rail increased only by 156%. As a result of the severe bottleneck of the railway infrastructure, only 43.5% of national coal output was transported by rail in 2009, a sharp decline from the rail share of 59.4% in 1985 and 65.1% in 1978. In comparison, since 1985, coal and coal products transported by major coastal ports and major inland ports have increased by nearly nineteen-fold and more than eleven-fold, respectively (MOT, various years; NBS, various years-b). Not surprisingly, rail capacity is so far the main constraint of coal transport in China.

To meet growing demand for traffic such as coal arising from the booming economy, the Ministry of Transport (MOT) promulgated the 11th Five Year Development Plan for Road and Water Transportation in 2006. The plan sought to (1) increase total distance of highways by 37%, reaching 2.3 million km in 2010, and (2) increase total length of highways in county and villages by 32%, reaching 1.8 million km in 2010. The above expansion is especially beneficial for small TVE mines, which rely heavily on road to truck their coal to end users due to lack of access to the national railway network (LFRC, 2007). Though the bottleneck of road infrastructure is expected to be eased over time, coal transport by road is likely to be increasingly constrained by costs and environmental concerns (e.g., dust emissions and air pollutant discharge of trucks).

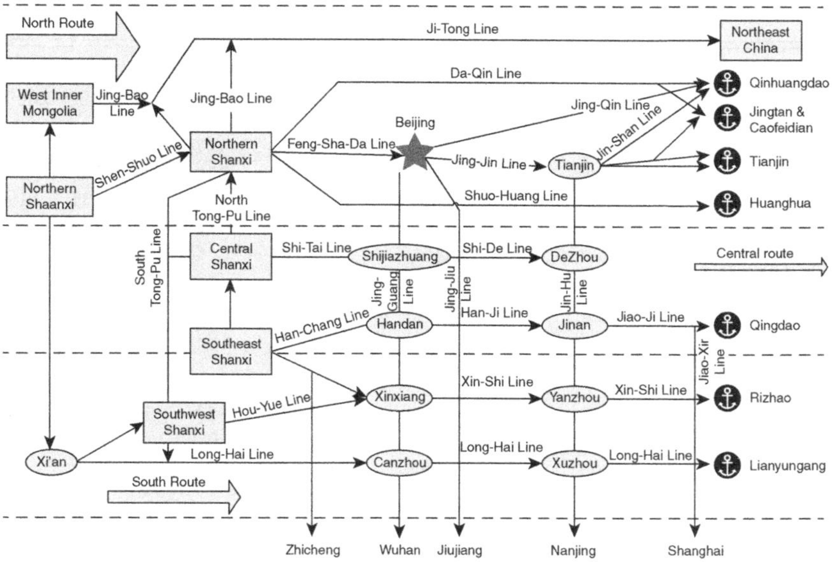

Figure A.16 Major coal transport routes in China. Sources: Sagawa et al. (2003), MOR (2008), and interviews with Chinese academia and officials (2009).

A.4.1 Major coal transport corridors in China

The 3-Xi[14] region including Shanxi, Shaanxi, and west Inner Mongolia is China's primary coal producing region. Together with east Ningxia, the 3-Xi region supplies most of China's out-of-province coal transport volume (coal that is exported outside of the province it is produced in). Given the majority of China's coal resources lie in the 3-Xi region with major consuming centers in the distant coastal regions, two key coal transport corridors exist in China to move coal from the 3-Xi region and east Ningxia to east and southeast China.

As illustrated in Figure A.16, these two corridors include (1) the west to east coal transport corridor and (2) the north to south coal transport corridor. Coal between north and south China accounts for 75% of national coal transport tonnage – of which 53.5% of coal moves from north to south and 21.3% of coal moves along the south to north direction. The remaining 25% of China's total coal transport tonnage moves between west and east China – of which 21% moves west to east and only 4% from east to west (Duan, 2007).

Railway transport in the 3-Xi region can be further divided into three interconnected routes: (1) the north route includes Feng-Sha-Da line,

Table A.9 *Coal throughput at the three coal transport routes (in Mt)*

Transport route / line	Start	End	2004	2005	2006	2007	2008	2009
North route			**309**	**351**	**427**	**522**	**565**	**564**
Da-Qin line	Datong	Qinhuangdao	153	203	250	304	340	330
Shuo-Huang line	Shenchinan	Huanghua	75	94	112	133	135	149
Jing-Yuan line	Beijing	Yuanping	18	11	17	22	20	20
Ji-Tong line	Jininnan	Tongliao	13	14	5	10	20	15
Feng-Sha-Da line	Fengtai	Datong	50	30	42	53	50	50
Middle route			**67**	**72**	**76**	**70**	**76***	**98****
Shi-Tai line	Shijiazhuang	Taiyuan	61	67	69	62	68	90
Han-Chang line	Handan	Changzhibei	6	5	7	8	8*	8*
South route			**89**	**109**	**175***	**190***	**210***	**210***
Tai-Jiao line	Xiuwen	Jiaozuobei	43	40	50	50	70	70
Hou-Yue line	Houma	Yueshan	24	52	105	120	120	120
Xi-Zheng of Long-Hai	Xi'an	Zhengzhou	13	9	11	11*	11*	11*
Xi-Kang line	Xi'an	Ankang	4	5	5*	5*	5*	5*
Ning-Xi line	Xi'an	Nanjing	5	4	4*	4*	4*	4*
TOTAL			**465**	**532**	**678***	**782***	**851***	**872***

* Estimated because starting in September 2010, the Han-Chang line was under construction to expand its capacity. Sources: Various industrial sources.

Da-Qin line, Shuo-Huang line, Jing-Yuan line, and Ji-Tong line and primarily transports coal produced in Datong, Pingshuo, Zhungeer, Hebaopian, Shenfu, Dongsheng, Wuda, and Haibowan; (2) the central route includes Shi-Tai line, Han-Chang line, and Tai-Jiao line and primarily transports coking coal and anthracite produced in Xishan, Yangquan, Jinzhong, and Lvliang and other coal produced in Lu'an, Jincheng, and Yangquan; and (3) the south route includes South Tong-Pu line, Long-Hai line, and Hou-Yue line. In addition, this route also transports small amounts of coal produced in Shaanxi through Xi-Kan line and Xiang-Yu line.

A.4.1.1 West to east corridor

Table A.9 details coal throughput at major rail lines in the three west to east coal transport routes in the 3-Xi region and east Ningxia. Since 2004, coal hauled by the north route increased by 83%, reaching 564 Mt in 2009. While the north route accounted for about 65% of coal throughput handled by the west to east coal transport corridor, Daqin is the most important coal rail transport route in China. Nevertheless, a bottleneck of railway infrastructure has long existed for this route, especially after coal demand in the coastal provinces spiked in the past decade. In comparison, the remaining two routes together accounted for only 35% of total coal throughput at the west to east corridor, thus the bottleneck of railway infrastructure has been an even more acute issue for these two routes. To tackle the infrastructure challenge, on December 22, 2009, the construction of a dedicated coal railway between Xinxian in Central South Shanxi and Rizhao port in Shandong officially started. The 1,260 km rail line will provide coal mines in central south Shanxi an additional 200 Mt/annum transport capacity once the rail line is completed in 2014.[15]

A.4.1.2 North to south corridors

Integrated rail-waterway transport is the primary mode to transfer coal from the 3-Xi region and east Ningxia to east and southeast China. The so-called Northern Seven (N7) ports are the primary linkage between rail lines traveling the west to east coal transport corridor and the north to south coastal marine transport corridor. According to the National Coastal Ports Layout Plan promulgated by the MOT in 2006, the national marine coal transport network primarily consists of N7 ports for coal delivery and dedicated ports owned by utilities and public coal ports of discharge in east and southeast China.

Figure A.17 shows that the N7 ports include (1) Qinhuangdao Port in Hebei, (2) Jingtang in Hebei, (3) Tianjin Port in Tianjin, (4) Huanghua in Hebei, (5) Qingdao in Shandong, (6) Rizhao in Shandong, and (7) Lianyungang in Shandong. These together account for about 90% of coal throughput handled by major Chinese coal coastal ports of delivery from which coal is first shipped to end users. The major coal ports of discharge in China include (1) Shanghai Ports, which have fifteen berths with design capacity at 45.7 Mt/annum in 2006; (2) Ningbo-Zhoushan Ports, which have four berths with design capacity at 34.2 Mt/annum in 2006; and (3) Guangzhou Port, which has ten berths with design capacity at 40.5 Mt/annum in 2006.[16]

Table A.10 lists capacity of the N7 ports since 2006 and their primary railway connection, through which these ports are interconnected with a complex railway network in the west to east corridor. In 2006, total coal hauled by the west to east rail routes reached 693 Mt. In comparison, the aggregated throughput of the N7 ports was 447 Mt, and most cargoes were destined to end users in east and southeast China. More specifically, coal from Shanxi and Inner Mongolia is primarily transferred at Qinhuangdao and Tianjin. Coal from Shaanxi is primarily transferred at

Throughput at Major Chinese Coal Ports of Delivery (Mt)

Ports	1995	2000	2005	2006	2007	2008
Major coastal ports	124.9	188.9	371.0	406.3	463.4	509.5
Northern seven	122.5	184.2	331.6	370.8	413.8	447.1
1. Qinhuangdao	64.9	83.8	144.7	176.4	209.7	214.8
2. Jingtan	2.3	7.5	13.6	13.8	14.2	36.9
3. Tianjin	28.3	43.2	68.9	74.1	74.8	80.6
4. Huanghua	0.4	0.3	66.9	59.9	81.7	78.0
5. Qingdao	6.7	13.4	8.0	15.0	11.0	13.0
6. Rizhao	12.0	21.4	19.5	16.9	14.3	15.5
7. Lianyuangang	8.0	14.6	10.0	14.8	7.9	8.2

Status of Coal Ports of Discharge in South China in 2006

	No. of berths	Throughput in 2006 (Mt)	Capacity (Mt)	Uitlization rate
Coal Ports of Discharge	80	328.0	262.2	125%
Yangtze River Delta	50	215.6	152.2	142%
Coastal Southeast	9	24.9	25.8	97%
Pearl River Delta	17	79.9	75.0	106%
Coastal Southwest	4	7.6	9.3	82%

Figure A.17 China's major coastal ports and marine transport of coal. Sources: CNCA (2006), CCTD, MOT (various years), and SXCOAL.

Tianjin and Huanghua. Coal from Shandong is primarily transferred at Rizhao.

Qinhuangdao is China's largest coal port of delivery and all coals in Qinhuangdao currently come from the Da-Qin line, which primarily serves mines owned by China National Coal, Shenhua, and Yitai Group in the 3-Xi region and east Ningxia. After the receipt of coal from Da-Qin, coal at Qinhuangdao is eventually sent to the following regions along the north to south marine transport corridor: 85% to east and southeast China (Shanghai and Zhejiang are the primary destination markets in east China), with remaining coal going to Jiangsu along the Yangtze River, and Guangzhou, Shenzhen, and Fujian are the primary destination markets in southeast China, with remaining coal going to Guangxi and Shandong (Wu, 2009).

The north to south inland coal transport corridor follows more complex routes: (1) major southward rail trunk lines such as Jing-Jiu and Jing-Hu can directly move coal from north China to central south and

Table A.10 *Overview of the linkage between the N7 ports and railways*

Port	Capacity (Mt) 2006	2007	2008	2009**	2010**	Railway connection
Qinhuangdao	193	193	193	193	193	Da-Qin line, Jing-Qin line
Tianjin	90	98	108	108	108	Huang-Wan line, Jing-Jin line
Jingtang	16	25	46	46	46	Jingtang Port line, Da-Qin line
Caofeidian*	-	-	50	50	80	Qian-Cao line, Da-Qin line
Huanghua	90	90	90	90	90	Shen-Shuo-Huang line
Qingdao	15	15	15	15	15	Jiao-Ji line
Rizhao	27	27	27	27	45	Xin-Shi line
Lianyungang	17	17	17	20	20	Long-Hai line

* Caofeidian is often considered part of Jingtang port due to their geographic approximation.

** Estimated capacity (based on interviews)

southeast China, and (2) rail lines in east China can connect with ports along the Jing-Hang Great Canal to move coal along the canal region and the Yangtze River Delta. Wanzai Port, located at the intersection of Jing-Hu line, Long-Hai line, and the Jing-Hang Great Canal, thus is an increasingly important regional coal transfer hub and southward rail lines can connect with ports along the Yangtze River such as Zhicheng, Hankou, Yuxikou, and Pukou.

A.4.2 Coal transport by mode

A.4.2.1 Rail

At the end of 2009, the total length of Chinese railways in operation reached 85,518 km, ranking third in the world after the United States and Russia and accounting for about 8% of the world total. In comparison, the freight throughput (tonne.km) handled by Chinese rail networks is estimated to represent more than a quarter of the world total, which implies an overly congested traffic pattern.[17] Table A.11 shows the composition of freight transport by Chinese railways in terms of ownership of infrastructure. Rail freight transport in China is currently dominated by the national railway network, which accounted for 83.3% of freight tonnage and 93.7% of freight t.km in 2009. In comparison, though the share of cargo tonnage handled by local railways increased from 2.4% in 1978 to 7.2% in 2009, the share of freight t.km handled by local railways

Table A.11 *Freight transport composition of Chinese railways*

	1978	1985	1995	1996	2000	2005	2006	2007	2008	2009
Freight (Gt)	1.10	1.31	1.66	1.71	1.79	2.69	2.88	3.14	3.30	3.33
National (%)	97.6	97.6	96.1	94.6	93.0	86.1	85.2	83.5	83.3	83.3
Local (%)	2.4	2.4	3.9	4.2	4.7	6.6	6.8	7.8	8.2	7.2
Joint venture (%)	—	—	—	1.2	2.3	7.3	8.0	8.7	8.5	9.6
Freight (trillion tonnes.km)	0.53	0.81	1.30	1.31	1.38	2.07	2.20	2.38	2.51	2.52
National (%)	99.8	99.8	99.7	98.7	97.6	94.2	93.6	92.9	94.2	93.7
Local (%)	0.2	0.2	0.3	0.4	0.3	0.5	0.5	0.6	0.6	0.5
Joint venture (%)	—	—	—	0.9	2.1	5.3	5.9	6.5	5.2	5.8

Source: NBS (various years-b).

grew only from 0.2% in 1978 to 0.5% in 2009, which indicates that local railways focus their business on short-distance cargo transport. The importance of joint ventures has increased steadily over time, but their share of freight transport has not exceeded 10% market share.

Lack of competition in the rail sector has long made the growth of railway networks in China lag far behind the national coal output, especially in recent years. To make the situation worse, the slow growth of rail line additions is especially acute in China's major coal-producing regions such as the 3-Xi region. In 2009, this region supplied about half of China's coal production but possessed only 17% of China's rail lines (NBS, various years-b). As a result, the heavily congested railway network in major coal-producing regions has long been a bottleneck for moving coal to coastal ports in north China and end users in central south China and coastal provinces.

The primary driving force underlying the composition of coal transport in China is transport cost by mode. Average freight rate by rail in China has increased from 7.95 cents/t.km in 2000 to 10.31 cents/t.km in 2009 (CCTD), and rail is still the most desirable mode to move large quantities of coal in China due to its large capacity, fuel efficiency, and relatively lower freight rate (especially when compared with much more expensive road transport). Rail freight rates in China are expected to increase to 15 to 20 cents/t.km by 2015 and 20 to 26 cents/t.km by 2020 (Li, 2008).

To tackle over-congestion of the railway network, the construction of rail lines in China reached its heights following the promulgation of the

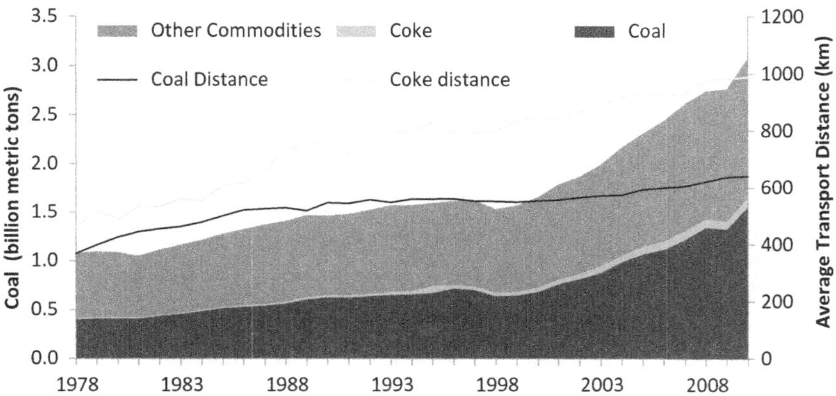

Figure A.18 Coal and coke transport versus freight tonnage in China, 1978–2009. Source: NBS (various years-b).

Mid- and Long-term Railway Development Plan by the Ministry of Railways (MOR) in 2004. A revised plan was issued in 2008, which indicates that China will continue to extend rail coverage, improve the network structure, lift the quality of services, boost transport capacity, and upgrade equipment rapidly. The total length of Chinese railways in operation is estimated to reach 100,000 km, with passenger lines and freight lines separated for busy routes. The MOR also set the goal of having 50% of railways electrified and 50% double-tracked by 2020. According to the plan, coal rail network development was identified as a top priority. Greenfield coal rail lines are planned to be added for the central west corridor in Inner Mongolia and central south corridor in Shanxi, which are expected to serve as new high-capacity coal transport corridors for the 3-Xi region. In addition, existing coal rail lines will be expanded in ten major coal export regions (MOR, 2008b).[18]

Figure A.18 shows the tonnage of coal and coke transported by rail over time. In 1978, the national rail network hauled 402 Mt of coal, which accounted for 37.4% of total tonnage handled by railway in China and 65% of national coal output. Since then, coal hauled by national railways more than tripled, reaching 1.343 Gt in 2008, and then declining slightly to 1.327 Gt in 2009. Due to the slower growth of coal hauled by railway compared with national coal output, coal handled by the national rail network accounted for only 44% of China's coal production in 2009. Nevertheless, coal's share of total cargo hauled by rail increased from 37% in 1978 to 48% in 2009. In addition, the average coal transport distance by national railways increased steadily from

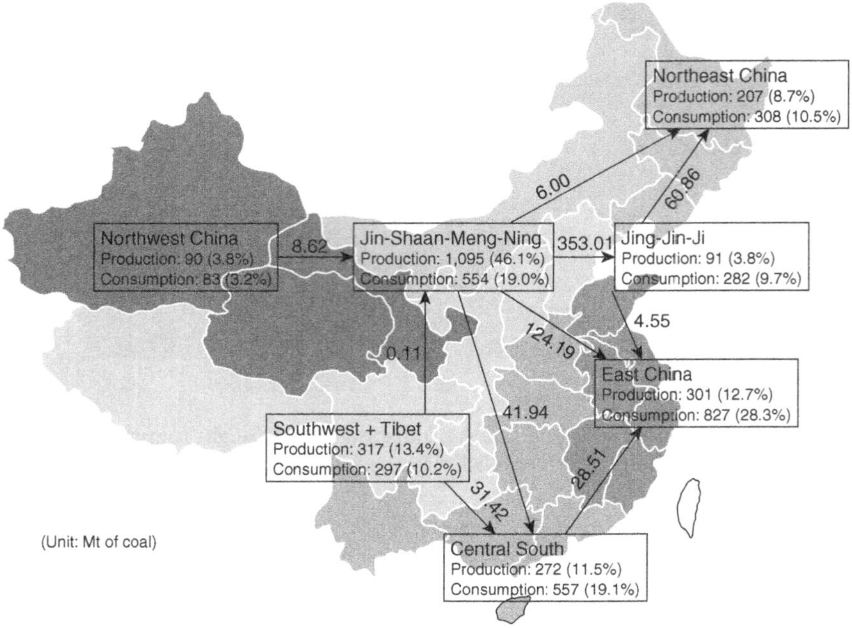

Figure A.19 China's inter-regional coal flow by rail in 2006, in Mt. Coal consumption reported at provincial level is much higher than coal production at a similar level provided by the NBS. Sources: MOT (various years), NDRC (2008), and CCTD.

358 km in 1978 to 639 km in 2009, which further aggravated the burdens on China's congested rail infrastructure. Local and joint venture railways play an increasingly important but still marginal role in China's rail transportation sector. In 2006, 1.39 Gt of coal was hauled by Chinese railways, with 1.12 Gt on the national network (IEA, 2009a; NBS, various years-b).

Figure A.19 shows China's inter-regional coal flow by rail. In 2006, China's inter-provincial coal flow by rail reached 1.12 Gt, which is 63% higher than the 2000 level. Shanxi, Shaanxi, Inner Mongolia (especially the western part), and Ningxia (especially the eastern part) are primary coal supply and delivery centers in China. This region alone exported 525 Mt of coal to the rest of China and the international market by rail. Central south, east China, and northeast China are the primary coal recipient regions in China. In 2006, east China received 157 Mt of coal by rail. Central south China received 73 Mt of coal by rail and sent 29 Mt

Table A.12 *Railways owned by Shenhua Group*

Railway	Year of completion	Capacity (Mt/annum)	Length (km)	Start	End
Bao-Shen line	March 1989	30	172	Wangshuiquan Station of Jing-Bao Line	Daliuta at Shenmu, Shaanxi
Shen-Shuo line	1996	150	270	Daliuta at Shenmu, Shaanxi	Shuozhou in Shanxi
Shuo-Huang line	October 2001	200	594	Shenchi S. Station at Shuozhou, Shanxi	Huanghua Port in Hebei
Da-Zhun line	1997	45	264	Zhungeer, Inner Mongolia	Datong, Shanxi
Huang-Wang line	October 2006	30	67	Huanghua S. Station of Shuo-Huang line	Wangjia Port Station in Tianjin
Dong-Wu line	March 2008	35	343	Wuhai, Inner Mongolia	Shashagetai Station of Bao-Shen line

Source: Wang (2008).

to east China in the same year. In 2006, northeast China received 67 Mt of coal by rail.

In 2006, 37% of coal transported by rail moved less than 500 km, 45% moved between 500 and 800 km, 13% moved between 800 and 1,200 km, and only about 5% of coal transported by rail moved beyond 1,200 km in China. As a result, 1,200 km is estimated to be the threshold for moving coal economically in China by rail (Zhang, 2009).

Due to the MOR's monopoly status, the majority of Chinese railways, especially those strategically important trunk lines, were firmly controlled by the MOR. Nevertheless, there is one exception. After Shenhua Group was founded in 1995, it was owned by the former State Planning Commission. Since then, preferential treatment from the central government has allowed the company to expand at a breakneck pace, and it has become the largest coal mining company in the world. With strong state favoritism, Shenhua has even been able to break into the state railway's monopoly turf and build its own captive transportation capacity as shown in Table A.12.

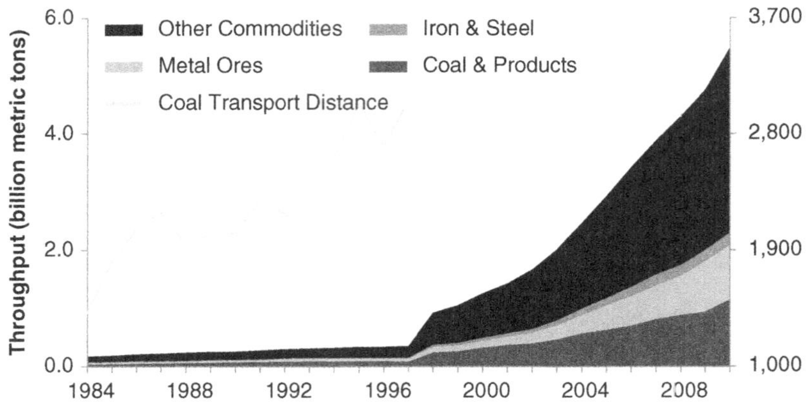

Figure A.20 Coal throughput at major coastal ports, from 1984 to 2009. Source: NBS (various years-b).

Shen-Shuo-Huang line is Shenhua's coal transport trunk line, carrying coal produced at Shendong and Wangli coalfields, primarily to Huanghua and Tianjin. In 2007, this line hauled 136 Mt of coal, which accounted for 72% of Shenhua's coal transport tonnage. In addition, Huanghua port is 70% owned by Shenhua with the Hebei provincial government. Completed in 2006, Shenhua's Tianjin coal port (55% interest) has a capacity of 35 Mt. Nevertheless, due to the state railway's monopoly status, Shenhua's success with captive coal rail line development is considered an exception in China.

A.4.2.2 Sea

Port sector competition, unlike for rail transport, has already been introduced by the Chinese government. Figure A.20 shows that coal and coal products transported by major coastal ports in China have increased by 12.6% annually since 2000, reaching 949 Mt in 2009. In comparison, national coal production grew at only about 9% annually during the same period. Although the high growth rate of coal transport by major coastal ports is expected to be sustained by market competition, the bottleneck of coastal port capacity on coal transport will continuously be eased in the years to come.

In 2006, average freight rates of coastal marine transport were 5 to 6 cents/tonne.km, with cargo handling charges ranging from 8 to 22 yuan/tonne. Since then, freight rates of coastal marine transport increased rapidly until late 2008, when the world economy dipped into a recession.

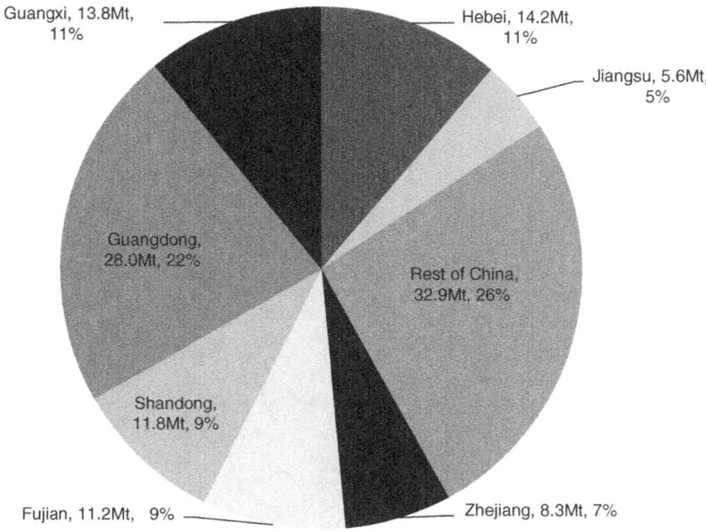

Figure A.21 China's coal imports by province in 2009. Source: China Customs.

Since late 2009, freight rates of coastal marine transport started to increase again but have been volatile. Freight rates of coastal marine transport are expected to grow to 10 to 14 cents/t.km by 2020. In comparison, cargo handling charges at coastal ports are projected to increase to 14 to 28 yuan/t by 2020 (Li, 2008). Even so, the relative cost competitiveness of different coal transport modes is expected to remain the same, with coastal marine transport the least expensive and road transport the most expensive.

China's coal import-export balance is an important driving force underlying coal transport at coastal ports. In the past, the N7 ports in north China not only transferred coal from the 3-Xi region and east Ningxia to coastal provinces in east and southeast China but also exported large amounts of coal to the international market. With quickly diminishing coal export levels and spiking coal imports, China first became a net coal importer in 2009.

As illustrated in Figure A.21, coal consumers in China's coastal provinces, especially Hebei, Shandong, Fujian, Jiangsu, Zhejiang, Guangdong, and Guangxi, are expected to keep importing moderate levels of coal from overseas suppliers, which will inevitably affect coal flows among different Chinese coastal ports. In addition, China's ongoing coal imports are expected to stimulate inland waterway transport, particularly in the Yangtze River Delta and Pearl River Delta regions.

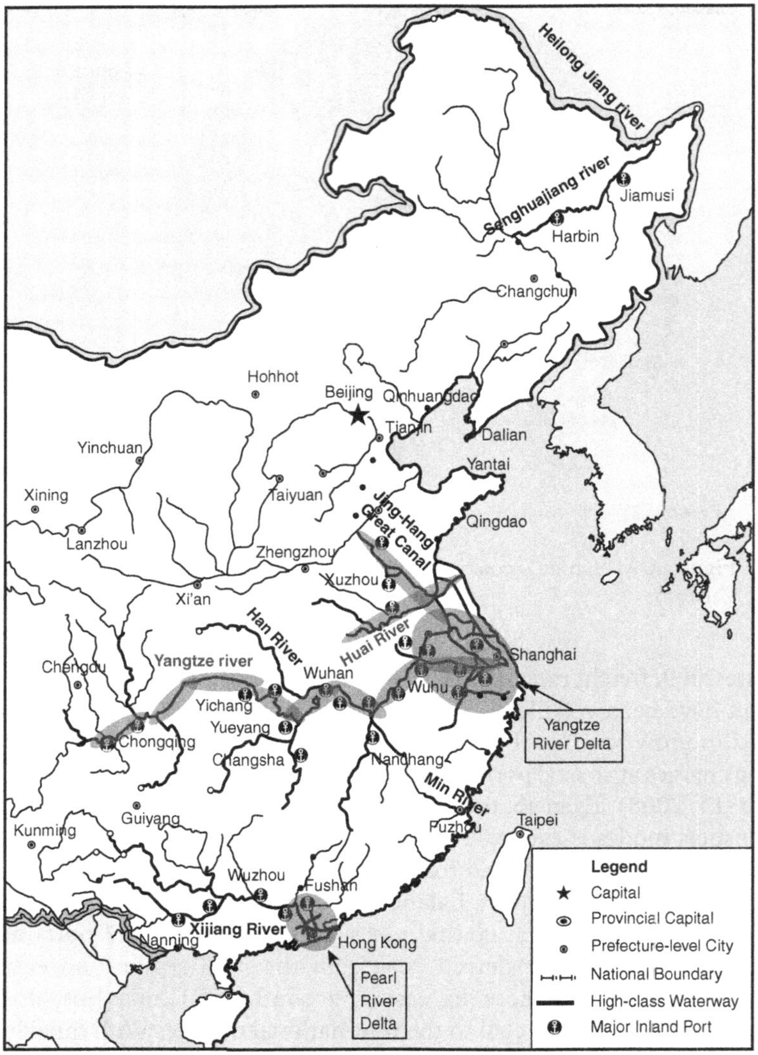

Figure A.22 High-class waterways and major inland ports in China. Source: MOT (various years).

A.4.2.3 River (barge)

Transportation on the inland waterways is an important element of the domestic coal distribution system, carrying 374 Mt of coal in 2009, which accounted for approximately 12% of China's coal output in the same year. Currently, coal represents an important share of shipping on the inland waterways, accounting for approximately 17% of total cargo tonnage (MOT, various years).

In 2007, the MOT announced the National Plan for Inland Waterways and Ports Layout, for the development of inland waterways and ports by 2020 (LFRC, 2009). According to the layout, the development of an inland waterway network includes the following main streams as illustrated in Figure A.22:

1) Yangtze River – In 2007, coal hauled by major ports along Yangtze River reached 179 Mt, accounting for 19.6% of total cargo tonnage at these ports.[19]
2) Beijing-Hangzhou (Jing-Hang) Grand Canal – The canal has become the most important inland waterway coal transport trunk line at the strategically important north to south coal transport corridor.
3) Yangtze River Delta – The high-class waterways network of the delta includes one of the largest cities in East China, Shanghai.
4) Pearl River Delta – In 2007, coal hauled through the Xijiang River and the Pearl River Delta reached 4.8 and 30 Mt, respectively, accounting for 11.6% of inland waterway cargo tonnage in this region.[20]
5) The waterway network also comprises another eighteen major tributaries of a number of rivers across the country.

Table A.13 *Selected major inland coal ports in China*

Port	Mainstream	Note
Zhicheng Port	Yangtze River	27 berths: The Shigu Main Port has 4 dedicated coal berths, with a capacity of handling above 3 kt dwt vessels. The capacity is expected to be increased to 15 Mt in five years.
Hankou Port	Yangtze River	250 berths: with maximum capacity of handling 5 kt dwt vessels.
Yuxikou Port	Yangtze River	China's first mechanized coal port: annual capacity at 3 Mt.
Pukou Port	Yangtze River	15 berths: with design capacity at 11.72 Mt/annum.
Wanzhai Port	Jing-Hang Great Canal	6 berths: annual capacity at 8.8 Mt with coal mixing facility.
Pizhou Port	Jing-Hang Great Canal	19 berths, 6.02 km of rail line: annual design capacity at 6.5 Mt with the maximum annual throughput of 4.38 Mt. Coal accounts for more than 95% throughput at this port.
Shuanglou Port	Jing-Hang Great Canal	Annual throughput has reached 3 Mt.
Mengjiagou Port	Jing-Hang Great Canal	Annual throughput has exceeded 3 Mt in 2005.

Source: CCTD, SXCOAL, tadercn.com, and Xuzhou Port Group.

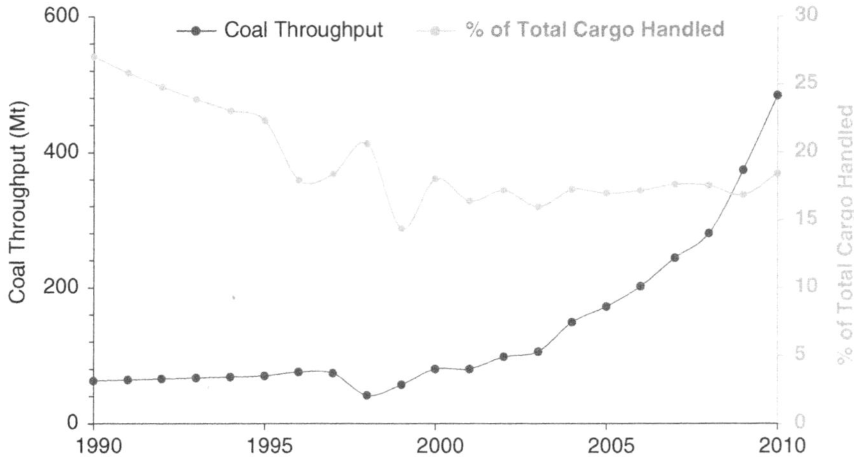

Figure A.23 Coal handled at major Chinese inland ports. Data in 1991–1994 was interpolated according to 1991 and 1995 data. Source: MOC (various years).

The inland waterway coal transport network in China primarily consists of ports along the Yangtze River, Pearl River, Jing-Hang Great Canal, and Huai River. The following ports are specialized in coal transport: (1) Zhicheng, Hankou, Yuxikou, and Pukou along the Yangtze River (in 2007, 16.4 Mt of coal were transferred from railways to these ports: Zhicheng 1.27 Mt, Hankou 1.91 Mt, Yuxikou 7.56 Mt, and Pukou 5.65 Mt); and (2) Wanzhai, Pizhou, Shuanglou, and Mengjiagou along the Jing-Hang Great Canal. Spot prices at these ports have been widely used as benchmark coal prices due to their critical location to bridge China's major coal producing and consuming centers.

Figure A.23 shows that coal handled at major Chinese inland coal ports grew by 17% between 1990 and 1997, then declined sharply by 44% in 1998, reaching 57 Mt in the same year, which is 34% lower than the 1990 level. Since then, coal handled at major Chinese inland ports rebounded strongly at 21% annually, reaching 374 Mt in 2009, which is 345% higher than the 1990 level. Nevertheless, coal's share of total cargo throughput handled at major inland ports still declined from 27% in 1990 to 17% in 2009.

In 2006, coal shipping rates on inland waterways were 5 to 7 cents/tonne.km with cargo handling charges ranging from 8 to 22 yuan/tonne. Since then, freight rates for coal transport on the Yangtze River fluctuated widely and declined sharply in 2008, following the worldwide economic slowdown, and then rebounded strongly after July 2009.[21] According to Li (2008), freight rates of coal transport by inland waterway are expected to increase to 10 to 14 cents/t.km by 2020, and cargo handling charges at inland ports are projected to grow to 14 to 28 yuan/t by 2020. Nevertheless, inland waterways will remain cost competitive to move large quantities of coal along long-distance transport corridors in China, and geographic suitability and availability of port facilities instead of freight rates are expected to be the primary constraint to sustaining growth of inland waterway coal transport in China.

A.4.2.4 Road (truck)

In this study, coal transport by road refers to coal tonnage hauled by truck from coal producers to coal consumers or transfer centers. In terms of coal transport, road is considered to be only a supplemental mode to rail and waterway due to the economy of scale of the latter two modes. Road is most likely to be used as an intermediate mode along the producer-rail-port-user transport chain. In addition, long-distance trucking of coal is both economically unattractive and environmentally undesirable; road is suitable only for short-distance coal delivery within a coal-producing province or between coal-producing and coal-consuming regions that are not far apart. Though when market conditions create high prices for delivered coal and rail capacity is tight, trucks have been profitably used to deliver "swing transport supplies" (which tend to be quite expensive as a substitute for rail; thus "swing truck capacity" is likely to enter the medium to long-distance transport market only when coal prices are high).

In 2005, the average freight rate by road was 55.3 cents/t.km in China, which increased to 56.2 cents/t.km in 2006. In the past, overloading was a widespread practice by Chinese truck drivers. The government initiatives to eliminate overloading by trucks have significant impacts on road freight rates. For instance, after the Shanxi government initiated its campaign to control overloading by trucks, the average short-distance freight rate for coal transport increased rapidly, by 62.3% between December 19, 2007 and May 28, 2008 (Yang et al., 2009).

Table A.14 shows detailed statistics of coal throughput by road during the 10th FYP period. In 2005, coal transport tonnage by road was about 400 Mt, with inter-provincial transport accounting for 184 Mt of coal and intra-province transport accounting for the remaining 216 Mt. During the 10th FYP period between 2001 and 2005, coal transport by road grew at 17% annually at the national level and 19% annually in the Jin-Shaan-Meng-Ning region (Li, 2008).

Table A.14 *Coal throughputs by road during the 10th FYP period (in Mt)*

	2001	2002	2003	2004	2005
National coal production	1,161	1,391	1,660	1,956	2,190
Coal transport by road	192	273	378	426	400
Ratio of road transport tonnage to production in China (%)	16.5	19.6	22.8	21.8	19.3
Coal production in Jin-Shaan-Meng-Ning region	411	391	731	777	840
Coal export tonnage in Jin-Shaan-Meng-Ning region	344	369	469	557	631
Coal export tonnage by road in Jin-Shaan-Meng-Ning region	50	65	101	120	95
Ratio of road transport tonnage to production in Jin-Shaan-Meng-Ning (%)	12.2	16.6	13.8	15.4	12.6
Coal transport by road in Shanxi			79.78	107.41	132.42
Out-of-province coal transport	55.34	66.05	68.35	90.29	109.28
Intra-provincial coal transport by road			11.43	17.12	23.14

Sources: Coal transport by road in Shanxi is based on NBS (various years-a) and Shanxi Bureau of Statistics (various years); the remaining data are from Li (2008).

At the end of the 10th FYP period, road transport authorities tightened their control on overloading to protect road infrastructure and conserve the environment. As a result, coal transport tonnage by rail increased continuously at the expense of trucking since 2004, with a higher percentage drop on inter-provincial coal transport. In 2005, inter-provincial coal exports by road were 153.1 Mt. Major coal export provinces by road include Shanxi (65.18 Mt in 2005), Shaanxi (35.23 Mt), Inner Mongolia (14.33 Mt), Hunan (8.59 Mt), Hebei (7.54 Mt), Guizhou (6.94 Mt), and Yunnan (6.00 Mt). Major coal import provinces by road include Henan (14.42 Mt), Ningxia (2.33 Mt), Jiangsu (1.99 Mt), Hubei (1.65 Mt), and Zhejiang (1.54 Mt) (Li, 2008).

In 2005, coal transport by road accounted for 19% of national coal production. In the Jin-Shaan-Meng-Ning region, coal export tonnage by road accounted for 13% of regional coal production in the same year, and Shanxi alone accounted for about 25% of national coal throughput by road during the 10th FYP period. In 2005, growth on coal transport by road in Shanxi was higher than coal hauled by rail, which again indicates the severity of the railway bottleneck in China.

In Table A.14, coal trucked between mine mouth and railway stations is excluded. If this category is incorporated, coal transport by road was as high as 2.7 Gt in 2008. According to Dong Yan, head of the Comprehensive Transport Research Center at the NDRC, inter-provincial coal transport by road in China was estimated to be close to 500 Mt in that year. With throughput at more than 400 billion t.km, coal accounted for about 12% to 15% of national cargo throughput by road (SWS Research, 2010).

Road normally serves as a supplemental mode for coal transport in China, but it can play a more important role in certain regional markets with severe bottlenecks of railway infrastructure. For instance, being both a major coal-producing and coal-consuming province, Henan has long been China's coal transport hub, especially given its geographic proximity to Shanxi. Since 1998, the bottleneck in the railway network propelled coal tonnage hauled by truck to increase significantly in Henan. The 278 Mt of coal hauled by trucks in 2009 was 21% higher than volume transported by rail in Henan in the same year. Average coal transport distance by road in Henan was below 100 km in the past, but a coal price spike in 2008 pushed average trucking distance from 96 km in 2007 to 256 km in 2008 and 279 km in 2009 (Henan Bureau of Statistics, various years).

A.4.2.5 Pipeline

Coal pipelines are used to transport coal from coal producers to end users. There are two types of coal pipelines. Slurry pipelines transport a slurry of water and pulverized coal. The ratio of coal to water is about 1:1. Coal log pipelines use coal that has been compressed into logs with a diameter 5% to 10% less than the diameter of the pipeline and a length about twice the diameter of the pipeline. The ratio of coal to water is about 3 or 4:1 (Liu et al., 1993). Given the rapid development of other coal transport modes, the long-term prospect of coal pipeline development is not promising in China, especially given the widespread water shortage across China's major coal-producing regions.

A.4.3 The role of government

Figure A.24 shows the trends of cargo tonnage by transport mode in China. Since 1978, cargo tonnage by air grew at a much faster pace than any other transport modes, which can be explained by the small freight tonnage handled by air in 1978 and the successful introduction of competition in this sector. In comparison, cargo tonnage hauled by both rail and pipeline grew the least during the study period; this coincides with the fact that both the railway and petroleum industries are dominated by state-owned monopolies in China.

The Chinese government has long regarded rail transport as a strategically important industry, and the MOR has been both the regulator of

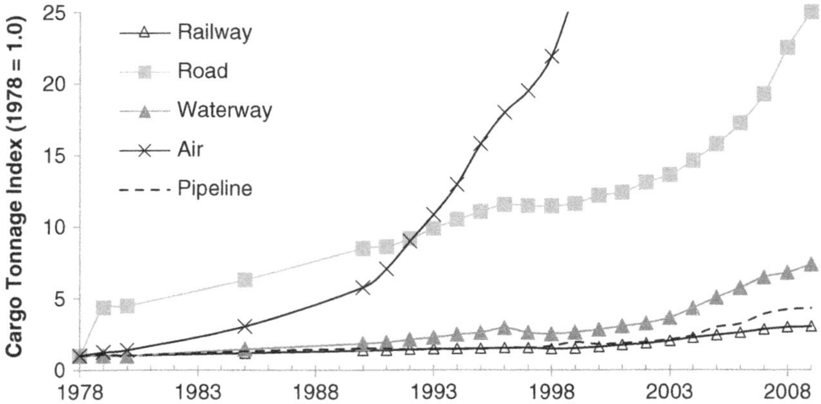

Figure A.24 Indices of cargo tonnage by transport mode, normalized around 1978. Source: NBS (various years-b).

the industry and the operator of the majority of the country's rail lines. Since its inception, the MOR's management style can be considered quasi-militarized, which is unique compared with other ministries under the State Council. The MOR has survived China's turbulent government restructuring in the past decades, and it has never become part of the MOT. As the Chinese economy has become increasingly market-oriented, decentralization and competition have been introduced to most parts of the transport industry – except the railways. In March 2013, it was announced that MOR would be dissolved, with its responsibilities taken by the MOT (safety and regulation), State Railways Administration (inspection), and China Railway Corporation (CRC, construction and management).[22] Currently, the primary investor in China's rail transport industry is CRC, and the monopoly status enjoyed by CRC has stymied investment from private sector, state-owned enterprises that do not belong to the CRC network, local governments, and even other ministries. Not surprisingly, the railway industry has the worst performance in the transport sector in terms of cargo tonnage growth since 1978 (NBS, various years-b).

A.4.4 The future of coal transport in China

Based on the reference coal production trajectory in this study, Figure A.25 presents a forecast of coal transport tonnages by mode in China by 2030. Rail is expected to continue to be the most dominant mode for moving coal from west China to the east and from the north to the south. In 2009, about

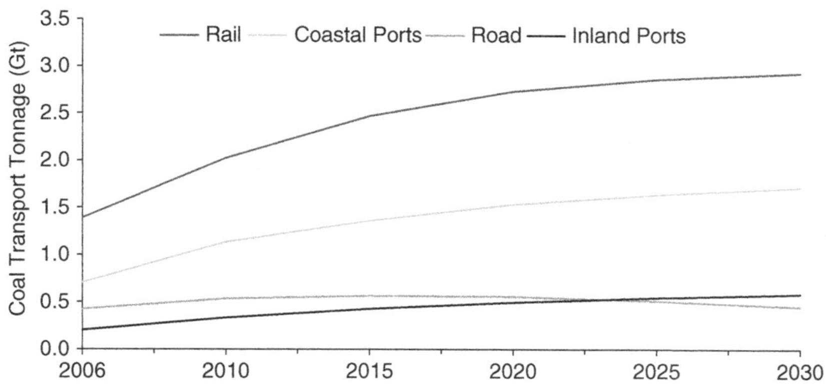

Figure A.25 Possible scenario of coal transport by mode in China. Only inter-provincial coal transport by road is included. Sources: NBS (various years-b), MOT (various years), Li (2008), and industrial sources.

1.7 Gt of coal were hauled by rail.[23] In future years, coal tonnage by rail is expected to increase by 2.3% annually, approaching 3 Gt by 2030. After out-of-province coal sales from Heilongjiang, Henan, Chongqing, Sichuan, Yunnan, Gansu, and Xinjiang are lowered due to either depleting resource bases or increasingly higher regional coal consumption levels, the average coal transport distance by rail in China is expected to increase over time. Shanxi, Shaanxi, Inner Mongolia, Ningxia, and Xinjiang are the primary coal export provinces, with major coal import provinces concentrated in the coastal region. The pattern of transporting coal from north to south and from west to east is expected to be further strengthened in the future. With high fuel costs and increasingly stringent regulatory countermeasures on truck overloading, railways are expected to haul most inter-provincial coal transport tonnages.

Except in a few regions, road is only a supplemental coal transport mode. In 2009, coal transported by road was about 0.5 Gt. Since then, coal transport tonnage by road is expected to first increase slightly and then gradually decline to the late 2010s level by 2030. After out-of-province coal sales from Heilongjiang, Henan, Chongqing, Sichuan, Yunnan, Gansu, and Xinjiang are reduced, inter-provincial coal transport tonnage by road is expected to be lowered accordingly. With spiking fuel costs and increasingly stringent regulatory countermeasures on truck overloading, coal transport tonnage by trucking in the future is unlikely to be significantly higher than the current level. Coal transport by road is expected to concentrate at major coal-producing provinces such as Shanxi, Shaanxi, Inner Mongolia, Ningxia, Aihui, and Guizhou. Road is expected to continue to be a supplemental coal transport mode.

In 2009, coal throughput at major inland ports was 0.4 Gt, which is expected to increase over time. After out-of-province coal sales from Chongqing, Sichuan, and Yunnan are reduced, coal throughput at the upstream and midstream sections of the Yangtze River is expected to be lowered accordingly. Following the expected coal output drop in Henan and Jiangsu and the forthcoming railway capacity expansion in China, the share of coal throughput by inland waterways is expected to grow over time to handle coal produced in Shanxi, Shaanxi, Inner Mongolia, Anhui, and Guizhou. Coal transport capacity of the Jing-Hang Grand Canal is expected to increase significantly after major waterway expansion projects. With spiking fuel costs and increasingly stringent regulatory countermeasures on truck overloading, inland waterway is expected to partially substitute for road for inter-provincial coal transport. Inland waterways will become increasingly cost competitive, and the integration of rail and inland waterway transport modes is expected to be strengthened in the future.

In 2009, coal throughput at major coastal ports was near 1.0 Gt, which is expected to grow at 2.7% annually, reaching 1.8 Gt in 2030. With coal exports primarily originating from Shanxi, Shaanxi, Ningxia, Anhui, Guizhou, and coal import provinces concentrating in the coastal region, coal handling capacity at coastal ports is expected to grow over time. With increasingly higher levels of coal imports, capacity of coal ports of discharge in southeast China is projected to increase over time. The geographic advantage of northern coal ports especially the N7 (close to the 3-Xi region and east Ningxia) will make them cost competitive in the long run, and the integration of rail and coastal marine transport modes is expected to be strengthened over time.

A.5 Grey markets and coal statistical distortion

A grey market is the trade of a commodity through distribution channels that, while legal, are unofficial, unauthorized, or unintended by the original producer. In contrast, a black market is the trade of goods and services that are themselves illegal and/or distributed through illegal channels. In the context of the Chinese coal industry, a grey market consists of coal sold by private and sometimes state-owned collieries at local or even regional markets without appropriate permits from governmental authorities. Illegal mines may supply much of the coal traded in grey markets. These so-called “illegal” mines often operate with recognition of local officials. Because Shanxi has produced 23% of China’s aggregate coal supply since 1949 and has long been the largest out-of-province coal exporter in China until 2010, Shanxi was selected as the case study to investigate the formation of grey coal markets in China.

Understanding the grey market – and thus the degree to which market statistics do not capture the reality of the Chinese coal market – is important for researchers, policy makers, and investors. Analyses for purposes ranging from investment planning to policy decision making to international climate model development that rely on Chinese coal statistics need to understand the potential for error and risks associated with making firm conclusions based on this data.

A.5.1 Introduction to Shanxi and its coal industry

Shanxi possesses 32% of China's proven coal reserves and more than half of its proven coking coal reserves. In 2008, Shanxi produced more than one quarter of China's coal output but only accounted for about 10% of China's coal consumption. Since Shanxi does not produce any oil and gas, which is in sharp contrast with its neighboring province Shaanxi, coal thus is the primary fossil fuel in its energy sector. Even with its high per capita coal endowment, economic development in Shanxi lags behind the

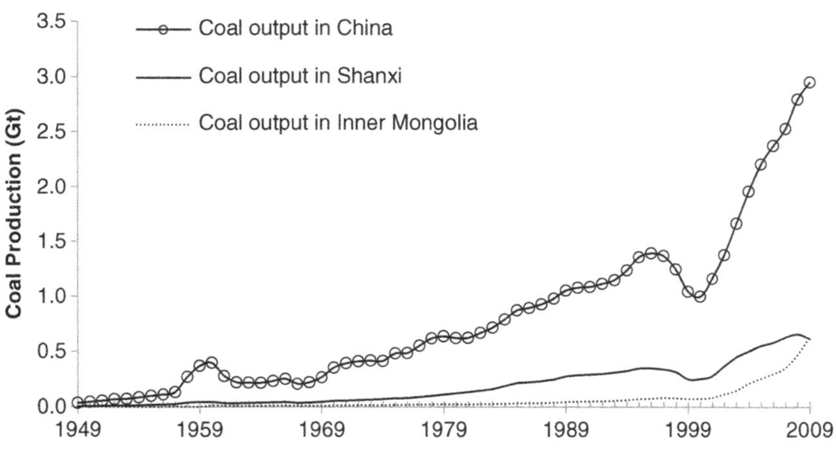

Figure A.26 Coal output of Shanxi, Inner Mongolia, and China from 1949 to 2009. Coal output figures around 2000 are based on original statistics reported by the NBS, as outputs at provincial levels have never been revised by the provincial bureaus of statistics in later years. Sources: NBS (2005 and various years-b), Shanxi Bureau of Statistics (various years), and CCTD.

national average. In 2008, Shanxi's GDP accounted for only 2.3% of the national total, and per capita GDP in Shanxi was only 91% of the national average, when coal prices had already increased significantly compared with historical levels (NBS, various years-b and Shanxi Bureau of Statistics, various years).

Figure A.26 presents the role of Shanxi's coal industry in China. After the inception of the PRC in 1949, Shanxi's coal output increased quickly from 8% of the national total in 1949 to 27% of the national total in 1993. Since then, Shanxi's share of national coal output fluctuated over time and decreased slightly to 23.5% in 2008. However, the Shanxi government's second resource consolidation campaign had severe impacts on the province's coal production capacity, with its share of national coal output falling significantly to 20.9% in 2009. Even so, due to the existence of a sizable coal grey market in Shanxi, Inner Mongolia did not overtake Shanxi as China's largest coal producer until 2010.

A.5.2 Grey markets and coal statistical distortion

In 1949, there were fewer than 100 state-owned mines, with an average capacity below 30 kt/annum. Though several thousand TVE mines existed, the extremely low capacities of these mines meant that most of them could supply only adjacent communities. Furthermore, the private coal industry in Shanxi had been stifled by the leftist fever until 1978 when the Chinese economy was opened to the outside world.

In 1980, TVE mines played an important role in Shanxi by supplying about one-third of the province's total coal output. Encumbered by heavy welfare obligations to their bloated workforces and millions of retired workers, China's state-owned mines were unable to meet the burgeoning coal demand, and the central government had to encourage private investment by issuing the "let water flow" policy in 1981. Since then, the number of registered TVE mines in Shanxi increased significantly and reached 6,659 in 1995. In comparison, the number of small coal mines across the country once exceeded 100,000 during this period (CCII, various years).

The existence of too many small TVE mines soon became a major policy concern for the Chinese government. After the outbreak of the Southeast Asia Financial Crisis in 1997/98, Beijing grasped the opportunity (lower coal demand) and started to exert greater control over local coal mine operations in the service of national efficiency, safety, and environmental goals, with a later target of reducing the number of small mines to 10,000 by 2010. However, as the aged state-owned coal mines became increasingly unable to keep pace with the demand spike, many of the small mines ordered for closure managed to survive. Attracted by tax revenue, local governments in coal mining regions especially Shanxi were unwilling to comply with the production cap and simply underreported the coal supply and demand within their geographic boundaries. Suffering a serious staffing shortage, NBS headquarters struggled just to compile the country's energy balance tables, let alone conduct investigations of suspicious claims. As a result, China's coal statistics since 1998 have been seriously underreported (Tu, 2006).

Grey markets actually existed long before the government campaign to shut down small coal mines in 1997/98. China's energy statistical collection system was initially developed and functioned well under a planned economy, with the assumption of complete reporting from all units producing, transforming, delivering, and consuming commercial energy. However, as China moved toward a market-oriented economy, the energy activities outside the state-owned enterprises grew rapidly. For example, the percentage of coal produced by TVE mines increased from 18% in 1980 to 46% in 1995 (CCII, various years). As there was no reliable mechanism for collecting coal statistics from small, non state-owned mines, the quality of China's coal statistics deteriorated over time.

To rein in its chaotic coal mining industry, Shanxi Province (2006b) issued the Guidelines for Coal Development and Exploration in 2006. The number of small coal mines in Shanxi was expected to be lowered to 2,172 by 2010 and 1,000 by 2020. The average capacity of small coal mines is expected to still be below 20 kt/annum during the planning period. However, in the same year, Shanxi also launched its first resource consolidation campaign. According to Shanxi Province (2006a), small coal mines with capacity below 90 kt/annum would be closed and mines with capacity below 300 kt/annum would not be allowed to acquire closed small collieries or undeveloped coal resources. As a result, the average capacity of coal mines in Shanxi showed an increase from 18.5 kt/annum in 2006 to 25 kt/annum in 2008. In addition, the number of operating coal mines in Shanxi decreased from 3,199 in 2006 to 2,598 by the end of 2008.

According to Shanxi Province (2008), the capacity of all coal enterprises in Shanxi was to be above 3 Mt/annum by 2010. In addition, the number of operating coal mines was to be reduced to 1,500 by 2010, which was significantly lower than the previous governmental target. In February 2009, a methane blast at the Tunlan Coal Mine in Shanxi killed 78 miners, which soon prompted Shanxi to launch the second round of a resource consolidation campaign (Tu, 2009). According to Shanxi

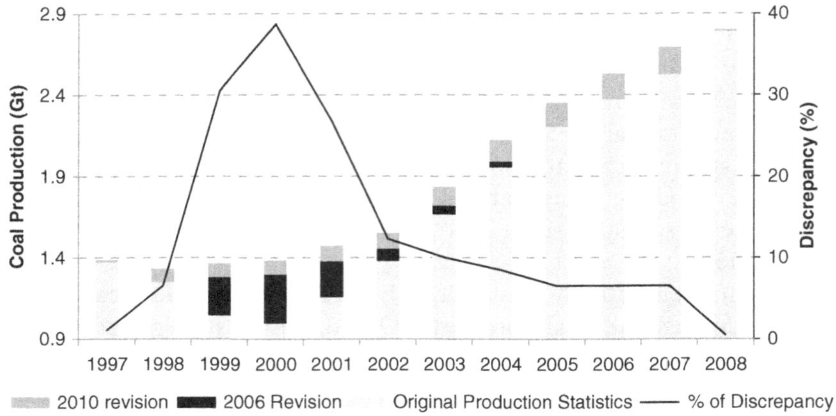

Figure A.27 Statistical revisions of China's coal output in 2006 and 2010. Percent of discrepancy is based on the ratio between the sum of the 2006 and 2010 revisions and the original coal production statistics. Source: NBS (various years-b).

(2009), the number of operating mines in Shanxi was to be lowered from the previous target of 1,500 to 1,000 by the end of 2010. In addition, capacity of any operating mine and a coal mining enterprise must be above 0.9 Mt/annum and 3.0 Mt/annum, respectively.

While the above administrative orders clearly show Shanxi government's determination to rein in its chaotic coal mining industry, the measurements specified in the 2009 order are apparently not compatible with the principle of Chinese laws (e.g., the Real Rights Law). The failure of Shanxi to govern its coal industry according to the principle of laws is expected to have very negative impacts in the long run by (1) further reducing the trust level of Chinese society as a whole, (2) lowering the efficiency of the Chinese coal industry by eliminating competition from the private coal mining industry, and (3) encouraging coal production underreporting in Shanxi.

Eventually, pressure on energy demand prompted NBS officials to revise historical coal statistics in 2006. Since both the extent and coverage of the 2006 statistical revisions were considered to be insufficient, the NBS revised China's coal supply statistics in 2010 again. Nevertheless, the revision methodology and procedure have not yet been revealed.

Figure A.27 compares the original statistics of the national coal output in China with the updated data revised in both 2006 and 2010. While the size of coal grey markets in 2000 was as high as 39% of originally reported national coal output in 2000, the discrepancy between original statistics

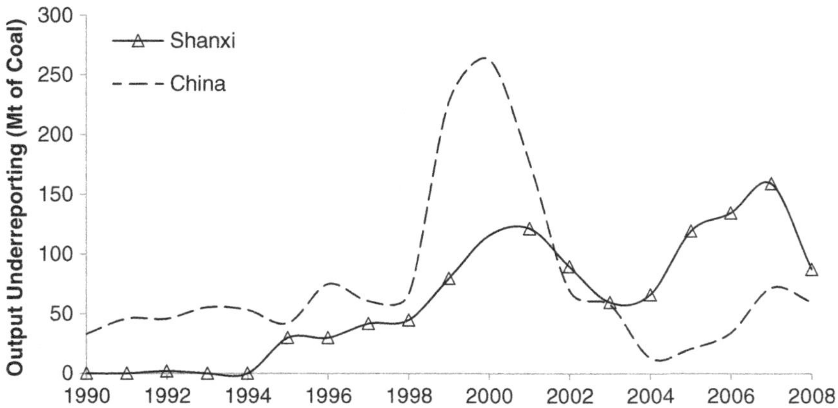

Figure A.28 Unexplainable coal usage: China vs. Shanxi. Trend lines are based on the "balance" category in coal balance tables for either Shanxi or China. The latest national statistics distortion quantification is based on *China Energy Statistical Yearbook 2009*, so the statistical revisions by the NBS in 2010 have not been incorporated. Sources: NBS (various years-b) and Shanxi Bureau of Statistics (various years).

and revised data diminished quickly thereafter. However, even after two major statistical revisions in 2006 and 2010, whether the size of coal grey markets has been fully identified by the Chinese government is still an open-ended question.

In an energy balance table, the "balance" category is used to capture energy consumption that cannot be balanced by either coal supply, storage change, or imports/exports. In this study, the balance category of coal in China is termed as "unexplainable coal usage," which is the amount of coal production beyond the grasp of Chinese statistical agencies and is considered to reflect the possible size of the coal grey market in China.

Figure A.28 compares the unexplainable coal usage in China and Shanxi. Between 1990 and 1997, national unexplainable coal usage fluctuated over time, ranging from 33 Mt to 75 Mt. The size of grey markets in China was generally below 5% of national coal output. Since 1998, national unexplainable coal usage increased abruptly and reached 228 Mt in 1999, 264 Mt in 2000, and 177 Mt in 2001, which accounted for 15% to 26% of national coal output originally reported by the NBS.

The national markets follow a similar trend as Shanxi. However, since 2003, the size of the coal grey market in Shanxi has increased significantly and was consistently above the national levels. In 2007, the "balance" category of coal in Shanxi was as high as 160 Mt, which suggests that 25%

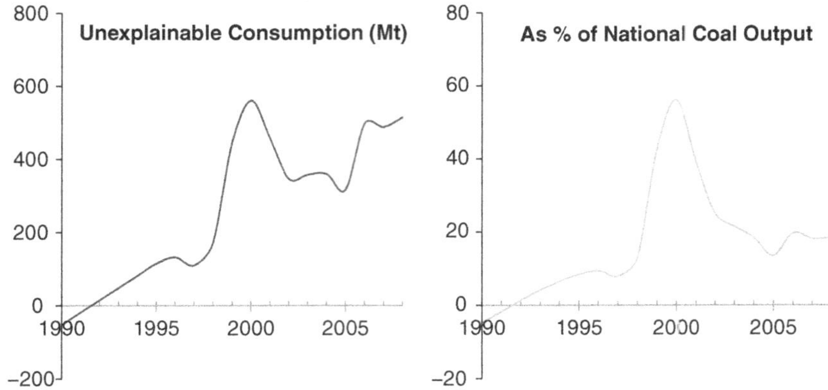

Figure A.29 Unexplainable coal consumption at the provincial level in China. Calculated as percent of national coal output in a specific year, which equals (unexplained coal consumption at provincial level in China) / (original national coal output reported by the NBS in respective year). Data for 1991 to 1994 are interpolated according to 1990 and 1995 figures (NBS, various years-b).

of coal usage in Shanxi cannot be reconciled with reported coal production in this province (with consideration of imports/exports and storage change).

Currently, national coal production in China equals the summation of coal output in all provinces and autonomous regions. While one would expect that a similar relation holds for national coal consumption, it is actually not the case in China. According to the China Energy Statistical Yearbook, the summation of coal consumption in all Chinese provinces and autonomous regions is significantly higher than aggregate coal consumption data at the national level reported by the NBS in recent years.

Taking coal imports/exports, storage change, and losses in coal washing and dressing into consideration, Figure A.29 shows that unexplainable amounts of coal consumption at the provincial level have increased rapidly since 1990 and first peaked at 561 Mt in 2001, a year for which underreporting of coal statistics was officially recognized as a particularly serious problem. Since then, the grey picture of coal statistical distortion in China improved marginally until 2005 and rebounded sharply thereafter, reaching 512 Mt in 2008. As a result, grey markets may still account for up to 18% of the national coal output reported by the NBS in 2008.

One important driving force underpinning coal grey markets in China is the historic and chronic difficulty of compelling local officials to obey central policies, as local officials often have flexibility to interpret regulations and administrative orders from higher-level authorities according to their departmental or personal interests. The statistical distortion on coal mine safety can best illustrate this phenomenon. The coal mine safety record in China is full of controversy. While the official coal mine fatality statistics at the beginning of the PRC range from 242 in 1951 to 6,036 in 1960, Guo et al. (2006) reported that annual fatalities were as high as 70,000. Since 1978, the official coal mine fatalities in China fluctuated over time but were generally between 5,000 and 7,000 deaths per year. Nevertheless, independent estimations suggest that annual coal mine fatalities were about 40,000 in the 1980s, 20,000 in the 1990s, and 10,000 at the beginning of this millennium (Guo et al., 2006, Elegant, 2007, and Tu, 2007a). The sharp contrast between official coal mine death tolls and independent estimations of coal mine fatalities in China again suggests the accuracy of China's coal statistics as a whole is problematic.

A.5.3 Hypotheses about coal reporting gaps

Based on the aforementioned assessment, the author proposed several hypotheses regarding the status of China's coal statistical reporting, and it is suggested that the hypotheses below should be assessed with rigor by researchers, ideally with encouragement from the Chinese government:

- Since the mid-1990s, coal statistical distortion has become a systematic error in China's statistical collection system. It is caused by the existence of grey coal markets in China.
- Up to 100–200 Mt of coal output may be underreported in Shanxi in recent years, which needs to be corrected by the Shanxi Bureau of Statistics.
- Up to 300–600 Mt of coal output may be underreported at the national level in recent years, which needs to be corrected by the National Bureau of Statistics.
- The coal supply statistical revisions by the NBS in 2006 and 2010 have not fully corrected the aforementioned statistical distortion.

In early 2015, NBS reported China's preliminary energy consumption in 2014 as 4.26 Gtce, the equivalent of a 13.6% spike compared with a similar indicator in 2013.[24] Nevertheless, as the reported year-over-year change of primary energy consumption between 2013 and 2014 was only 2.2%, Chinese coal statistics in the past are expected to be revised upward significantly again, with profound impacts on China's 13th Five Year Plan, legitimacy of past energy-economy

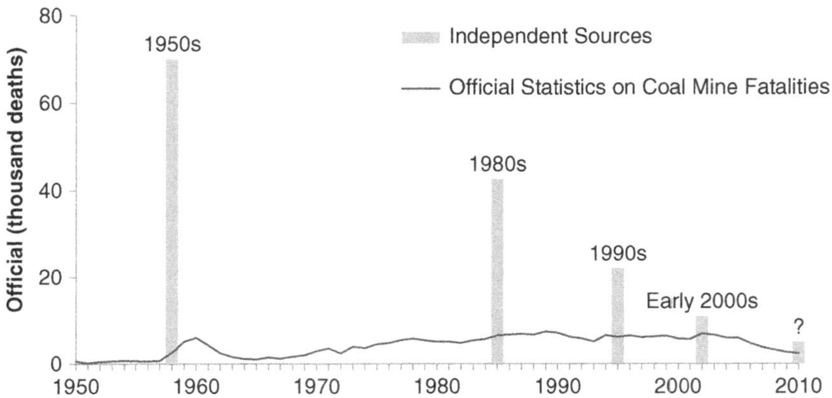

Figure A.30 Official statistics versus independent estimations of coal mining safety in China. Sources: Guo et al. (2006), Tu (2007a), Elegant (2007), Safety Bureau of Former Ministry of Coal Industry (1998), CCII (various years), and the State Administration of Work Safety.

modeling exercises in China and beyond, and international climate negotiation.

A.6 The future of coal in China and the climate factor

A.6.1 Background

While the coal resource, because of its abundance and low cost, has been a sustained contributor to China's economic and social development, its extraction, processing, transport, and utilization are all associated with major sustainability challenges. The predominance of underground coal mining leads to significant amounts of fugitive methane emissions. In addition, coal combustion accounts for 82% of CO_2 emissions, 75% of SO_2 emissions, 85% of NOx emissions, and 70% of particulate emissions in China (TFSUC, 2009). For more than a decade, China has been the world's largest SO_2 emitter, and Figure A.31 shows that China first overtook the United States for the world lead in CO_2 emissions in 2006. Deteriorating air quality is a major environmental issue in many Chinese urban centers, and GHG emissions abatement becomes an increasingly important policy concern for Chinese decision makers.

Given the necessity of balancing economic growth with portraying itself as a responsible power, China understands that it cannot escape the responsibility of curbing its spiking carbon emissions forever, so it started

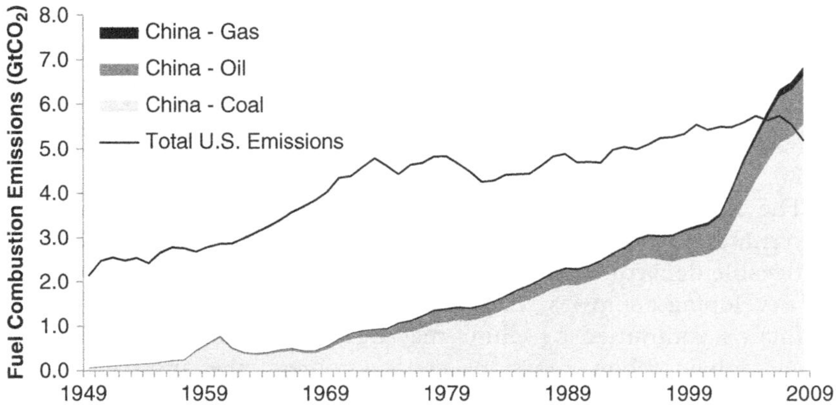

Figure A.31 Fuel combustion and carbon emissions: China vs the United States. China's updated energy statistics released in 2010 indicated that China had already passed the United States as the world's largest carbon emitter in 2006, which is one year earlier than what the previous energy statistics implied. Sources: NBS (various years-b), ERI (2007), ONCCCC (2007), US EPA, and the author's own estimation.

developing an increasingly proactive and comprehensive climate policy. During the United Nations Climate Summit in September 2009, former Chinese president Hu Jintao first announced that China will cut its carbon emissions per unit GDP by an unspecified "notable margin" by 2020 from the 2005 level. In November 2009, the question of notable margin was specified by the State Council as a 40% to 45% reduction of 2005 levels by 2020.[25]

A.6.2 China's carbon intensity abatement approach

Between 1990 and 2005, China's national GDP increased 326%. In comparison, the country's fuel combustion carbon emissions increased 129%. China's carbon intensity per unit GDP declined significantly, by 46% for the period from 1990 to 2005 (IEA, 2009b). In its 11th FYP (2006–2010), the Chinese government set a goal of cutting national energy intensity per unit GDP by 20% between 2005 and 2010, translating into an annual reduction of more than 1.5 $GtCO_2$ by 2010, making the largely energy security and local pollution-based effort one of the most significant carbon mitigation initiatives in the world (Lin et al. 2008). Since China is pursuing a 16% intensity improvement target in the subsequent FYP periods, a carbon intensity reduction of 40% to 45% between 2005 and 2020 is achievable in the absence of the aforementioned commitment. Nevertheless, with significant upward revisions of China's national energy consumption in 2015, it is uncertain whether China could still meet the 40%–45% carbon intensity reduction target by 2020.

The carbon intensity abatement approach proposed by the Chinese government before the 2009 Copenhagen conference represents a noticeable departure from past stances on climate change negotiation by developing countries. Though the percentage of carbon emissions reduction committed by China may be considered as insufficient to meet global carbon emissions control targets, the approach itself, coupled with Chinese president Xi Jinping's commitment to peak the country's carbon emissions by 2030 during the APEC Summit in Beijing in November 2014, may open a window of opportunity to bring at least key emerging economies into the fold of global efforts on GHG emissions abatement.

A.6.3 Summary and recommendation for future research

Chinese decision makers will continue to rely on carbon-intensive coal to fuel the country's booming economy in the decades to come, primarily because of coal's low cost, resource abundance, and the increasing availability of "cleaner coal" technology that can be applied to alleviate environmental impacts ("cleaner coal" in the context of the Chinese power sector primarily implies controls for SO_2, NOx, and particulate emissions). New policies promoting a reduction in both the energy intensity of GDP and the carbon intensity of GDP are likely to gradually reduce the "coal intensity of GDP." But because China's dependence on coal will likely last decades into the future, understanding the complex coal value chain that delivers the vast majority of energy in the world's largest energy consuming economy is essential for understanding global energy markets and climate change.

At the time of writing, the size of grey markets in the Chinese coal industry seems to have grown to dangerous levels that are too significant to be ignored; it is recommended that the Chinese government should consider assessing the current situation and fixing any inconsistency within its statistical reporting system. Otherwise, ongoing coal statistical distortion is likely not only to severely undermine China's policy initiatives on energy conservation and carbon intensity abatement in the years to come but also make it difficult for the international community to verify policy achievements claimed by the Chinese government. Further, statistical distortions impact the investment, policy, and climate communities' ability to make sound, accurate decisions in the Chinese energy sector.

Finally, the complex Chinese value chain of Chinese coal production, transport, and power generation is now undergoing major reforms. While this study has presented a comprehensive review of the Chinese coal value chain including resources, production, transport, end uses, and environmental impacts, a follow-up study is recommended to thoroughly explore the future of coal in China under an energy security-, air quality-, and carbon-constrained world.

Notes

1. Statistics from industrial sources suggest 3.41 Gt as the coal output in 2010 (National Bureau of Statistics and BP, 2010).
2. Official statistics. Independent sources suggest much higher fatalities; see Section A.5 for more detail.
3. See Figure A.31 for more detail.
4. See http://www.china5e.com/show.php?contentid=77252 [2010-10-18].
5. Classification of coals in China (GB/T 5751–2009 Chinese Classification of Coals) is not consistent with the international standards. For instance, ASTM D 388–2005 Standard Classification of Coals by Rank classifies coals as meta-anthracite, anthracite, semianthracite, low volatile bituminous, medium volatile bituminous, high volatile (A, B, C) bituminous, sub-bituminous (A, B, C), and lignite (A, B). In comparison, ISO 11760:2005 – Classification of Coals categorizes coals as anthracite (A, B, C), bituminous (A, B, C, D), sub-bituminous (A), and lignite/brown coal (B, C).
6. http://kbs.cnki.net/forums/50150/ShowThread.aspx [July 1, 2009].
7. The difference was coal bought by Shenhua from other producers (e.g., TVE mines) primarily due to the latter's lack of access to the rail transport network.
8. This category includes both indigenous coke ovens and modified-indigenous coke ovens. Indigenous coke ovens are the oldest and most simple indigenous methods of making coke, with significant heat loss and pollutant emissions during the production process. Since the 1970s, the Chinese coke manufacturers have continuously made improved versions of coke ovens (the so-called "modified indigenous"). Though this process can improve the quality of coke product, it can still result in serious air pollution and groundwater contamination.
9. Interview with several industrial contacts in 2009.
10. Liquid products of direct CTL plant include diesel, naphtha, and LPG.
11. See http://cccmc.mofcom.gov.cn/aarticle/zhongyswhd/200806/200806056 30580.html [August 09, 2009].

12. CIMS is an energy-economy simulation model developed by the Energy and Materials Research Group at Simon Fraser University; please refer to http://www.emrg.sfu.ca/Our-Research/Policy-Modelling.
13. For more detail, please refer to (1) Bataille, C., J. Tu, and M. Jaccard (2008), Permit sellers, permit buyers: China and Canada's role in a global low-carbon society, *Climate Policy* 2008: 93–107, and (2) Tu (2007), The Application of a Hybrid Energy-Economy Model to a Key Developing Country – China, *Energy for Sustainable Development*, XI(1): 35–47.
14. "Xi" in Chinese stands for "west."
15. Based on *Shanxi Daily*, December 23, 2009.
16. See http://www.tadercn.com/mtxy/mtlt/hy/644361.shtml [August 1, 2009].
17. Statistics for China are based on data from the National Bureau of Statistics. Statistics for the rest of world are based on the World Bank's online database – World Development Indicators.
18. Datong including west Inner Mongolia, Shenfu; Taiyuan including Southern Shanxi, Southeast Shanxi, Shaanxi, Guizhou, Henan, Gunzhou, Huainan and Huaibei, East Heilongjiang and Xinjiang.
19. See http://www.tadercn.com/mtxy/mtlt/hy/644349.shtml [July 15, 2009].
20. See http://www.tadercn.com/mtxy/mtlt/hy/644347.shtml [July 15, 2009].
21. Source: Changjiang River Administration of Navigational Affairs.
22. See http://www.china-railway.com.cn [March 8, 2015].
23. This figure includes coal hauled by local railways and Shenhua.
24. See http://www.stats.gov.cn/tjsj/zxfb/201502/t20150226_685799.html [March 8, 2015].
25. See http://www.chinadaily.com.cn/china/2009-11/26/content_9058731.htm [December 1, 2009].

References

Bataille, C., J. Tu, and M. Jaccard (2008). Permit sellers, permit buyers: China's and Canada's role in a global low-carbon society. *Climate Policy* 2008: 93–107.

BP (2010). BP statistical review of world energy 2010. British Petroleum, London.

Cai, J., J. Hao, Z. Lu, G. Li, W. Wang, and L. Kong (2002). Analysis of energy saving and energy consumption in Chinese steel industry for last 20 years and next 5 years. *Iron & Steel* 37: 68–73.

CCII (various years). *China Coal Industry Yearbook*. Beijing: China Coal Information Institute.

CEC (various years). *Express Statistics of the Chinese Electricity Industry*. Beijing: China Electricity Council.

CNCA (2006). *Statistics of the Chinese Coal Industry: 1949–2004*. Beijing: Coal Industry Press.

Duan, Q. (2007). Status of China's coal transport corridors and study on optimized policies. *Research and Approach* 29: 27–32.

Editorial Board of CCCI (1999). *Chronicles of the Chinese Coal Industry: The Comprehensive Volume*. Beijing: Coal Industry Press.

Editorial Board of CEIYB (various years). *China Electricity Industry Yearbook*. Beijing: China Electric Power Press.

Elegant, S. (2007). Where the coal is stained with blood. Time, *March 2*.

Fan, Y., Z. Lu, J. Cheng, Z. Zhou, and G. Wu (2003). Major ecological and environmental problems and the ecological reconstruction technologies of the coal mining areas in China. *Acta Ecologica Sinica* 23(10): 2144–2152.

Guo, G., M. Lu, H. Zhang, Y. Liu, and G. Ma (2006). *Coal Mine Safety Technology and Management*. Beijing: Metallurgical Industry Press.

Guo, X. (2008). *The Experience on Developing China's Transport Infrastructure. Experience-Sharing Program on Development between China and Africa*. Beijing: World Bank.

Henan Bureau of Statistics (various years). *Henan Statistical Yearbook*. Beijing: China Statistics Press.

Huang, S., W. Liu, and Q. Dou (2008). *China Coal Outlook 2008*. Beijing: National Insititute of Occupational Health.

IEA (2009a). *Cleaner Coal in China*. Paris: International Energy Agency.

IEA (2009b). *World Energy Outlook 2009*. Paris: International Energy Agency.

IEA (2010). *Coal Information: 2010 Edition*. Paris: International Energy Agency.

IEA (2013). *World Energy Outlook 2013*. Paris: International Energy Agency.

Jiang, K., X. Hu, X. Zhuang, and Q. Liu (2009). China's low carbon scenarios and low carbon development paths to 2050. *Sino-Global Energy* 14(6): 1–7.

LFRC (2007). *Transport Infrastructure Development in China, Part I: Road – Highway and Expressway*. Hong Kong: Li & Fung Research Center.

LFRC (2009). *An Update on the Transport Infrastructure Development in China: Water Transportation*. Hong Kong: Li & Fung Research Center.

Li, H. (2008). *Coal Transport in China: Capacity, Consumption and Prices*. Beijing: China Market Publishing House.

Li, S., Y. Liu, Z. Xu, and J. He (2010). *Outlook of China's Economic Development between the 12th Five-Year Plan Period and 2030*. Beijing: Development Research Center of the State Council.

Lin, J., N. Zhou, M. Levine, and D. Fridley (2008). Taking out 1 billion tons of CO_2: The magic of China's 11th Five-Year Plan. *Energy Policy* 36: 954–970.

Liu, H., R. Zuniga, J. L. Richards (1993). Economic analysis of coal log pipeline transportation of coal, CPRC report no. 93–1. University of Missouri, Columbia, MO.

Minchener, A. J. (2007). *Coal Supply Challenges for China*. London: IEA Clean Coal Centre.

MLR (1999). *Forecast and Evaluation of China's Coal Resources*. Beijing: Ministry of Land and Resources.

MOC (various years). *Report on China's Shipping Development*. Beijing: Ministry of Communications.

MOR (2008). *Illustrative Maps of Operating Stations of the National Railway Bureaus*. Beijing: China Railway Publishing House.

MOT (various years). *Report on China's Shipping Development*. Beijing: Ministry of Transport.

NBS (2005). *Comprehensive Statistical Data and Materials on 55 Years of New China*. Beijing: National Bureau of Statistics.

NBS (various years-a). *China Energy Statistical Yearbook*. Beijing: National Bureau of Statistics, Beijing.
NBS (various years-b). *China Statistical Yearbook*. Beijing: National Bureau of Statistics.
NDRC (2006). Draft Mid- to Long-Term Development Plan for the Coal Chemical Industry. Beijing: National Development and Reform Commission.
NDRC (2008). *Medium- to Long-Term Railway Network Plan* (2008 revision). Beijing: National Development and Reform Commission.
Office of the National Coordination Committee on Climate Change, Energy Research Institute (2007). *The People's Republic of China: National Greenhouse Gas Inventory*. Beijing: China Environmental Science Publishing House.
Peabody (2009). *Peabody Energy: 2008 Corporate & Social Responsibility*. St. Louis, MO: Peabody Energy.
Safety Bureau of the MCI (1998). *Assembly of Statistical and Analytic Documents for Coal Mine Injuries and Fatal Accidents in China*. Beijing: Coal Industry Press.
Sagawa, A., A. Fukushima, Y. Mimuroto, and C. C. Siang (2003). *Prospects for the Supply and Demand of Coal and Related Coal Transportation Issues in China*. Tokyo: Institute of Energy Economics.
SERC (various years). *Annual Report on Electricity Regulation*. Beijing: State Electricity Regulatory Commission, People's Republic of China.
SETC (2000). *50 Years of the Chinese Industry*. Beijing: China Economic Publishing House.
Shanxi Bureau of Statistics (various years). *Shanxi Statistical Yearbook*. Beijing: China Statistics Press.
Shanxi Government (2008). Opinions of Shanxi government on implementation of accelerating the merger and acquisition amongst coal mining enterprises. Taiyuan: Shanxi Development and Reform Commission, [2008]23. Available at http://www.chinaacc.com/new/63_74_/2009_2_27_wa11593547441722900216425.shtml.
Shanxi Province (2006a). Administrative order on coal resource consolidation and compensated use in Shanxi, People's Government Order of Shanxi – No. 187. Shanxi Provincial Government, Taiyuan.
Shanxi Province (2006b). Guidelines for coal production and exploration between 2006 and 2020 in Shanxi. Shanxi Provincial Government, Taiyuan.
Shanxi Province (2009). Notice of Shanxi government on issues regarding further accelerating the merger, acquisition and consolidation amongst coal mining enterprises, Shanxi Government [2009] No. 10. Shanxi Provincial Government, Taiyuan.
SWS Research (2010). Meeting minutes for SWS research coal market forecast conference 2011. SWS Research, Shanghai.
TFSUC (2009). Conference proceedings of the Third Working Meeting of the CCICED Task Force on Sustainable Use of Coal. CCICED Task Force on Sustainable Use of Coal, Beijing.
Thomson, E. (2003). *The Chinese Coal Industry: An Economic History*. London, New York: RoutledgeCurzon.
Tu, J. (2006). China's botched coal statistics? *China Brief* 6(21). Available at http://www.jamestown.org/single/?tx_ttnews[tt_news]=32168%26no_cache=1#.VQQ2ROF1wUM.

Tu, J. (2007). Coal mining safety: China's Achilles' heel. *China Security* 3(2): 36–53.

Tu, J. (2009). The Shanxi coal mine blast and the failure of safety governance in China. *China Brief* 9(10). Available at http://www.jamestown.org/programs/chinabrief/single/?tx_ttnews[tt_news]=34991%26tx_ttnews[backPid]=459%26no_cache=1#.VQQ1JOF1wUM.

Tu, J., M. Jaccard, and J. Nyboer (2007). The application of a hybrid energy-economy model to a key developing country – China. *Energy for Sustainable Development* XI: 35–47.

US EIA (2009). *Annual Coal Report 2008*. Washington, DC: Energy Information Administration.

Wang, H., and N. Horii (2008). *Chinese Energy Market & Reform on Pricing Mechanisms – Case Study and Analysis of Shanxi's Experiment*. Chiba: Institute of Developing Economies. Available at http://www.ide.go.jp/English/Publish/Download/Jrp/pdf/146_1.pdf.

Wang, W. (2009). *Scientific Assessment on the Difference between Energy Intensities of the Chinese Iron and Steel Industry and International Best Performance*. Beijing: China Iron and Steel Enterprises' Network.

Wang, Y., and W. Pan (2009). Analysis of the coal economy in 2008. *China Energy* 31(3): 13–17.

Wang, Z. (2008). *Research on Improving Operation Efficiency of Shenhua Group's Huanghua Port*. Beijing: Beijing Jiaotong University.

Wu, R. (1996). Development and trends of China's coal production. *China Coal* 1996(12): 11–14. Available at http://lib.cqvip.com/qk/95382A/199612/2310770.html.

Wu, S., and M. Wang (2006). Position of coal in Chinese energy supply. *Coal Processing & Comprehensive Utilization* 2006: 2–7.

Wu, Y. (2009). *Investigating the economy through coal transport – grass-root study on coal trade market at Qinhuangdao*. Wuhan: Research Division of Changjiang Securities Co. Available at http://finance.ifeng.com/stock/hybg/20090428/592840.shtml.

Yang, X., C. Wu, and D. Wang (2009). *2009 Coal Marketing Handbook*. Beijing: China University of Mining and Technology Press.

Zhang, P. (2009). Brief assessment on selection of rail routes to export coal in Shanxi. *Road and Bridge Economy* 2009: 59–60.

Index

Lightning Source UK Ltd.
Milton Keynes UK
UKOW06n2132210715

255597UK00003B/101/P

9 781107 092426